CLASSICAL MECHANICS

CLASSICAL MECHANICS

A. DOUGLAS DAVIS
Department of Physics
Eastern Illinois University

ACADEMIC PRESS COLLEGE DIVISION
Harcourt Brace Jovanovich, Publishers
Orlando   San Diego   San Francisco   New York
London   Toronto   Montreal   Sydney   Tokyo   São Paulo

To

Bruce and Lillian,

*my parents,
from whom I received unbounded love*

and

Judy, Amy, and Emily,

*my wife and daughters,
to whom I give unending love*

Copyright © 1986 by Academic Press, Inc.
All rights reserved.

No part of this publication may be reproduced or transmitted in any form or by any means, electronic or mechanical, including photocopying, recording, or any information storage and retrieval system, without permission in writing from the publisher.

Academic Press, Inc.
Orlando, Florida 32887

United Kingdom Edition published by
Academic Press, Inc. (London) Ltd.
24/28 Oval Road, London NW1

ISBN 0-15-507630-2

Library of Congress Catalog Card Number: 85-71119

Printed in the United States of America

CONTENTS

Preface ix

1 Conversational BASIC 1

1.1 Getting Started 1
1.2 Elegant Output 8
1.3 Sometimes More Is Really Less 12
1.4 Into the Wild Blue... 14

2 One-Dimensional Motion 21

2.1 Kinematics in One Dimension 21
2.2 Dynamics in One Dimension 22
2.3 Constant Force 25
2.4 Force as a Function of Time 31
2.5 Force as a Function of Position 33
2.6 Force as a Function of Velocity 39

3 The Harmonic Oscillator 50

3.1 Introduction 50
3.2 Simple Harmonic Oscillator 52
3.3 Power Series Representation of an Arbitrary Function 56
3.4 Damped Harmonic Oscillator 58
3.5 Forced Harmonic Oscillator 64
3.6 Application to AC Circuits 69
3.7 Simple Pendulum 71
3.8 Physical Pendulum 73

4 Vectors 79

4.1 Introduction   79
4.2 Vector Algebra   79
4.3 Vector Multiplication   86
4.4 Coordinate Systems   89
4.5 Vector Calculus   91
4.6 Vector Differential Operators   95

5 Motion in Three Dimensions 114

5.1 Introduction   114
5.2 Separable Forces   115
5.3 Three-Dimensional Harmonic Oscillator   119
5.4 Potential Energy Function   122
5.5 Motion in Electromagnetic Fields   133

6 Coordinate Systems 144

6.1 Introduction   144
6.2 Plane Polar Coordinates   144
6.3 Cylindrical Polar Coordinates   147
6.4 Spherical Polar Coordinates   148
6.5 Moving Coordinate Systems   153
6.6 Rotating Coordinate Systems   156
6.7 Motion Observed on the Rotating Earth   164
6.8 Foucault Pendulum   168

7 Central Forces 176

7.1 Introduction   176
7.2 Potential Energy and Central Forces   177
7.3 Angular Momentum and Central Forces   179
7.4 Inverse-Square Force   184
7.5 Kepler's Laws   189
7.6 Orbital Transfers and "Gravitational Boosts"   191
7.7 Radial Oscillations about a Circular Orbit   200
7.8 Gravity   204
7.9 Rutherford Scattering   210

8 Systems of Particles 222

8.1 N Particles, the General Case   222
8.2 Momentum   225

8.3 Motion with a Variable Mass — Rockets   227
8.4 Motion with a Variable Mass — Conveyor Belts   230
8.5 Collisions   231
8.6 Center of Mass Frame   236

9 Rigid Bodies 246

9.1 Center of Mass   246
9.2 Angular Momentum   247
9.3 Rotation about an Axis   251
9.4 Moment of Inertia Theorems   256
9.5 The Inertia Tensor   260
9.6 Principal Axes   262
9.7 Kinetic Energy   268
9.8 Euler's Equations   269

10 Lagrangian Mechanics 282

10.1 Introduction   282
10.2 Generalized Coordinates   283
10.3 Generalized Forces   285
10.4 Lagrange's Equations   287
10.5 Elementary Examples   289
10.6 Systems with Constraints   292
10.7 Applications   295
10.8 Ignorable Coordinates   297
10.9 Lagrangian Mechanics and a Rotating Top   298
10.10 Hamilton's Equations   304
10.11 Hamilton's Principle   306

11 Statics 314

11.1 Introduction   314
11.2 Plane Trusses   315
11.3 Method of Joints   320
11.4 Method of Sections   325
11.5 Cables under Distributed Loads   327
11.6 Parabolic Cables   329
11.7 Catenary Cables   332
11.8 Cables with Concentrated Loads   335

12 Fluid Mechanics 346

12.1 Introduction 346
12.2 Hydrostatics: Fluids at Rest 346
12.3 Moving with the Flow 355
12.4 Hydrodynamics 359

13 Special Relativity 370

13.1 Introduction 370
13.2 Galilean Relativity 371
13.3 Historical Background 373
13.4 Einsteinean Relativity/Spacetime Coordinates 374
13.5 Simultaneity 376
13.6 Lorentz Transformations 378
13.7 Application of the Lorentz Transformations 382
13.8 Minkowski Diagrams 385
13.9 Velocity Transformation 392
13.10 Doppler Effect 394
13.11 Momentum and Mass-Energy 395
13.12 Relativistic Mass 403

APPENDICES

A Conversational Pascal 407

B Calculus Review 424

C Multiple Integrals 432

D Matrix Multiplication 444

Index 447

PREFACE

Mechanics is usually the first course following introductory physics taken by students of physics and engineering. For many students of mathematics, chemistry, and other sciences, mechanics may be their only additional physics course. Most practicing physicists remember this as a tough course. Undergraduate physics majors may well be convinced it is their toughest course.

Indeed, mechanics is a tough course. For it is here that most students first *apply* the concepts of calculus to another field. Students leave a typical introductory physics course with an appreciation for the basic ideas and concepts of physics; in concurrent mathematics courses they have developed an appreciation for functions, differentiation, integration, and perhaps even differential equations. Now the object is to put these two together—to turn calculus into an effective *tool* to be used in physics.

This text is written to help students learn to use calculus effectively to solve problems in classical mechanics. It is intended to be an aid to understanding new ideas, not a reference book to be used only after the material is thoroughly understood. Nor is it written to impress colleagues with the author's elegance and sophistication. I have diligently, though not always successfully, attempted to avoid such ominous phrases as "clearly, it can be shown that..."; "after a little algebra, we have..."; or "as is well known...."

For the same reasons I have not lumped all of the necessary mathematical ideas into a first chapter for future reference nor relegated them to appendices, which I fear few people ever read. The details of vector algebra and vector calculus appear in Chapter 4, Vectors, just before they are used extensively in the next chapter, Motion in Three Dimensions. The details of spherical coordinates and moving coordinate systems appear in Chapter 6, Coordinate Systems, just before they are used in the following chapter, Central Forces. Mathematics is so important to this course that it can sometimes obscure the physics. Care must be taken to ensure that the necessary mathematics is always seen as a valuable tool and not as an end in itself.

The harmonic oscillator plays a significant role throughout the first half of this text, as it does in any typical mechanics course. There is much to be learned there, in terms of both physical concepts and mathematical tools. Students interested in electricity or electronics are often unhappy when so much time is devoted to "a block on a spring." Yet the ideas, equations, and techniques developed here are directly applicable to electric circuits and electronic oscillators (to say nothing of such other diverse endeavors as nuclear physics or optics).

Chapter 1, Conversational BASIC, is a crash course (or brief refresher) in the BASIC computer language and its immediate *application* to solving the harmonic oscillator. This text is *not* intended to be a computer course with a bit of physics on the side; it is not even intended to be a computer-based mechanics course. Rather, it is meant to be a solid, fundamental physics course that recognizes the usefulness of a new tool, the computer. In my own classroom I begin the semester with about two lectures devoted to developing the computer program in Chapter 1, which solves for the motion of a harmonic oscillator. During the next couple of weeks my students write and run programs that illustrate the behavior of a harmonic oscillator under a wide variety of conditions, concurrently with their study of the more usual analytical solutions. Just as with classroom demonstrations, these graphical results from the computer provide yet another bridge between the real, physical world of blocks and springs and the mathematical world that describes it.

It has been my observation that the two days I spend developing the simple ideas of numerical analysis are more than repaid throughout the course. Later material is understood better and mastered more quickly because of this additional tool at the disposal of my students. Further, many interesting problems in physics do not have analytical solutions in closed form; numerical solutions are their *only* solutions. Such problems no longer need to be avoided. Problems involving the use of computers are included throughout the text, usually at the ends of the problem sets. These are indicated by a star (★). Thus I find the numerical approach in Chapter 1 to be both efficient and pedagogically important. Appendix A, Conversational Pascal, is included for those who prefer Pascal.

One can reasonably ask what Chapter 13, Relativity, is doing in a text on *classical* mechanics: after all, relativity is certainly *modern* physics (along with quantum mechanics). It is included partly out of tradition. A first course in introductory physics often touches on relativity—and it should. But time constraints and the newness of the material often mean that even good students leave a course in introductory physics with little or no understanding of relativity. Relativity is important, but it is also interesting and even intriguing. Professor Ohanian describes relativity as doing "violence to our intuition." We must be willing to travel with Alice down yet another rabbit hole—one where our common sense is often wrong. But that is part of the fun of relativity. Chapter 13 is but a brief introduction to special relativity (that is, high velocities but no accelerations). It is included as an extension of the ideas developed earlier for lower speeds, but may be omitted as time constraints dictate.

This text is designed for the usual course in theoretical mechanics, classical mechanics, or analytical mechanics that follows a course in introductory physics. It may be taken concurrently with the end of that course, which usually deals with modern physics. If freshmen take introductory physics they will be ready for this course as sophomores; if introductory physics is delayed until the sophomore year, this will be a junior course.

In a one-year course, I would like to cover all of the first ten chapters and, perhaps, Chapter 13 on relativity. However, I've never had enough time to do that. I usually settle for covering the first eight chapters in depth and quickly surveying Chapters 9 and 10. Chapters 2 through 10 contain the basics, the "heart and soul," of classical mechanics. Chapter 11 on statics and Chapter 12 on hydrodynamics may be considered optional. They are included as interesting and serious applications. These chapters are quite important for students going into engineering or applied physics, and may even supplant Chapters 9 or 10.

For a one-semester course, I would still attempt to cover the first eight chapters. For any course, Sections 4.1, 4.2, and 4.3 may well be treated as a review of vectors. The vector operations of divergence and curl could also be omitted for a one-semester course. Chapter 6 on coordinates and Chapter 7 on central forces could be treated briefly. Orbital Transfers, Section 7.6, and Rutherford Scattering, Section 7.9, could be omitted altogether.

Books are special to me. Therefore, authoring a book is quite special. Only a little reflection reminds me that this book is the indirect product of many, many people. My mother and father gave me a very early appreciation of books (the summer I was six, we checked out *White Prince* every time the bookmobile came to town), of numbers, and of questions and learning in general. Only recently have I realized how important a part my family physician, Dr. Louis Morgan, has played in my interest in science and technology. Professors Henry Unruh, Gerald Loper, and Henry Pronko had much to do with molding my view of science and leading me into a career in University teaching. I appreciate Professor Steve Moszkowski at UCLA and the trust and patience he exhibited.

Regarding the text itself, I especially appreciate the encouragement offered from the beginning by my friend and colleague, Professor Scott Smith. My friend and department chairman, Professor William Butler, has also been very supportive. Thanks are extended to Ken Bowman and Dick Morel for their encouragement. Special thanks go to Jeff Holtmeier, senior editor at Academic Press, for his enthusiastic support, encouragement, and efficiency. Many thanks are also due Pam Robertson and Lenn Holland at Academic Press for their efforts with this book. George Morris and his able crew at Scientific Illustrators are greatly appreciated for their fine illustrations. I want to thank Greg Hubit and his staff at Bookworks, who handled the production of this book. Writing a book is great—but it is also a long and tedious process. I greatly appreciate the support and encouragement that my wife Judy has given me throughout this project. For their help in typing early versions of this text, I extend my thanks to Linda Hall and Rebecca Wheeler. For typing the later versions, I appreciate all the efforts of Rebecca Wheeler, Sheri Smith, Marie Moline, Yvette Carey, Mary Houston, and, especially, Mary Ostendorf and Chris Sims. I appreciate, too, my students who have used this text, my friends and colleagues, Professors William Franz, Michael McInerney, and Scott Smith, who have taught from this text, and the reviewers. I want to thank fellow physicists John Reed, Jr. and Andy Jones for their skill and efforts in helping so greatly with the *Instructor's Solutions Manual*. All these people have offered many useful suggestions from which the text has benefited.

1

CONVERSATIONAL BASIC

1.1 Getting Started

In any physics course, it's often easy for the physics of the situation to become obscured by the mathematics. To avoid this (for the present), let's look at a harmonic oscillator and use the simplest mathematical tools possible.

For our harmonic oscillator, we'll look at a mass, m, attached to one end of a spring having spring constant k. The other end of the spring is firmly attached to a rigid support, as shown in Figure 1.1.1. When the mass is moved some distance, x, away from equilibrium, the spring exerts a force to restore it—to push or pull it back to equilibrium. This force is found to be proportional to the distance the mass is from equilibrium (the constant of proportionality being the spring constant k). This behavior of a linear restoring force is known as Hooke's Law and, as an equation, can be written as

$$F = -kx \qquad (1.1.1)$$

From Newton's Second Law of Motion we also know

$$F = ma \qquad (1.1.2)$$

where a is the acceleration experienced by mass m when acted upon by force F. Therefore,

$$ma = -kx$$

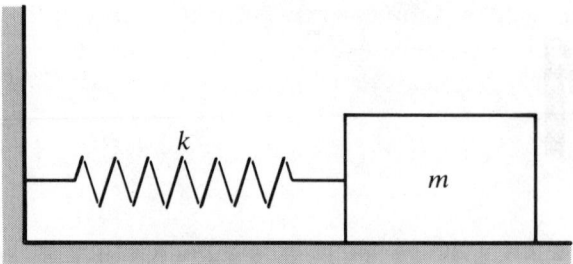

Figure 1.1.1

or

$$a = -\frac{kx}{m} \tag{1.1.3}$$

If the mass is moved to some initial position, x_0, and released with some initial velocity, v_0, Eq. 1.1.3 should enable us to locate the mass at any later time—*in principle*.

For simpler cases, such as $a = a_0$ (a constant), you already know that

$$\begin{aligned} v &= a_0 t + v_0 \\ x &= \tfrac{1}{2} a_0 t^2 + v_0 t + x_0 \end{aligned} \tag{1.1.4}$$

and for $a = a(t)$, we can integrate directly to get

$$\begin{aligned} v &= \int a(t)\, dt + v_0 \\ x &= \int v(t)\, dt + x_0 \end{aligned} \tag{1.1.5}$$

But our present situation doesn't lend itself to such direct methods. So, we shall omit integration entirely and solve this problem *by hand*! Using a hand-held calculator will make the calculation go much faster.

To begin, label some columns on a sheet of paper as follows:

Time	Position	Velocity	Acceleration
0	x_0	v_0	$-\dfrac{k}{m} x_0$

Usually we know where the mass is and how fast it is going at the beginning. These are the *initial conditions*. We can write x_0 and v_0 for these initial conditions when $t = 0$. Knowing the position allows us to calculate the force and then the acceleration. Acceleration for this mass and spring is $-(k/m)x_0$. All these values are entered in the previous table.

Although we begin by looking at the motion of a mass attached to a spring—a simple harmonic oscillator—the ideas and techniques we develop here will be useful in many different systems. This simple harmonic oscillator is just an example, but an important one.

The acceleration changes as x changes. If we observe this system for a very brief time, Δt, so that x doesn't change much, then acceleration doesn't change much either. For that matter, the velocity stays nearly the same during this time, too.

If we let time increase to $t_1 = \Delta t$, where Δt is small enough that the change in acceleration is negligible, we may treat the acceleration as a constant. Then,

$$v(t_1) = v_0 + a\,\Delta t \tag{1.1.6}$$

Likewise, the velocity will be nearly constant over such a small time interval. So we can write

$$x(t_1) = x_0 + v\,\Delta t \tag{1.1.7}$$

Now the data table looks like this:

Time	Position	Velocity	Acceleration
0	x_0	v_0	$-\dfrac{k}{m}x_0$
t_1	$x(t_1)$	$v(t_1)$	

Now let's try this with some actual numbers. Suppose we use a spring with spring constant $k = 1000$ N/m attached to a body with a mass of $m = 5$ kg. Pull it to one side a distance of 10 cm or $x_0 = 0.10$ m. Let it go from rest so that $v_0 = 0.00$ m/s. As you release it, its acceleration is $-(k/m)x_0 = -(1000/5)0.1 = -20$; that is, -20 m/s^2. For the time increment Δt, use 0.01. From Eq. 1.1.6 the velocity at $t = 0.01$ s must be $v(0.01) = 0 + (-20)(0.01) = -0.2$; that is, -0.2 m/s. This velocity and Eq. 1.1.7 allow us to calculate the position at $t = 0.01$ s. We find $x(0.01) = 0.10 + (-0.2)(0.01) = 0.0998$, which is really 0.0998 m.

With actual numbers, then, our data table looks like this:

Time	Position	Velocity	Acceleration
0.00	0.10000	0.00	-20
0.01	0.0998	-0.20	

Since we now know the current (or "new") position, we can calculate the current (or "new") value of the acceleration from

$$a(t_1) = \frac{k}{m}x(t_1) \tag{1.1.8}$$

and include it in our table.

Let time increase by Δt again, this time to t_2. Just as before, assume that Δt is so small that v and a are almost constant. Then we can write

$$v(t_2) = v(t_1) + a(t_1)\,\Delta t \tag{1.1.9}$$

$$x(t_2) = x(t_1) + v(t_2)\,\Delta t \tag{1.1.10}$$

Or we can write this as

$$v(\text{new}) = v(\text{old}) + a(\text{old})\,\Delta t \tag{1.1.11}$$

$$x(\text{new}) = x(\text{old}) + v(\text{old})\,\Delta t \tag{1.1.12}$$

We can then calculate a new value for the acceleration:

$$a(\text{new}) = -\frac{k}{m} x(\text{new}) \tag{1.1.13}$$

Using this method over and over again, we can continue to fill in our table of values. The smaller we make Δt, the more accurate our assumptions (and our results) become. The larger we make Δt, the quicker we're finished, but with less accurate results.

In this way, we reduce this problem to a particular pattern of "plug and crank arithmetic." But we fear you will quickly tire of this. So think. Is there anyone on campus you can get to do these simple arithmetical operations and give you the finished table, all neatly filled in?

Of course there is—your friendly campus computer! It speaks several languages, if not English or algebra. So we have to take a crash course in what we'll call "conversational BASIC." This isn't a course in computer programming, so we're not going to learn *real* BASIC—rather, just enough to solve our current problem.

The spring constant, k, and the mass, m, have numerical values. We can inform the computer of this by telling it

LET K = k (1.1.14)

LET M = m (1.1.15)

K and M are locations—just like a scratch pad, a chalkboard, or an envelope. k and m are the actual numbers we then store in the respective locations—or write on the scratch pad or chalkboard or put in the envelope.

To make this example more explicit, suppose the spring has a spring constant of 5.23 N/m and is attached to a mass of 0.750 kg. Eqs. 1.1.14 and 1.1.15 then become

LET K = 5.23 (1.1.16)

and

LET M = 0.750 (1.1.17)

Please note that the computer has no way of keeping track of units, so that's up to you. The computer works only on the numbers.

SECTION 1.1 / GETTING STARTED

When $t = 0$, the initial position is x_0 and the initial velocity is v_0. We can tell the computer this by saying

$$\text{LET } T = 0.0 \tag{1.1.18}$$

$$\text{LET } X = x_0 \tag{1.1.19}$$

$$\text{LET } V = v_0 \tag{1.1.20}$$

x_0 and v_0 are just numbers—like 1.50 (m) and 2.35 (m/s). T, X, and V are locations inside the computer—or scratch pads, chalkboards, or envelopes, if you please—that now contain the numbers 0.0, x_0, and v_0, respectively.

We must now determine a value for Δt. For the time being, let's choose $\Delta t = 0.05$. Tell the computer

$$\text{LET } D = 0.05 \tag{1.1.21}$$

The location D now contains our choice for Δt.

The computer can now calculate the acceleration if we tell it

$$\text{LET } A = -K*X/M \tag{1.1.22}$$

Note that the asterisk (*) is used for multiplication and the slash (/) for division. The computer doesn't understand "LET $A = -KX/M$."

To start printing our table, we could now instruct the computer to

$$\text{PRINT } T; X; V; A \tag{1.1.23}$$

Although we need A for the calculations, we're usually more interested in X and V. So even though the computer will be keeping A's value inside, let's ask for a simplified printout:

$$\text{PRINT } T; X; V \tag{1.1.24}$$

We need one more thing. In BASIC, even "conversational BASIC," every statement must be numbered. The statements need not be consecutive, but they must be sequential. Therefore, if we were to number our statements so far and put them together, it would look like this (since we don't need A just yet, we omit its calculation for the moment):

$$10 \quad \text{LET } K = k \tag{1.1.25}$$

$$20 \quad \text{LET } M = m \tag{1.1.26}$$

$$30 \quad \text{LET } T = 0.0 \tag{1.1.27}$$

$$40 \quad \text{LET } X = x_0 \tag{1.1.28}$$

$$50 \quad \text{LET } V = v_0 \tag{1.1.29}$$

$$60 \quad \text{LET } D = \Delta t \tag{1.1.30}$$

$$70 \quad T; X; V \tag{1.1.31}$$

Now we can reinsert our instruction to calculate the acceleration:

80 LET A = −K∗X/M (1.1.32)

90 LET V = V + A∗D (1.1.33)

Whoops! That doesn't seem to be correct unless A∗D is equal to zero. But "=" in BASIC isn't quite what you would expect. A good English paraphrase of this BASIC statement might be:

Take the numerical value written at location A and multiply it by the numerical value written at location D. Now take the numerical value written at location V and add it to this product. Finally, put this *new* value at location V.

The *original* value stored at location V is erased, forgotten, lost, destroyed. The values at A and D remain.

Likewise, Eq. 1.1.12, which we used to calculate the new position, readily translates into

100 LET X = X + V∗D (1.1.34)

These are the *new* values of X, V, and A, corresponding to a new time. To find the new time, we can instruct the computer to

110 LET T = T + D (1.1.35)

So that we can see the results, we tell it to

125 PRINT T; X; V (1.1.36)

to get a new set of entries on our data table.

We now need to repeat these calculations over and over again. But instead of making more and more copies of the same statements, we add

130 GO TO 80 (1.1.37)

and the computer will do exactly that. It will GO TO Statement 80. There it will calculate a new A. At 90, it calculates a new V, at 100 it calculates a new X. 110 then changes T to its new value. In this way, a new set of entries have been calculated and are printed out by Statement 125. 130 directs the computer to GO TO 80 and calculate another set of entries for the data table.

And what a data table this will produce! Like the sorcerer's apprentice, we've created a helpful servant, but we need a way to stop it. To do this, let's insert

115 IF T > 10.0 THEN 140 (1.1.38)

SECTION 1.1 / GETTING STARTED

after Statement 110. Then, following Statement 130, we add

$$140 \quad \text{END} \tag{1.1.39}$$

Statement 115 doesn't change a thing *until* the time gets larger than 10.0 (seconds). Then the computer skips to Statement 140 and a quick END. We're back in control of the situation.

What do we choose for Δt? For accuracy, we want it small. We might try 0.01; or 0.001 would be even more accurate. But that will generate 1000 or 10,000 entries on our data table! Increasing Δt to 0.1 or 0.5 gives us a reasonable number of entries, but these data are unreliable. How can we have both accuracy (from a small Δt) and a reasonable number of data entries?

What if you were still processing Eqs. 1.1.11, 1.1.12,. and 1.1.13 by hand? You could use $\Delta t = 0.001$ for accuracy, but only use the results for $t = 0$, $t = 0.1$, $t = 0.2$, and so on. In this way, all the calculations are made every 0.001 seconds, but the results are *printed* only every 0.1 seconds—one printout every 100 calculations.

To tell the computer to do this, we must first establish a method of counting. So early in our program we should define a location—we'll call it N—and set the contents equal to zero:

$$75 \quad \text{LET N} = 0 \tag{1.1.40}$$

This statement, then, belongs between Statements 70 and 80. Every time T is incremented we'll also increment N by adding Statement 112

$$112 \quad \text{LET N} = \text{N} + 1 \tag{1.1.41}$$

Instead of printing every calculation, we want to print only when $N = 100$. We can ensure this by adding Statement 120:

$$120 \quad \text{IF N} < 100 \text{ THEN } 80 \tag{1.1.42}$$

Just as with our IF statement in Statement 115, when "IF $N < 100$" can be answered by "yes" or "true," THEN the right-hand command is followed and control returned to line 80. When N finally does reach 100, the "IF $N < 100$" is no longer true and the computer will go on to the very next statement.

$$125 \quad \text{PRINT T; X; V} \tag{1.1.43}$$

Once $N = 100$, we must reinitialize $N = 0$ and proceed. Following the print statement, we add

$$130 \quad \text{GO TO } 75 \tag{1.1.44}$$

Statement 75, you remember, initialized $N = 0$ the first time through.
Now the final program looks something like this:

```
10   LET K = k
20   LET M = m
30   LET T = 0.0
40   LET X = x₀
50   LET V = v₀
60   LET D = 0.001
70   PRINT T; X; V
75   LET N = 0
80   LET A = −K∗X/M
90   LET V = V + A∗D
100  LET X = X + V∗D
110  LET T = T + D
112  LET N = N + 1
115  IF T > 10.0 THEN 140
120  IF N < 100 THEN 80
125  PRINT T; X; V
130  GO TO 75
140  END
```

This program works quite well. If we interchange lines 90 and 100, calculating the new position before the new velocity, then the program would use a numerical integration method developed by Euler over two centuries ago. But the order of the lines in your program greatly increases the accuracy of the calculations, an important result only recently recognized by Professor Alan Cromer of Northeastern University[1].

1.2 Elegant Output

We now have written a thoroughly functional program. It can do calculations and print out an exhaustive data table. Can we ask for anything more? Of course we can!

There are often times when it is helpful to make a remark in the program to remind ourselves of something. We can do this in BASIC with a REM statement. For example, we can add a heading like

5   REM   CALCULATION OF POSITION FOR HARMONIC OSCILLATOR   (1.2.1)

---

[1] Alan Cromer, "Stable solutions using the Euler approximation," *American Journal of Physics*, Vol. 49 (May 1981), pp. 455–459.

to our program. Or, between Lines 75 and 80 we could insert

    78   REM   MAJOR CALCULATIONS OCCUR HERE   (1.2.2)

The computer ignores these lines, but they may be helpful when we read the program.

The **PRINT** statement in BASIC also can be used to print more than just numbers. The statement

    7   PRINT   "HARMONIC OSCILLATOR CALCULATIONS"   (1.2.3)

tells the computer to print everything *within* the quotations. We can use this feature and print numerical values at the same time, too.

As an example, suppose the spring constant is $k = 125$ N/m and the spring is attached to a mass of $m = 0.75$ kg. We can tell the computer this by

    10   LET K = 125   (1.2.4)

    20   LET M = 0.75   (1.2.5)

We could then add

    25   PRINT "SPRING CONST ="; K; "MASS ="; M   (1.2.6)

The computer will then print the expression in the first quotation,

    SPRING CONST =

print the numerical value stored in location K,

    SPRING CONST = 125

and then print the expression in the second set of quotation marks:

    SPRING CONST = 125   MASS =

Finally, it will print the numerical value stored in M:

    SPRING CONST = 125   MASS = 0.75

Things could be spruced up a bit if we had the units printed out as well. To do this we change Statement 25 to

    25   PRINT "SPRING CONST ="; K; "N/M MASS ="; M; "KG"   (1.2.7)

The material inside the quotation marks will be printed out *exactly* as written. The material outside the quotation marks will be used as a command to look up and then print a numerical value.

Graphs are often better at displaying information than data tables. For example, it's easier to find the period, or see if it's varying, or determine what's happening to the amplitude in a graph.

BASIC has a TAB ( ) function that operates like a tabulator on a typewriter and that can be used to generate a graph (albeit rough) of data.

Instead of just printing the position, the value of X, TAB ( ) can direct the printer to a certain column and then to print a symbol—X, *, or., or whatever you like. For instance, with

  200 PRINT TAB (10); "X" (1.2.8)

the printer moves to column 10 and then prints X.

  210 PRINT "V"; TAB (35); "Z" (1.2.9)

will print V in column 1, and then move to column 35 and print Z.

  220 PRINT A; B; TAB (57); "X" (1.2.10)

will print numerical values for A and B and then move to column 57 and print X. (Note that it does *not* move 57 spaces from the numerical value of B.)

  The TAB function can also contain an arithmetic function inside it. Look at

  210 LET A = 5.5 (1.2.11)

  215 LET X = 10.3 (1.2.12)

  225 PRINT A; TAB (4*X); "*" (1.2.13)

In this case, the computer will print the numerical value of A, 5.5, and then calculate 4*X or 4*10.3 = 41.2. Then it will move to column 41 (the decimal part is ignored), where it prints an asterisk, *. In BASIC, up to 80 columns can be printed per line. Let's reserve the first 10 or 15 to continue to print numerical values for T and X. This leaves 65 or so columns in which to make a graph. As X varies from its minimum to maximum values, we want, say, an asterisk (*) to be printed between columns 16 and 80 and printed in the middle—column 45—when $X = 0$.

  How large is X going to get? That depends upon the initial conditions, x_0 and v_0. Suppose we don't expect X to ever get greater than 1 or 1.5. Then we might multiply X by 20 to get a position P to use with TAB. But we want the asterisk printed in column 45 when X is zero. We can do that by

  190 LET P = 45 + 20*X (1.2.14)

  200 PRINT T; X; TAB (P); "*" (1.2.15)

But we can simplify this to just

  200 PRINT T; X; TAB (45 + 20*X); "*" (1.2.16)

since we can place an arithmetic expression inside TAB. If the range of X is generally larger, or smaller, we'll have to change this expression.

  Now look back at our original program. Our PRINT statement is line number 125, which we can replace with a new statement:

  125 PRINT T; X; TAB (45 + 20*X); "*" (1.2.17)

This modification should now result in a reasonable graph with time plotted downward and position horizontally. Rotating the computer's printout by 90° yields a more conventional distance/time graph.

So far, we've only used $A = -K*X/M$ in our program. We *could* have done this explicitly by writing

$$80 \quad \text{LET } F = -K*X \qquad (1.2.18)$$

$$85 \quad \text{LET } A = F/M \qquad (1.2.19)$$

These statements hold for a simple harmonic oscillator and assume no friction at all. But suppose our apparatus is submerged in water—or corn syrup or molasses. Then there would be a frictional force opposing the motion. This force is proportional to the velocity, so our *net* force would be

$$80 \quad \text{LET } F = -K*X - C*V \qquad (1.2.20)$$

where C is as yet undefined (it's small for water and very large for molasses). We can go back to the beginning of the program and define C by

$$25 \quad \text{LET } C = c \qquad (1.2.21)$$

where c is any value we want.

Ordinarily, we will be concerned with the units of anything we use; C must have units of N/(m/s) or kg/s. But we shall ignore such reasonable considerations in this chapter so we can concentrate on the computer program and the motion it describes.

Adding this velocity-dependent term to the force means that we are describing a *damped harmonic oscillator*.

Now we can add still another component or term to the force, an external driving force like

$$F = f_0 \sin \omega t \qquad (1.2.22)$$

that varies with time. This is now a *driven harmonic oscillator*.

Once we define f_0 and ω as

$$27 \quad \text{LET } E = f_0 \qquad (1.2.23)$$

$$28 \quad \text{LET } W = \omega \qquad (1.2.24)$$

we can rewrite the force equation of Line 80:

$$80 \quad \text{LET } F = -K*X - C*V + E*\text{SIN}(W*T) \qquad (1.2.25)$$

This means that we're trying to move the mass with this external driving force. How will it respond? Certainly increasing f_0 (or E) will increase the amplitude of the motion. But the motion's dependence upon the angular frequency ω or W is of more interest. What frequency causes the largest amplitude? This short program allows immediate investigation of this driven harmonic oscillator. We'll return to it later and find analytical solutions to this problem.

1.3 Sometimes More Is Really Less

We're finished with the physics of the problem for the time being. After all our modifications, our program now looks like this:

```
   5   REM CALCULATION OF POSITION FOR HARMONIC OSCILLATOR
   7   PRINT "HARMONIC OSCILLATOR CALCULATIONS"
*10    LET K = k
*20    LET M = m
*25    LET C = c
*27    LET E = f₀
*28    LET W = ω
*30    LET X = x₀
*40    LET V = v₀
  50   LET T = 0.0
  60   LET D = 0.001
  70   PRINT T; X; TAB (45 + 20*X); "*"
  75   LET N = 0
  78   REM MAJOR CALCULATIONS OCCUR HERE
  80   LET F = −K*X − C*V + E*SIN(W*T)
  85   LET A = F/M
  90   LET V = V + A*D
 100   LET X = X + V*D
 110   LET T = T + D
 112   LET N = N + 1
 115   IF T > 10.0 THEN 140
 120   IF N < 100 THEN 080
 125   PRINT T; X; TAB (45 + 20*X); "*"
 130   GO TO 075
 140   END
```

The seven lines marked with an asterisk define the parameters of our system and the initial conditions. Numerical values here determine the numbers calculated by the program.

So far, we have made a separate run to the computer every time we needed to change any of these values. There must be something we can do to decrease this workload (of course there is!). Rather than assigning each value in a LET statement, we can use the following READ and DATA statements:

$$10 \quad \text{READ K, M, C, E, W, X, V} \tag{1.3.1}$$

$$20 \quad \text{DATA k, m, c, } f_0, \omega, x_0, v_0 \tag{1.3.2}$$

This READ statement tells the computer to go to all the numbers listed on all the DATA statements and store the first value in K, the next in M, and so on. The DATA statement can be placed anyplace in the program—anyplace before END, that is. An example might be:

$$10 \quad \text{READ K, M, C, E, W, X, V} \tag{1.3.3}$$

$$20 \quad \text{DATA 1.0, 1.0, 0.01, 0.0, 0.0, 1.0, } -0.1 \tag{1.3.4}$$

This is equivalent to

```
LET K = 1.0
LET M = 1.0
LET C = 0.01
LET E = 0.00, no external force
LET W = 0.00, but after E = 0, the value of W doesn't matter
LET X = 1.0, for its initial position
LET V = −0.1, having started at X = x₀ = 1.0 m, the 1 kg mass has initial
     velocity of V = v₀ = −0.1 m/s.
```

Now any new runs can be made by just retyping that *single* DATA statement on Line 20. That's easier than it was.

Let's change Line 140 to

$$140 \quad \text{GO TO 10} \tag{1.3.5}$$

and add

$$150 \quad \text{END} \tag{1.3.6}$$

After the program has finished its first run and control moves to Line 140, it now returns to Line 10 and tries to read more values. If these values are supplied on subsequent DATA statements, our wondrous computer will continue to produce beautiful graphs.

To make three runs of the simple harmonic oscillator ($C = 0$, $E = f_0 = 0$) with varying initial conditions and three runs of the damped harmonic oscillator ($C \neq 0$, $E = f_0 = 0$) with varying values of the damping constant, C, and three runs for the forced harmonic oscillator, we might submit a single computer program as follows:

```
 5  REM HARMONIC OSCILLATOR PROGRAM
 7  PRINT "HARMONIC OSCILLATOR CALCULATIONS"
10  READ K, M, C, E, W, X, V
19  REM DATA FOR SIMPLE HARMONIC MOTION (SHM)
20  DATA 1, 1, 0, 0, 0, 1, −0.25
21  DATA 1, 1, 0, 0, 0, 1, 0
22  DATA 1, 1, 0, 0, 0, 1, 0.25
23  REM DATA FOR DAMPED HARMONIC MOTION
24  DATA 1, 1, 0.5, 0, 0, 1, 0
25  DATA 1, 1, 1.0, 0, 0, 1, 0
26  DATA 1, 1, 1.5, 0, 0, 1, 0
27  REM DATA FOR FORCED HARMONIC MOTION
28  DATA 1, 1, 0.25, 0.1, 0.8, 1.0, 0.0
29  DATA 1, 1, 0.25, 0.1, 0.9, 1.0, 0.0
30  DATA 1, 1, 0.25, 0.1, 1.0, 1.0, 0.0
```

```
50   LET T = 0.0
60   LET D = 0.001
      ⋮
REST OF THE PROGRAM
      ⋮
140  GO TO 10
150  END
```

Each time $T > 10.0$, the calculations are complete; control at Line 140 is turned back to Line 10 and seven *new* values are read from DATA statements. As currently written, after the last data statement is read and control is returned to Line 10, the computer will search in vain for more data. Finding no data, the program will terminate with an error message something like

DATA NOT FOUND AT LINE 10   (1.3.7)

This is somewhat like ending a telephone conversation by simply hanging up—effective, but not as nice as it might be. But it will suffice for our purposes.

Of course there is always a chance of a typographical error in the DATA statements. A good way to check this is to follow the READ statement immediately by a PRINT statement, thereby verifying that it is actually using the data you expect. This can simply be

```
10   READ K, M, C, E, W, X, V                                    (1.3.8)

11   PRINT K; M; C; E; W; X; V                                   (1.3.9)
```

Or you can spruce it up a bit like this:

```
10   READ K, M, C, E, W, X, V
11   PRINT "HARMONIC OSCILLATOR WITH"
12   PRINT "SPRING CONST =" K; "N/M and MASS ="; M; "KG"
13   PRINT "DAMPING CONST, C ="; C
14   PRINT "EXTERNAL FORCE ="; E; "*SIN ("; W; "*T)"
15   PRINT "INITIAL CONDITIONS ARE x₀ ="; X; "v₀ ="; V
```

1.4   Into The Wild Blue . . .

We can now accurately (and elegantly) calculate the rather complex motion of an oscillator. As you shall see later, this is far from a trivial problem. It's the computer and BASIC that make our calculations fairly easy. Because of that, *you* can expend your current energies on understanding the physics of the motion—how it moves and why.

Now let's apply our powerful tool to some other situation. First, consider the very familiar case of free fall with no air resistance. This is a case of constant acceleration; the results are described by

$$v = v_0 - gt \tag{1.4.1}$$

$$x = x_0 + v_0 t - \tfrac{1}{2}gt^2 \tag{1.4.2}$$

with which you should be very familiar. Such motion is a result of a force

$$F = -mg \tag{1.4.3}$$

that is proportional to the mass of the body. (This was one of Galileo's big contributions to physics.) To use your BASIC program, all we need do is change the force Statement 080 to:

$$\text{80 \quad LET F} = -\text{M} * \text{G} \tag{1.4.4}$$

We must define G earlier ($G = 9.8$ for SI units) and do not need to define K, C, E, or W since they have been removed from our force equation.

We can now turn to air resistance, and find that for small, smooth objects at low speeds it is nearly proportional to the speed of the object. We accommodate for this by changing Statement 80 to

$$\text{80 \quad LET F} = -\text{M} * \text{G} - \text{C} * \text{V} \tag{1.4.5}$$

This slight change is trivial as far as the computer is concerned, but it makes our force a much closer approximation of real forces encountered in nature. As you shall see later, this slight change also greatly increases the complexity of the analytic solution in closed form.

With such a force, it becomes very interesting to look at the behavior of the speed. We therefore change Statement 125 to include the speed again. If dropped from rest, the speed increases initially, but reaches (or approaches) some value and then remains constant. This is called the *terminal speed* of the object. A constant speed requires zero acceleration or zero force, so

$$F = -mg - cv_t = 0 \tag{1.4.6}$$

$$v_t = -\frac{mg}{c} \quad \text{or} \quad |v_t| = \frac{mg}{c} \tag{1.4.7}$$

where v_t is the terminal velocity ($|v_t|$ is the terminal speed). Terminal speed for a man falling without a parachute might be about 120 mph. A parachute increases the drag (increases c) so the terminal speed is (greatly) reduced. Even before deploying a parachute, a sky diver can change his speed by extending his arms and legs; this changes his drag (and, thus, c) and, in turn, the terminal velocity.

Air resistance is complex. As we shall see later, considering it as a force proportional to velocity (or speed) is a reasonable approximation and allows us to solve problems without too much difficulty using calculus. For some objects—and especially for higher speeds—a force proportional to the *square* of the velocity is a better approximation.

Our first attempt might be to change our definition of the force to

$$80 \quad \text{LET F} = -\text{M}*\text{G} - \text{C2}*\text{V}*\text{V} \tag{1.4.8}$$

where $C2$ is our proportionality constant. $C2$ is a coefficient just like C for the linear case, but it has different units. Therefore, we can avoid difficulties later by remembering that it is different and not writing it simply as C.

But wait just a minute. Whether V is positive when the object is going upward or negative when the object is going downward, the term $-C2*V*V$ is always negative (assuming $C2 > 0$). A negative force is *downward*. Air resistance, being a frictional force, is always opposed to the motion. When V is positive, the air resistance force is negative, and vice versa.

We could take care of this with IF statements checking on whether V is directed upward or downward. But a shorter method is to use the absolute value function, ABS ( ), available in BASIC. With that we can correctly define our force in Statement 80 by

$$80 \quad \text{LET F} = -\text{M}*\text{G} - \text{C2}*\text{V}*\text{ABS (V)} \tag{1.4.9}$$

ABS (V) is the absolute value of V. It is V whenever V is positive and $-V$ whenever V is negative. That gives the correct sign relations we need on V and the air resistance force.

Again, if we look at velocity as a function of time, we will find a *terminal speed* as in the earlier case. As then, a terminal speed occurs when the net force is zero (the downward gravitational component is exactly balanced by the upward air resistance component). Thus,

$$F = -mg + c_2 v_t^2 = 0 \tag{1.4.10}$$

$$v_t^2 = \frac{mg}{c_2} \tag{1.4.11}$$

$$v_t = \sqrt{\frac{mg}{c_2}} \tag{1.4.12}$$

To calculate the motion of a body under the influence of this force, our entire BASIC program looks something like this:

```
 5  REM FREE FALL WITH AIR RESISTANCE
10  PRINT "AIR RESISTANCE CALCULATION"
20  READ C2, X, V, M
30  DATA 0.05, 0.0, 50.0, 1.0
40  LET G = 9.8
50  LET T = 0.0
60  LET D = 0.001
70  PRINT T; X; V
75  LET N = 0
```

SECTION 1.4 / INTO THE WILD BLUE ...

```
78   REM MAJOR CALCULATIONS OCCUR HERE
80   LET F = -M*G - C2*V*ABS (V)
85   LET A = F/M
90   LET V = V + A*D
100  LET X = X + V*D
110  LET T = T + D
112  LET N = N + 1
115  IF T > 25.0 THEN 140
120  IF N < 1000 THEN 80
121  REM RESULTS WILL BE PRINTED FOR EACH SECOND
125  PRINT T; X; V
130  GO TO 75
140  END
```

This program, as well as all the others, has handled motion in one dimension only. But it is easy to alter it to handle projectile motion—or any motion in two dimensions under the influence of any two-dimensional force. The general case of three dimensions can easily be handled as well. Instead of dealing only with $F = ma$, we must look at $F_x = ma_x$ and $F_y = ma_y$. In some restricted versions of BASIC we cannot define FX or FY, but we can always define $F1$ and $F2$, $V1$ and $V2$, and $A1$ and $A2$ for the x- and y-components of force, velocity, and acceleration. Even in older, restricted versions of BASIC we may use variable names that start with a letter and are then followed by a number. Most current versions of BASIC are not this restrictive.

To look at simple projectile motion, we replace the force calculations of Statement 80 by

$$80 \quad \text{LET F1} = 0 \tag{1.4.13}$$

$$81 \quad \text{LET F2} = -M*G \tag{1.4.14}$$

The components of the acceleration must be handled separately, so Statement 85 becomes

$$85 \quad \text{LET A1} = \text{F1}/M \tag{1.4.15}$$

$$86 \quad \text{LET A2} = \text{F2}/M \tag{1.4.16}$$

The velocity calculation of Statement 90 becomes

$$90 \quad \text{LET V1} = \text{V1} + \text{A1}*D \tag{1.4.17}$$

$$91 \quad \text{LET V2} = \text{V2} + \text{A2}*D \tag{1.4.18}$$

and both the horizontal and vertical positions must now be calculated:

$$100 \quad \text{LET X} = X + \text{V1}*D \tag{1.4.19}$$

$$101 \quad \text{LET Y} = Y + \text{V2}*D \tag{1.4.20}$$

(Always be sure that *you* control the computer—*not* vice versa.)

PROBLEMS

1.1 Run the harmonic oscillator program with $m = 1$ kg, $\Delta t = 0.001$ s, $k = 1$ N/m, and initial conditions of $x_0 = 1.0$ m and $v_0 = 0$ (i.e., it is released from rest). From the output, determine the amplitude and period. What relation can you find between position and velocity?

1.2 With the same spring constant k and the same initial conditions x_0 and v_0 of Problem 1.1, run the harmonic oscillator program with $m = 1, 2, 3, 4, 9,$ and 16 kg. How does the mass affect the period and amplitude?

1.3 With the same mass (m) and same initial conditions (x_0 and v_0) of Problem 1.1, run the harmonic oscillator with $k = 1, 2, 3, 4, 9,$ and 16 N/m. How does the spring constant affect the period and amplitude?

1.4 Setting $m = 1$ kg and $k = 1$ N/m, vary the initial conditions
  (a) $x_0 = 0, v_0 = 1$ m/s
  (b) $x_0 = 0, v_0 = 0.5$ m/s
  (c) $x_0 = 0.5$ m, $v_0 = 0.5$ m/s
  (d) $x_0 = 0.5$ m, $v_0 = -0.5$ m/s
  (e) $x_0 = -1.0$ m, $v_0 = +0.25$ m/s
For each case, describe the motion. Specifically, what are the amplitude and period?

1.5 With $m = 1.0$ kg, $k = 1.0$ N/m, and initial conditions of $x_0 = 1.5$ m and $v_0 = 0$, make runs for the damped harmonic oscillator with damping constant values of $C = 0.0, 0.05, 0.1, 0.5, 1.0, 1.5, 2.0, 2.5,$ and 4.0. What happens to the period and amplitude as C varies?

1.6 With $m = 1.0$ kg, $k = 1.0$ N/m, $C = 0.05$, $E = 0.25$, and initial conditions of $x_0 = 0.5$ and $v_0 = 0$, make runs for the forced harmonic oscillator with the external driving frequency $W = 0.5, 0.75, 0.95, 1.0, 1.05, 1.10, 1.5,$ and 2.0. Look at the output for time *later* than 50 seconds. How does the driving frequency W affect the motion?

1.7 Consider free fall with no air resistance. Make data tables and graphs of time versus distance and time versus velocity. Compare these to the results obtained from the equations $x = x_0 + v_0 t - \frac{1}{2}gt^2$ and $v = v_0 - gt$ (they should be identical). Do this for two sets of initial conditions: (a) $x_0 = v_0 = 0$, an object released from rest, and (b) $x_0 = 0, v_0 = 100$, an object fired directly upward at 100 m/s.

1.8 Consider an object moving under the influence of air resistance proportional to velocity ($F_r = -cV$). Make data tables and graphs of time versus distance and time versus velocity for $c = 0.1, 0.25,$ and 0.5. What is the terminal speed? Do this for two sets of initial conditions: (a) $x_0 = v_0 = 0$, an object released from rest, and (b) $x_0 = 0$, $v_0 = 100$, an object fired directly upward at 100 m/s.

1.9 Consider an object moving under the influence of air resistance proportional to the *square* of the velocity ($F_r = \pm c_2 v^2$ or $F_r = -c_2 v|v|$). Make data tables and graphs of time versus distance and time versus velocity for $c_2 = 0.1, 0.25,$ and 0.5. What is the terminal speed from your data tables? What would you *expect* it to be?

Do this for two sets of initial conditions: (a) $x_0 = v_0 = 0$, an object released from rest, and (b) $x_0 = 0$, $v_0 = 100$, an object fired directly upward at 100 m/s.

1.10 Body A falls with air resistance proportional to its velocity with $c = 0.25$. Body B falls with air resistance proportional to the *square* of the velocity. What value of c_2 will give B the same terminal velocity as A? Make graphs of time versus distance and time versus velocity with A and B shown on the *same* graph. Do this for two sets of initial conditions: (a) $x_0 = v_0 = 0$, an object released from rest, and (b) $x_0 = 0$, $v_0 = 100$, an object fired directly upward at 100 m/s.

1.11 Plot the trajectory for an object fired at 45° above the horizontal with an initial speed of 100 m/s by computing its x and y positions at one-second intervals and then plotting these on a graph. Ignore air resistance.

1.12 Repeat Problem 1.11 with initial firing angles of 15°, 30°, 45°, 60°, and 75°. Plot all the trajectories on a single, clearly labeled graph.

1.13 Repeat Problem 1.12 including the air resistance term proportional to the first power of velocity. Make calculations for $c = 0.1, 0.25$, and 0.5.

1.14 Repeat Problem 1.12 including the air resistance term proportional to velocity *squared*. Make calculations for $c_2 = 0.05, 0.1$, and 0.2.

1.15 The inverse-square force, $F = K/r^2$, is important as it describes the gravitational force that holds the solar system together as well as the electrostatic force that holds atoms together. Write a BASIC program to calculate the position of a body moving under the influence of this inverse-square force. Choose units so $K = -1.0$. Let the initial position be $x_0 = 1$, $y_0 = 0$. Be careful to determine the x- and y-components of the force for calculating the x- and y-components of the motion. Calculate the positions and plot the path taken by the body for the following initial velocities:

(a) $v_{x0} = 0$, $v_{y0} = 1.0$
(b) $v_{x0} = 0$, $v_{y0} = 1.1$
(c) $v_{x0} = 0$, $v_{y0} = 1.25$
(d) $v_{x0} = 0$, $v_{y0} = 1.50$
(e) $v_{x0} = 0.7$, $v_{y0} = 0.7$
(f) $v_{x0} = 0$, $v_{y0} = 0.9$

1.16 Consider central forces that are *not* inverse square forces. Consider a force of the form $F = Kr^n$. Thus, $n = -2$ will provide the common inverse square force. Write a computer program to calculate the position of a body moving under the influence of this general central force. Graph the motion resulting from the following initial conditions:

$x_0 = 1$ $v_{x0} = 0$

$y_0 = 0$ $v_{y0} = 1, 1.05, 1.1$, and 1.2

Do this for $n = -1.9, -2.0, -2.1, -3.0$, and $+1.0$. Use $K = -1.0$.

1.17 Consider a binary star system with stars of some unit mass located at $x = 1$ and $x = -1$ (in arbitrary units). As viewed from a particular reference frame, these remain fixed. The force felt by any other mass is the *vector sum* of the two forces from the two stars, each of which is, of course, an inverse-square force. Write a computer program to calculate the position or motion of a mass "orbiting" in this system. Start with initial conditions of $x_0 = 2$, $y_0 = 0$, $v_{x0} = 0$, and $v_{y0} = 1$, but vary these initial conditions and see what types of "orbits" are possible.

1.18 The only way the binary stars in the previous problem can appear to remain fixed is if they are viewed from a *rotating* reference frame. Unexpected "inertial forces" are observed in this rotating frame (see Chapter 6). Include these additional forces and redo Problem 1.17.

2

ONE-DIMENSIONAL MOTION

The subject of mechanics can be divided into *kinematics*, *dynamics*, and *statics*. Kinematics is a study of the concise *description* of a body in motion, whereas dynamics seeks to explain the *cause* of that motion. Statics seeks to understand the details of forces that hold a body in equilibrium. We shall defer our discussion of statics until later.

Sections 2.1, 2.2, and 2.3 may well be review. If they are, look at them quickly just to refresh the ideas and become familiar with the notation we use in this book.

2.1 Kinematics in One Dimension

The location of an object can often be fully specified by a single number—the distance from an origin to that object. For example, we can locate a train by giving its distance from the station. We know it must be on the track, so the additional information of its distance from the station, positive for one direction and negative for the other, fully locates the train. Similarly, a car moving down a straight road, a bead sliding on a wire, an ant climbing a piece of string, a bucket being lowered into a well, and a mass attached to a spring can all be located fully and completely by a single number.

This specific location, *x*, is the *displacement* from some reference point, or origin, to the object. An *operational definition* of this distance involves comparing it with some standard. We might move a meterstick over and over to see how many times it fits between the two positions. Or we might pull out a steel tape with distances marked on it. Or we might stand at an origin, shine a light on our object, and see how much time elapses before the reflected light returns.

Now we have to worry about an operational definition of *time*. We can begin with a crude water clock and define time as that which elapses between drops from the water clock. Or we can say that time is what elapses between oscillations of a pendulum. From there, it's easy to say that time is what elapses between ticks of a clock. Or we can be quite contemporary and define time as that which elapses between oscillations of a cesium atom.

Physicists work with operational definitions to make measurements precise and reproducible. Philosophers might say that we still haven't answered the questions of what distance really *is* or what time really *is*, and that's true. But we can *measure* these quantities and that is all we shall (or can) do. If we can't find an operational definition for something, then we really can't discuss it in physics. That doesn't mean it's not real or important —like love, truth, beauty, right, and wrong. These subjects just aren't physics. Time and distance are more complex (subtle may be a better description) than we commonly think. But in this book on classical mechanics, we shall treat them as absolutes (at least until Chapter 13).

As an object moves, we need to discuss how fast it's moving. So we define its *velocity*, *v*, as the time rate of change of its position. That is,

$$v = \frac{dx}{dt} \tag{2.1.1}$$

Velocity may also change, so we need to discuss the time rate of change of the velocity, which we shall call the *acceleration*, *a*.
That is,

$$a = \frac{dv}{dt} \tag{2.1.2}$$

or

$$a = \frac{d^2x}{dt^2} \tag{2.1.3}$$

Until now, in your introductory physics course, you may have considered only cases involving *constant* acceleration. For that special case you have already found the following important relationships:

$$v = v_0 + at \tag{2.1.4}$$

$$x = x_0 + v_0 t + \tfrac{1}{2}at^2 \tag{2.1.5}$$

and

$$v^2 = v_0^2 + 2a(x - x_0) \tag{2.1.6}$$

Describing motion is interesting and useful. But more interesting and more useful is understanding the *cause* of the motion. And that, exactly, is the subject of *dynamics*.

2.2 Dynamics in One Dimension

When you push a cart, it's quite easy to understand that it moves *because* you push it. As you push, you may increase its velocity. Now stop pushing it. Why does it *continue* to move?

This simple question caused great difficulties for the ancient Greeks. To explain this continued movement, they invented complicated systems and methods and actions of which Rube Goldberg could have been quite proud. Something must *continue* to cause the motion, they insisted.

It remained for Galileo and Newton to carry out experiments and verify their new ideas in order to understand such motion. In 1687, Newton published his *Principia*, in which he put forth his three Laws of Motion. The First Law of Motion describes an object in the *absence* of a net force. (The net force is the vector sum of *all* the forces acting on an object.) Newton's First Law of Motion states:

> *In the absence of a net force, a body at rest remains at rest and a body in motion continues in motion along the same straight line and at constant speed.*

We often call this the Law of Inertia and describe this characteristic of matter to remain in its particular state of motion as *inertia*.

Thus, a moving cart continues to move simply because it is already moving. It needs nothing external to help it continue its motion. In fact, influences from the outside—namely, the force of friction—will cause it to come to rest.

But is this Law of Inertia valid for all observers? Put your sunglasses on the dash of your car and then turn a corner or accelerate or brake to a stop. Your glasses, which seemed to be at rest, certainly don't remain at rest. Why is that? We shall discuss that later, in Section 6.5, "Moving Coordinate Systems."

All measurements of position or motion must be made relative to some reference point. A complete system for describing position or motion in three dimensions is called a *coordinate system* or a *reference frame*. Three meter sticks held together at right angles provide a ready example of a reference frame. A reference frame in which the Law of Inertia holds is called an *inertial frame*. A reference frame in which the Law of Inertia does not hold is called a *noninertial frame*. The car turning a corner was a noninertial frame. For practical purposes, any frame firmly attached to the Earth's surface is an inertial frame. The Earth's own rotation causes it to be a noninertial frame if we look *very* closely. But it doesn't really differ very much. Experiments carried out in a laboratory attached to a building attached to solid Earth beneath are carried out in an inertial frame. We expect the Law of Inertia—indeed, all of Newton's Laws—to be valid in such a laboratory.

Newton's Second Law explains what happens when external forces are present:

The net *force on a body causes that body to accelerate. The acceleration is in the direction of the force, proportional to the force and inversely proportional to the mass of the body.*

We can write this in equation form as

$$\mathbf{a} = \frac{\mathbf{F}}{m} \tag{2.2.1}$$

Or, we usually solve for the force and write this as

$$\mathbf{F} = m\mathbf{a} \tag{2.2.2}$$

(Note that $\mathbf{F} = m\mathbf{a}$ is a vector equation, "... the acceleration is in the *direction* of the force,..." We shall cover vectors in detail in Chapter 4 and look at the general case of $\mathbf{F} = m\mathbf{a}$ in Chapter 5. In this chapter, we will limit ourselves to considering forces and motions that occur in one direction only, such as back and forth along a track.

Equation 2.2.2 means that if we know the cause of the motion—that is, the force $\mathbf{F}$—then we know the change in the motion—that is, the acceleration. If we know what some object was doing originally—its initial position, initial velocity—and we know the force, Eq. 2.2.1 tells us its acceleration and Eqs. 2.1.1 and 2.1.2 allow us to use this information to predict the body's future motion—to find $v(t)$ and $x(t)$. That is, we can find velocity and position as functions of time. Briefly stated, that is the prime goal of *mechanics*.

We can also use Newton's Second Law as a definition of the mass of an object. Mass is a measure of the resistance to *change* in motion. If identical forces are applied to two bodies of masses m_1 and m_2, and two different accelerations are measured, a_1 and a_2, then Eq. 2.2.2 can be used to relate the two masses by

$$F = m_1 a_1 = m_2 a_2$$

$$m_2 = \frac{a_1}{a_2} m_1$$

Newton's Third Law describes the forces involved when two bodies interact with each other:

> Let $\mathbf{F}_{12}$ be the force exerted by body 1 on body 2. There is also a force exerted by body 2 on body 1, $\mathbf{F}_{21}$. These forces lie along the same straight line and in opposite directions and are of identical magnitude. That is, $\mathbf{F}_{12} = -\mathbf{F}_{21}$.

This has often been referred to as the principle of action and reaction.

As we seek various methods to "solve for the motion"—to predict the future positions and velocities, $x(t)$ and $v(t)$—of a body acted upon by some particular force F, we shall use Newton's Laws of Motion. We will develop and discuss new quantities that describe particular aspects of motion, but that are always consistent with these laws.

From Newton's Second Law, $F = ma$, we readily know the acceleration if the force is known. Now we must solve Eq. 2.1.3: ($a = d^2x/dt^2$). If the force—and, hence, the acceleration—is constant, the motion is fully described by Eqs. 2.1.4, 2.1.5, and 2.1.6. But what if the acceleration is not constant? What if it varies with time, position, or velocity? We've already seen how we can handle these situations numerically on the computer. Now we shall use a more explicit tool for understanding these situations—*calculus*.

Appendix B contains a review of calculus and is recommended reading for all students. Calculus is not merely an interesting subject in mathematics, it is a necessary tool in physics. Calculus is vital.

2.3 Constant Force

We're now ready to apply the ideas of calculus to the problems we encounter in mechanics. You should already be familiar with problems involving a constant force (and hence a constant acceleration). So, let's begin here by using the ideas of calculus on already familiar ground:

$$F = \text{constant}$$
$$F = ma$$
$$a = \frac{F}{m} = \text{constant}$$

But the acceleration is just the time derivative of the velocity:

$$a = \frac{dv}{dt} \qquad (2.3.1)$$

We shall adopt the usual convention of indicating a derivative with respect to time by a dot, $\dot{v} = dv/dt$. Hence,

$$a = \dot{v} \tag{2.3.1a}$$

Likewise, a second derivative with respect to time shall be indicated by two dots, $\ddot{x} = d^2x/dt^2$.

From Eq. 2.3.1, we readily have

$$dv = a\,dt \tag{2.3.2}$$

[handwritten: a = constant]

The small change in velocity, dv, is equal to the acceleration, a, multiplied by the short length of time, dt, over which it acts. We now integrate both sides of Eq. 2.3.2, making sure that both sets of limits correspond to the same configuration:

$$\int_{v_0}^{v} dv = \int_{t_0}^{t} a\,dt \tag{2.3.3}$$

(Formally, of course, the variable of integration is only a "dummy variable" and is not—and cannot be—the same as one of the limits. So this equation should really be written as

$$\int_{v_0}^{v} dv' = \int_{t_0}^{t} a\,dt'$$

However, most practicing physicists and engineers will write it in the previous form. Therefore, I have chosen to use that form throughout the text.)

Our object has velocity v at time t and velocity v_0 at time t_0. We shall "start our clock" at zero so that $t_0 = 0$. Since a is just a constant, this can be written as

$$\int_{v_0}^{v} dv = a \int_{0}^{t} dt \tag{2.3.4}$$

$$\int_{v_0}^{v} dv = v \bigg|_{v_0}^{v} = v - v_0 \tag{2.3.5}$$

$$\int_{0}^{t} dt = t \bigg|_{0}^{t} = t - 0 = t \tag{2.3.6}$$

Therefore, Eq. 2.3.3 becomes

$$v = v_0 + at \tag{2.3.7}$$

As we have just done for the acceleration, we can also write the velocity as the time derivative of the position and integrate. That is,

$$v = \dot{x} = \frac{dx}{dt} \tag{2.3.8}$$

$$\int_{x_0}^{x} dx = \int_{0}^{t} v \, dt$$

$$\int_{x_0}^{x} dx = \int_{0}^{t} (v_0 + at) \, dt$$

$$\int_{x_0}^{x} dx = v_0 \int_{0}^{t} dt + a \int_{0}^{t} t \, dt$$

$$x \Big|_{x_0}^{x} = v_0 \left(t \Big|_{0}^{t}\right) + a \left(\tfrac{1}{2}t^2 \Big|_{0}^{t}\right)$$

$$x = x_0 + v_0 t + \tfrac{1}{2} a t^2 \tag{2.3.9}$$

The kinematic equations, Eqs. 2.3.7 and 2.3.9, describe all the constant acceleration problems you've seen previously—free fall, inclined plane, Atwood's machine, friction, and so on. They should be quite familiar.

Such a constant acceleration is the result of a constant force since $F = ma$. Determining which acceleration to use in Eqs. 2.3.7 and 2.3.9 begins by determining the net force—the sum of all the forces—acting upon a body. Therefore, we apply Newton's Second Law.

Near Earth's surface, objects are observed to fall with a constant acceleration of 9.8 m/s² if air resistance is neglected. This constant acceleration means that we can use Eqs. 2.3.7 and 2.3.9 to describe such motion.

EXAMPLE 2.3.1   A tennis ball is thrown upward with initial velocity v_0. How high does it go? As sketched in Figure 2.3.1, it momentarily comes to rest at the top of its path, so $v = 0$. We can use Eq. 2.3.7 to find the time t:

$$v = v_0 + at$$

$$0 = v_0 + at$$

$$t = -\frac{v_0}{a} = -\frac{v_0}{(-9.8 \text{ m/s}^2)}$$

$$t = \frac{v_0}{g}$$

where $g = 9.8$ m/s² is the acceleration due to gravity. Since the acceleration is downward, it is negative. That is, $a = -g = -9.8$ m/s².

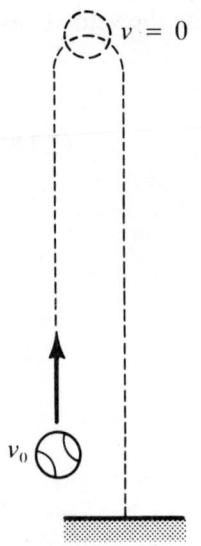

Figure 2.3.1 A ball is thrown upward.

Knowing the time, Eq. 2.3.9 lets us determine the position:

$$x = x_0 + v_0 t + \tfrac{1}{2}at^2$$

$$x = 0 + v_0\left(\frac{v_0}{g}\right) + \frac{1}{2}(-g)\left(\frac{v_0}{g}\right)^2$$

$$x = \frac{1}{2}\frac{v_0^2}{g}$$

Note, again, that the acceleration is $-g$; the negative sign indicates *downward*.

The previous example of constant acceleration or constant force was very simple and straightforward. Physical situations can become more complex (and more interesting), yet still involve only constant acceleration, as we shall see in the following example.

EXAMPLE 2.3.2 Consider a block of mass m initially projected up a rough plane inclined at angle θ with respect to the horizontal. The coefficient of kinetic friction between the block and the plane is μ. If the initial velocity of the block up the plane is v_0, describe its motion. Specifically, find how far up the plane it goes and how long it requires to return to its starting position.

Any physics problem begins with a sketch or diagram. This mass moving up an inclined plane is shown in Figure 2.3.2. From that sketch, we should also draw a "free-body diagram," showing *all* the forces acting *upon that body*. This is also shown in Figure 2.3.2. Gravity exerts a force on the block directly downward and of magnitude mg; this is the *weight* of the block. The plane supports the block with a "normal force," N, that is perpendicular to the plane's surface and also exerts a friction force, F_f, that is proportional to N ($F_f = \mu N$, defines the coefficient of friction μ), parallel to the plane's surface, and opposite to the direction of motion. The block will move only

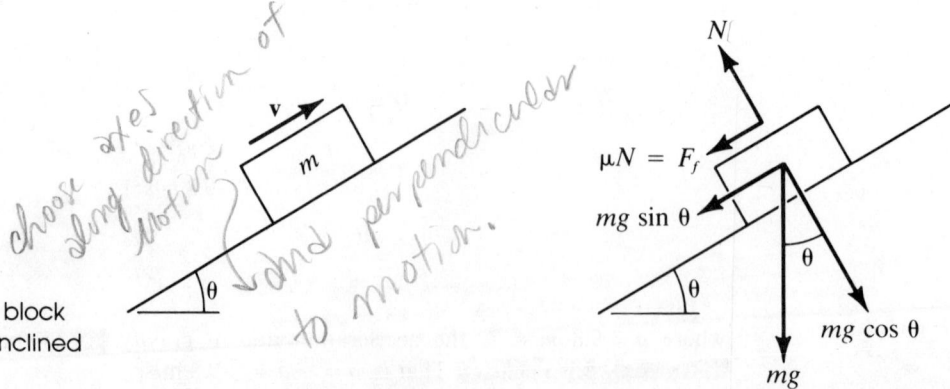

Figure 2.3.2 A block moving up an inclined plane.

SECTION 2.3 / CONSTANT FORCE

along the plane—it isn't going to tunnel itself into the plane, or hop up off of the plane—so we need to look at the components of force that lie in this direction. Therefore, we choose axes along this direction and perpendicular to it.

F_f and N already meet this criteria. But we must resolve the weight, mg, into components parallel and perpendicular to the direction of motion. This is shown in the free-body diagram; the two components are $mg \sin \theta$ and $mg \cos \theta$. $F = ma$ must be satisfied for each direction. Since there is no motion *perpendicular* to the plane, the acceleration in that direction must certainly be zero. So the *net* force in that direction is also zero.

Forces in one direction (*up* from the plane) are considered positive and forces in the opposite direction (*down* from the plane) are considered negative. Thus,

$$F_{net})_\perp = N - mg \cos \theta = 0$$

or

$$N = mg \cos \theta$$

The net force parallel to the plane is

$$F_{net})_\parallel = -mg \sin \theta - \mu N = ma$$

where the positive direction has been taken to be toward the right along the inclined plane. From this we can readily solve for the acceleration in Eqs. 2.3.7 and 2.3.9:

$$a = \frac{-mg \sin \theta - \mu N}{m}$$

or

$$a = -g(\sin \theta + \mu \cos \theta) \qquad (2.3.10)$$

The negative sign means an acceleration to the left, or down the plane.

How long does it take it to stop? Set $v = 0$ in Eq. 2.3.7:

$$v = 0 = v_0 + at$$

$$0 = v_0 - g(\sin \theta + \mu \cos \theta)t$$

$$t = \frac{v_0}{g(\sin \theta + \mu \cos \theta)} \qquad (2.3.11)$$

How far does it move in that time? Evaluate x from Eq. 2.3.9:

$$x = v_0 \frac{v_0}{g(\sin \theta + \mu \cos \theta)}$$

$$+ \tfrac{1}{2}[-g(\sin \theta + \mu \cos \theta)]\left[\frac{v_0}{g(\sin \theta + \mu \cos \theta)}\right]^2$$

$$x_{max} = \frac{v_0^2}{2g(\sin \theta + \mu \cos \theta)} \qquad (2.3.12)$$

Of course we could also have obtained that in one easy step with Eq. 2.1.6.

Now, how long does it take to return to its initial position? Once it stops and begins to slide back, the forces are no longer the same. We need a new free body diagram to show the new forces; see Figure 2.3.3.

Figure 2.3.3 A block sliding down an inclined plane. Note that the direction of the friction force, N, has now changed.

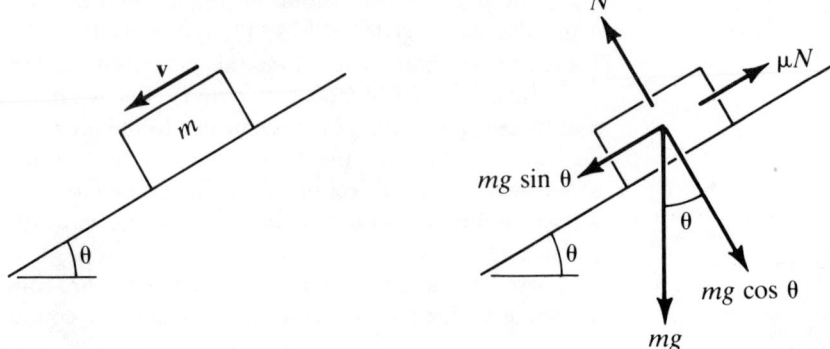

The friction force *always* acts in a direction opposite the motion. Thus, when the block slides down the plane, the friction force, F_f, points up the plane as shown in Figure 2.3.3. Since the net force *perpendicular* to the plane must still be zero, we find just as before that

$$N = mg \cos \theta$$

The net force *parallel* to the plane is now

$$F_{net} = -mg \sin \theta + \mu N$$

As always, Newton's Second Law can be applied to find the acceleration. Thus,

$$F_{net} = -mg \sin \theta + \mu \, mg \cos \theta = ma$$

from which the acceleration can readily be found:

$$a = g(-\sin \theta + \mu \cos \theta) \qquad (2.3.13)$$

Notice that there will be no acceleration (the block will remain at rest at X_{max}) for

$$\mu \cos \theta = \sin \theta$$

or

$$\mu = \tan \theta \qquad (2.3.14)$$

For larger values of μ, our *equation* would predict a *positive* value of the acceleration—an acceleration to the right, *up* the plane. Is that reasonable? Of course not! These equations describe a block starting from rest and sliding down the plane to the left. Any result inconsistent with that must be wrong. In this instance a value of μ larger than $\mu = \tan \theta$ means that the block will *not* return—it remains stopped at X_{max}. At rest the friction force need not be as large as μN. So don't blindly trust the mathematics. Always ask what a certain result means. Have correct mathematics been applied to the wrong situation? Always question a result. Is it what you expect? Is it reasonable?

SECTION 2.4 / FORCE AS A FUNCTION OF TIME

Returning to our example, if friction is small enough to allow the mass to slide down the plane, its acceleration is

$$a = -g(\sin\theta - \mu\cos\theta) \qquad (2.3.13')$$

and its speed as it returns to its initial position can be found from Eq. 2.1.6:

$$v^2 = v_0^2 + 2a(x - x_0)$$

$$v^2 = 2g(\sin\theta - \mu\cos\theta)\frac{v_0^2}{2g(\sin\theta + \mu\cos\theta)}$$

$$v = -v_0\sqrt{\frac{\sin\theta - \mu\cos\theta}{\sin\theta + \mu\cos\theta}}$$

2.4 Force as a Function of Time

Even if the force (or acceleration) is not constant, but it is known to be an explicit function of time, then finding the equation of motion—solving for the position—is as straightforward as before. Eq. 2.3.3 then becomes

$$v = v(t) = v_0 + \int_0^t a(t)\,dt \qquad (2.4.1)$$

and, instead of Eq. 2.3.9, we obtain

$$x = x(t) = x_0 + \int_0^t v(t)\,dt \qquad (2.4.2)$$

Once the function $a(t)$ is known, these two integrals can be carried out in a straightforward manner.

EXAMPLE 2.4.1 Consider an acceleration (or force) that increases linearly with time as

$$a = bt \qquad (2.4.3)$$

Eq. 2.4.1 then becomes:

$$v = v_0 + \int_0^t (bt)\,dt$$

$$= v_0 + b\int_0^t t\,dt$$

$$= v_0 + b\tfrac{1}{2}t^2\Big|_0^t$$

$$v = v_0 + \tfrac{1}{2}bt^2 \qquad (2.4.4)$$

This can now be used to evaluate Eq. 2.4.2:

$$x = x_0 + \int_0^t (v_0 + \tfrac{1}{2}bt^2)\, dt$$

$$= x_0 + v_0 \int_0^t dt + \tfrac{1}{2}b \int_0^t t^2\, dt$$

$$= x_0 + v_0(t|_0^t) + \tfrac{1}{2}b(\tfrac{1}{3}t^3|_0^t)$$

$$x = x_0 + v_0 t + \tfrac{1}{6}bt^3 \tag{2.4.5}$$

This expression entirely describes the motion of the object by giving its position as a function of time. Throughout mechanics, this is usually our goal.

EXAMPLE 2.4.2 Consider a force that varies *sinusoidally* with time. What motion does it produce? Let

$$F = A \sin wt \tag{2.4.6}$$

$$a = \frac{F}{m} = \frac{A}{m} \sin wt \quad = \frac{dv}{dt}$$

$$v = v_0 + \int_0^t \frac{A}{m} \sin wt\, dt \quad v = v_0 + \int a\, dt$$

$$= v_0 + \frac{A}{m} \int_0^t \sin wt\, dt$$

$$= v_0 + \frac{A}{m}\left(-\frac{1}{w}\cos wt\right)\bigg|_0^t$$

$$= v_0 - \frac{A}{mw}[\cos wt - \cos(0)]$$

$$v = v_0 + \frac{A}{mw} - \frac{A}{mw}\cos wt \qquad v = \frac{dx}{dt} \tag{2.4.7}$$

$$v\, dt = dx$$

$$x = x_0 + \int_0^t \left[v_0 + \frac{A}{mw} - \frac{A}{mw}\cos wt\right] dt$$

$$= x_0 + \left(v_0 + \frac{A}{mw}\right)\bigg|_0^t dt - \frac{A}{mw}\int_0^t \cos wt\, dt$$

$$= x_0 + \left(v_0 + \frac{A}{mw}\right)\left(t\bigg|_0^t\right) - \frac{A}{mw}\left(\frac{1}{w}\sin wt\bigg|_0^t\right)$$

$$= x_0 + \left(v_0 + \frac{A}{mw}\right)(t - 0) - \frac{A}{mw}\left(\frac{1}{w}\sin wt - \frac{1}{w}\sin 0\right)$$

$$x = x_0 + \left(v_0 + \frac{A}{mw}\right)t - \frac{A}{mw^2}\sin wt \tag{2.4.8}$$

2.5 Force as a Function of Position

Suppose we know the force (and, thus, the acceleration) as a function of position; i.e., $F = F(x)$. We can no longer proceed with straightforward integrations as we did in Eqs. 2.4.1 and 2.4.2.

We know

$$F = F(x) = ma \qquad (2.5.1)$$

Using the chain rule from calculus, we can rewrite a as

$$a = \frac{d^2x}{dt^2} = \frac{dv}{dt} = \frac{dv}{dx} \cdot \left(\frac{dx}{dt}\right) = \frac{dv}{dx} \cdot v \qquad (2.5.2)$$

Therefore,

$$F(x) = mv\frac{dv}{dx} \qquad (2.5.3)$$

This can still be rewritten as

$$F(x) = mv\frac{dv}{dx} = m\frac{1}{2}\frac{d}{dx}(v^2) \qquad (2.5.4)$$

or

$$F(x) = \frac{d}{dx}\left(\frac{1}{2}mv^2\right) \qquad (2.5.5)$$

assuming the mass doesn't change.

But you should recall that $\frac{1}{2}mv^2$ is by definition the kinetic energy T; that is

$$T \equiv \tfrac{1}{2}mv^2 \qquad (2.5.6)$$

Therefore,

$$F(x) = \frac{dT}{dx} \qquad (2.5.7)$$

We can now multiply by dx and integrate to get

$$\int_{x_0}^{x} F(x)\,dx = \int_{T_0}^{T} dT = T - T_0 \qquad (2.5.8)$$

The left-hand side represents the *work* done on the object by the force as it moves from x_0 to x. This work is equal to the change in the kinetic energy.

We can get more meaning out of this equation if we define a *potential energy* $V = V(x)$ such that

$$F(x) = -\frac{dV}{dx} \tag{2.5.9}$$

or

$$V(x) = \int_x^{x_s} F(x)\, dx \tag{2.5.10}$$

That is, the potential energy is the work done by the force as the object moves *from x to* some standard reference position x_s. You should recall from your introductory physics course that in talking about the gravitational potential energy of a brick raised above a desk when you wrote

$$V = mgx \tag{2.5.11}$$

you had to specify a reference point. Was x the distance above the desk or above the floor? Only the *difference* in potential energy ever enters a problem.

Equation 2.5.10 can, of course, be turned around and written as

$$V(x) = -\int_{x_s}^{x} F(x)\, dx \tag{2.5.10a}$$

We always choose x_s to make $V(x)$ as simple as possible. If we can, we choose x_s so that $V(x_s)$ equals 0. Therefore, Eq. 2.5.10a is often written merely as the indefinite integral:

$$V(x) = -\int F(x)\, dx \tag{2.5.10b}$$

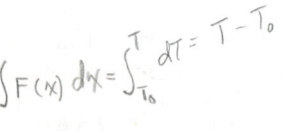

We can use the potential energy to evaluate the left-hand side of Eq. 2.5.8. Then

$$-V(x) + V(x_0) = T - T_0 \tag{2.5.12}$$

or

$$T_0 + V(x_0) = T + V(x) \tag{2.5.13}$$

This means that

$$T + V(x) = E = \text{constant} \tag{2.5.14}$$

The sum of the kinetic energy and the potential energy remains constant throughout the motion; this is called the *total energy*. We describe this by saying that the total energy is *conserved*. This is true only for forces definable by Eq. 2.5.9 or 2.5.10, which we call *conservative forces*.

Before actually solving for $x(t)$, we can learn a lot by looking at a graph of $V(x)$ as shown in Figure 2.5.1.

SECTION 2.5 / FORCE AS A FUNCTION OF POSITION

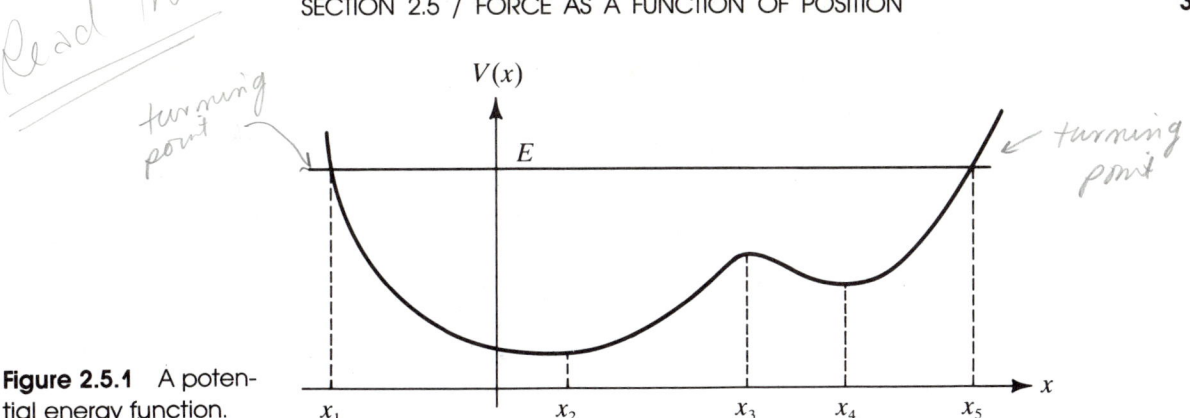

Figure 2.5.1 A potential energy function.

Figure 2.5.1 shows the potential energy a body has as a function of position for this particular case. It also shows the total energy, E. The difference between E and V is the kinetic energy

$$T = E - V \qquad (2.5.15)$$

If the potential energy is equal to the total energy, then there is no kinetic energy. The body must be at rest. In Figure 2.5.1, this occurs at $x = x_1$ and $x = x_5$. These are called *turning points* because the body (with total energy E as shown) stops and changes direction there.

By definition, kinetic energy must always be positive or zero. Motion to the left of $x = x_1$ or to the right of $x = x_5$ requires a *negative* value for the kinetic energy. This is forbidden in classical mechanics, so these are referred to as *forbidden regions*. All motion of the particle will be confined to the *allowed region* between $x = x_1$ and $x = x_5$.

Starting at $x = x_1$, we can *qualitatively describe* the motion of the body just by looking at the graph of $V(x)$ in Figure 2.5.1:

$x = x_1$: The body stops here since $T = 0$ and $E = V$ here.
$x_1 < x < x_2$: The body accelerates to the right since the speed and kinetic energy increase as V continues to decrease from E.
$x = x_2$: The body reaches its maximum speed as V reaches its minimum.
$x_2 < x < x_3$: The body decelerates (still moving to the right) since the speed and kinetic energy decrease as V increases.
$x_3 < x < x_4$: The body accelerates again as V decreases.
$x_4 < x < x_5$: The body decelerates again as V increases. Finally, as $V = E$ at $x = x_5$, the body stops, and then begins to accelerate to the left.
$x = x_5$: The body stops since $T = 0$ and $E = V$.

All of this is an application of the energy conservation ideas contained in Eq. 2.5.14. The ideas of forces in Eq. 2.5.9 could also be used. From *that* equation, we find that the force is just the *negative of the slope* of the potential energy function. We can use this to qualitatively describe the motion:

$x_1 < x < x_2$: The slope, dV/dx, is negative (i.e., as x increases, V decreases) so the force is positive—to the right. Therefore, the body *accelerates to the right*.

$x = z_2$: The slope is zero so the speed is unchanging as it passes through x_2.

$x_2 < x < x_3$: The slope is positive (i.e., as x increases, V also increases) so the force is negative—to the left. An acceleration to the left means the body loses speed.

$x_3 < x < x_4$: As in the region $x_1 < x < x_2$, a negative slope indicates a positive acceleration and thus an increasing speed.

$x_4 < x < x_5$: As in the region $x_2 < x < x_3$, a positive slope means a negative acceleration and decreasing speed.

We should take time to appreciate the preceding *qualitative* description of a motion. The turning points and positions of minimum or maximum velocity are all very important. In fact, such determinations may generate more *understanding* than an eventual, explicit solution of $x(t)$. But to find such an explicit solution, we can rewrite Eq. 2.5.15 as

$$\tfrac{1}{2}mv^2 = E - V(x)$$

and solve for v:

$$v = \frac{dx}{dt} = \sqrt{\frac{2}{m}[E - V(x)]} \qquad (2.5.16)$$

We can now separate variables and integrate to obtain

$$t = \int_0^t dt = \int_{x_0}^x \frac{dx}{\sqrt{2[E - V(x)]/m}} \qquad (2.5.17)$$

This equation is actually $t = t(x)$, a solution of the time as a function of x. In principle, we can then find, from this, $x = x(t)$, the position as a function of time. This approach—and Eq. 2.5.17 in particular—may be rather unruly, but it still is necessary.

Before we apply this method to a more complicated force, let's look at the example of free fall:

$$F = -mg \qquad (2.5.18)$$

This isn't a very *interesting* function of position, but because we understand this force thoroughly, it will be *useful* in understanding a new method.

SECTION 2.5 / FORCE AS A FUNCTION OF POSITION

From the force we can define the potential energy

$$V(x) = \int_x^{x_s} F(x)\, dx = \cdots = mgx - mgx_s \qquad (2.5.19)$$

We simplify this by choosing $x_s = 0$ so that

$$V(x) = mgx \qquad (2.5.20)$$

and

$$E = \tfrac{1}{2}mv^2 + mgx \qquad (2.5.21)$$

If we start at $x = 0$ and throw an object upward with velocity v_0, then

$$E = \tfrac{1}{2}mv_0^2 \qquad (2.5.22)$$

At the turning point, $v = 0$ and $x = x_{\max}$, and Eq. 2.6.21 becomes

$$E = mgx_{\max} \qquad (2.5.23)$$

or

$$x_{\max} = \frac{E}{mg} = \frac{\tfrac{1}{2}mv_0^2}{mg} = \frac{v_0^2}{2g} \qquad (2.5.24)$$

Back to finding $x(t)$ using Eq. 2.5.16, we have

$$t = \int_{x_0}^{x} \frac{dx}{\sqrt{2(E - mgx)/m}}$$

or

$$t = \sqrt{\frac{m}{2}} \int_0^x \frac{dx}{\sqrt{E - mgx}} \qquad (2.5.25)$$

This integral may not be one with which you're familiar. Evaluating it with a table of integrals we have

$$t = \sqrt{\frac{m}{2}} \left[\frac{2\sqrt{E - mgx}}{(-mg)} \right]\Big|_0^x$$

$$= -\sqrt{\left(\frac{m}{2}\right)\frac{(-4)(\tfrac{1}{2}mv_0^2 - mgx)}{m^2 g^2}}\Big|_0^x$$

$$= -\frac{1}{g}\sqrt{v_0^2 - 2gx}\,\Big|_0^x$$

$$t = \frac{1}{g}(v_0 - \sqrt{v_0^2 - 2gx}) \qquad (2.5.26)$$

This is now $t = t(x)$. Solving this for x yields $x = x(t)$, which we want:

$$x = v_0 t - \tfrac{1}{2}gt^2 \qquad (2.5.27)$$

EXAMPLE 2.5.1 Now that we have seen that this method does indeed work, we can apply it to a less trivial problem; namely, the simple harmonic oscillator. As discussed earlier, the classic example is a mass attached to the end of a spring obeying Hooke's Law.

$$F = -kx \qquad (2.5.28)$$

From this we can determine the *potential energy* by

$$V(x) = \int_x^{x_s} -kx\, dx = -\tfrac{1}{2}kx_s^2 + \tfrac{1}{2}kx^2$$

Choosing $x_s = 0$ simplifies this to

$$V(x) = \tfrac{1}{2}kx^2 \qquad (2.5.29)$$

The spring has *no* potential energy at its equilibrium position but gains potential energy as the spring is *either* compressed *or* extended. In either case, work must be expended to compress or extend the spring.

Now we can write the total energy as

$$E = V + T = \tfrac{1}{2}kx^2 + \tfrac{1}{2}mv^2 \qquad (2.5.30)$$

Proceeding as before, we can solve for t:

$$v = \sqrt{2/m(E - \tfrac{1}{2}kx^2)} = \frac{dx}{dt} \qquad (2.5.31)$$

$$t = \int_{x_0}^{x} \frac{dx}{\sqrt{2/m(E - \tfrac{1}{2}kx^2)}} \qquad (2.5.32)$$

To make the calculus easy, let us limit this example to the case of $x_0 = 0$ with some initial speed v_0. From integral tables we find that

$$\int \frac{dx}{\sqrt{a^2 - x^2}} = \sin^{-1}\left(\frac{x}{a}\right)$$

To get Eq. 2.5.32 in that form, we make the following substitutions:

$$u^2 = \tfrac{1}{2}kx^2 \quad \text{or} \quad x = \sqrt{\frac{2}{k}}\, u$$

$$a^2 = E \qquad dx = \sqrt{\frac{2}{k}}\, du$$

Therefore,

$$t = \sqrt{\frac{m}{2}} \int_0^u \frac{(\sqrt{2/k}\,du)}{\sqrt{a^2 - u^2}} = \sqrt{\frac{m}{k}} \int_0^u \frac{du}{\sqrt{a^2 - u^2}} \quad (2.5.33)$$

$$t = \sqrt{\frac{m}{k}} \sin^{-1}\left(\frac{u}{a}\right)\Big|_0^u = \sqrt{\frac{m}{k}} \sin^{-1}\left(\frac{\sqrt{k/2}\,x}{\sqrt{E}}\right) \quad (2.5.34)$$

As always by this method, we have solved for $t = t(x)$. We now need to solve this for x to get our desired result, $x = x(t)$:

$$x = \sqrt{\frac{2E}{k}} \sin\sqrt{\frac{k}{m}}\,t \quad (2.5.35)$$

This result *looks* a lot better if we write it as

$$x = A \sin \omega t \quad (2.5.36)$$

where $\omega = \sqrt{k/m}$ is the angular frequency of oscillation (a stiffer spring, with larger spring constant k means a larger frequency, or faster oscillation; a larger mass m means a lower frequency or slower oscillation) and $A = \sqrt{2E/k}$ is the *amplitude* or maximum displacement. (The *amplitude* is the maximum displacement from equilibrium. The amplitude describes how big the swing is. Pull the oscillator away from zero and release it. That distance from zero—from the equilibrium position—is the *amplitude*. It almost seems strange that the amplitude would not affect the period. Move it a little or give it a tiny push, and our mass on the spring oscillates back and forth over a small distance. Move it more or give it a large shove, and it rushes back and forth over a large distance, but *always* with the same period or frequency! This is true for *every* system undergoing simple harmonic motion—the period is independent of the amplitude.)

You should compare this analytical result with the numerical solution in Chapter 1. We shall return to glean more information from this simple harmonic oscillator later on.

2.6 Force as a Function of Velocity

Consider a body moving in air or some other fluid. A friction force, which varies with the body's speed through the fluid, slows it down (the force is usually larger for larger velocities).

In many cases, we can reasonably approximate the frictional force—or viscous damping force—as being *directly* proportional to the velocity. This can be expressed as

$$F = F(v) = -cv \quad (2.6.1)$$

Such a force, which depends on velocity, doesn't fit our previous techniques. To solve for the motion under the influence of this force, we can write $F = ma$ as

$$F(v) = m\frac{dv}{dt} \tag{2.6.2}$$

Separating variables gives us

$$dt = m\frac{dv}{F(v)} \tag{2.6.3}$$

This looks almost as awkward as the integrals we encountered with position-dependent forces. Solving this yields

$$t = m\int_{v_0}^{v} \frac{dv}{F(v)} \tag{2.6.4}$$

which is $t = t(v)$. We then solve *that* for v to obtain $v = v(t)$. This we can then solve in a straightforward manner for $x = x(t)$ by integration with respect to time.

An alternate way of solving velocity-dependent forces is to start by writing $F = ma$ as

$$F(v) = m\frac{dv}{dx}\cdot\frac{dx}{dt} = m\frac{dv}{dx}v \tag{2.6.5}$$

Separation of variables then yields

$$dx = m\frac{v\,dv}{F(v)} \tag{2.6.6}$$

Integration yields

$$x = x_0 + m\int_{v_0}^{v} \frac{v\,dv}{F(v)} \tag{2.6.7}$$

This is $x = x(v)$, which we can solve to obtain $v = v(x)$. We can then write

$$v = \frac{dx}{dt} = v(x) \tag{2.6.8}$$

Separation of variables once more yields

$$dt = \frac{dx}{v(x)} \tag{2.6.9}$$

Integration yields $t = t(x)$, which can be solved to provide $x = x(t)$, our final goal.

The work, effort, or understanding involved in these two methods is similar. The second method can provide a connection between x and v if we don't need the full details of $v(t)$ or $x(t)$.

EXAMPLE 2.6.1 Let's return to Eq. 2.6.1 and demonstrate these techniques. $F = -cv$ describes the force acting on, say, a raft launched with some initial velocity v_0:

$$F = -cv = m\frac{dv}{dt}$$

$$t = \int_0^t dt = -\frac{m}{c}\int_{v_0}^v \frac{dv}{v}$$

$$t = -\frac{m}{c}\ln v\Big|_{v_0}^v = -\frac{m}{c}(\ln v - \ln v_0)$$

$$t = \frac{m}{c}\ln\frac{v_0}{v} \tag{2.6.10}$$

This is $t = t(v)$; but we need $v = v(t)$. We can write $\ln(v_0/v) = (c/m)t$. We can then raise e to the power of each side of this equation (or take the antilogarithm of each side of this equation) to get

$$\frac{v_0}{v} = e^{\ln(v_0/v)} = e^{(c/m)t}$$

or

$$\frac{v}{v_0} = \frac{1}{e^{(c/m)t}} = e^{-(c/m)t}$$

or, finally,

$$v = v_0 e^{-(c/m)t} \tag{2.6.11}$$

The velocity decays exponentially from an initial velocity of v_0 with a time constant of m/c. Such behavior is found throughout physics—from the electric current in a circuit to the number of nuclei remaining in a radioactive sample. Therefore, it is important to have a solid understanding of this behavior.

Since $e = 2.7183$, if we wait until $t = m/c$ (define $\tau \equiv m/c$ and call this a *characteristic time* of the system), the speed will have decreased to e^{-1} or 36.7% of its initial value v_0. When $t = 2m/c$ (or $t = 2\tau$), the speed has decreased to e^{-2}, or 13.5% of its initial value. For $t = 3m/c$ (or $t = 3\tau$), it is e^{-3} or 5% of its initial value. For $t = 4m/c$ (or $t = 4\tau$), it's a tiny e^{-4} or 1.8% of v_0. A quick glance at Eq. 2.6.11 might leave the unsettling feeling that the boat will *never* come to a stop (that worried the ancient Greeks!), but the speed will get as slow as you care to measure.

Since Eq. 2.6.11 gives $v = v(t)$, we can proceed in a straightforward manner with the integration to get $x = x(t)$:

$$\int_{x_0}^x dx = v_0 \int_0^t e^{-(c/m)t}\, dt$$

$$x = x_0 - \frac{mv_0}{c} e^{-(c/m)t}\Big|_0^t$$

$$x = x_0 + \frac{mv_0}{c}(1 - e^{-(c/m)t}) \tag{2.6.12}$$

So how far does the boat travel? For very large t, the exponential term is essentially zero, so

$$x_{max} = x_0 + \frac{mv_0}{c} \tag{2.6.13}$$

EXAMPLE 2.6.2 If we throw an object straight up or down, we find both the force of gravity *and* a viscous force due to the friction of the air resisting its motion. We might describe this force as

$$F = -mg - cv \tag{2.6.14}$$

which is only slightly more complicated than Eq. 2.6.1. We can solve it exactly as before:

$$t = \int_{v_0}^{v} \frac{m \, dv}{F(v)} = -m \int_{v_0}^{v} \frac{dv}{mg + cv}$$

$$= \cdots = -\frac{m}{c} \ln(mg + cv) \Big|_{v_0}^{v}$$

$$t = \frac{m}{c} \ln \frac{mg + cv_0}{mg + cv} \tag{2.6.15}$$

$$\frac{mg + cv}{mg + cv_0} = e^{-(c/m)t}$$

$$mg + cv = (mg + cv_0)e^{-(c/m)t}$$

$$v = \left(\frac{mg}{c} + v_0\right)e^{-(c/m)t} - \frac{mg}{c} \tag{2.6.16}$$

For very large values of t, this exponential term becomes negligibly small; this final velocity is called the *terminal velocity* and is given by

$$v_t = -\frac{mg}{c} \tag{2.6.17}$$

Looking back to Eq. 2.6.14, we see that for this velocity the *force* is zero—the downward force of gravity is just balanced by the upward force of air resistance.

We can proceed with the solution for the position as before to obtain

$$x = x_0 - \frac{mg}{c}t + \left(\frac{m^2 g}{c^2} + \frac{mv_0}{c}\right)(1 - e^{-(c/m)t}) \tag{2.6.18}$$

EXAMPLE 2.6.3 Accurately describing air resistance is *not* a trivial task. For higher speeds, a better approximation of the force of air resistance is proportional to the *square* of the speed:

$$F = -mg \pm \alpha v^2 \tag{2.6.19}$$

SECTION 2.6 / FORCE AS A FUNCTION OF VELOCITY

For the *rising* case, when velocity is upward and the friction force downward, this becomes

$$F = -mg - \alpha v^2 \tag{2.6.19a}$$

and for the *falling* case, when velocity is downward and the friction force upward, it becomes

$$F = -mg + \alpha v^2 \tag{2.6.19b}$$

Since v^2 is always positive, we must explicitly choose which sign to use for the frictional viscous drag force, $\pm \alpha v^2$. Note that this was not necessary for linear viscous force, $-cv$, since this term changes sign when the velocity changes sign.

Let us look at an object thrown *upward* with initial velocity v_0.

We can begin to solve this as before:

$$t = -m \int_{v_0}^{v} \frac{dv}{mg + \alpha v^2} \tag{2.6.20}$$

With the help of a table of integrals, we can evaluate this as

$$t = \left[-m \frac{1}{\sqrt{mg\alpha}} \tan^{-1} \frac{v\sqrt{mg\alpha}}{mg} \right]_{v_0}^{v}$$

$$t = -\sqrt{\frac{m}{\alpha g}} \left[\tan^{-1} v \sqrt{\frac{\alpha}{mg}} - \tan^{-1} v_0 \sqrt{\frac{\alpha}{mg}} \right] \tag{2.6.21}$$

We now define $\tau \equiv \sqrt{m/\alpha g}$ and call it the *characteristic time*.

$$\tan^{-1} v \sqrt{\frac{\alpha}{mg}} = \tan^{-1} v_0 \sqrt{\frac{\alpha}{mg}} - \frac{t}{\tau}$$

$$v = \sqrt{\frac{mg}{\alpha}} \tan\left[\tan^{-1} v_0 \sqrt{\frac{\alpha}{mg}} - \frac{t}{\tau} \right] \tag{2.6.22}$$

This velocity finally becomes zero at the top of its trajectory. We can then find the time it requires to reach the top by setting the velocity equal to zero. This gives

$$t_{\text{top}} = \tau \tan^{-1} v_0 \sqrt{\frac{\alpha}{mg}} \tag{2.6.23}$$

With $v = v(t)$, we can proceed to $x = x(t)$ by direct integration. Again, the integrals are not too common, so we use a table of integrals. To simplify our equations, we define $v_t \equiv \sqrt{mg/\alpha}$. We shall see later that this becomes the *terminal speed* for the falling case.

$$x - x_0 = v_t \int_0^t \tan\left[A - \frac{t}{\tau} \right] dt \tag{2.6.24}$$

where

$$A = \tan^{-1}\frac{v_0}{v_t}$$

$$x = x_0 + v_t\tau\left[\ln\cos\left(A - \frac{t}{\tau}\right)\right]_0^t$$

$$x = x_0 + v_t\tau\ln\frac{\cos\left(A - \dfrac{t}{\tau}\right)}{\cos A}$$

or

$$x = x_0 + v_t\tau\ln\frac{\cos\left(\tan^{-1}\dfrac{v_0}{v_t} - \dfrac{t}{\tau}\right)}{\cos\left(\tan^{-1}\dfrac{v_0}{v_t}\right)} \tag{2.6.25}$$

At the top of the trajectory, when $v = 0$,

$$\tan^{-1}\frac{v_0}{v_t} - \frac{t}{\tau} = 0$$

from Eq. 2.6.22. So the maximum height reached is

$$x_{\max} = x_0 + v_t\tau\ln\frac{1}{\cos\left(\tan^{-1}\dfrac{v_0}{v_t}\right)} \tag{2.6.26}$$

Now, as the body *falls*—either from rest or if thrown downward with some initial velocity, it experiences the force of Eq. 2.6.19b. We can solve this using the same techniques as before:

$$t = \int_{v_0}^v \frac{m\,dv}{F(v)} = m\int_{v_0}^v \frac{dv}{-mg + \alpha v^2}$$

$$= \cdots = -\sqrt{\frac{m}{g\alpha}}\tanh^{-1}\sqrt{\frac{\alpha}{mg}}\,v\bigg|_{v_0}^v$$

$$t = -\tau\left[\tanh^{-1}\left(\frac{v}{v_t}\right) - \tanh^{-1}\left(\frac{v_0}{v_t}\right)\right] \tag{2.6.27}$$

For $v_0 = 0$, this simplifies to

$$t = -\tau\tanh^{-1}\left(\frac{v}{v_t}\right) \tag{2.6.28}$$

or

$$v = -v_t \tanh \frac{t}{\tau} \tag{2.6.29}$$

The minus sign comes about because v_t is the *terminal speed* or the magnitude of the terminal velocity and is positive. But the velocity is really downward, so v is negative.

We can continue to solve for x using

$$x = x_0 + \int_0^t v\, dt$$

Restricting ourselves to $v_0 = 0$ simplifies the form of v in the integral. For that case we have

$$x = x_0 - v_t \int_0^t \tanh \frac{t}{\tau}\, dt$$

$$= x_0 - v_t \tau \ln \cosh \frac{t}{\tau}\bigg|_0^t$$

$$x = x_0 - v_t \tau \ln \cosh \frac{t}{\tau} \tag{2.6.30}$$

Figure 2.6.1 compares the velocity of two bodies, one experiencing air resistance directly proportional to the velocity, and the other proportional to the *square* of the velocity. Both are dropped from rest. C and α differ, of course, but have been chosen so the terminal speed, v_t, is the same for both.

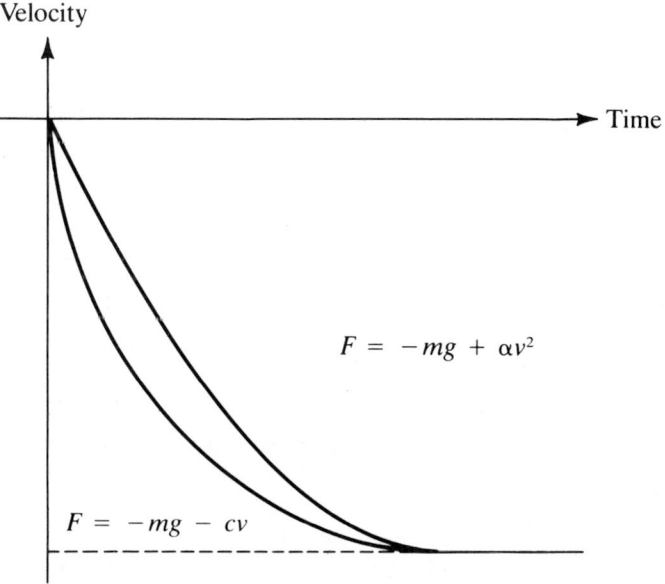

Figure 2.6.1 Velocities as functions of time for two bodies with a common terminal velocity. Curve 1 describes a body falling with a *linear* viscous force, $F = -mg - cv$, and curve 2 describes a *quadratic* viscous force, $F = -mg + \alpha v^2$.

PROBLEMS

2.1 Find the area of a quadrant of a circle (of radius $r = 1$) by constructing a sum of rectangles fitting underneath the circle as shown in the accompanying figure. That is, evaluate

$$A = \sum f(x_i)\, \Delta x$$

for $\Delta x = 0.25$ and 0.1. Compare these to the actual value of $\pi/4 = 0.7854$.

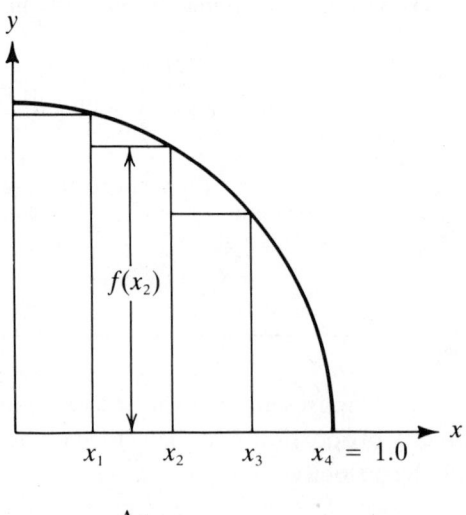

*2.2 Continue the evaluation in the accompanying illustration with smaller values of x such as 0.05, 0.01, 0.005, 0.001. You will need a computer, a programmable calculator, or a great deal of patience.

2.3 A body with mass m is initially at rest at the origin ($v_0 = 0$, $x_0 = 0$). At time $t = 0$, a force is applied that increases quadratically with time, $F = ct^2$. Solve for the position $x = x(t)$.

2.4 A body with mass m is initially at rest at the origin. At time $t = 0$, a force is applied that increases quadratically, $F = ct^2$. This continues until time t_1, when the force has reached a value of F_1. The force then remains constant.
  (a) Find a value for c in terms of F_1 and t_1.
  (b) Find $x = x(t)$ for $t > t_1$.
  (c) Determine the speed and position of the mass for time $t_2 = 2t_1$.

2.5 A body of mass m is initially at rest at the origin. Beginning at $t = 0$, a constant force F_1 acts on the body for time t_1. The force then increases linearly with time so that at time $t_2 = 2t_1$, the force is $2F_1$.
  (a) Find an explicit expression for the force during $t_1 < t < t_2$.
  (b) Find the position and velocity of the body at time t_2.

PROBLEMS

2.6 A body of mass m is initially at rest at the origin. Beginning at $t = 0$, a constant force F_1 acts on the body for time t_1. The force then decreases linearly with time so that it vanishes at time $t_2 = 2t_1$.
    (a) Find an explicit expression for the force during $t_1 < t < t_2$.
    (b) Find the position and velocity of the body at time t_2.

2.7 A tug of war is held between two teams of five men each. Each man has a mass of 80 kg and can initially pull on the rope with a force of 1000 newtons. At first the teams are evenly matched, of course, but as the men tire, the force with which each man pulls decreases according to $F = (1000 \text{ newtons}) e^{-t/T}$, where T is the mean tiring time. The team on the left has a mean tiring time of $T_L = 10$ s while the one on the right has a mean tiring time of $T_R = 20$ s.
    (a) Find the motion (i.e., $x(t)$).
    (b) What is their final velocity?
    (c) What assumptions lead to the foolish result of part (b)?

2.8 A particle of mass m is at rest at the origin (i.e., $x_0 = 0$, $v_0 = 0$). At time $t = 0$, a force begins. This force is proportional to the *square* of the time. At time $t = t_1$, the force has reached a value of F_1. The force then remains constant until time $t = 2t_1$. Then it decreases linearly until it disappears at time $t = 3t_1$. Write explicit mathematical expressions for the force as a function of time. Find the location of the particle for $t = t_1$, $t = 2t_1$, and $t = 3t_1$. How fast is it going at those times?

2.9 A body of mass m is initially at rest. Starting at time $t = 0$, it is subject to a force

$$F = F_0 e^{-\alpha t} \cos(\omega t + \theta_0)$$

Solve for its motion, $x = x(t)$. How does its final velocity depend on the initial phase angle θ_0? *Hint*: Writing the cosine as a sum of complex exponentials will simplify the algebra.

2.10 A rope of mass m and length l lies on a frictionless table with a length l_0 initially hanging over the edge. How fast is the rope moving when the bitter end of the rope leaves the table?

2.11 What would the coefficient of static friction between the rope and table in Problem 2.10 need to be so that the rope wouldn't slip?

*2.12 A body of mass m is acted upon by a *quadratic* restoring force, $|F| = Kx^2$. If $v_0 \neq 0$, and $x_0 \neq 0$, find (a) the total energy, (b) the turning point, (c) the velocity at any position, and (d) the motion, $x = x(t)$. You will best be able to do this part with numerical methods (i.e., on a computer).

2.13 Suppose a spring worked "in reverse." Instead of a linear restoring force, consider a linear *repelling* force, $F = +Kx$. Find the general solution for the motion $x = x(t)$ with general initial conditions of x_0 and v_0.

2.14 A particle of mass m is repelled from the origin by a force inversely proportional to the cube of its distance from the origin. Find the motion, $x(t)$, for a particle released from rest at x_0.

2.15 A particle of mass m moves in a region where the potential energy is $V(x) = A(e^{-bx}/x^2)$ (A and b are both positive).

(a) What is the force on the particle as it approaches the origin from the positive x-direction? From the negative x-direction?
(b) Make a rough sketch of the potential energy function and describe the possible kinds of motion.

2.16 An alpha particle is held inside a nucleus by a potential whose shape is roughly given in the accompanying diagram.
    (a) Describe the motion qualitatively.
    (b) Write a function that has the general form shown for $V(x)$—that is, vanishing but positive as $x \to \pm \infty$ and equal to $-V_0$ for $x = 0$.
    (c) Adjust the parameters so that $V(0) = -V_0$ and $V(\pm x_1) = V_1$.
    (d) Find the force associated with this potential.

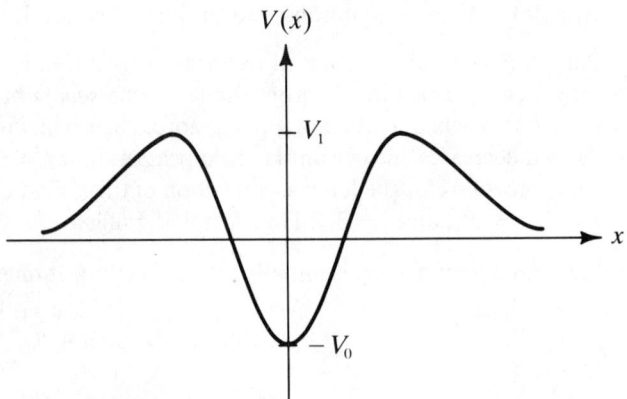

2.17 A jet engine that develops a maximum thrust F is used to power a plane with frictional drag proportional to the square of the velocity. If the plane starts at rest and the thrust is kept constant at its maximum, find $v(t)$ and $x(t)$. This should be familiar to you by now.

2.18 Find the speed as a function of position for a body released from rest and that is subject only to gravity and air resistance proportional to (a) the velocity and (b) the square of the velocity.

2.19 Find the maximum height an object thrown upward with initial speed v_0 will reach if it is subject to gravity and to air resistance proportional to (a) the velocity and (b) the square of the velocity.

2.20 The velocity of a certain body varies inversely with the displacement. That is, $v = b/x$. Find the force as a function of the displacement that must be acting on this body of mass m.

*2.21 Write a computer program to do a numerical integration (this only involves a slight modification of the program from Chapter 1) to determine the velocity and position as functions of time for an object falling from rest with air resistance proportional to the (a) velocity and (b) square of velocity. For both, adjust the parameters to give a terminal velocity of 50 m/s. Construct graphs of v versus t and x versus t for both cases.

*2.22 Write a computer program to determine velocity and position as functions of time for an object thrown *upward* with velocity $v_0 = 20$ m/s with air resistance proportional to (a) velocity and (b) square of velocity. For both, adjust the parameters so that the terminal velocity is 40 m/s. Construct graphs of v versus t and x versus t for both cases.

3

THE HARMONIC OSCILLATOR

3.1 Introduction

We have already seen harmonic oscillators in two forms—the driven, damped harmonic oscillator that we solved numerically in Chapter 1 and the simple harmonic oscillator that we solved explicitly in Chapter 2 using energy considerations. We shall now look at harmonic oscillators in more detail.

Harmonic oscillators are important for several reasons. First, they occur throughout nature. Indeed, *any* motion about a stable equilibrium is harmonic as long as it is small. Second, the equations involved offer challenging yet solvable examples of second-order differential equations. Third, as we shall see later, the same *equations* that describe a mechanical body under oscillations also describe electrical circuits.

$F = ma$, where $a = d^2x/dt^2$, is an example of a second-order differential equation; entire books and courses are devoted to the solution of such equations. Let's begin here by looking at the *simple* harmonic oscillator—a mass attached to a spring in a friction*less* environment. We shall assume a spring that obeys Hooke's Law that the stretch of the spring, x, and the force of the spring, F, are proportional. We label the proportionality constant k and call it the *spring constant*. Thus,

$$F = m\ddot{x} = -kx \qquad (3.1.1)$$

SECTION 3.1 / INTRODUCTION

By definition, a *second*-order differential equation involves taking *two* derivatives of the position, $\ddot{x}$. Every time a derivative is taken, some information is lost since the derivative involves only the slope. Thus, the functions $f_1 = x^2$ and $f_2 = x^2 + 7$ both have the *same* derivative, $2x$. If we start with the derivative $df/dx = 2x$, and wish to reconstruct the original function, we know it must be $x^2 + C$. C is an arbitrary constant. We need more information than *just* the differential equation to determine the value of C.

A *second*-order differential equation always has *two* arbitrary constants. Our additional information usually includes the *initial conditions*—the initial position x_0 and the initial velocity v_0. With the differential equation *and* these initial conditions, we can then determine the later position $x(t)$ for any time t in the future.

All the differential equations we will encounter describe actual physical systems that really do something. Therefore, we know that a solution *exists* and that the system is only going to do *one* thing. Therefore we also know that there is only one *unique* solution. When we have found a solution to our differential equation—by any means—and satisfied the initial conditions, we're through. The problem is completely solved; there cannot be another, different solution.

Exponentials (Euler's Relations)

Since we shall find that a harmonic oscillator moves sinusoidally with time, we begin by reviewing the ideas of a power series in connection with some useful functions. The trigonometric functions sine and cosine can be written as a power series. This form will prove to be very useful.

$$\sin x = x - \frac{x^3}{3!} + \frac{x^5}{5!} - \frac{x^7}{7!} + \cdots \quad (3.1.2)$$

and

$$\cos x = 1 - \frac{x^2}{2!} + \frac{x^4}{4!} - \frac{x^6}{6!} + \cdots \quad (3.1.3)$$

Another very useful function is e^y. This, too, can be written as a power series:

$$e^y = 1 + y + \frac{y^2}{2!} + \frac{y^3}{3!} + \frac{y^4}{4!} + \frac{y^5}{5!} + \cdots \quad (3.1.4)$$

If we now let $y = ix$ (where $i^2 = -1$), this becomes

$$e^{ix} = 1 + ix - \frac{x^2}{2!} - i\frac{x^3}{3!} + \frac{x^4}{4!} + i\frac{x^5}{5!} + \cdots$$

$$e^{ix} = \left(1 - \frac{x^2}{2!} + \frac{x^4}{4!} - \frac{x^6}{6!} + \cdots\right) + i\left(x - \frac{x^3}{3!} + \frac{x^5}{5!} - \frac{x^7}{7!} + \cdots\right)$$

that is

$$e^{ix} = \cos x + i \sin x \tag{3.1.5}$$

which will shortly prove to be extremely useful in solving for the motion of harmonic oscillators. Eq. 3.1.5 is known as the *Euler relation*.

3.2 Simple Harmonic Oscillator

A *simple* harmonic oscillator is a system that obeys our equation of motion:

$$\ddot{x} = -\frac{k}{m}x \tag{3.2.1}$$

We developed this equation for x to measure the position of a mass m attached to a spring with spring constant k. With appropriate interpretation, this also describes the current flowing in a circuit with a capacitor, an inductor (coil), and no resistance (just as we have no friction in the mechanical case).

We can begin to solve Eq. 3.2.1 by remembering the solution from Chapter 2, Eq. 2.5.36:

$$x = A \sin \omega t \tag{3.2.2}$$

where $\omega^2 = k/m$, ω is the angular frequency of oscillation. We can check this easily by finding

$$\dot{x} = A\omega \cos \omega t$$

$$\ddot{x} = -A\omega^2 \sin \omega t$$

and putting it back into Eq. 3.2.1 to see if Eq. 3.2.2 is correct.

$$m(-A\omega^2 \sin \omega t) = -k(A \sin \omega t)$$

$$m\omega^2 = k$$

$$\omega^2 = \frac{k}{m}$$

which is certainly true from our definition of ω. But Eq. 3.2.2 cannot be the general or complete solution of a second-order differential equation because it doesn't contain *two* arbitrary constants (like the A in Eq. 3.2.2).

If the sine function is a solution, let's try the cosine function, too:

$$x = B \cos \omega t \tag{3.2.3}$$

then

$$\dot{x} = -B\omega \sin \omega t$$

SECTION 3.2 / SIMPLE HARMONIC OSCILLATOR

and

$$\ddot{x} = -B\omega^2 \cos \omega t$$

This can be put back into Eq. 3.2.1 to determine that it, too, is a valid (although incomplete) solution.

A complete solution, then, containing the *two* necessary arbitrary constants, might be found by taking the *sum* of these as

$$x = A \sin \omega t + B \cos \omega t \qquad (3.2.4)$$

This is a general, or complete, solution. Substitution into Eq. 3.2.1 will demonstrate that it is *a* solution. But because it contains two arbitrary constants, A and B, it is also the general or *complete* solution.

With two arbitrary constants, we need two more pieces of data to solve for A and B. Usually, these data take the form of *initial conditions*. If the initial position is $x(0) = x_0$ and the initial velocity is $v(0) = v_0$, we can then write two equations for the two unknowns—the arbitrary constants:

$$x(0) = A \sin 0 + B \cos 0$$

$$x_0 = B \qquad (3.2.5)$$

$$v(0) = A\omega \cos 0 - B\omega \sin 0$$

$$v_0 = A\omega$$

$$A = \frac{v_0}{\omega} \qquad (3.2.6)$$

An alternate way of obtaining a general solution to our equation is to introduce another arbitrary constant into either Eq. 3.2.2 or 3.2.3 by writing

$$x = A \sin(\omega t + \phi) \qquad (3.2.7)$$

This is still a valid solution to Eq. 3.2.1, as substitution will show. Now A and ϕ can be evaluated from the two equations using our initial conditions. (*Note*: This A is *not* the same as the A solved in Eq. 3.2.6. A is used here simply by convention.)

$$x(0) = x_0 = A \sin \phi \qquad (3.2.8)$$

$$v(0) = v_0 = A\omega \cos \phi \qquad (3.2.9)$$

Dividing Eq. 3.2.7 by Eq. 3.2.8 yields

$$\tan \phi = \frac{\omega x_0}{v_0}$$

or

$$\phi = \tan^{-1} \frac{\omega x_0}{v_0} \qquad (3.2.10)$$

which can then be used with Eq. 3.2.7 to solve for A:

$$A = \frac{x_0}{\sin \phi} = \frac{x_0}{\sin\left(\tan^{-1}\dfrac{\omega x_0}{v_0}\right)}$$

or

$$A = \frac{\sqrt{\omega^2 x_0^2 + v_0^2}}{\omega} \qquad (3.2.11)$$

Still another method involves the exponential function. Since we already know, from Eq. 3.1.5, that certain exponentials can be written in terms of sines and cosines, it seems reasonable to expect that a solution to the simple harmonic oscillator can be written in terms of exponentials. Of more importance, though, is the ease and simplicity with which derivatives of exponentials may be written since

$$\frac{de^{ay}}{dy} = ae^{ay} \qquad (3.2.12)$$

Let's begin by trying

$$x = Ae^{qt} \qquad (3.2.13)$$

Then

$$\dot{x} = qAe^{qt} \qquad (3.2.14)$$

and

$$\ddot{x} = q^2 Ae^{qt} \qquad (3.2.15)$$

Substituting these back into our differential equation, Eq. 3.2.1, we have

$$m\ddot{x} = mq^2 Ae^{qt} = -kAe^{qt} = -kx$$

or

$$mq^2 = -k \qquad (3.2.16)$$

This *algebraic* equation is called the *secular equation*. Its solution is

$$q = \pm i\sqrt{\frac{k}{m}} = \pm i\omega \qquad (3.2.16a)$$

Therefore, our general solution can be written as

$$x = Ae^{i\omega t} + Be^{-i\omega t} \qquad (3.2.17)$$

(*Note*: This A and B are *not* the same as those in Eq. 3.2.4. A and B are used here again simply by convention.)

SECTION 3.2 / SIMPLE HARMONIC OSCILLATOR

Since such exponential terms can always be expressed as sine and cosine terms, this is totally equivalent to the other forms we've seen in Eqs. 3.2.4 and 3.2.7. As before, the arbitrary constants A and B must be solved from the initial conditions.

$$x(0) = x_0 = Ae^0 + Be^0$$

$$A + B = x_0 \qquad (3.2.18)$$

$$V(0) = i\omega A e^0 - i\omega B e^0 = v_0$$

$$i\omega(A - B) = v_0 \qquad (3.2.19)$$

In this way, Eqs. 3.2.18 and 3.2.19 form a system of simultaneous equations that can be used to solve for A and B:

$$A = \frac{x_0 - i\dfrac{v_0}{\omega_0}}{2} \qquad (3.2.20)$$

and

$$B = \frac{x_0 + i\dfrac{v_0}{\omega_0}}{2} \qquad (3.2.21)$$

Of course, A and B are now complex numbers.

From all three methods we know that a simple harmonic oscillator like the mass and spring sketched in Figure 3.2.1 will move back and forth—oscillate sinusoidally—with an angular frequency of $\omega = \sqrt{k/m}$.

This is the *simple* harmonic oscillator with no friction present; we will refer back to this again and again as we continue. Therefore, we shall call this the *natural* frequency and indicate this with a subscript, $\omega_0 = \sqrt{k/m}$.

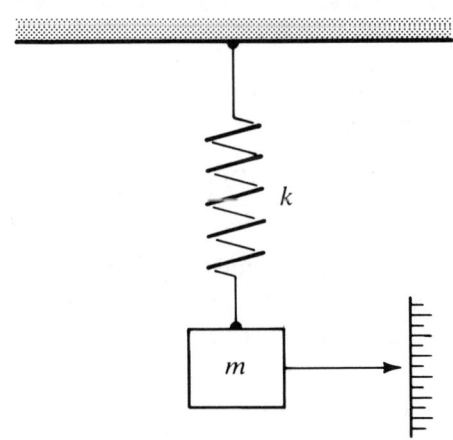

Figure 3.2.1 A simple harmonic oscillator.

This is the *angular frequency* in radians per second. Since there are 2π radians in one cycle, the *frequency*, f, in cycles per second (or hertz) is

$$f_0 = \frac{1}{2\pi}\omega_0 = \frac{1}{2\pi}\sqrt{\frac{k}{m}} \qquad (3.2.22)$$

The period, then, is

$$T_0 = \frac{1}{f_0} = 2\pi\sqrt{\frac{m}{k}} \qquad (3.2.23)$$

This period is independent of the *amplitude*.

We have now solved the simple harmonic oscillator in three different—yet totally equivalent—forms. In each of these cases, we have essentially guessed a solution to the differential equation, and then checked carefully to see if our guess was correct. We checked to see if the functional form of our proposed solution (i.e., the "guess") satisfied the differential equation. If it was satisfied, then we evaluated some of the constants (*e.g.*, $\omega^2 = k/m$). Since we were dealing with a second-order differential equation, we knew that its solution must contain two arbitrary constants. When these conditions were all met, the solution was complete.

At first thought, guessing a solution may seem a very poor or crude—or even silly—method. There are more rigorous ways to solve a differential equation. In fact, you've already solved this one rigorously in Chapter 2. But this isn't a guess in the usual sense. Rather, it's a reasoned proposal built upon your physical observations of how you *expect* the system to behave. Don't let the mathematics overpower the physics. Since you know how the system behaves, you can begin to write down the solution immediately. Then the mathematics can guide you on to the complete solution. This method—using a reasoned proposal and making corrections or completions as guided by the mathematics—is not unique to the harmonic oscillator. It is a valuable method used throughout physics.

3.3 Power Series Representation of an Arbitrary Function

Any "well behaved" function can be expressed as an infinite power series by using a Taylor series expansion such as

$$f(x) = f(a) + \left.\frac{df}{dx}\right|_a (x-a) + \frac{1}{2!}\left.\frac{d^2f}{dx^2}\right|_a (x-a)^2 + \frac{1}{3!}\left.\frac{d^3f}{dx^3}\right|_a (x-a)^3 + \cdots \qquad (3.3.1)$$

By *well behaved*, we mean that the function and all its derivatives are continuous. Since this is true for *any* function, it is certainly true for a function defining the potential energy, $V(x)$:

$$V(x) = V(a) + \left.\frac{dV}{dx}\right|_a (x-a) + \frac{1}{2!}\left.\frac{d^2V}{dx^2}\right|_a (x-a)^2 + \frac{1}{3!}\left.\frac{d^3V}{dx^3}\right|_a (x-a)^3 + \cdots$$

$$(3.3.1a)$$

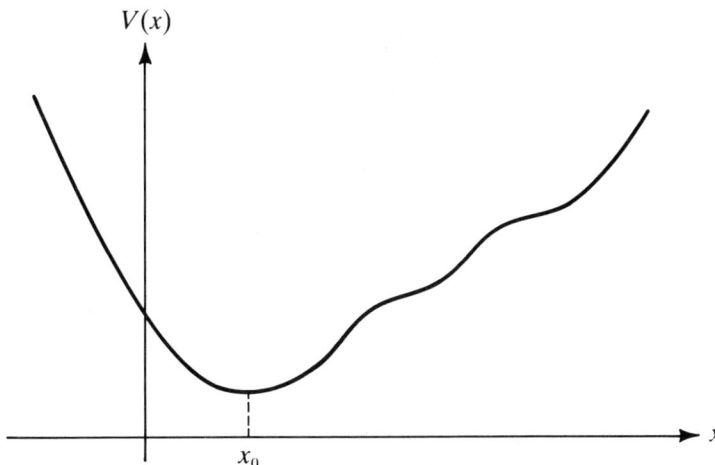

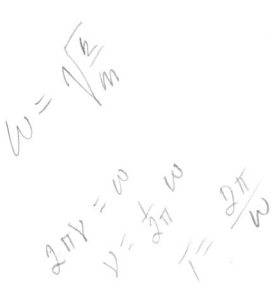

Figure 3.3.1 A graph of a particular potential energy, V(x).

Now consider some arbitrary potential energy function, $V(x)$, as sketched in Figure 3.3.1.

The value x_0 locates a minimum of the function $V(x)$. If we replace a in Eq. 3.3.1a with x_0, the Taylor series for $V(x)$ can be written as

$$V(x) = V(x_0) + \frac{dV}{dx}\bigg|_{x_0}(x-x_0) + \frac{1}{2!}\frac{d^2V}{dx^2}\bigg|_{x_0} + \frac{1}{3!}\frac{d^3V}{dx^3}\bigg|_{x_0}(x-x_0)^3 + \cdots \quad (3.3.2)$$

$V(x_0)$ is a constant; let's call it V_0. $\frac{dV}{dx}\bigg|_{x_0}$ is the slope of the curve at x_0. Our choice of x_0 at the minimum requires that the slope be zero for $x = x_0$. So this term is zero. We define $\frac{d^2V}{dx^2}\bigg|_{x_0}$ as equal to k. Each term involves $(x - x_0)$ to a power. As long as x remains near the equilibrium position x_0, then $(x - x_0)$ is small and the higher the power of n, the smaller $(x - x_0)^n$ will be.

Therefore, for x very close to x_0, we may neglect all higher ordered terms containing $(x - x_0)$. Since $\frac{dV}{dx}\bigg|_{x_0} = 0$, we shall retain only the next term, in $(x - x_0)^2$. We can now write the potential energy as

$$V(x) = V_0 + \tfrac{1}{2}k(x - x_0)^2 \quad (3.3.3)$$

But any potential energy is only defined to within an *additive* constant. So we can just as well describe our system by

$$V'(x) = V(x) - V_0 \quad (3.3.4)$$

as by $V(x)$ itself. To make things *look* nice, we change the coordinate system to

$$X = x - x_0 \quad (3.3.5)$$

Now we can write the potential energy function as

$$V'(X) = \tfrac{1}{2}kX^2 \quad (3.3.6)$$

And *that* is just the potential energy for a simple harmonic oscillator. The constant k plays the same role here as in the spring/mass simple harmonic oscillator, so we shall call it the *effective spring constant*.

$$k_{\text{eff}} = \left.\frac{d^2V}{dx^2}\right|_{x_0} \tag{3.3.7}$$

This means that, for small oscillations near equilibrium, our system has an angular frequency of

$$\omega_0 = \sqrt{\frac{k_{\text{eff}}}{m}} = \sqrt{\left.\frac{1}{m}\frac{d^2V}{dx^2}\right|_{x_0}} \tag{3.3.8}$$

Therefore, such a system undergoes simple harmonic oscillations when displaced only a small amount from a stable equilibrium position. This applies to huge suspension bridges or to atoms in a crystal, to a tuning fork or to ionized particles in the upper atmosphere. Thus understanding the simple harmonic oscillator helps us learn a little about all of the universe. Any potential energy function describing any imaginable system acted upon by horribly complex forces *looks* like a plain old simple harmonic oscillator as long as we keep the amplitude small (provided it and its derivatives are continuous at x_0).

3.4 Damped Harmonic Oscillator

Now you thoroughly understand the motion of a simple harmonic oscillator as characterized by a mass on a spring. Such a system oscillates forever with an undiminished amplitude. Consider as a more realistic system an oscillator immersed in a fluid (like air, water, warm syrup or, eventually, cold molasses) as shown in Figure 3.4.1.

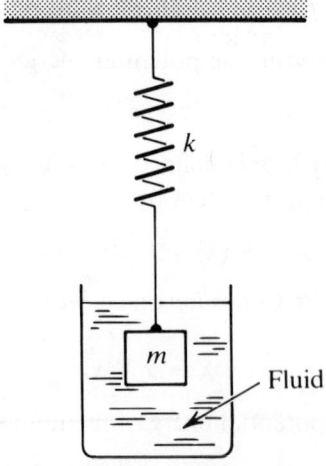

Figure 3.4.1 A damped harmonic oscillator.

SECTION 3.4 / DAMPED HARMONIC OSCILLATOR

Such a mass may have a force exerted on it by the spring as before. But it also feels a force from the fluid. This force varies with the speed, and all friction forces always oppose the motion; as the mass changes direction, so does this friction force. We can write it as simply $-cv$ or $-c\dot{x}$. Our total force then becomes

$$F = -kx - c\dot{x} = m\ddot{x} \tag{3.4.1}$$

It is more convenient to rearrange this as

$$m\ddot{x} + c\dot{x} + kx = 0 \tag{3.4.1a}$$

To find the solution to this *damped* harmonic oscillator equation, we might look back at our method of attack on the simple harmonic oscillator we solved earlier. What sort of motion do you *expect*? This is an important part of developing *physical intuition*. We might expect some sort of oscillation with a decaying amplitude. We could describe that by

$$x = Ae^{pt}\sin(\omega t + \theta_0) \tag{3.4.2}$$

Pardon the pun, but if we're very careful with our *p*'s and *q*'s, we can describe the same solution with

$$x = Ae^{qt} \tag{3.4.3}$$

if we allow q to be complex. Now we can take time derivatives of this and substitute them back into Eq. 3.4.1 to see what values (if any) q must have for this to be a solution.

$$\dot{x} = Aqe^{qt} \tag{3.4.4}$$

$$\ddot{x} = Aq^2 e^{qt} \tag{3.4.5}$$

As before, these can be substituted into our second-order *differential* equation, Eq. 3.4.1, to obtain the secular equation, a quadratic *algebraic* equation:

$$mq^2 + cq + k = 0 \tag{3.4.6}$$

From the familiar quadratic formula, the solution is

$$q = \frac{-c \pm \sqrt{c^2 - 4mk}}{2m} \tag{3.4.7}$$

These solutions naturally fall into three categories:

Case I (overdamping): When the frictional force is large, so that

$$c^2 > 4mk$$

then the square root term is a real number and we can write

$$q_1 = \frac{-c + \sqrt{c^2 - 4mk}}{2m} \quad \text{and} \quad q_2 = \frac{-c - \sqrt{c^2 - 4mk}}{2m} \tag{3.4.8}$$

Both of these are negative, so we can define

$$\gamma_1 = -q_1 = \frac{+c - \sqrt{c^2 - 4mk}}{2m} \quad \text{and} \quad \gamma_2 = -q_2 = \frac{+c + \sqrt{c^2 - 4mk}}{2m}$$
(3.4.9)

and write our solution as

$$x = A_1 e^{-\gamma_1 t} + A_2 e^{-\gamma_2 t} \tag{3.4.10}$$

This has the *two* arbitrary constants, A_1 and A_2, required of any second-order differential equation. A_1 and A_2 will be determined from, say, the initial conditions x_0 and v_0. Both of these terms decay exponentially with time so the motion will be as shown in Figure 3.4.2.

This case is called, appropriately enough, the *overdamped* case. Once moved from equilibrium and released, the mass returns to equilibrium without overshoot, slowing down as it nears equilibrium so that it requires a long (actually, infinite) time to come to rest there.

Case II (underdamping): When the frictional force is small so that

$$c^2 < 4mk$$

the square root terms in Eq. 3.4.7 become imaginary and the q's become complex:

$$q = \frac{-c \pm \sqrt{(-1)(4mk - c^2)}}{2m}$$

$$q = \frac{-c}{2m} \pm i\sqrt{\frac{k}{m} - \left(\frac{c}{2m}\right)^2} \tag{3.4.11}$$

Remembering that $k/m = \omega_0^2$, we can also define $\gamma = c/2m$ and simplify Eq. 3.4.11 to

$$q = -\gamma \pm i\sqrt{\omega_0^2 - \gamma^2} \tag{3.4.12}$$

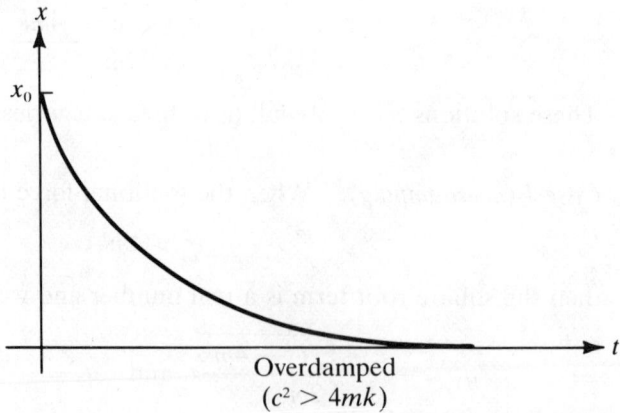

Figure 3.4.2 Motion of an overdamped harmonic oscillator.

Overdamped ($c^2 > 4mk$)

SECTION 3.4 / DAMPED HARMONIC OSCILLATOR

We define

$$\omega = \sqrt{\omega_0^2 - \gamma^2} \tag{3.4.13}$$

and can write

$$q_1 = -\gamma + i\omega \quad \text{and} \quad q_2 = -\gamma - i\omega \tag{3.4.14}$$

That means the position is given by

$$x = A_1 e^{(-\gamma + i\omega)t} + A_2 e^{(-\gamma - i\omega)t} \tag{3.4.15}$$

or

$$x = (e^{-\gamma t})(A_1 e^{i\omega t} + A_2 e^{-i\omega t}) \tag{3.4.16}$$

The first term is a decaying exponential that forms an *envelope* for the behavior of the second term. From our earlier discussions of the complex numbers and Eq. 3.1.5, you should be familiar with this second term. It describes a sinusoidal oscillation. In fact, with careful choice of A_1 and A_2, this can be written as

$$x = (e^{-\gamma t}) B \sin(\omega t + \theta) \tag{3.4.17}$$

It requires some trigonometric manipulations, but the required substitutions are

$$A_1 = \tfrac{1}{2} B(\sin\theta + i\cos\theta)$$

and

$$A_2 = \tfrac{1}{2} B(\sin\theta - i\cos\theta)$$

This decaying oscillation is shown in Figure 3.4.3.

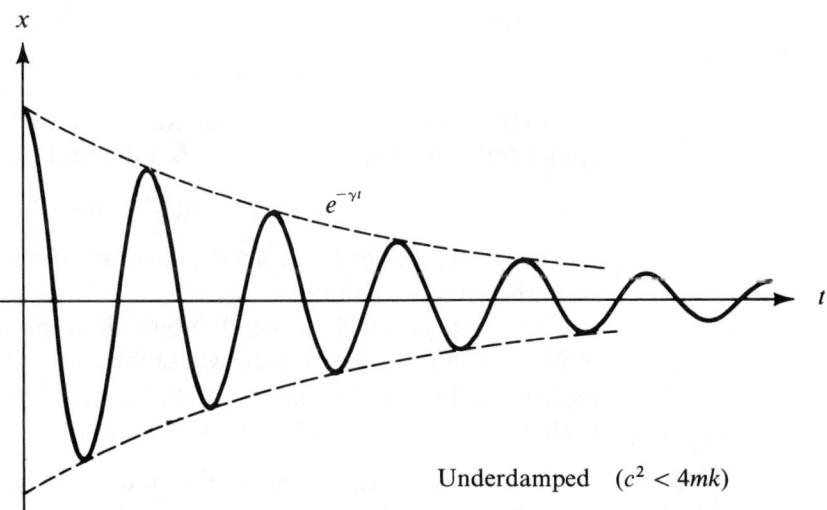

Figure 3.4.3 Motion of an underdamped harmonic oscillator.

Underdamped $(c^2 < 4mk)$

This is the case of *under*damped motion. In the limit that c is zero, γ becomes zero and we have oscillation with constant amplitude, as we must, since $c = 0$ brings us back to the simple harmonic oscillator. Notice also in this underdamped case that as c increases, the frequency *decreases* since $\omega = \omega_0^2 - \gamma^2$. That is, the frictional drag slows down the motion. Again, this is as we would expect from experimental observation.

Case III (critical damping): When the frictional force is between the two cases just discussed so that

$$c^2 = 4mk$$

there is only one value of q:

$$q = \frac{-c}{2m} = -\gamma$$

or

$$x = Ae^{-\gamma t} \tag{3.4.18}$$

But because this contains a single arbitrary constant, A, it cannot be the complete solution to our second-order differential equation. We must find another solution with another arbitrary constant. For this other solution, we might try

$$x = Bte^{-\gamma t}$$
$$\dot{x} = Be^{-\gamma t} - B\gamma te^{-\gamma t} \tag{3.4.19}$$
$$\ddot{x} = -\gamma 2Be^{-\gamma t} + B\gamma^2 te^{-\gamma t}$$

Substituting these into Eq. 3.4.1 yields

$$m(\gamma^2 t - 2t) + c(1 - \gamma t) + kt = 0 \tag{3.4.20}$$

or, rearranging terms,

$$t(m\gamma^2 - c\gamma + k) + (c - 2m\gamma) = 0 \tag{3.4.20a}$$

Both terms in parentheses are identically zero for $\gamma = c/2m$. Therefore, our choice for a solution in Eq. 3.4.19 is valid and our general solution is

$$x = Ae^{-\gamma t} + Bte^{-\gamma t} \tag{3.4.21}$$

As always, the values of the arbitrary constants A and B must be determined from the initial conditions.

This case is called *critical* damping. It is the smallest value of c (i.e., the weakest frictional drag) that will allow, at most, a single overshoot or oscillation. Figure 3.4.4 shows a sketch of critically damped motion for a body released from rest. Figure 3.4.5 shows all three cases superposed.

The mass returns to its equilibrium position more rapidly with critical damping than with overdamping. Yet, there is a lack of oscillation or going

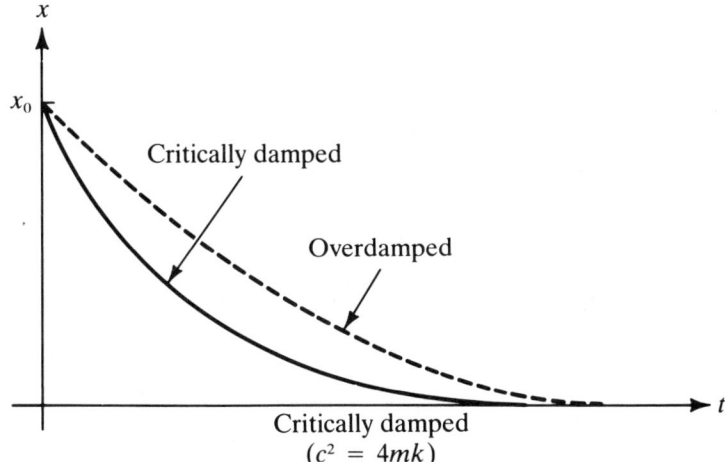

Figure 3.4.4 Motion of a critically damped harmonic oscillator.

past the equilibrium position, which has important practical applications. A simple spring on a wooden screen door pulls the door shut with a linear restoring force (a simple harmonic oscillator). This means that the door reaches maximum speed as it passes through (or tries to pass through) its equilibrium position, the door jamb. *Wham*! Putting some sort of shock absorber on the door closer saves both door and jamb from damage. But to close the door as quickly as possible, and yet have its velocity essentially zero as it eases into equilibrium at the jamb, we choose the strength of the return spring, friction drag of the shock absorber, and the mass of the door such that the system is *critically damped*. Likewise, the suspension system of a car is designed for *critical damping*. Worn shock absorbers have little frictional drag and allow oscillations due to *underdamping*. Shocks that are too heavy prevent a quick return to equilibrium and cause a harsh, bumpy ride from *overdamping*.

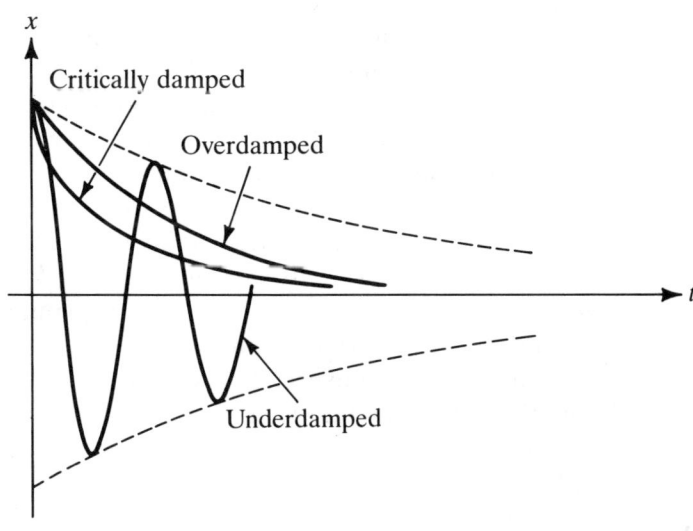

Figure 3.4.5 Harmonic oscillators of all three types.

3.5 Forced Harmonic Oscillator

Let us now impress an external force upon our oscillator. We can do this by putting a steel object in a varying magnetic field, an electron in a varying magnetic field, an electron in a varying electric field, by driving a car over joints in a highway, or by causing the top support of the spring to move up and down. To keep things (relatively) simple, we shall restrict our consideration to an external force that varies sinusoidally with time as in

$$F_{ex} = F_0 \cos(\omega t + \theta) \quad (3.5.1)$$

Now our force equation becomes

$$F_{net} = -kx - c\dot{x} + F_{ex} = m\ddot{x} = ma \quad (3.5.2)$$

We have already seen that exponentials with complex exponents are a powerful tool when dealing with a harmonic oscillator. So we can write

$$F_{ex} = \text{Re}(F_0 e^{i(\omega t + \theta)}) \quad (3.5.3)$$

where Re means "the real part" of the expression that follows it. Likewise, if the need arises, we can write

$$\sin(\omega t + \theta) = \text{Im}(e^{i(\omega t + \theta)}) \quad (3.5.4)$$

where Im means "the imaginary part" of the expression following it. In fact, we can simply write Eq. 3.5.2 as

$$-kx - c\dot{x} + F_0 e^{i(\omega t + \theta)} = m\ddot{x} \quad (3.5.5)$$

with the *understanding* that *only* the real—or only the imaginary—part of any solution x is to be used. We can then rearrange the terms as

$$m\ddot{x} + c\dot{x} + kx = F_0 e^{i(\omega t + \theta)} \quad (3.5.5a)$$

We already know the *general* solution to the *homogeneous* differential equation corresponding to this; that is, when the right-hand side is zero. When $F_0 = 0$, this is simply the equation of the damped harmonic oscillator that we've just solved. It has two arbitrary constants that depend on the initial *conditions*. This general solution dies out with time. We now need to find a *particular* solution to Eq. 3.5.5 with the external force present. There will be no arbitrary constants. Since the force continues with time, this particular solution will also continue in time. This is called the *steady state solution*. The solution to the homogeneous equation is called the *transient solution* since it decays with time.

What might we expect this steady state solution to be? The force is sinusoidal so the motion it produces will be, too—with the same frequency but, perhaps, shifted in time. We can write that as

$$x = A e^{i(\omega t + \delta)} \quad (3.5.6)$$

SECTION 3.5 / FORCED HARMONIC OSCILLATOR

Now we need to find the time derivatives and substitute back into the differential Eq. 3.5.5 to solve for A and δ:

$$\dot{x} = iA\omega e^{i(\omega t + \delta)} \tag{3.5.7}$$

$$\ddot{x} = -A\omega^2 e^{i(\omega t + \delta)} \tag{3.5.8}$$

$$Ae^{i(\omega t + \delta)}[-m\omega^2 + ic\omega + k] = F_0 e^{i(\omega t + \theta)}$$

$$-m\omega^2 + ic\omega + k = \frac{F_0}{A} e^{i[(\omega t + \theta) - (\omega t + \delta)]} = \frac{F_0}{A} e^{i(\theta - \delta)}$$

A single complex equation, just like a vector equation, is actually two equations. The real parts must be equal and the imaginary parts must be equal. Thus, rearranging terms after expanding the exponentials, we have

$$(k - m\omega^2) + ic\omega = \frac{F_0}{A}\cos(\theta - \delta) + i\frac{F_0}{A}\sin(\theta - \delta) \tag{3.5.9}$$

which really means

$$k - m\omega^2 = \frac{F_0}{A}\cos(\theta - \delta) \tag{3.5.10}$$

and

$$c\omega = \frac{F_0}{A}\sin(\theta - \delta) \tag{3.5.11}$$

We can divide Eq. 3.5.11 by Eq. 3.5.10 and define $\phi = \theta - \delta$ to get

$$\tan\phi = \frac{c\omega}{k - m\omega^2} \tag{3.5.12}$$

where ϕ is the *difference* in phase between the external force and the resulting steady state motion. This equation is more meaningfully written in terms of the natural frequency, $\omega_0^2 = k/m$, and the parameter, $\gamma = c/2m$. Thus,

$$\tan\phi = \frac{2\gamma\omega}{\omega_0^2 - \omega^2}$$

or

$$\phi = \tan^{-1}\frac{2\gamma\omega}{\omega_0^2 - \omega^2} \tag{3.5.13}$$

To solve for the amplitude, we can square Eqs. 3.5.10 and 3.5.11 and add them together. This yields

$$(k - m\omega^2)^2 + c^2\omega^2 = \frac{F_0^2}{A^2} = \frac{F_0^2}{A^2}(\cos^2\phi + \sin^2\phi) \tag{3.5.14}$$

This can be solved for A, the *amplitude* of the steady state motion, to give

$$A = \frac{F_0}{\sqrt{(k - m\omega^2)^2 + c^2\omega^2}} \tag{3.5.15}$$

Again recalling $\omega_0^2 = k/m$ and $\gamma = c/2m$, cosmetic algebra can reduce this somewhat to

$$A = \frac{F_0}{m}\left[\left(\frac{k}{m} - \omega^2\right)^2 + \frac{c^2\omega^2}{m^2}\right]^{-1/2}$$

or

$$A = \frac{F_0}{m}\left[(\omega_0^2 - \omega^2)^2 + 4\gamma^2\omega^2\right]^{-1/2} \tag{3.5.16}$$

Now to find the condition that gives maximum steady state amplitude, we can differentiate A with respect to ω, the frequency of the external force, and set that equal to zero:

$$\frac{dA}{d\omega} = -\frac{1}{2}\frac{F_0}{m}\left[(\omega_0^2 - \omega^2)^2 + 4\gamma^2\omega^2\right]^{-3/2}\left[2(\omega_0^2 - \omega^2)(-2\omega) + 8\gamma^2\omega\right] = 0 \tag{3.5.17}$$

Therefore,

$$-4\omega(\omega_0^2 - \omega^2) + 8\gamma^2\omega = 0$$

or

$$\omega^2 - \omega_0^2 + 2\gamma^2 = 0$$

or

$$\omega^2 = \omega_0^2 - 2\gamma^2$$

Maximum steady state amplitude occurs, then, when the external force has a frequency of

$$\omega = \omega_r = \sqrt{\omega_0^2 - 2\gamma^2} \simeq \omega_0 - \frac{\gamma^2}{\omega_0} \tag{3.5.18}$$

This is known as the *resonance* frequency. For systems with very little damping, this resonance frequency ω_r is very nearly equal to the undamped natural frequency ω_0. But in general it is different from the undamped natural frequency *and* from the frequency the damped harmonic oscillator has in the absence of an external driving force.

The *maximum* amplitude is found by evaluating Eq. 3.5.16 for the resonance frequency. The result is

$$A_{max} = \frac{F_0}{m2\gamma\sqrt{\omega_0^2 - \gamma^2}} \tag{3.5.19}$$

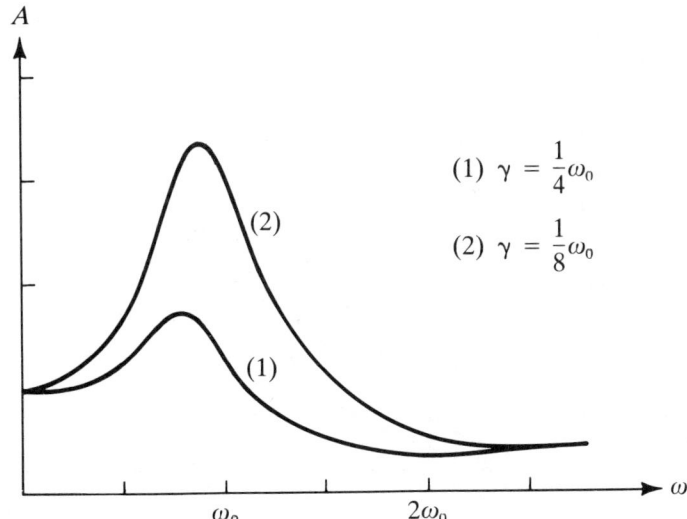

Figure 3.5.1 Steady state amplitude versus driving frequency. (from Fowles, p. 68, 3rd ed.)

Figure 3.5.1 shows the steady state amplitude A as a function of ω for two different values of γ or two different cases of damping.

Notice that as the damping decreases, the *maximum* amplitude increases (as might be expected). Such a decrease in damping also causes the resonance curve to be more sharply peaked; an increase in damping broadens or flattens the resonance curve. A good example of this is a chime, tuning fork, or piano string that has nearly zero damping. A musical note of *exactly* its natural frequency will cause the chime, fork, or string to vibrate and sound itself. But there will be no noticeable response to another note that differs by only a very few hertz (cycles per second).

As the frequency of the external driving force increases from zero, the amplitude will increase to some maximum and then decrease. While this is happening, what is the *phase* relation between the motion and the external force? The phase angle ϕ is given by Eq. 3.5.13. Figure 3.5.2 shows ϕ graphed as a function of driving frequency.

For low frequencies, ϕ is nearly zero; the external force and motion are *in phase*. When the force is at its maximum in one direction, so is the displacement. When the force is zero, the displacement is zero. When the force is at its maximum in the other direction, so is the displacement. As the driving frequency increases from zero, approaching ω_0, the phase angle ϕ approaches $\pi/2$. For $\omega = \omega_0$, the force and motion are $\pi/2$ or 90° out of phase. When the force is maximum, the displacement is zero, and vice versa. As the driving force increases further, the phase angle increases toward π or 180° so that external force and resulting motion become entirely *out of phase*. Under those conditions, when the force is maximum in one direction, the displacement is at a maximum in the *opposite* direction. You can demonstrate such behavior by suspending a mass on a spring while holding the top end of

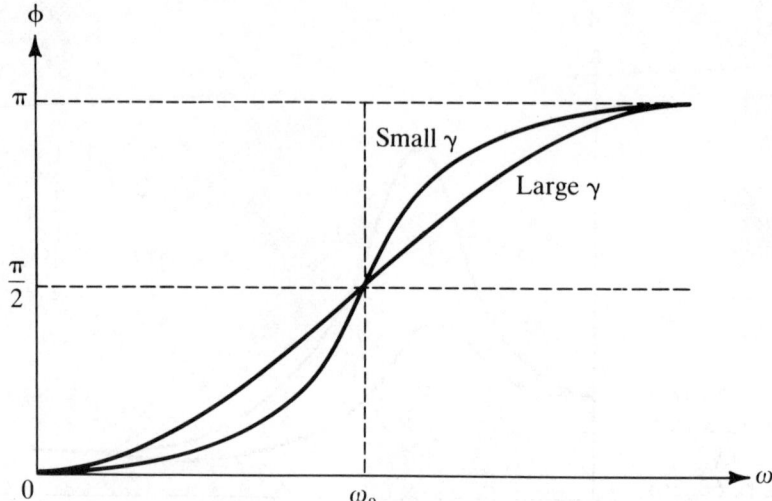

Figure 3.5.2 Phase angle versus driving frequency.

the spring. If you move your hand up and down very slowly, the mass follows in phase. Now move your hand up and down rather rapidly; that is, increase the frequency of the external driving force, ω. If you watch closely, you will see the mass moving up as your hand moves down and vice versa; the force and motion are π or 180° out of phase.

This phase difference is quite important if we look at the average power supplied by the external force or absorbed by the system (these should be the same, of course). Power supplied equals the force times the velocity:

$$P = F_{ex}\dot{x} \tag{3.5.20}$$

From Eqs. 3.5.7 and 3.5.16, we can write this as the following, if we remember to use *only* the imaginary part of $\dot{x}$ and F_{ex}:

$$P = [F_0 \sin(\omega t + \theta)] \frac{\omega F_0 \cos(\omega t + \delta)}{m\sqrt{(\omega_0^2 - \omega^2)^2 + 4\gamma^2\omega^2}} \tag{3.5.21}$$

Recalling that $\phi = \theta - \delta$ or $\delta = \theta - \phi$, this can be expanded as

$$P = \frac{\omega F_0^2 \sin(\omega t + \theta)[\cos(\omega t + \theta)\cos(-\phi) - \sin(\omega t + \theta)\sin(-\phi)]}{m\sqrt{(\omega_0^2 - \omega^2)^2 + 4\gamma^2\omega^2}} \tag{3.5.22}$$

Taken over a complete cycle, the average value of the first term, with $\sin(\omega t + \theta)\cos(\omega t + \theta)$, is zero so it gives no contribution. But the average of the $\sin^2(\omega t + \theta)$ from the second term over a cycle is $\frac{1}{2}$. Thus, the *average power* supplied by the external force is

$$\langle P \rangle = \frac{\omega F_0^2 \sin \phi}{2m\sqrt{(\omega_0^2 - \omega^2)^2 + 4\gamma^2\omega^2}} \tag{3.5.23}$$

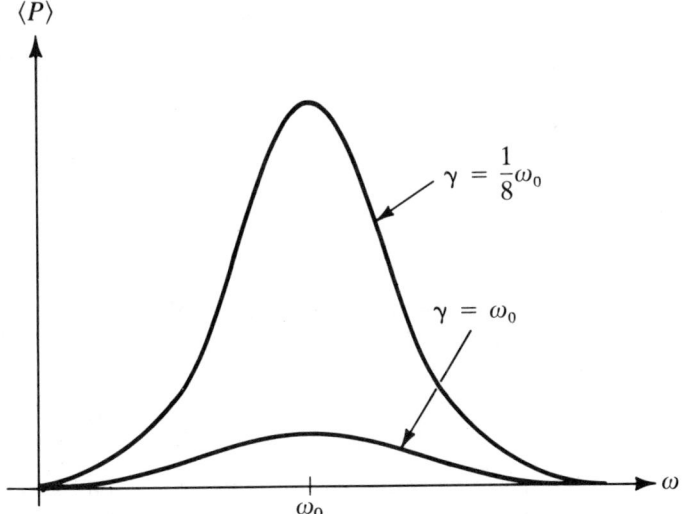

Figure 3.5.3 Power versus driving frequency.

We have already seen part of this expression before in Eq. 3.5.16 for the amplitude. We can write the maximum velocity as $\dot{x}_m$ where

$$\dot{x}_m = \omega A \qquad (3.5.24)$$

Then the average power can be written as

$$\langle P \rangle = \tfrac{1}{2} F_0 \dot{x}_m \sin \phi \qquad (3.5.25)$$

The $\sin \phi$ in this expression is called the *power factor*. Here ϕ is the phase angle between the applied force and the resulting motion. In discussing the power dissipated in an electric circuit, a similar term occurs with ϕ being the phase angle between the applied voltage (emf) and the resulting voltage across the ends of a particular circuit element.

Figure 3.5.3 is a graph of Eq. 3.5.23, showing power as a function of frequency. Note, of course, the rise in power as resonance is reached. Note also the broadening of this resonance peak with increased damping. Figure 3.5.3 is similar to Figure 3.5.1 (as should be expected).

3.6 Application to AC Circuits

Figure 3.6.1 is a schematic drawing of an *LRC* series circuit—*L* is an inductor (or coil) with inductance *L*, *R* is a resistor of resistance *R*, *C* is a capacitor (or so-called condenser) with capacitance *C*, *S* is a source of an external applied voltage (or emf) $E(t)$.

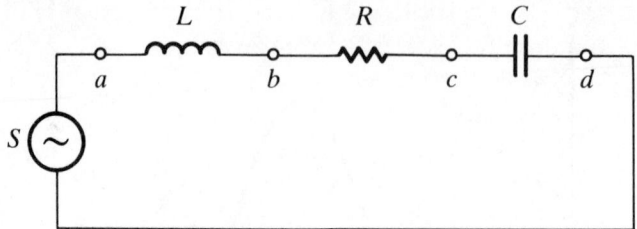

Figure 3.6.1 LCR series circuit schematic diagram.

The voltage across the ends ab of the inductor is proportional to the time rate of change of current through this (perfect) inductor. So

$$V_{ab} = V_L = L\frac{di}{dt} \tag{3.6.1}$$

where i is the current. But the current is itself the time rate of change of charge q. That is, $i = \dot{q}$. So this can then be written as

$$V_{ab} = V_L = L\ddot{q} \tag{3.6.2}$$

The voltage across the ends bc of the resistor is proportional to the current flowing through it, or

$$V_{bc} = V_R = Ri = R\dot{q} \tag{3.6.3}$$

The voltage across the ends cd of the capacitor is proportional to the charge stored on the capacitor. Thus,

$$V_{cd} = V_C = \frac{1}{C}q \tag{3.6.4}$$

The sum of all of these voltages must be equal to the applied voltage from Kirchhoff's rules. Thus, the charge q and, since $i = \dot{q}$, the current throughout the circuit, is governed by

$$L\ddot{q} + R\dot{q} + \frac{1}{C}q = E(t) \tag{3.6.5}$$

And this equation is identical in *form* to Eq. 3.5.2 or 3.5.5a for the forced harmonic oscillator. All of our previous solutions are *immediately* applicable to this electric circuit.

The externally applied voltage $E(t)$ is analogous to the external driving force F_{ex}; the inductance L, to the mass m; the resistance R, to the frictional constant; one over the capacitance $1/C$, to the spring constant k. This analogy is physical as well as mathematical. Energy is dissipated by the resistor R through ohmic heating just as energy is dissipated in the friction due to damping in the harmonic oscillator. Energy is stored in the capacitor just as it is in the compressed or elongated spring. A little thought will show that an inductor has characteristics of inertia. It offers no impedance to direct current, but opposes or resists any *change* in current by offering an impedance proportional to the time rate of change of the current.

The electrical circuit experiences resonance, like the mechanical system. Suppose the externally applied voltage is sinusoidal; that is,

$$E(t) = E_0 \sin(\omega t + \theta) \qquad (3.6.6)$$

Now the analogy with the forced harmonic oscillator is complete: As ω increases, the current will increase to some maximum value and then decrease, identical to the behavior of the mechanical system shown in Figure 3.5.1. Likewise, the phase angle between current and voltage will vary, undergoing a change at the resonance frequency. The power absorbed will undergo a maximum as shown by Figure 3.5.3.

This resonant behavior is used in tuning a radio. Circuit elements (most usually the capacitance) can be changed to vary the resonant frequency ω_r. Energy of many different frequencies impinges upon the radio's antenna. Power absorbed (and then amplified) is quite small except for frequencies very close to ω_r. Figure 3.5.3 shows two resonance curves—one strongly peaked and the other fairly broad. A radio tuning circuit characterized by the latter will absorb energy from a wide range of frequencies about ω_r. The result will be that several signals with several somewhat different frequencies will be absorbed and amplified. That is, several stations will be picked up simultaneously. A circuit corresponding to the former curve will absorb energy over a very narrow range so that only one station's signal will be absorbed and amplified.

3.7 Simple Pendulum

As a simple pendulum, consider a point mass m attached to a massless string of length l attached to a rigid support, as sketched in Figure 3.7.1. Pull the mass over to one side so that the string makes an angle θ with the vertical (its

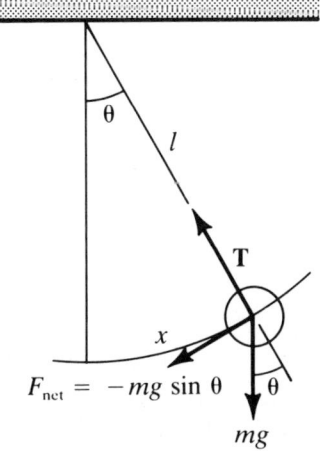

Figure 3.7.1 Simple pendulum.

equilibrium position). What motion does this simple pendulum undergo when released? It moves toward its equilibrium position, gains speed, overshoots, and continues on the other side. It certainly *resembles* simple harmonic motion.

To understand the pendulum's motion in detail, we must look at the forces in detail. The net force is the *vector* sum of the tension in the string and the weight

$$\mathbf{F}_{net} = \mathbf{T} + m\mathbf{g} \tag{3.7.1}$$

We can best understand this by resolving the forces into components along the string and perpendicular to it. The tension adjusts itself to provide the centripetal force necessary since this mass is moving in a circle. But the tangential motion is caused—or, at least, altered—by the component of the force in *that* direction, $-mg \sin \theta$ (the minus sign because it is a restoring force). If we define x as the *arc length along the path* the mass m travels, then

$$F_x = m\ddot{x} = -mg \sin \theta$$

or

$$\ddot{x} = -g \sin \theta \tag{3.7.2}$$

And that doesn't look like Eq. 3.1.1 for the simple harmonic oscillator at all. It isn't.

We've written $\sin \theta$ as a power series already. If we restrict ourselves to *small angles* θ, then we can use the approximation that

$$\sin \theta \approx \theta \tag{3.7.3}$$

and, by the definitions of angular measurements,

$$\theta = \frac{x}{l} \tag{3.7.4}$$

So, Eq. 3.7.2 now can be written as

$$\ddot{x} = -g\left(\frac{x}{l}\right) = -\left(\frac{g}{l}\right)x \tag{3.7.5}$$

Now this has *exactly* the same form as the now-familiar simple harmonic oscillator equation,

$$\ddot{x} = -\frac{k}{m}x \tag{3.2.1}$$

Therefore, we can conclude that as long as the angle θ is small enough that $\sin \theta \approx \theta$ (or, to the degree this is true), then the simple pendulum is, indeed, a simple harmonic oscillator. We know that our "standard" simple harmonic

SECTION 3.8 / PHYSICAL PENDULUM

oscillator (the mass and spring system) has an angular frequency of $\omega = \sqrt{k/m}$ or $\omega^2 = k/m$, so Eq. 3.2.1 could be written as

$$\ddot{x} = -\omega^2 x \qquad (3.7.6)$$

valid for all simple harmonic oscillators. Comparing this with Eq. 3.7.5, we can see that the simple pendulum has a frequency given by

$$\omega = \sqrt{\frac{g}{l}} \qquad (3.7.7)$$

This means that the period is independent of the mass. As for all simple harmonic oscillators, the period is also independent of the amplitude. Only the length of the pendulum and the acceleration due to gravity determine the period. Of course, these statements are only true to the degree that the amplitude is small enough that $\sin\theta \approx \theta$, for that is the approximation we used.

3.8 Physical Pendulum

Instead of a point mass suspended from a massless string, we might more realistically pivot some real body about a point O as sketched in Figure 3.8.1. The center of mass is now located a distance l from the pivot O. This time it's

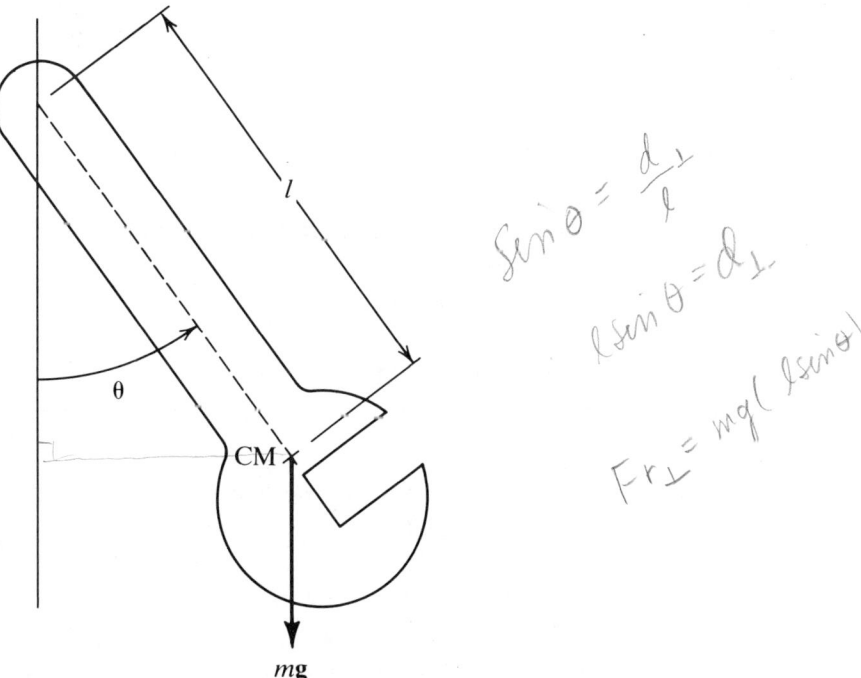

Figure 3.8.1 Physical pendulum.

easier to describe the motion entirely in terms of the angle θ, moment of inertia *about the pivot* I, and the *net* torque $\tau = -mgl \sin \theta$. You should recall that torque exerted about some point is equal to the force applied (the weight mg in this case), multiplied by the distance from the pivot to the point of application (l), multiplied by the sine of the angle between this distance and the force; that is, the force times its moment arm, the perpendicular distance from the pivot to the line of action of the force, $l \sin \theta$. The negative sign indicates it's a *restoring* torque. We'll discuss torque and angular motion in detail later, but for the present discussion of a physical pendulum, we'll assume you remember these ideas from previous work in your introductory physics course.

Just as $F = m\ddot{x}$, we can write the equivalent $\tau = I\ddot{\theta}$ for the rotational motion. In this case,

$$\tau = -mgl \sin \theta = I\ddot{\theta} \tag{3.8.1}$$

If we restrict ourselves to small angles so that $\sin \theta \simeq \theta$, then

$$\ddot{\theta} = -\frac{mgl\theta}{I} \tag{3.8.2}$$

so we can see that this physical pendulum will oscillate with a frequency of

$$\omega = \sqrt{\frac{mgl}{I}} \tag{3.8.3}$$

This is the same frequency as a *simple* pendulum of length k where

$$\omega^2 = \frac{g}{k} = \frac{mgl}{I}$$

or

$$k = \frac{I}{ml} \tag{3.8.4}$$

This k, the length of an equivalent simple pendulum, is known as the radius of oscillation.

PROBLEMS

3.1 A simple harmonic oscillator consists of a spring and a 100-gm (0.1-kg) mass. When the mass is suspended from the end of the spring, the spring stretches 10 cm (0.1 m). Find the following: (a) the spring constant k, (b) the natural angular frequency ω_0, and (c) the period T.

3.2 The simple harmonic oscillator in Problem 3.1 is pulled down an additional 5 cm (0.05 m) beyond its equilibrium position. Determine the following: (a) its amplitude, (b) its total energy, and (c) the velocity of the mass as it passes through the equilibrium position.

3.3 The simple harmonic oscillator in Problems 3.1 and 3.2 is set in motion by giving it an initial velocity of 2 cm/s (0.02 m/s) at its equilibrium position. Determine the following: (a) its total energy, (b) its maximum potential energy, and (c) its amplitude.

3.4 Two bodies of mass m_1 and m_2, respectively, are observed to execute simple harmonic motion with amplitudes A_1 and A_2. Their total energies are identical. Find the ratio of their periods.

3.5 A simple harmonic oscillator has a speed v_1 and displacement x_1 at time t_1 and a speed v_2 and displacement x_2 at another time t_2. Find the period and amplitude of this oscillator in terms of x_1, v_1, x_2, and v_2.

3.6 Two springs with spring constants k_1 and k_2 are attached to a body of mass m. Find the *effective spring constant* when they are attached in the following manners:

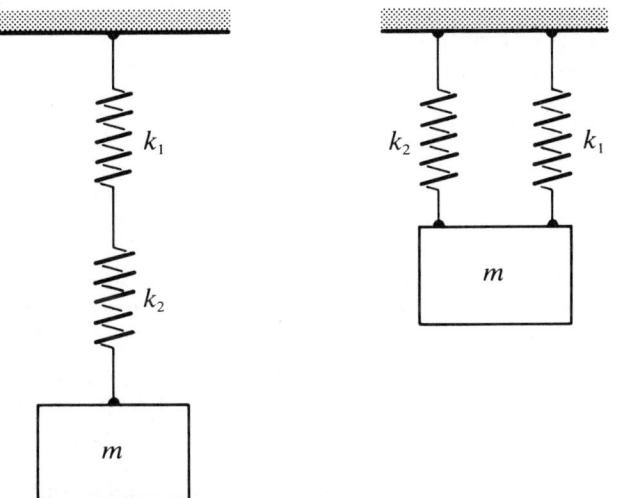

3.7 Consider a spring with spring constant k attached to a body of mass m. When the spring is horizontal and the mass is supported by a frictionless horizontal surface, the net force exerted on the mass is clearly $F = -kx$, where x is the displacement from the position when the spring is unstretched (i.e., x is the displacement from the equilibrium position and the equilibrium position is the location of the unstretched spring). Now *suspend* the same mass by the same spring. There is now an additional downward force due to the weight, $-mg$. Show that the vertical motion this now undergoes, once disturbed vertically from equilibrium, is still simple harmonic motion. *Hint*: Look for a new equilibrium position.

3.8 A spring with spring constant k is attached to a box of mass M in which is placed a small body of mass m ($m \ll M$). The system is displaced a distance A from equilibrium and released. Find the normal force between the box and small mass m as a function of time. For what initial displacement A will the small mass m just begin to lose contact with the box?

3.9 Sketch the following potential energy functions. Explicitly locate any stable equilibrium positions by calculation. Find the period of small oscillations for a 1-kg body disturbed from those equilibria.
   (a) $V(x) = 3x^3 - 4x + 5$
   (b) $V(x) = 5x^5 - 3x^3 + 1$
   (c) $V(x) = 5x^5 - 12x^3$

3.10 The suspension system of your car has (supposedly) been carefully "tuned" to provide critical damping. If this is true when you, the driver, ride alone in the car, what sort of damping do you have when four additional people and 100 kg of luggage are added to the car? How does this affect the ride?

3.11 Suppose, instead, that the suspension system is tuned to be critically dampened when your car is fully loaded. Now what sort of damping do you experience when you drive by yourself? What do you experience in the ride?

3.12 A body of mass m is attached to a spring so it experiences a linear restoring force, $-kx$. It also experiences a linear damping force, $-c\dot{x}$. The body is initially displaced from equilibrium a distance x_0 and released from rest. Find analytical expressions for the motion in the undamped, critically damped, and overdamped cases.

*3.13 Solve Problem 3.12 for the case when the body initially leaves the equilibrium position with initial velocity v_0. In addition, make a sketch of displacement versus time for the cases (this is an interesting one to investigate numerically, too).

3.14 A body of mass m is subject to our usual linear restoring force, $-kx$, and to a damping force due to dry sliding friction, $\pm \mu mg$. Solve for its motion. Show that the period does not depend on the amplitude (*hint*: First determine the effect of an additional *constant* force. When the force changes, as it does here, you must solve for the motion separately for each case and then carefully match the boundary conditions.)

3.15 After n cycles, the amplitude of a certain damped harmonic oscillator drops to $1/e$ of its initial value. Show that the frequency of this damped oscillator is approximately $[1 - (1/8^2\pi^2 n^2)]$ times the "natural frequency" of the corresponding undamped oscillator.

3.16 Show that the ratio of two successive displacement maxima for the damped harmonic oscillator is constant.

3.17 The terminal speed of a freely falling object is v_t (assume a *linear* form for air resistance). When the object is suspended by a spring, the spring stretches an amount x_0. Show that the frequency of oscillation is

$$\omega = \sqrt{\frac{g}{x_0} - \frac{g^2}{4v_t^2}}$$

3.18 Starting with $e^{i2\theta} = (e^{i\theta})^2$, obtain formulas for $\sin 2\theta$ and $\cos 2\theta$ in terms of $\sin \theta$ and $\cos \theta$.

3.19 In each of the following cases, verify that x_p is a particular solution to the given inhomogeneous differential equation:

(a) $x + \dot{x} = 2e^t$ $x_p = e^t$
(b) $x - \dot{x} = 3e^{2t}$ $x_p = e^{2t}$
(c) $x + \dot{x} = -3 \sin 2t$ $x_p = \sin 2t$
(d) $x + \dot{x} - 2\ddot{x} = 14 + 25 + 2t^2$ $x_p = t^2 - 6$
(e) $x + \dot{x} = -6 \sin 2t$ $x_p = 2 \sin 2t$
(f) $x - \dot{x} = 2e^t$ $x_p = te^t$
(g) $x + 4\dot{x} = -12 \sin 2t$ $x_p = 3t \cos 2t$
(h) $x - 4\dot{x} + 3\ddot{x} = 4e^{3t}$ $x_p = 2te^{3t}$

3.20 A force $F_0 \cos(\omega t + \theta_0)$ acts on a damped harmonic oscillator beginning at $t = 0$.

  (a) What must x_0 and v_0 be so there is no transient?
  (b) If $x_0 = v_0 = 0$, find the complete solution (transient plus steady state).

3.21 Find the driving frequency ω that imparts the greatest *speed* to a forced harmonic oscillator.

3.22 Show that the driving frequency, ω, for which the amplitude of a driven harmonic oscillator is one-half the amplitude at the resonant frequency, is approximately $\omega_0 \pm \gamma\sqrt{3}$.

3.23 An underdamped harmonic oscillator is subject to an external force given by $F = F_0 e^{-\alpha t} \cos(\omega t + \theta)$. Find a particular (steady state) solution by expressing F as the real part of a complex exponential function and looking for a solution, x, having the same form.

*3.24 In the text we have solved the simple pendulum with small oscillations and found it to be a simple harmonic oscillator. In Eqs. 3.7.2 and 3.7.3 we replaced the *chord* length by the *arc* length and the sine of the angle by just the angle to obtain Eq. 3.7.5. What is the "true motion"? That is, what is the motion of a simple pendulum if these approximations are not made? Or, how valid are these approximations?

Go back to your computer program of Chapter 1 and use Eq. 3.7.2 to write the equation for the actual acceleration *without approximation*—

           85 LET A = −G∗SIN (X/L).

See what the motion looks like for various amplitudes.

In particular, what is the *period* and how does it compare to your expected value for $2\pi\sqrt{l/g}$? Start it from rest ($v_0 = 0$) and use initial positions of 0.1, 0.25, 0.5, 0.75, 0.9, and 1.0 times the length l. For convenience, you may as well take $l = 1.0$ m.

*3.25 A mass of 1000 kg drops from a height of 10 m onto a platform of negligible mass. We want to design a spring and dashpot on which to mount the platform so that it will settle to a new equilibrium position 0.2 m below its original position as quickly as possible after impact, *without overshooting*. Find the spring constant k and the damping coefficient c of the dashpot. Analytical solution to this is rather messy; it is better to use a computer.

*3.26 An underdamped harmonic oscillator moves with velocity v_0 through position x_0 at time $t = 0$. Find its equation of motion. That is, use these initial conditions to solve for the arbitrary constants and, thus, fully specify $x(t)$. This is very interesting to investigate using the *numerical* techniques from Chapter 1. Use $x_0 = 10$ cm and $v_0 = 2$ cm/s.

***3.27** A critically damped harmonic oscillator has an initial position of x_0. What is the maximum initial *downward* velocity, $-v_0$, that it can have and not overshoot the origin. This is a rather interesting problem to solve with a computer. Start with $M = 1$ and $K = 1$. Then critical damping demands $C = 2$. Begin with an initial position of, say, $x_0 = 1$ and look at the resulting motion for *various* values of the initial velocity. Try the following values: $v = 1.0, 0.5, 0.0, -0.1, -0.25, -0.5, -0.75, -1.0$.

4

VECTORS

4.1 Introduction

When we classify physical quantities, it is important to differentiate between *scalars* and *vectors*. *Scalars* are fully specified by a number (how many?) and a unit (of what?). Some examples are temperature (100°C), mass (14.54 kg), time (62.1 s), and energy (32.6 J). *Vectors*, on the other hand, require a *direction* in addition to a number and a unit. Some examples of vectors are force (12 Newtons applied at 30° above a plane), displacement (25 meters 45° W of N), and velocity (110 km/hr, due east).

4.2 Vector Algebra

A vector can be represented graphically by a directed line segment—an arrow, if you will—as shown in Figure 4.2.1. We shall indicate that a quantity is a vector by writing it in boldface type as **F**. In handwritten notes, a vector is usually indicated by writing an arrow above it, as $\vec{F}$. In a graphical representation, a vector is defined by its length (which we shall call its *magnitude*) and its direction.

In Figure 4.2.1, the three vectors are parallel and of equal length. Thus, they are identical, or equal. That is, **A** = **B** = **C**. It is useful to talk of the

Figure 4.2.1 Three graphical representations of the same vector (**A** = **B** = **C**) and scalar multiplication of a vector.

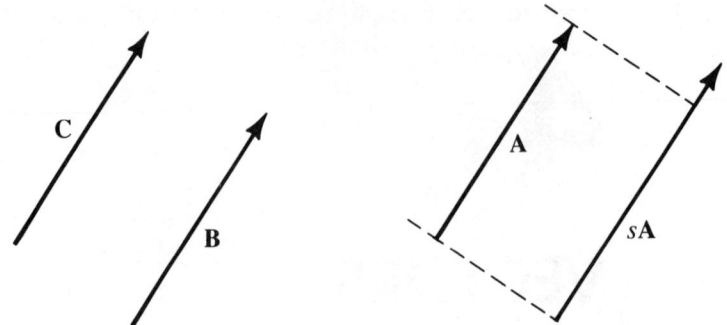

magnitude of a vector. This is just the scalar part of a vector—a number and a unit—without the information about its direction. You may think of magnitude as the length of the line segment in Figure 4.2.1 without a direction associated with it. We will indicate the magnitude by an ordinary italic symbol (no arrow or boldface type). To emphasize this, we may also use absolute value bars on the vector symbol, as

$$A = |\mathbf{A}| \tag{4.2.1}$$

A vector can be multiplied by a scalar, as $s\mathbf{A}$. The result is another vector parallel to the first, but whose magnitude is the magnitude of the original vector, A, multiplied by the scalar. That is,

$$|s\mathbf{A}| = (s)(A) \tag{4.2.2}$$

This is illustrated in Figure 4.2.1 as well.

Vector addition is quite important—and quite different from ordinary, scalar addition. Consider a displacement of 3 m due north and then a displacement of 4 m due east (perhaps as directions on a pirate's treasure map). Or consider an airplane heading due west at 125 km/hr while being blown off course by a southern gale blowing at a hearty 60 km/hr. These are two brief situations involving *vector addition*.

It is easiest to understand vector addition by looking at a diagram. Consider the two vectors **A** and **B** shown in Figure 4.2.2. These might

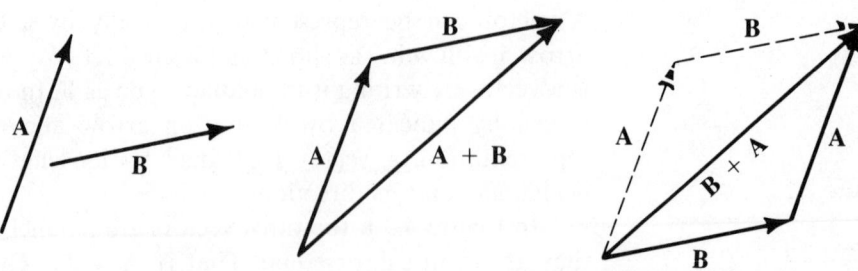

Figure 4.2.2 Vector addition.

represent velocities, forces, or any vector quantity at all. For the moment, let them represent two displacements:

$$\mathbf{A} = 40 \text{ paces } 10° \text{ E of N}$$

$$\mathbf{B} = 50 \text{ paces } 60° \text{ E of N}$$

What is the net displacement of these two displacements together (that is, what is the sum of **A** and **B**)? What would you do if you were following directions on a treasure hunt? Starting at some origin O, you would take 40 paces in the direction of 10° E of N. That's like drawing vector **A**. Then, at that point, you would take 50 paces 60° E of N. Right? That's like drawing vector **B** *starting from* the arrow end of vector **A**. The net displacement, then, is a vector drawn from the origin O, or the *beginning* of **A**, to the *end* of **B**. This is shown in Figure 4.2.2.

All vector addition can be represented this way. You first draw one vector, and then draw the second vector starting at the end of the first. The vector sum of these two is another vector drawn from the beginning of the first to the end of the last vector. This is sometimes referred to as constructing a *vector triangle*. As you can see from Figure 4.2.2, the two vectors also form two sides of a parallelogram, with their vector sum being the diagonal between them. This is referred to as the *parallelogram method* of vector addition.

As we can already see from Figure 4.2.2, vector addition is commutative; that is,

$$\mathbf{A} + \mathbf{B} = \mathbf{B} + \mathbf{A} \qquad (4.2.3)$$

The vector triangle for two vectors can be extended to a *vector polygon* for several vectors—each successive vector is placed at the tip of the preceding one. The resultant, the sum of all the vectors, is a new vector again drawn from the tail of the first vector to the head of the final one. This is illustrated in Figure 4.2.3.

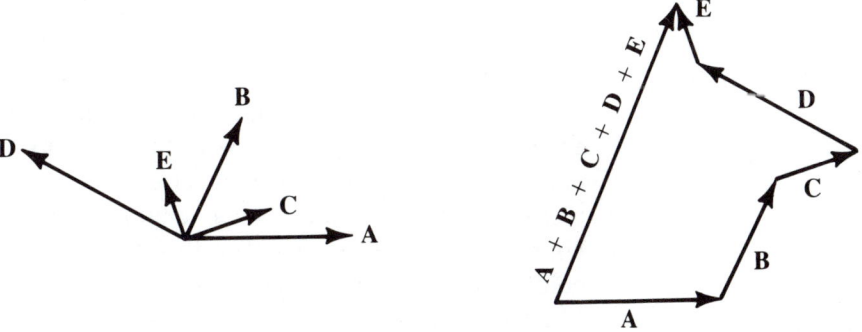

Figure 4.2.3 Vector polygon method of vector addition.

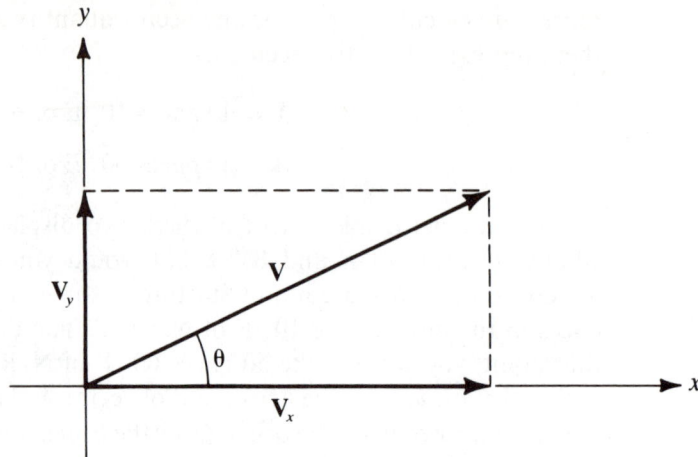

Figure 4.2.4
Rectangular components of a vector.

We sometimes find it useful to describe a vector in relation to a coordinate system. For example, consider some vector, **V**, as sketched in Figure 4.2.4. Vector **V** can also be represented as the vector *sum* of two other vectors, each of which lies parallel to the *x*- and *y*-axes, $\mathbf{V}_x$ and $\mathbf{V}_y$. That is,

$$\mathbf{V} = \mathbf{V}_x + \mathbf{V}_y \tag{4.2.4}$$

If we consider a vector in *three* dimensions instead of two, as sketched in Figure 4.2.4, this can be extended to

$$\mathbf{V} = \mathbf{V}_x + \mathbf{V}_y + \mathbf{V}_z \tag{4.2.4a}$$

$\mathbf{V}_x$, $\mathbf{V}_y$, and $\mathbf{V}_z$ are called the vector *components* of the vector **V**. We can also represent a vector by writing the magnitudes of these three vector components *in order* and referring to them as the scalar *components* of a vector. Hence,

$$\mathbf{V} = (V_x, V_y, V_z) \tag{4.2.4b}$$

Such vector notation is fully complete and satisfactory. *Note*: we can always find the component along any particular direction by multiplying the magnitude of the vector by the cosine of the angle between the vector and the direction of interest; for example, $V_x = V\cos\theta$ in Figure 4.2.4.

However, another notation system is more frequently used in science and engineering. Figure 4.2.5 shows a three-dimensional coordinate system with mutually perpendicular axes labeled *x*, *y*, and *z*. Near each axis is a *unit vector*, or a vector that has a *unit* length and a *direction* along one of the axes. Unit vectors will be indicated with a caret (^) above them in addition to the boldface type used on all other vectors. We could label these unit vectors $\hat{\mathbf{e}}_x$, $\hat{\mathbf{e}}_y$, and $\hat{\mathbf{e}}_z$ ("**e**" from the German word *Einheitsvektor* meaning unit vector).

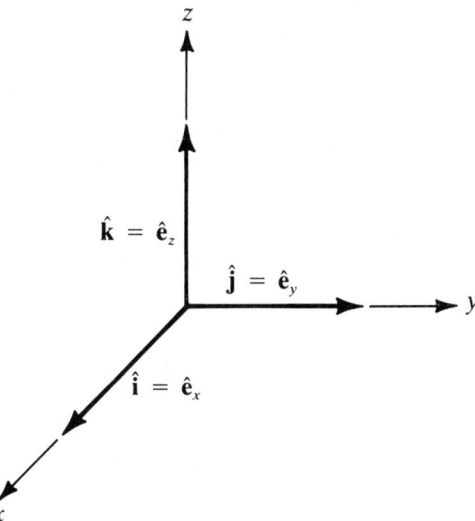

Figure 4.2.5 Unit vectors.

Instead, we shall simply label them $\hat{\imath}$, $\hat{\jmath}$, and $\hat{k}$ and refer to the following as *the ĳk notation*. Clearly,

$$\hat{e}_x = \hat{\imath} = (1, 0, 0)$$
$$\hat{e}_y = \hat{\jmath} = (0, 1, 0) \qquad (4.2.5)$$
$$\hat{e}_z = \hat{k} = (0, 0, 1)$$

Our general vector, **V**, can now be written in ĳk notation using these unit vectors:

$$\mathbf{V} = (V_x, V_y, V_z)$$
$$\mathbf{V} = V_x(1, 0, 0) + V_y(0, 1, 0) + V_z(0, 0, 1)$$
$$\mathbf{V} = V_x\hat{\imath} + V_y\hat{\jmath} + V_z\hat{k} \qquad (4.2.6)$$

Since these components are mutually perpendicular, the magnitude of a vector is just the length of a diagonal of a rectangular solid (that is, the longest rod that will fit in a shoe box) whose sides are V_x, V_y, and V_z. That is

$$V = |\mathbf{V}| = \sqrt{V_x^2 + V_y^2 + V_z^2} \qquad (4.2.7)$$

We have already defined equality of two vectors in terms of their graphical representation. An equivalent definition in our ĳk notation is that two vectors are equal if and only if the three components of one are equal to the three corresponding components of the second. Thus a single vector equation like

$$\mathbf{A} = \mathbf{B} \qquad (4.2.8)$$

is simply a shorthand notation for *three* equations:

$$A_x = B_x$$
$$A_y = B_y \quad (4.2.9)$$
$$A_z = B_z$$

In the $\hat{\imath}\hat{\jmath}\hat{k}$ notation, vector addition is likewise carried out by the addition of the respective scalar components. Thus,

$$\mathbf{C} = \mathbf{A} + \mathbf{B} = (A_x\hat{\imath} + A_y\hat{\jmath} + A_z\hat{k}) + (B_x\hat{\imath} + B_y\hat{\jmath} + B_z\hat{k})$$
$$C_x\hat{\imath} + C_y\hat{\jmath} + C_z\hat{k} = (A_x + B_x)\hat{\imath} + (A_y + B_y)\hat{\jmath} + (A_z + B_z)\hat{k} \quad (4.2.10)$$

Since

$$\mathbf{C} = \mathbf{A} + \mathbf{B} \quad (4.2.10a)$$

really means

$$C_x = A_x + B_x$$
$$C_y = A_y + B_y \quad (4.2.10b)$$
$$C_z = A_z + B_z$$

which are simply scalar equations, we see again that vector addition is commutative—just as ordinary addition is. That is,

$$\mathbf{A} + \mathbf{B} = \mathbf{B} + \mathbf{A} \quad (4.2.11)$$

Multiplication of a vector by a scalar leaves the direction of the vector the same, but alters its length, as shown in Figure 4.2.1. Such multiplication can also be done by multiplication of *each* component by the scalar as

$$s\mathbf{A} = (sA_x)\hat{\imath} + (sA_y)\hat{\jmath} + (sA_z)\hat{k} \quad (4.2.12)$$

Subtraction—whether vector or scalar—is really just addition after the quantity *to be subtracted* has been multiplied by minus one (-1). For a vector, such multiplication results in a new vector, parallel to the original, of the same length, but opposite in direction. (This is also referred to as being *antiparallel*). Thus,

$$\mathbf{A} - \mathbf{B} = \mathbf{A} + (-\mathbf{B}) \quad (4.2.13)$$

as shown in Figure 4.2.6.

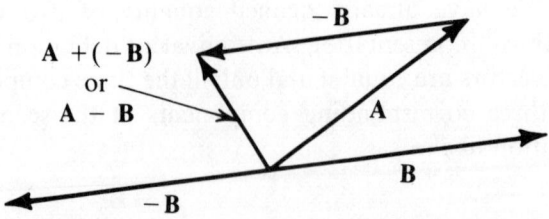

Figure 4.2.6 Vector subtraction.

SECTION 4.2 / VECTOR ALGEBRA

In terms of the $\hat{\mathbf{i}}\hat{\mathbf{j}}\hat{\mathbf{k}}$ notation, we need only subtract respective components. That is,

$$\mathbf{C} = \mathbf{A} - \mathbf{B} \qquad (4.2.14)$$

implies three separate equations for three components:

$$\begin{aligned} C_x &= A_x - B_x \\ C_y &= A_y - B_y \\ C_z &= A_z - B_z \end{aligned} \qquad (4.2.14a)$$

There is a zero (or null) vector, each of whose three components is identically zero:

$$\mathbf{0} = (0, 0, 0) = 0\hat{\mathbf{i}} + 0\hat{\mathbf{j}} + 0\hat{\mathbf{k}} \qquad (4.2.15)$$

It follows that $\mathbf{A} + \mathbf{0} = \mathbf{A}$ and $\mathbf{A} - \mathbf{A} = \mathbf{0}$. When there is (little or) no ambiguity, we shall usually simply use $0 = \mathbf{0}$.

Since vector addition really is equivalent to three separate scalar additions, the characteristics of scalar addition also hold for vector addition. We have already seen its *commutative* nature in Eq. 4.2.11 and Figure 4.2.2. It is also *associative* in that

$$(\mathbf{A} + \mathbf{B}) + \mathbf{C} = \mathbf{A} + (\mathbf{B} + \mathbf{C}) \qquad (4.2.16)$$

We can see this by looking at the scalar components:

$$\begin{aligned}{}[(\mathbf{A} + \mathbf{B}) + \mathbf{C}]_x &= (A_x + B_x) + C_x \\ &= A_x + (B_x + C_x) \\ &= [\mathbf{A} + (\mathbf{B} + \mathbf{C})]_x \end{aligned} \qquad (4.2.16a)$$

Identical equations hold, of course, for the y- and z-components.

Vector addition is also *distributive* for multiplication by a scalar as in

$$s(\mathbf{A} + \mathbf{B}) = s\mathbf{A} + s\mathbf{B} \qquad (4.2.17)$$

This is clearly evident if we look at the equation in component form:

$$\begin{aligned}{}[s(\mathbf{A} + \mathbf{B})]_x &= s(A_x + B_x) \\ &= s(\mathbf{A})_x + s(\mathbf{B})_x \\ &= [s\mathbf{A} + s\mathbf{B}]_x \end{aligned} \qquad (4.2.17a)$$

When vector equations are written in their component form, each component forms a separate, ordinary, scalar, algebraic equation. This is the beauty and utility of writing vectors in terms of their components.

4.3 Vector Multiplication

So far we have only discussed vector addition. What about vector multiplication? It turns out that there are two different types of vector multiplication. First we shall consider the *dot product*, also called the *inner product* or the *scalar product*. It is defined by

$$\mathbf{A} \cdot \mathbf{B} \equiv AB \cos \theta \tag{4.3.1}$$

where θ is the angle between vectors A and B. Note that this dot product—or scalar product—of vectors **A** and **B** is itself a scalar. The dot product is commutative; that is,

$$\mathbf{A} \cdot \mathbf{B} = \mathbf{B} \cdot \mathbf{A} \tag{4.3.2}$$

Note that we can write Eq. 4.3.1 as $\mathbf{A} \cdot \mathbf{B} = (A)(B \cos \theta)$ and then interpret it as the magnitude of one vector multiplied by the component of the second one parallel to the first. This is shown in Figure 4.3.1. This interpretation is useful in discussing *work*. Work can be written in the simple form of force times distance only when the force applied and distance moved lie along the same straight line. In more general cases, it is only the component of the force *along* the direction of motion that contributes to the work. That is,

$$W = (F \cos \theta)(D) = \mathbf{F} \cdot \mathbf{D} \tag{4.3.3}$$

Work is the dot product of the vector **F** applied and the vector displacement **D** through which an object moves.

It is interesting and worthwhile to look at the dot products of the unit vectors $\hat{\mathbf{i}}, \hat{\mathbf{j}}$, and $\hat{\mathbf{k}}$. They are mutually perpendicular, so the dot product of any two different vectors is zero since $\cos(\pi/2) = 0$. Each one "dotted into" itself is unity because

$$|\hat{\mathbf{i}}| = |\hat{\mathbf{j}}| = |\hat{\mathbf{k}}| = 1$$

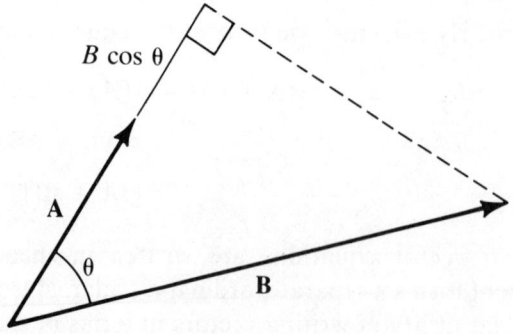

Figure 4.3.1
Graphical interpretation of the dot product.

and $\cos(0) = 1$. That is,

$$\hat{\imath} \cdot \hat{\imath} = 1 \qquad \hat{\imath} \cdot \hat{\jmath} = 0 \qquad \hat{\imath} \cdot \hat{k} = 0$$
$$\hat{\jmath} \cdot \hat{\imath} = 0 \qquad \hat{\jmath} \cdot \hat{\jmath} = 1 \qquad \hat{\jmath} \cdot \hat{k} = 0 \qquad (4.3.4)$$
$$\hat{k} \cdot \hat{\imath} = 0 \qquad \hat{k} \cdot \hat{\jmath} = 0 \qquad \hat{k} \cdot \hat{k} = 1$$

We can now use this to write vectors **A** and **B** in their component form and take the dot product:

$$\mathbf{A} \cdot \mathbf{B} = (A_x \hat{\imath} + A_y \hat{\jmath} + A_z \hat{k}) \cdot (B_x \hat{\imath} + B_y \hat{\jmath} + B_z \hat{k})$$

This yields

$$\mathbf{A} \cdot \mathbf{B} = A_x B_x + A_y B_y + A_z B_z \qquad (4.3.5)$$

since all of the cross terms involving different unit vectors are identically zero. Equation 4.3.5 is often the most useful form of the dot product. Letting $\mathbf{A} = \mathbf{B}$ and applying it immediately gives

$$\mathbf{A} \cdot \mathbf{A} = A_x A_x + A_y A_y + A_z A_z = A^2 \qquad (4.3.6)$$

The second form of vector multiplication is usually called the *cross product*, and, sometimes, the *outer product* or *vector product*. We can indicate the cross product of vectors **A** and **B** by

$$\mathbf{C} = \mathbf{A} \times \mathbf{B}, \qquad (4.3.7)$$

and then define the magnitude and direction of **C**. Its magnitude is given by

$$C = |\mathbf{A} \times \mathbf{B}| = AB \sin \theta \qquad (4.3.7a)$$

The magnitude of the cross product is a product of the magnitudes of the two vectors and the *sine* of the angle between them. This is the *area* of a parallelogram bounded on two sides by the vectors of the product. The direction of C is defined as perpendicular to the plane defined by **A** and **B**. That still leaves two possible directions for **C**, so we completely specify its direction by using the right-hand rule: the direction of **C** is the direction of advance of a right-hand screw as it is rotated *from* the first vector (**A**) *into* the second vector (**B**).

Now order of multiplication *is* important! The cross product is *not* commutative. In fact,

$$\mathbf{A} \times \mathbf{B} = -\mathbf{B} \times \mathbf{A} \qquad (4.3.8)$$

As for the dot product, it is very instructive to apply this *cross product* to the unit vectors. Since they are mutually perpendicular, the sine of the angle between two different ones will be unity; therefore, the cross product of two perpendicular unit vectors is another unit vector determined by the right-hand rule. What of the cross product of a unit vector with itself? In our definition, Eq. 4.3.7, the angle θ must be zero and the sine must be zero. Thus, the cross product of *any* vector with itself (or another parallel vector) is zero. We can list the cross products of the unit vectors:

$$\hat{\imath} \times \hat{\imath} = 0 \qquad \hat{\imath} \times \hat{\jmath} = \hat{k} \qquad \hat{\imath} \times \hat{k} = -\hat{\jmath}$$
$$\hat{\jmath} \times \hat{\imath} = -\hat{k} \qquad \hat{\jmath} \times \hat{\jmath} = 0 \qquad \hat{\jmath} \times \hat{k} = \hat{\imath} \qquad (4.3.9)$$
$$\hat{k} \times \hat{\imath} = \hat{\jmath} \qquad \hat{k} \times \hat{\jmath} = -\hat{\imath} \qquad \hat{k} \times \hat{k} = 0$$

We can now use this to write the cross product in terms of components:

$$\begin{aligned}
\mathbf{A} \times \mathbf{B} &= (A_x\hat{\imath} + A_y\hat{\jmath} + A_z\hat{k}) \times (B_x\hat{\imath} + B_y\hat{\jmath} + B_z\hat{k}) \\
&= A_xB_x(\hat{\imath} \times \hat{\imath}) + A_xB_y(\hat{\imath} \times \hat{\jmath}) + A_xB_z(\hat{\imath} \times \hat{k}) \\
&\quad + A_yB_x(\hat{\jmath} \times \hat{\imath}) + A_yB_y(\hat{\jmath} \times \hat{\jmath}) + A_yB_z(\hat{\jmath} \times \hat{k}) \\
&\quad + A_zB_x(\hat{k} \times \hat{\imath}) + A_zB_y(\hat{k} \times \hat{\jmath}) + A_zB_z(\hat{k} \times \hat{k}) \\
&= A_xB_x(0) + A_xB_y(\hat{k}) + A_xB_z(-\hat{\jmath}) \\
&\quad + A_yB_x(-\hat{k}) + A_yB_y(0) + A_yB_z(\hat{\imath}) \\
&\quad + A_zB_x(\hat{\jmath}) + A_zB_y(-\hat{\imath}) + A_zB_z(0)
\end{aligned} \qquad (4.3.9a)$$

$$\mathbf{A} \times \mathbf{B} = \hat{\imath}(A_yB_z - A_zB_y) + \hat{\jmath}(A_zB_x - A_xB_z) + \hat{k}(A_xB_y - A_yB_x)$$

It is probably easier to remember this very useful form of the cross product as the expansion of a determinant:

$$\mathbf{A} \times \mathbf{B} = \begin{vmatrix} \hat{\imath} & \hat{\jmath} & \hat{k} \\ A_x & A_y & A_z \\ B_x & B_y & B_z \end{vmatrix} \qquad (4.3.9b)$$

Forces on charged objects moving in a magnetic field are peculiar in that they are proportional to velocity and act perpendicular to the velocity (and perpendicular to the magnetic field **B**). These magnetic forces can be written in terms of a cross product:

$$\mathbf{F}_{\text{mag}} = q\mathbf{v} \times \mathbf{B} \qquad (4.3.10)$$

Torque, the moment of a force or the rotational equivalent of a force, can also be written as a cross product. You already know, from the everyday example of opening doors, that the torque exerted (or the rotational effect a certain force has) depends both on where the force is located and what direction it has. Figure 4.3.2 shows a force **F** being applied to an object at point P. Vector **r** locates P with respect to point O (point O can be any point, but it is especially useful if it happens to be the axis of rotation). Only the component of **F** perpendicular to **r** causes the body to rotate about O, so we might write

$$\tau = rF_\perp = r(F \sin \theta)$$

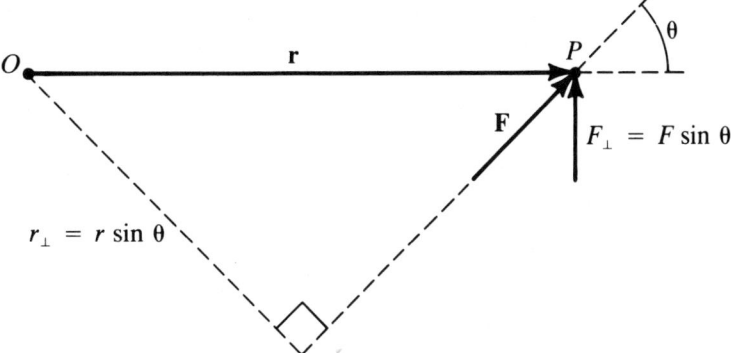

Figure 4.3.2 Torque.

An alternate definition of torque is *force multiplied by moment arm*, $r_\perp$, the perpendicular distance from the origin O to the line of action of the force applied at P. So we would write

$$\tau = r_\perp F = (r \sin \theta) F$$

You can see that these two are identical. So we can omit the parentheses and write

$$\tau = rF \sin \theta$$

If we stop here, we must restrict ourselves to forces **F** and positions defined by **r** lying in a plane. We also have to label all torques that cause a clockwise rotation with an arrow (↷), or call them *negative*, and label counterclockwise torques (↶), or call them *positive*. An easier and more general alternative is to treat torque as the vector it really is and use the following definition:

$$\boldsymbol{\tau} = \mathbf{r} \times \mathbf{F} \tag{4.3.11}$$

The direction of $\boldsymbol{\tau}$ defines an axis about which **F** applied at **r** would cause the object to rotate. The sense about this axis is given by the right-hand rule.

4.4 Coordinate Systems

A vector can be defined in terms of its components along mutually perpendicular coordinate axes. If we know the components of a vector in one reference frame, say V_x, V_y, and V_z, and we know the relation between the xyz coordinate system and another $x'y'z'$ system, how would we describe this vector in the other system? That is, what are its components $V_{x'}$, $V_{y'}$, and $V_{z'}$ along the mutually perpendicular x'-, y'-, and z'-axes? Note that we've written $V_{x'}$ and *not* V'_x. It is the coordinate system that has changed, not the vector.

Figure 4.4.1 shows a two-dimensional vector **V** that can be described in terms of its components along (or projections onto) the x- and y-axes by

$$\mathbf{V} = V_x \hat{\mathbf{i}} + V_y \hat{\mathbf{j}} \tag{4.4.1}$$

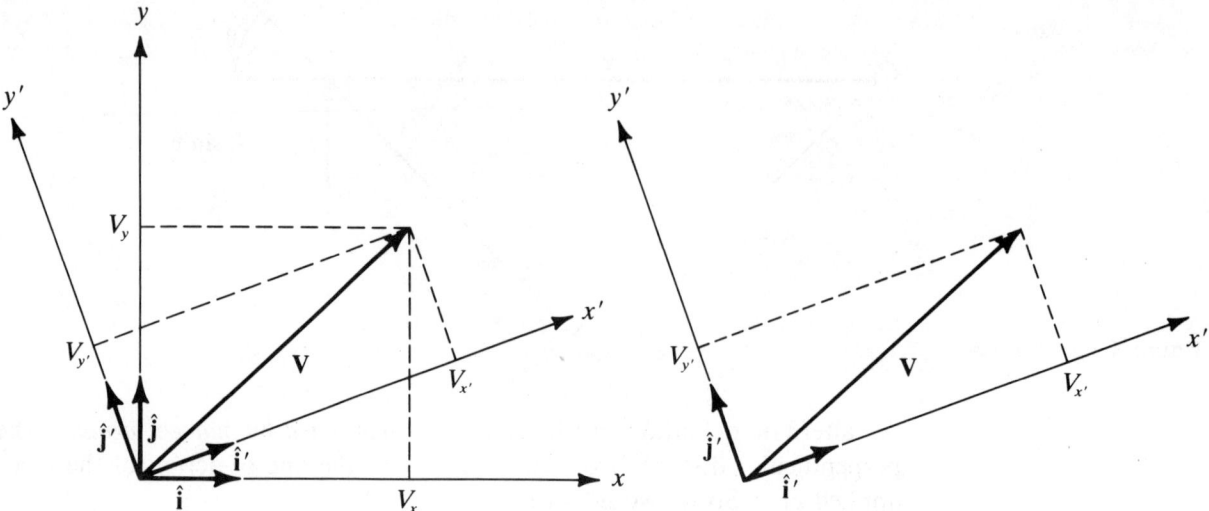

Figure 4.4.1 A vector represented in two coordinate systems.

We simply pretend that the x- and y-axes aren't present. This same vector **V** could just as easily be represented by

$$\mathbf{V} = V_{x'}\hat{\mathbf{i}}' + V_{y'}\hat{\mathbf{j}}' \tag{4.4.2}$$

where $\hat{\mathbf{i}}'$ is the unit vector in the direction of increasing x', and $\hat{\mathbf{j}}'$, in the direction of increasing y'. This can be generalized to three dimensions.

Cumbersome though it appears, we could get the components V_x, V_y, and V_z by taking dot products of $\hat{\mathbf{i}}$, $\hat{\mathbf{j}}$, and $\hat{\mathbf{k}}$, respectively, with the vector **V**. Each dot product is the component of **V** parallel to the direction of the unit vector used. Hence,

$$V_x = \hat{\mathbf{i}} \cdot \mathbf{V}$$
$$V_y = \hat{\mathbf{j}} \cdot \mathbf{V} \tag{4.4.3}$$
$$V_z = \hat{\mathbf{k}} \cdot \mathbf{V}$$

Likewise, we can find the components $V_{x'}$, $V_{y'}$, and $V_{z'}$ that describe this vector V as seen in the $x'y'z'$ system by taking dot products with $\hat{\mathbf{i}}'$, $\hat{\mathbf{j}}'$, and $\hat{\mathbf{k}}'$:

$$V_{x'} = \hat{\mathbf{i}}' \cdot \mathbf{V} = \hat{\mathbf{i}}' \cdot (V_x\hat{\mathbf{i}} + V_y\hat{\mathbf{j}} + V_z\hat{\mathbf{k}}) = V_x(\hat{\mathbf{i}}' \cdot \hat{\mathbf{i}}) + V_y(\hat{\mathbf{i}}' \cdot \hat{\mathbf{j}}) + V_z(\hat{\mathbf{i}}' \cdot \hat{\mathbf{k}})$$
$$V_{y'} = \hat{\mathbf{j}}' \cdot \mathbf{V} = \hat{\mathbf{j}}'(V_x\hat{\mathbf{i}} + V_y\hat{\mathbf{j}} + V_z\hat{\mathbf{k}}) = V_x(\hat{\mathbf{j}}' \cdot \hat{\mathbf{i}}) + V_y(\hat{\mathbf{j}}' \cdot \hat{\mathbf{j}}) + V_z(\hat{\mathbf{j}}' \cdot \hat{\mathbf{k}})$$
$$V_{z'} = \hat{\mathbf{k}}' \cdot \mathbf{V} = \hat{\mathbf{k}}' \cdot (V_x\hat{\mathbf{i}} + V_y\hat{\mathbf{j}} + V_z\hat{\mathbf{k}}) = V_x(\hat{\mathbf{k}}' \cdot \hat{\mathbf{i}}) + V_y(\hat{\mathbf{k}}' \cdot \hat{\mathbf{j}}) + V_z(\hat{\mathbf{k}}' \cdot \hat{\mathbf{k}})$$

$$\tag{4.4.4}$$

That is, we can find expressions for $V_{x'}$, $V_{y'}$, and $V_{z'}$ in terms of V_x, V_y, and V_z and relationships between the two coordinate systems contained in the dot products of the various unit vectors. In similar fashion, we could make the inverse transformation and find V_x, V_y, and V_z in terms of $V_{x'}$, $V_{y'}$, and $V_{z'}$:

$$V_x = \hat{\imath} \cdot \mathbf{V} = \hat{\imath} \cdot (V_{x'}\hat{\imath}' + V_{y'}\hat{\jmath}' + V_{z'}\hat{k}') = V_{x'}(\hat{\imath}\cdot\hat{\imath}') + V_{y'}(\hat{\imath}\cdot\hat{\jmath}') + V_{z'}(\hat{\imath}\cdot\hat{k}')$$

$$V_y = \hat{\jmath} \cdot \mathbf{V} = \hat{\jmath} \cdot (V_{x'}\hat{\imath}' + V_{y'}\hat{\jmath}' + V_{z'}\hat{k}') = V_{x'}(\hat{\jmath}\cdot\hat{\imath}') + V_{y'}(\hat{\jmath}\cdot\hat{\jmath}') + V_{z'}(\hat{\jmath}\cdot\hat{k}')$$

$$V_z = \hat{k} \cdot \mathbf{V} = \hat{k} \cdot (V_{x'}\hat{\imath}' + V_{y'}\hat{\jmath}' + V_{z'}\hat{k}') = V_{x'}(\hat{k}\cdot\hat{\imath}') + V_{y'}(\hat{k}\cdot\hat{\jmath}') + V_{z'} = (\hat{k}\cdot\hat{k}')$$

(4.4.5)

A very useful and, indeed, elegant way of writing these two transformations is in terms of matrix multiplication:

$$\begin{bmatrix} V_{x'} \\ V_{y'} \\ V_{z'} \end{bmatrix} = \begin{bmatrix} \hat{\imath}'\cdot\hat{\imath} & \hat{\imath}'\cdot\hat{\jmath} & \hat{\imath}'\cdot\hat{k} \\ \hat{\jmath}'\cdot\hat{\imath} & \hat{\jmath}'\cdot\hat{\jmath} & \hat{\jmath}'\cdot\hat{k} \\ \hat{k}'\cdot\hat{\imath} & \hat{k}'\cdot\hat{\jmath} & \hat{k}'\cdot\hat{k} \end{bmatrix} \begin{bmatrix} V_x \\ V_y \\ V_z \end{bmatrix} \quad (4.4.6)$$

$$\begin{bmatrix} V_x \\ V_y \\ V_z \end{bmatrix} = \begin{bmatrix} \hat{\imath}\cdot\hat{\imath}' & \hat{\imath}\cdot\hat{\jmath}' & \hat{\imath}\cdot\hat{k}' \\ \hat{\jmath}\cdot\hat{\imath}' & \hat{\jmath}\cdot\hat{\jmath}' & \hat{\jmath}\cdot\hat{k}' \\ \hat{k}\cdot\hat{\imath}' & \hat{k}\cdot\hat{\jmath}' & \hat{k}\cdot\hat{k}' \end{bmatrix} \begin{bmatrix} V_{x'} \\ V_{y'} \\ V_{z'} \end{bmatrix} \quad (4.4.7)$$

A *coordinate transformation matrix* makes life easier if multiple transformations are involved. For instance, we may know V_x, V_y, and V_z, the components along x, y and z, *and* know how the new coordinates x', y', and z' have been rotated. We may also know how another set of axes, x'', y'', and z'', have been rotated *from* x', y', and z'. Reapplication of Eq. 4.4.4 to obtain $V_{x''}$, $V_{y''}$, and $V_{z''}$ is straightforward, but may prove messy. Reapplying Eq. 4.4.6 yields

$$\begin{bmatrix} V_{x''} \\ V_{y''} \\ V_{z''} \end{bmatrix} = \begin{bmatrix} \hat{\imath}''\cdot\hat{\imath}' & \hat{\imath}''\cdot\hat{\jmath}' & \hat{\imath}''\cdot\hat{k}' \\ \hat{\jmath}''\cdot\hat{\imath}' & \hat{\jmath}''\cdot\hat{\jmath}' & \hat{\jmath}''\cdot\hat{k}' \\ \hat{k}''\cdot\hat{\imath}' & \hat{k}''\cdot\hat{\jmath}' & \hat{k}''\cdot\hat{k}' \end{bmatrix} \begin{bmatrix} V_{x'} \\ V_{y'} \\ V_{z'} \end{bmatrix}$$

$$\begin{bmatrix} V_{x''} \\ V_{y''} \\ V_{z''} \end{bmatrix} = \begin{bmatrix} \hat{\imath}''\cdot\hat{\imath}' & \hat{\imath}''\cdot\hat{\jmath}' & \hat{\imath}''\cdot\hat{k}' \\ \hat{\jmath}''\cdot\hat{\imath}' & \hat{\jmath}''\cdot\hat{\jmath}' & \hat{\jmath}''\cdot\hat{k}' \\ \hat{k}''\cdot\hat{\imath}' & \hat{k}''\cdot\hat{\jmath}' & \hat{k}''\cdot R' \end{bmatrix} \begin{bmatrix} \hat{\imath}'\cdot\hat{\imath} & \hat{\imath}'\cdot\hat{\jmath} & \hat{\imath}'\cdot\hat{k} \\ \hat{\jmath}'\cdot\hat{\imath} & \hat{\jmath}'\cdot\hat{\jmath} & \hat{\jmath}'\cdot\hat{k} \\ \hat{k}'\cdot\hat{\imath} & \hat{k}'\cdot\hat{\jmath} & \hat{k}'\cdot\hat{k} \end{bmatrix} \begin{bmatrix} V_x \\ V_y \\ V_z \end{bmatrix} \quad (4.4.8)$$

which may *look* overpowering at first, but is usually quite manageable. (Multiply the two transformation matrices first; you can now use the resulting transformation on *any* V_x, V_y, and V_z.)

4.5 Vector Calculus

You have previously discussed the derivative of a scalar function such as df/dx, dg/du, dx/dt, or dv/dt. We can similarly define the derivative of a vector function by

$$\frac{d\mathbf{r}}{dt} = \lim_{\Delta t \to 0} \frac{\mathbf{r}(t + \Delta t) - \mathbf{r}(t)}{\Delta t} \quad (4.5.1)$$

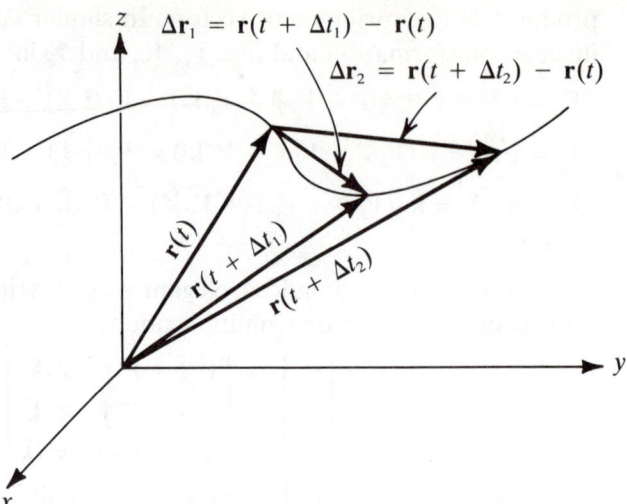

Figure 4.5.1 Vector differentiation.

While this definition is true for *any* vector **r**, and *t* can be any variable, Figure 4.5.1 gives an interpretation of this derivative if **r** is the *position* vector locating a particle at time *t*. As Δt approaches zero (shown in Figure 4.5.1 where $\Delta t_2 > \Delta t_1$), $\Delta \mathbf{r}$ and $\Delta \mathbf{r}/\Delta t$ become tangent to the path of the motion. $\Delta \mathbf{r}/\Delta t$ is just the net displacement divided by the time, or the average velocity. The expression $d\mathbf{r}/dt$, then, is the *instantaneous velocity*.

Since we can write vectors in component form, we can also write derivatives of vectors in component form. Thus, if we continue to focus our attention on the position vector,

$$\mathbf{r} = x\hat{\mathbf{i}} + y\hat{\mathbf{j}} + z\hat{\mathbf{k}} \tag{4.5.2}$$

Then

$$\mathbf{v} = \dot{\mathbf{r}} = \frac{d\mathbf{r}}{dt} = \frac{d}{dt}(x\hat{\mathbf{i}}) + \frac{d}{dt}(y\hat{\mathbf{j}}) + \frac{d}{dt}(z\hat{\mathbf{k}})$$

$$= \frac{dx}{dt}\hat{\mathbf{i}} + x\frac{d\hat{\mathbf{i}}}{dt} + \frac{dy}{dt}\hat{\mathbf{j}} + y\frac{d\hat{\mathbf{j}}}{dt} + \frac{dz}{dt}\hat{\mathbf{k}} + z\frac{d\hat{\mathbf{k}}}{dt} \tag{4.5.3}$$

$$\mathbf{v} = \dot{\mathbf{r}} = \dot{x}\hat{\mathbf{i}} + \dot{y}\hat{\mathbf{j}} + \dot{z}\hat{\mathbf{k}}$$

where we have carefully applied the chain rule and then seen that the time derivatives of the unit vectors are zero since they do not change as time changes. Unit vectors in our familiar cartesian or rectangular coordinate system do not change with time. But in Chapter 6 on coordinate systems, we shall encounter unit vectors that *are* functions of time (and of position, as well). Remember that **v** is the velocity. If we want only the *speed*, which is v, the magnitude, we can find it as usual from

$$v = |\mathbf{v}| = \sqrt{\dot{x}^2 + \dot{y}^2 + \dot{z}^2} \tag{4.5.4}$$

We can likewise define a vector acceleration by

$$\mathbf{a} = \dot{\mathbf{v}} = \ddot{\mathbf{r}} = \ddot{x}\hat{\mathbf{i}} + \ddot{y}\hat{\mathbf{j}} + \ddot{z}\hat{\mathbf{k}} \tag{4.5.5}$$

Differentiation of vectors follows rules similar to those we have already seen for differentiation of scalars. In particular, we have

$$\frac{d}{du}[\mathbf{A} + \mathbf{B}] = \frac{d\mathbf{A}}{du} + \frac{d\mathbf{B}}{du} \tag{4.5.6}$$

$$\frac{d}{du}[s(u)\mathbf{A}(u)] = \frac{ds}{du}\mathbf{A} + s\frac{d\mathbf{A}}{du} \tag{4.5.7}$$

$$\frac{d}{du}[\mathbf{A} \cdot \mathbf{B}] = \frac{d\mathbf{A}}{du} \cdot \mathbf{B} + \mathbf{A} \cdot \frac{d\mathbf{B}}{du} \tag{4.5.8}$$

$$\frac{d}{du}[\mathbf{A} \times \mathbf{B}] = \frac{d\mathbf{A}}{du} \times \mathbf{B} + \mathbf{A} \times \frac{d\mathbf{B}}{du} \quad \text{(watch the order!)} \tag{4.5.9}$$

where u can be any variable, $\mathbf{A}$ and $\mathbf{B}$ are any vector functions of u, and s is any scalar function of u.

A good example of the use of, say, Eq. 4.5.6 is in the idea of relative position or relative velocity. Figure 4.5.2 shows point P_1 located by position vector $\mathbf{r}_1$ and point P_2 located by $\mathbf{r}_2$. These position vectors are defined *with respect to* some origin O. They describe the position of points P_1 and P_2 "as seen by" the origin O. It may be of interest (and importance) to describe the position of P_2 "as seen by P_1" or to ask "where is P_2 with respect to P_1?" Its location is given by another position vector beginning with P_1 and terminating at P_2. We shall call this the relative position of P_2 with respect to P_1 and write it as the vector $\mathbf{r}_{21}$; it is defined by

$$\mathbf{r}_{21} = \mathbf{r}_2 - \mathbf{r}_1 \tag{4.5.10}$$

An observer located at P_1 will be moving with velocity $\mathbf{v}_1 = \dot{\mathbf{r}}_1$. Likewise, P_2 moves with $\mathbf{v}_2 = \dot{\mathbf{r}}_2$. These are velocities measured from the origin (measurable by observers at rest with respect to the origin). How does an observer at

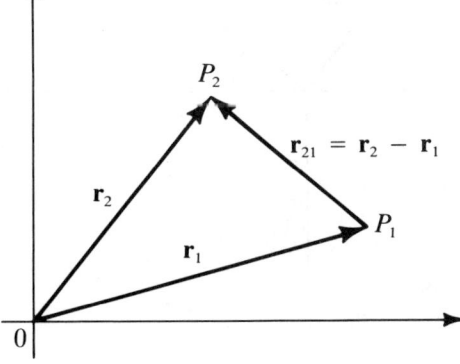

Figure 4.5.2 Position vectors for points P_1 and P_2 and the *relative* position of P_2 as seen from P_1.

P_1 (which is itself moving) describe the velocity of P_2? That is, what is the velocity of P_2 with respect to P_1? We shall call this $\mathbf{v}_{21}$ and it is defined by

$$\mathbf{v}_{21} = \dot{\mathbf{r}}_{21} = \dot{\mathbf{r}}_2 - \dot{\mathbf{r}}_1 \tag{4.5.11}$$

If this relative velocity varies, then an observer at P_1 will see P_2 accelerate. This relative acceleration will be labeled $\mathbf{a}_{21}$ and is given by

$$\mathbf{a}_{21} = \dot{\mathbf{v}}_{21} = \ddot{\mathbf{r}}_{21} = \ddot{\mathbf{r}}_2 - \ddot{\mathbf{r}}_1 \tag{4.5.12}$$

To see this better, consider two airplanes circling an airport in a holding pattern. They are flying in circles with a 5-mile radius at a uniform speed of 100 km/hr, and are separated by 90° in their circular flight path, as shown in Figure 4.5.3. We can write the airplanes' position vectors by

$$\mathbf{r}_1 = \hat{\mathbf{i}}b \cos \omega t + \hat{\mathbf{j}}b \sin \omega t$$

$$\mathbf{r}_2 = -\hat{\mathbf{i}}b \sin \omega t + \hat{\mathbf{j}}b \cos \omega t \tag{4.5.13}$$

where $b = 5$ km is the radius of their circular path and $\omega = v/b = 100/5 = 20$ radians/hour is the angular velocity of their circular motion.

Where is plane 2 as seen by the pilot in plane 1?

$$\mathbf{r}_{21} = \mathbf{r}_2 - \mathbf{r}_1 = -\hat{\mathbf{i}}b(\sin \omega t + \cos \omega t) + \hat{\mathbf{j}}b(\cos \omega t - \sin \omega t) \tag{4.5.14}$$

After playing with the trig for a bit, this can be written as

$$\mathbf{r}_{21} = -\hat{\mathbf{i}}b\sqrt{2} \cos\left(\omega t - \frac{\pi}{4}\right) + \hat{\mathbf{j}}b\sqrt{2} \sin\left(\omega t - \frac{\pi}{4}\right) \tag{4.5.15}$$

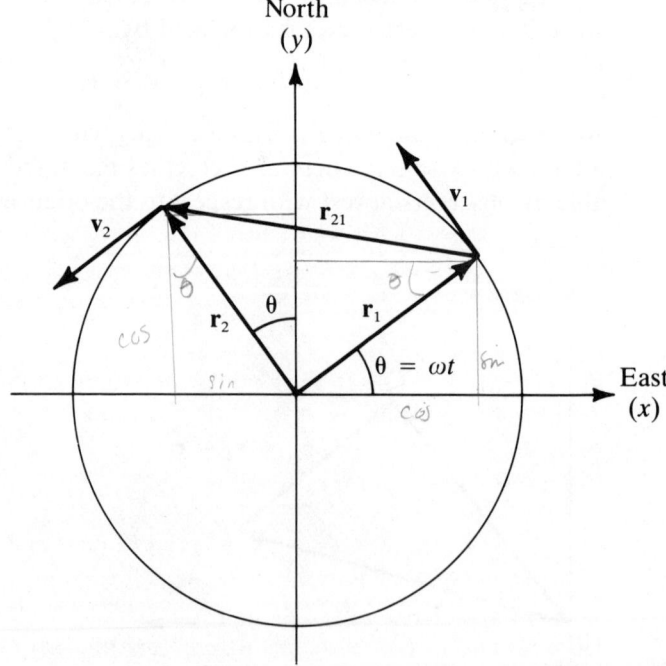

Figure 4.5.3 Relative motion in a airport's holding pattern.

That is, the pilot in plane 1 sees plane 2 moving around him in uniform circular motion at a constant radius of $b\sqrt{2}$, or 7.07 km. He sees plane 2 flying with a relative velocity of

$$\mathbf{v}_{21} = -\hat{\mathbf{i}}b\omega(\cos wt - \sin \omega t) - \hat{\mathbf{j}}bw(\sin \omega t + \cos \omega t) \qquad (4.5.16)$$

or

$$\mathbf{v}_{21} = \hat{\mathbf{i}}b\omega\sqrt{2}\sin\left(\omega t - \frac{\pi}{4}\right) + \hat{\mathbf{j}}bw\sqrt{2}\cos\left(\omega t - \frac{\pi}{4}\right) \qquad (4.5.17)$$

Our observer in plane 2 would conclude that plane 2—since it undergoes uniform circular motion—experiences a relative acceleration *with respect to* plane 1 of

$$\mathbf{a}_{21} = -\hat{\mathbf{i}}bw^2(\sin \omega t + \cos wt) + \hat{\mathbf{j}}bw^2(\sin \omega t - \cos \omega t)$$

or

$$\mathbf{a}_{21} = \hat{\mathbf{i}}bw^2\sqrt{2}\cos\left(\omega t - \frac{\pi}{4}\right) - \hat{\mathbf{j}}bw^2\sqrt{2}\sin\left(\omega t - \frac{\pi}{4}\right) \qquad (4.5.18)$$

4.6 Vector Differential Operators

There is an old adage, attributed to the Washington bureauracy, that says, "When you're up to your hips in alligators, it's hard to remember that you started out to drain the swamp." You may be so busy straightening out the ideas of vector analysis or coordinate representations that you've forgotten that your primary goal is to learn physics. Be warned now that this section can worsen the situation; don't let it happen!

The physical principles underlying mechanics can be summarized in Newton's three laws as previously stated. From them, we can develop ideas such as conservation of energy and momentum. All of these *ideas* should already be very clear to you from your introductory physics course. We don't want to introduce more *basic* principles now; rather, we are trying to apply these few principles to more and more complicated situations. Hence, we need new tools or methods of applying our old, familiar ideas. These tools are mathematical ideas, methods, and processes. We shall now proceed to add to our list of "tools of the trade" by investigating the vector differential operators known as the *gradient*, the *divergence*, and the *curl*.

Consider a two-dimensional function $f(x, y)$ as sketched in Figure 4.6.1. A good example for this function f might be the height of a hill as a function of position. A point P in the xy plane defines some value of x for that position. Now we can reasonably ask how f, or the height of the hill, varies as we move from this point P. That's like asking for the *slope* of the hill; it depends upon

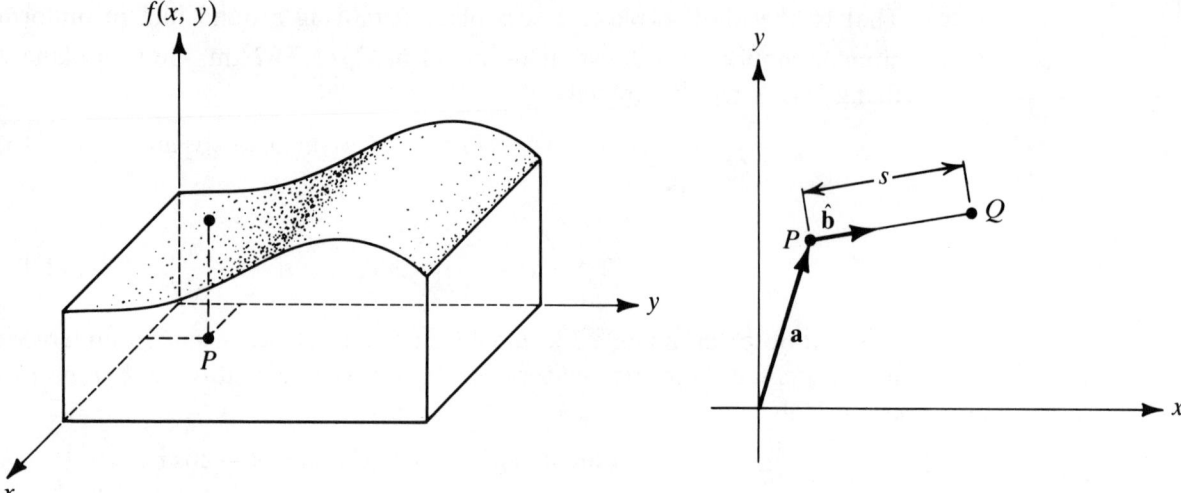

Figure 4.6.1 The gradient of a two-dimensional function.

which *direction* we move from P. If we move a distance s from P toward point Q, this new position is given by

$$\mathbf{r} = \mathbf{a} + s\hat{\mathbf{b}} \tag{4.6.1}$$

where $\hat{\mathbf{b}}$ is a unit vector pointing from P toward Q. The vector $\mathbf{a}$ locates point P. We can always write

$$\mathbf{r} = x\hat{\mathbf{i}} + y\hat{\mathbf{j}}$$

Therefore, we can now write

$$\mathbf{r} = x(s)\hat{\mathbf{i}} + y(s)\hat{\mathbf{j}} \tag{4.6.2}$$

where $x(s)$ and $y(s)$ are such that Eqs. 4.6.1 and 4.6.2 are consistent; that is, they describe the same $\mathbf{r}$.

Using the chain rule, we can write

$$\frac{df}{ds} = \frac{\partial f}{\partial x}\frac{dx}{ds} + \frac{\partial f}{\partial y}\frac{dy}{ds} \tag{4.6.3}$$

where $\partial/\partial x$ is the *partial* derivative—the rate of change of f as x changes and y (or anything else upon which f depends) remains constant.

From Eq. 4.6.1, we can see that

$$\frac{d\mathbf{r}}{ds} = \hat{\mathbf{b}} \tag{4.6.4}$$

From Eq. 4.6.2, we can write

$$\frac{d\mathbf{r}}{ds} = \frac{dx}{ds}\hat{\mathbf{i}} + \frac{dy}{ds}\hat{\mathbf{j}} \tag{4.6.5}$$

SECTION 4.6 / VECTOR DIFFERENTIAL OPERATORS

The expression *df/ds* is the slope of *f* along *s* (or in direction $\hat{\mathbf{b}}$) or the *directional derivative* of the function *f* along $\hat{\mathbf{b}}$ (the direction of increasing *s*). Using Eqs. 4.6.4 and 4.6.5, we can write *df/ds* in a very useful form:

$$\frac{df}{ds} = \left(\frac{\partial f}{\partial x}\hat{\mathbf{i}} + \frac{\partial f}{\partial y}\hat{\mathbf{j}}\right) \cdot \left(\frac{dx}{ds}\hat{\mathbf{i}} + \frac{dy}{ds}\hat{\mathbf{j}}\right)$$

$$\frac{df}{ds} = \left(\frac{\partial f}{\partial x}\hat{\mathbf{i}} + \frac{\partial f}{\partial y}\hat{\mathbf{j}}\right) \cdot \hat{\mathbf{b}} \tag{4.6.6}$$

If we extend these ideas to three dimensions, to a function $f(x, y, z)$, we can write

$$\mathbf{r} = \mathbf{a} + s\hat{\mathbf{b}} \tag{4.6.1a}$$

$$\mathbf{r} = x(s)\hat{\mathbf{i}} + y(s)\hat{\mathbf{j}} + z(s)\hat{\mathbf{k}} \tag{4.6.2a}$$

$$\frac{df}{ds} = \frac{\partial f}{\partial x}\frac{dx}{ds} + \frac{\partial f}{\partial y}\frac{dy}{ds} + \frac{\partial f}{\partial z}\frac{dz}{ds} \tag{4.6.3a}$$

$$\frac{d\mathbf{r}}{ds} = \hat{\mathbf{b}} \tag{4.6.4a}$$

$$\frac{d\mathbf{r}}{ds} = \frac{dx}{ds}\hat{\mathbf{i}} + \frac{dy}{ds}\hat{\mathbf{j}} + \frac{dz}{ds}\hat{\mathbf{k}} \tag{4.6.5a}$$

$$\frac{df}{ds} = \left(\frac{\partial f}{\partial x}\hat{\mathbf{i}} + \frac{\partial f}{\partial y}\hat{\mathbf{j}} + \frac{\partial f}{\partial z}\hat{\mathbf{k}}\right) \cdot \hat{\mathbf{b}} \tag{4.6.6a}$$

Eq. 4.6.6a can be rewritten as

$$\frac{df}{ds} = \nabla f \cdot \hat{\mathbf{b}} \tag{4.6.7}$$

where

$$\nabla f = \frac{\partial f}{\partial x}\hat{\mathbf{i}} + \frac{\partial f}{\partial y}\hat{\mathbf{j}} + \frac{\partial f}{\partial z}\hat{\mathbf{k}} \tag{4.6.8}$$

is defined as the *gradient* of the scalar function *f*. It is read as *gradient f, del f*, or sometimes *grad f*.

The symbol ∇ is called the *del operator*:

$$\nabla = \frac{\partial}{\partial x}\hat{\mathbf{i}} + \frac{\partial}{\partial y}\hat{\mathbf{j}} + \frac{\partial}{\partial z}\hat{\mathbf{k}} \tag{4.6.9}$$

Like any other operator (such as d/dx), ∇ is not complete by itself. It makes sense only after it has operated on a function.

Consider a surface of constant value; that is, values of *x*, *y*, and *z* that give

$$f(x, y, z) = c$$

A good two-dimensional example might be the *isotherms* on a weather map that show all the places with the same temperature.

Now consider taking the directional derivative as you move about this surface of constant value. That is, find the directional derivative while restraining $\hat{b}$ or s to lie in this surface of constant value. With that restriction, $df/ds = 0$, but from Eq. 4.6.7,

$$\frac{df}{ds} = \nabla f \cdot \hat{b} = 0 \tag{4.6.10}$$

and we conclude that ∇f *is a vector perpendicular to surfaces of constant value.*

Figure 4.6.2 shows a rough sketch of two surfaces, each of a different constant value. The change in the value of the function, Δf, is

$$\Delta f = c_2 - c_1 = \Delta c$$

If we move from P to Q along a vector $\Delta \mathbf{s}$ *and* $\Delta \mathbf{s}$ is very small (implying Δf or Δc is also very small), we can write

$$\Delta f = \frac{df}{ds} \Delta \mathbf{s} \tag{4.6.11}$$

from the very idea of a derivative. From Eq. 4.6.7, we have

$$\Delta f = (\nabla f \cdot \hat{b}) \Delta \mathbf{s} \tag{4.6.12}$$

or

$$\Delta f = \nabla f \cdot \Delta \mathbf{s} \tag{4.6.12a}$$

where $\Delta \mathbf{s} = \hat{b} \Delta \mathbf{s}$. The expression $\Delta f = \Delta c = c_2 - c_1$ is a constant. The magnitude Δf has a maximum value when ∇f and $\Delta \mathbf{s}$ are parallel due to the definition of the dot product. Therefore, we can conclude that ∇f *points in the direction of maximum increase in f. ∇f is the maximum directional derivative.*

In the one-dimensional case, we write the force as $-dV/dx$. Now, in three dimensions, we can write in similar fashion:

$$\mathbf{F} = -\nabla V \tag{4.6.13}$$

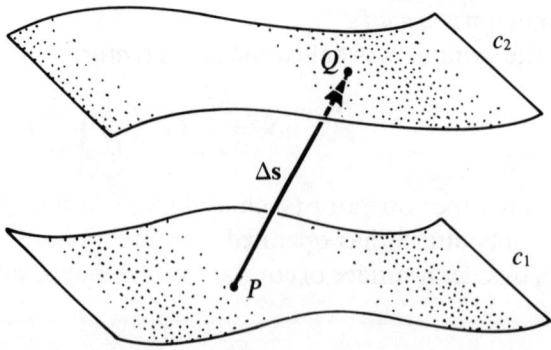

Figure 4.6.2 Two surfaces of (different) constant value.

where V is a potential energy function, and $\mathbf{F}$ the associated force. ∇V gives the direction of maximum *increase* in V. The force $\mathbf{F}$ pushes a body towards ever *lower* potential energy V.

Divergence

The del operator, ∇, acts like a vector in many respects so we may take a dot product with it and some vector function (or vector field). We will call this the *divergence* of the vector field:

$$\text{div } \mathbf{v} = \nabla \cdot \mathbf{v} = \frac{\partial v_x}{\partial x} + \frac{\partial v_y}{\partial y} + \frac{\partial v_z}{\partial z} \tag{4.6.14}$$

Note that this divergence of the vector field is itself a *scalar*, as is the case in any ordinary dot product.

To better understand the meaning or use of the divergence, consider the small rectangular solid with sides of Δx, Δy, and Δz that is shown in Figure 4.6.3.

Consider a momentum field where $\mathbf{u} = \rho \mathbf{v}$ gives the momentum carried per unit volume (ρ is the mass density and $\mathbf{v}$, the velocity). This can represent the flow of water down a river, air moving inside a wind tunnel, or air passing through your sinuses and out of your nostrils. Put a tiny rectangular solid in this momentum field with its "origin" located at position (x, y, z) and look at the mass that flows into and out of it. As long as the box is small enough that we can consider the flow $\mathbf{u}$ constant over each face (or use an equivalent average value of $\mathbf{u}$), we can write

$$I_x = \text{Rate of flow } in \text{ at } x \text{ (across back face)} = \rho v_x(x) \Delta y \Delta z \tag{4.6.15}$$

$$O_x = \text{Rate of flow } out \text{ at } x + x \text{ (across front face)} = \rho v_x(x + \Delta x) \Delta y \Delta z \tag{4.6.15a}$$

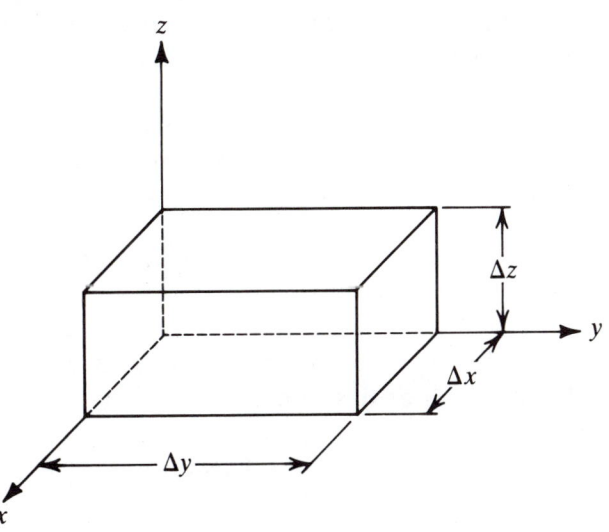

Figure 4.6.3
Rectangular solid $\Delta x \Delta y \Delta z$.

The *net* outward flow in the x-direction, then, is simply $O_x - I_x$. To evaluate this, we need $v_x(x + \Delta x)$. But as long as our box is very small, we can write

$$v_x(x + \Delta x) = v_x + \frac{\partial v_x}{\partial x} \Delta x \qquad (4.6.16)$$

This allows us to evaluate $O_x - I_x$ and, similarly for the y- and z-directions,

$$O_x - I_x = \rho \frac{\partial v_x}{\partial x} \Delta x \Delta y \Delta z \qquad (4.6.17)$$

$$O_y - I_y = \rho \frac{\partial v_y}{\partial y} \Delta x \Delta y \Delta z \qquad (4.6.17a)$$

$$O_z - I_z = \rho \frac{\partial v_z}{\partial z} \Delta x \Delta y \Delta z \qquad (4.6.17b)$$

Summing these, we have

$$O - I = \text{Net flow out of box} = \rho \left(\frac{\partial v_x}{\partial x} + \frac{\partial v_y}{\partial y} + \frac{\partial v_z}{\partial z} \right) \Delta x \Delta y \Delta z \qquad (4.6.18)$$

and the term in parentheses should be recognized as the divergence. $\Delta x \Delta y \Delta z$ is just the volume of the box under consideration. If we call that $\Delta \dot{V}$, then we can write

$$\text{Net flow out} = \rho (\nabla \cdot \mathbf{v}) \Delta V \qquad (4.6.19)$$

But the net flow out is just equal to the density multiplied by the component of velocity *normal* to the surface multiplied by the area of the surface; that is,

$$\text{Net flow out} = \sum \rho \hat{\mathbf{n}} \cdot \mathbf{v} \, \Delta S \qquad (4.6.20)$$

where $\hat{\mathbf{n}}$ is a unit vector normal to the surface, pointing outward, and ΔS is the area of the surface. This expression is summed over all the sides of our small rectangular solid that has volume ΔV.

Thus, the divergence of a vector field multiplied by a volume is equal to the *net* flow of that vector field across the surface bounding that volume. This can be stated mathematically by Gauss's Theorem (also called the divergence theorem).

$$\iiint_v \nabla \cdot \mathbf{A} \, dV = \iint_s \hat{\mathbf{n}} \cdot \mathbf{A} \, dS \qquad (4.6.21)$$

where **A** is any vector. In our previous example and discussion we used $\mathbf{A} = \mathbf{u} = \rho \mathbf{v}$. For this case, then,

$$\iiint_v \nabla \cdot (\rho \mathbf{v}) \, dV = \iint_n \hat{\mathbf{n}} \cdot (\rho \mathbf{v}) \, dS$$

SECTION 4.6 / VECTOR DIFFERENTIAL OPERATORS

For an incompressible fluid, like water, with a constant density, this becomes

$$\rho \iiint_v (\nabla \cdot \mathbf{v}) \, dV = \rho \iint_s \hat{\mathbf{n}} \cdot \mathbf{v} \, dS \qquad (4.6.22)$$

A positive divergence, then, means net flow outward from the volume—the volume contains a "source." A negative divergence, likewise, means net flow into the volume—the volume contains a "sink"; material comes into the volume and remains there.

A numerical example may help clarify this concept. Consider a box with dimensions $x = 2$ cm, $y = 3$ cm, and $z = 5$ cm, as illustrated in Figure 4.6.4. Simply for convenience in this example, there is no flow of fluid in the z direction and thus no fluid crosses the front and back faces of our box.

Fluid does flow across the other four sides with velocities as given in Figure 4.6.4. Only the normal component of the velocity carries fluid (and mass or momentum) *across* a surface. Only across surface 1 is fluid brought in. We can write by inspection

$$\text{Flow in} = v_1 A_1 = v_1 \Delta y \Delta z = 10 \, \frac{\text{cm}}{\text{s}} (3 \text{ cm})(5 \text{ cm}) = 150 \, \frac{\text{cm}^3}{\text{s}} \qquad (4.6.23)$$

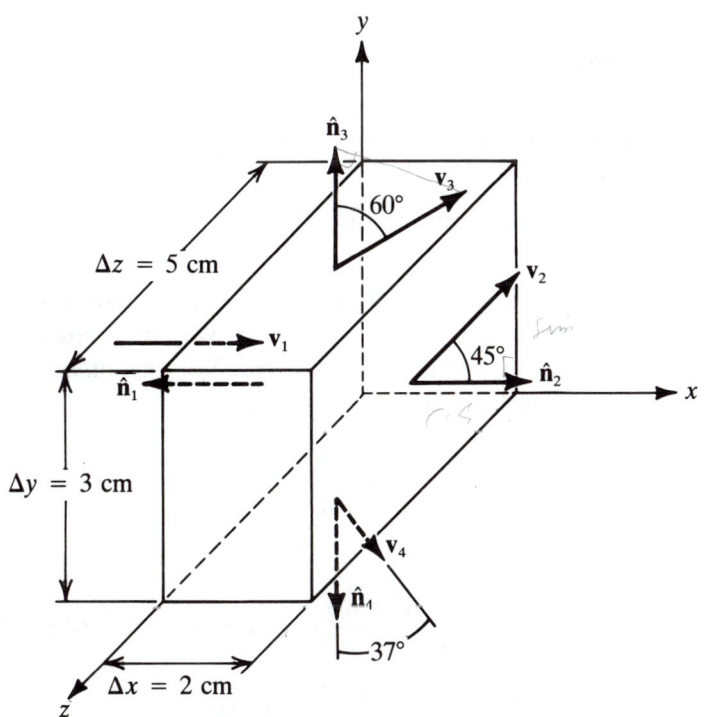

$\Delta x = 2$ cm
$\Delta y = 3$ cm
$\Delta z = 5$ cm
$\mathbf{v}_1 = 10$ cm/s $\hat{\mathbf{i}}$, $\hat{\mathbf{n}}_1 = -\hat{\mathbf{i}}$
$\mathbf{v}_2 = 7.1$ cm/s $\hat{\mathbf{i}}$ + 7.1 cm/s $\hat{\mathbf{j}}$, $\hat{\mathbf{n}}_2 = \hat{\mathbf{i}}$
$\mathbf{v}_3 = 8.67$ cm/s $\hat{\mathbf{i}}$ + 5 cm/s $\hat{\mathbf{j}}$, $\hat{\mathbf{n}}_3 = \hat{\mathbf{j}}$
$\mathbf{v}_4 = 3$ cm/s $\hat{\mathbf{i}}$ − 4 cm/s $\hat{\mathbf{j}}$, $\hat{\mathbf{n}}_4 = -\hat{\mathbf{j}}$

Figure 4.6.4 Fluid flow across various faces of a rectangular solid.

which we *could* have obtained explicitly from

$$\text{Flow } out = \hat{\mathbf{n}}_1 \cdot \mathbf{v}_1 A_1 = (-\hat{\mathbf{i}}) \cdot \left(10\,\frac{\text{cm}}{\text{s}}\,\hat{\mathbf{i}}\right)(15\text{ cm}^2) = -150\,\frac{\text{cm}^3}{\text{s}} \qquad (4.6.23\text{a})$$

but, a negative flow *out* is a positive flow *in*. For the other three surfaces we can directly write

$$\text{Flow out} = \hat{\mathbf{n}}_2 \cdot \mathbf{v}_2 A_2 + \hat{\mathbf{n}}_3 \cdot \mathbf{v}_3 A_3 + \hat{\mathbf{n}}_4 \cdot \mathbf{v}_4 A_4$$

$$= \hat{\mathbf{n}}_2 \cdot \mathbf{v}_2\, \Delta y \Delta z + \hat{\mathbf{n}}_3 \cdot \mathbf{v}_3\, \Delta x \Delta z + \hat{\mathbf{n}}_4 \cdot \mathbf{v}_4\, \Delta x \Delta z$$

$$= (\hat{\mathbf{i}}) \cdot \left(7.1\,\frac{\text{cm}}{\text{s}}\,\hat{\mathbf{i}} + 7.1\,\frac{\text{cm}}{\text{s}}\,\hat{\mathbf{j}}\right)(3\text{ cm})(5\text{ cm})$$

$$+ (\hat{\mathbf{j}}) \cdot \left(8.67\,\frac{\text{cm}}{\text{s}}\,\hat{\mathbf{i}} + 5\,\frac{\text{cm}}{\text{s}}\,\hat{\mathbf{j}}\right)(2\text{ cm})(5\text{ cm})$$

$$+ (-\hat{\mathbf{j}}) \cdot \left(3\,\frac{\text{cm}}{\text{s}}\,\hat{\mathbf{i}} - 4\,\frac{\text{cm}}{\text{s}}\,\hat{\mathbf{j}}\right)(2\text{ cm})(5\text{ cm})$$

$$= (7.1)(3)(5) + (5)(2)(5) + (4)(2)(5)\,\frac{\text{cm}^3}{\text{s}}$$

$$\text{Flow out} = 196.5\,\frac{\text{cm}^3}{\text{s}} \qquad (4.6.24)$$

Thus the *net* outward flow, flow out *minus* flow in, is

$$\text{Net outward flow} = 46.5\,\frac{\text{cm}^3}{\text{s}} \qquad (4.6.25)$$

That is, 46.5 cm³ of material is generated *within* this volume and passes across its surfaces every second to have velocities as indicated. We can relate this to our definition of the *divergence* of this velocity field. For this *numerical* example, we must replace the partial derivatives with Δ's:

$$\boldsymbol{\nabla} \cdot \mathbf{V} \approx \frac{\Delta v_x}{\Delta x} + \frac{\Delta v_y}{\Delta y} + \frac{\Delta v_z}{\Delta z} \qquad (4.6.26)$$

where

$$\frac{\Delta v_x}{\Delta x} = \frac{v_{2x} - v_{1x}}{\Delta x} = \frac{7.1\text{ cm/s} - 10\text{ cm/s}}{2\text{ cm}} = -1.45\text{ s}^{-1} \qquad (4.6.27)$$

$$\frac{\Delta v_y}{\Delta y} = \frac{v_{3y} - v_{4y}}{\Delta y} = \frac{5\text{ cm/s} - (-4\text{ cm/s})}{3\text{ cm}} = 3\text{ s}^{-1} \qquad (4.6.27\text{a})$$

$$\frac{\Delta v_z}{\Delta z} = 0 \qquad (4.6.27\text{b})$$

Therefore,

$$(\nabla \cdot \mathbf{v})(\Delta V) = (1.55 \text{ s}^{-1})(30 \text{ cm}^3) = 46.5 \frac{\text{cm}^3}{\text{s}} \qquad (4.6.28)$$

That is,

$$\text{\textit{Net} outward flow} = (\nabla \cdot \mathbf{v})(\Delta V) \qquad (4.6.29)$$

which is simply Gauss's Theorem once again.

Curl

We have taken a dot product with the del operator. If we take the *cross* product of the del operator with some vector function (or field), the result is called the *curl* (some texts, usually German, also call this the *rotation* of the vector field and write it as *rot* **v**):

$$\text{curl } \mathbf{v} = \nabla \times \mathbf{v} = \hat{\mathbf{i}}\left(\frac{\partial v_z}{\partial y} - \frac{\partial v_y}{\partial z}\right) + \hat{\mathbf{j}}\left(\frac{\partial v_x}{\partial z} - \frac{\partial v_z}{\partial x}\right) + \hat{\mathbf{k}}\left(\frac{\partial v_y}{\partial x} - \frac{\partial v_x}{\partial y}\right) \qquad (4.6.30)$$

The first thing to notice is that the curl, the result of this $\nabla \times$ operating on a vector, is another *vector* (whereas the divergence, $\nabla \cdot \mathbf{v}$, was a scalar).

The curl of a vector field—a velocity field is a good, almost tangible example—is a measure of the rotation (or curl) of that field. What in the world does *that* mean? Consider a small paddle wheel tossed into a fluid. As it is carried along, does it also *rotate*? If so, the curl of the vector field describing the fluid lines up with its axis of rotation and is related to its angular speed. The curl is zero for flow that would cause no rotation of our paddle wheel as it carries it along.

The curl of a vector can also be defined by Stokes's Theorem:

$$\iint_S (\nabla \times \mathbf{v}) \cdot \hat{\mathbf{n}} \, dS = \oint \mathbf{v} \cdot d\mathbf{r} \qquad (4.6.31)$$

where $\hat{\mathbf{n}}$ is a unit vector normal to the surface element dS and the integral on the right is taken around a path that forms the boundary of the surface used in the integral on the left-hand side. The direction used in evaluating this line integral is given by the right-hand rule with respect to $\hat{\mathbf{n}}$.

To make this more meaningful, let us look at another crude numerical example. Consider a 6-cm × 6-cm path located in the velocity field sketched in Figure 4.6.5.

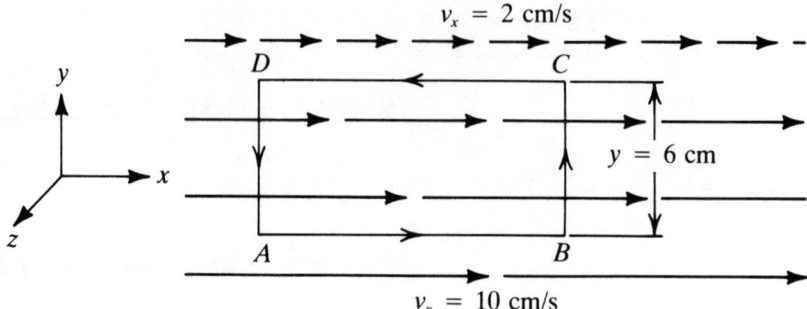

Figure 4.6.5
Rectangular path in a velocity field.

For such a numerical example, we must rewrite Stokes's Theorem, Eq. 4.6.31, in a crude fashion using summations and averages instead of integrals:

$$(\nabla \times \mathbf{v}) \cdot \hat{\mathbf{n}} \, \Delta S = \sum \mathbf{v} \cdot \Delta \mathbf{r} \tag{4.6.32}$$

The only term that contributes to the curl is the z-component. The other components are identically zero (as you should verify for yourself). If we replace the derivatives with the ratio of differences as

$$\frac{\partial v_x}{\partial y} \quad \text{becoming} \quad \frac{\Delta v_x}{\Delta y}$$

the curl is

$$\nabla \times \mathbf{v} = -\hat{\mathbf{k}}\left(\frac{\partial v_x}{\partial y}\right) \to \hat{\mathbf{k}}\left(\frac{\Delta v_x}{\Delta y}\right) = -\hat{\mathbf{k}}\,\frac{2\ \text{cm/s} - 10\ \text{cm/s}}{6\ \text{cm}} = +\frac{4}{3}\frac{1}{\text{s}}\hat{\mathbf{k}}$$

$\hat{\mathbf{n}}$ is the unit vector normal to ΔS *and* related to the direction taken in the line integral by the right-hand rule. If the line integral (or our sum $\sum \mathbf{v} \cdot \Delta \mathbf{r}$) is taken in the direction shown in Figure 4.6.5, then

$$\hat{\mathbf{n}} = \hat{\mathbf{k}}$$

$$\Delta S = 36\ \text{cm}^2$$

so

$$(\nabla \times \mathbf{v}) \cdot \hat{\mathbf{n}} \, \Delta S = \left(+\frac{4}{3}\frac{1}{\text{s}}\hat{\mathbf{k}}\right) \cdot (\hat{\mathbf{k}})(36\ \text{cm}^2) = +48\,\frac{\text{cm}^2}{\text{s}}$$

The right-hand side of Eq. 4.6.30 is a measure of whether the flow is going *with* or *against* a rotation. The two vertical sides, BC and DA, offer no contribution since the flow is perpendicular to the side $\Delta \mathbf{r}$. But the two horizontal sides *do* contribute:

$$\sum \mathbf{v} \cdot \Delta \mathbf{r} = (\mathbf{v} \cdot \Delta \mathbf{r})_{AB} + (\mathbf{v} \cdot \Delta r)_{CD}$$

$$= \left(10 \frac{\text{cm}}{\text{s}} \hat{\mathbf{i}}\right) \cdot (6 \text{ cm } \hat{\mathbf{i}}) + \left(2 \frac{\text{cm}}{\text{s}} \hat{\mathbf{i}}\right) \cdot (-6 \text{ cm } \hat{\mathbf{i}})$$

$$= (60 - 12) \frac{\text{cm}^2}{\text{s}}$$

$$= +48 \frac{\text{cm}^2}{\text{s}}$$

And, indeed, Stokes's Theorem is shown to be valid for this example.

It is worth ending this section with a quotation from K. R. Symon's classic text, *Mechanics* (Reading, Mass.: Addison-Wesley, third edition, 1971, p. 101):

> The reader should not be bothered by the difficulty of fixing these ideas in his [or her] mind. Understanding of new mathematical concepts like these comes to most people only slowly, as they are put to use. The definitions are recorded here for future use. One cannot be expected to be familiar with them until he [or she] has seen how they are used in physical problems.

PROBLEMS

4.1 Given two vectors $\mathbf{A} = \hat{\mathbf{i}} + 2\hat{\mathbf{j}} + 3\hat{\mathbf{k}}$ and $\mathbf{B} = 3\hat{\mathbf{i}} - 2\hat{\mathbf{j}} - \hat{\mathbf{k}}$, find the following:
(a) $|\mathbf{A}|$ and $|\mathbf{B}|$
(b) $\mathbf{A} + \mathbf{B}$ and $|\mathbf{A} + \mathbf{B}|$
(c) $\mathbf{A} - \mathbf{B}$ and $|\mathbf{A} - \mathbf{B}|$
(d) $\mathbf{B} - \mathbf{A}$ and $|\mathbf{B} - \mathbf{A}|$
(e) $\mathbf{A} \cdot \mathbf{B}$
(f) $\mathbf{A} \times \mathbf{B}$ and $|\mathbf{A} \times \mathbf{B}|$

4.2 Given three vectors $\mathbf{A} = \hat{\mathbf{i}} + 2\hat{\mathbf{j}} + 3\hat{\mathbf{k}}$, $\mathbf{B} = 3\hat{\mathbf{i}} - 2\hat{\mathbf{j}} - \hat{\mathbf{k}}$, and $\mathbf{C} = 2\hat{\mathbf{i}} + 3\hat{\mathbf{j}} - \hat{\mathbf{k}}$, find the following:
(a) $\mathbf{A} + \mathbf{B} + \mathbf{C}$
(b) $\mathbf{A} - \mathbf{B} + \mathbf{C}$
(c) $\mathbf{A} \cdot (\mathbf{B} + \mathbf{C})$
(d) $(\mathbf{A} + \mathbf{B}) \cdot \mathbf{C}$
(e) $\mathbf{A} \cdot (\mathbf{B} \times \mathbf{C})$
(f) $(\mathbf{A} \times \mathbf{B}) \cdot \mathbf{C}$
(g) $(\mathbf{A} \times \mathbf{B}) \times \mathbf{C}$

4.3 What is the angle between $\mathbf{A} = \hat{\mathbf{i}} + \hat{\mathbf{j}} + \hat{\mathbf{k}}$ and $\mathbf{B} = \hat{\mathbf{i}} + \hat{\mathbf{j}}$? Vector $\mathbf{A}$ is the diagonal *through* a cube and $\mathbf{B}$ is the diagonal *along* one face.

4.4 Given the vectors $\mathbf{A} = x_0 e^{-\lambda t}\,\hat{\mathbf{i}} + 2\hat{\mathbf{j}} + 3\hat{\mathbf{k}}$ and $\mathbf{B} = \hat{\mathbf{i}}\sin wt + \hat{\mathbf{j}}\cos wt$, find the following:

(a) $\dfrac{d\mathbf{A}}{dt}$ and $\left|\dfrac{d\mathbf{A}}{dt}\right|$

(b) $|\mathbf{A}|$ and $\dfrac{d}{dt}|\mathbf{A}|$

(c) $\dfrac{d}{dt}(\mathbf{A}\cdot\mathbf{B})$

(d) $\dfrac{d}{dt}(\mathbf{A}\times\mathbf{B})$

4.5 For what values of p are vectors $\mathbf{A}$ and $\mathbf{B}$ perpendicular? $\mathbf{A} = p\hat{\mathbf{i}} + \hat{\mathbf{j}} + p\hat{\mathbf{k}}$ and $\mathbf{B} = \hat{\mathbf{i}} - 2p\hat{\mathbf{j}} + p\hat{\mathbf{k}}$.

4.6 Use the dot product to obtain the Law of Cosines from plane trigonometry ($A^2 = B^2 + C^2 - 2BC\cos\gamma$, where A, B, C are the sides of a triangle).

4.7 Use the cross product to obtain the Law of Sines from plane trigonometry ($A/\sin\alpha = B/\sin\beta = C/\sin\gamma$; that is, the ratio of one side of a triangle and the sine of the angle opposite that side is a constant).

4.8 Show that the triple scalar product $(\mathbf{A}\times\mathbf{B})\cdot\mathbf{C}$ can be written in the form of a determinant; that is, as

$$(\mathbf{A}\times\mathbf{B})\cdot\mathbf{C} = \begin{vmatrix} A_x & A_y & A_z \\ B_x & B_y & B_z \\ C_x & C_y & C_z \end{vmatrix}$$

4.9 Show that the triple scalar product is independent of the order of the terms as long as they are in the same *cyclic* order. That is,

$$(\mathbf{A}\times\mathbf{B})\cdot\mathbf{C} = \mathbf{A}\cdot(\mathbf{B}\times\mathbf{C}) = \mathbf{B}\cdot(\mathbf{C}\times\mathbf{A})$$

Thus, the triple scalar product may be written as $\mathbf{ABC}$ with no ambiguity.

4.10 Calculate the volume of a parallelepiped whose sides are defined by the vectors $\mathbf{A}$, $\mathbf{B}$, and $\mathbf{C}$. How does this result compare to the value of the triple scalar product $\mathbf{ABC} = \mathbf{A}\cdot(\mathbf{B}\times\mathbf{C})$?

4.11 Show that the area of a parallelogram is given by $|\mathbf{A}\times\mathbf{B}|$ where $\mathbf{A}$ and $\mathbf{B}$ form adjoining sides of the parallelogram.

4.12 Prove the rather useful identity

$$\mathbf{A}\times(\mathbf{B}\times\mathbf{C}) = (\mathbf{A}\cdot\mathbf{C})\mathbf{B} - (\mathbf{A}\cdot\mathbf{B})\mathbf{C}$$

4.13 A body moves in an elliptic orbit with a position vector given by

$$\mathbf{r} = \hat{\mathbf{i}}a\cos wt + \hat{\mathbf{j}}b\sin wt.$$

Find its speed as a function of time.

PROBLEMS

4.14 Define (or describe) the *gradient* in words.

4.15 Show that $\nabla(\phi\psi) = \phi\nabla\psi + \psi\nabla\phi$.

4.16 Consider a potential energy function $V = k/r$ where r is given by

$$r^2 = x^2 + y^2 + z^2$$

Using $\mathbf{F} = -\nabla V$, find the components of the force associated with this potential energy.

4.17 Consider the function $S = 1/r^3$.
  (a) Find ∇S. Use $r^2 = x^2 + y^2 + z^2$.
  (b) Find a unit vector that points in the direction of maximum increase in S at the position $\mathbf{r} = \hat{\imath} + 2\hat{\jmath} + 3\hat{k}$.

4.18 Find the gradient of the following functions:
  (a) $f = x + y + z$
  (b) $g = xy + xz$
  (c) $h = x^2 + y^2$
  (d) $\psi = x^2y + xy^2 + z^2$
  (e) $\phi = xy^2 + xyz + yz^2$

4.19 Find the divergence of the following vector fields:
  (a) $\mathbf{v} = y\hat{\imath} + x\hat{\jmath}$
  (b) $\mathbf{v} = x\hat{\imath} + y\hat{\jmath}$
  (c) $\mathbf{v} = xy\hat{\imath} + xy\hat{\jmath} + z\hat{k}$
  (d) $\mathbf{v} = x^2\hat{\imath} - xy\hat{\jmath} - xz\hat{k}$
  (e) $\mathbf{v} = 2\hat{\imath} - y\hat{\jmath} + z\hat{k}$

4.20 Evaluate the divergence of $\mathbf{r}$, div $\mathbf{r} = \nabla \cdot \mathbf{r}$.

4.21 Find the divergence of the gravitational force

$$F = \frac{GMm}{r^2}\mathbf{r} = \frac{GMm}{r^3}\mathbf{r}$$

4.22 A cube, 1 cm on a side, is placed in a fluid and has the flow as shown in the sketch and table on page 108 (velocities are in cm/sec). For the four cases, use Gauss's Theorem (that is, Eqs. 4.6.19 and 4.6.20) to find whether there is a sink or source within the cube—and its magnitude. All flows given are perpendicular to the side of the cube.

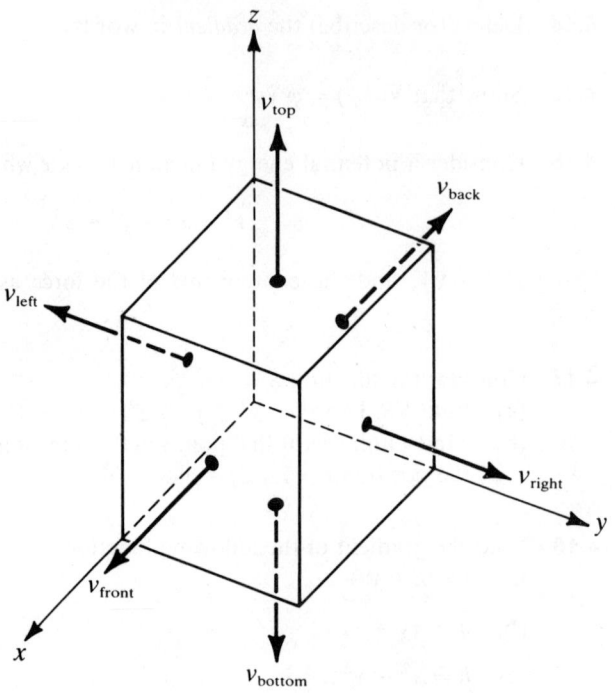

Case	A	B	C	D
Top	$v_z =$ 10 cm/s	10	10	−10
Bottom	$v_z =$ 20 cm/s	−10	10	−10
Right	$v_y =$ 10 cm/s	10	10	−10
Left	$v_y =$ −100 cm/s	−10	10	−10
Front	$v_x =$ −20 cm/s	10	10	20
Back	$v_x =$ 40 cm/s	−10	10 cm/s	−30

4.23 For the flow sketched at the top of page 109, use Gauss's Theorem to determine if there is a source or sink within the volume shown. All *velocities* lie in the xy plane.

$$\Delta x = 2 \text{ cm} \qquad v_1 = 10 \text{ cm/s}$$
$$\Delta y = 3 \text{ cm} \qquad v_2 = 10 \text{ cm/s}$$
$$\Delta z = 5 \text{ cm} \qquad v_3 = 10 \text{ cm/s}$$
$$\qquad\qquad\qquad\quad v_4 = 5 \text{ cm/s}$$

PROBLEMS

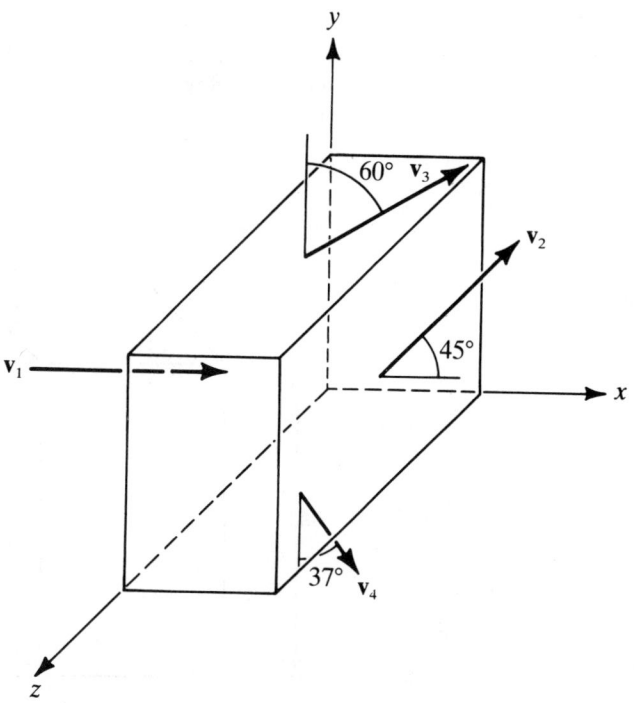

4.24 Using the numerical techniques of this chapter and Gauss's Theorem, find the average divergence of the flow within the box sketched below if it has the values shown for the average value of flow across each face.

Face	v_x	v_y	v_z
Front	2 m/s	2 m/s	−2 m/s
Right	3	−1	2
Left	−1	2	−3
Top	3	5	−4
Bottom	5	−3	3
Back	−1.5	3	2 m/s

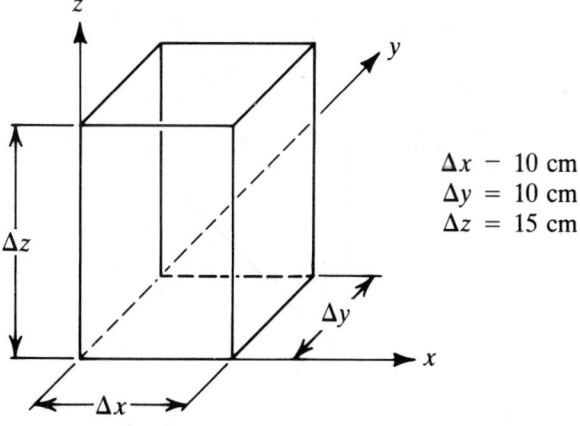

$\Delta x = 10 \text{ cm}$
$\Delta y = 10 \text{ cm}$
$\Delta z = 15 \text{ cm}$

4.25 Using the numerical techniques of this chapter and Gauss's Theorem, find the average divergence of the flow within the volume sketched below if it has the values shown for the average value of the flow across each face.

Face	v_x	v_y	v_z
Front	3 m/s	1 m/s	4 m/s
Right	1	1	2
Left	2	2	3
Top	−2	2	2
Bottom	3	−1	1
Back	4	−1	−3

$\Delta x = 3$ cm
$\Delta y = 4$ cm
$\Delta z = 5$ cm

4.26 For the flow sketched below, use Gauss's Theorem to find the average divergence of the flow within the volume. All *velocities* line in the xy plane.

$\Delta x = 2$ cm $v_1 = 10$ cm/s
$\Delta y = 3$ cm $v_2 = 10$ cm/s
$\Delta z = 5$ cm $v_3 = 10$ cm/s
            $v_4 = 5$ cm/s

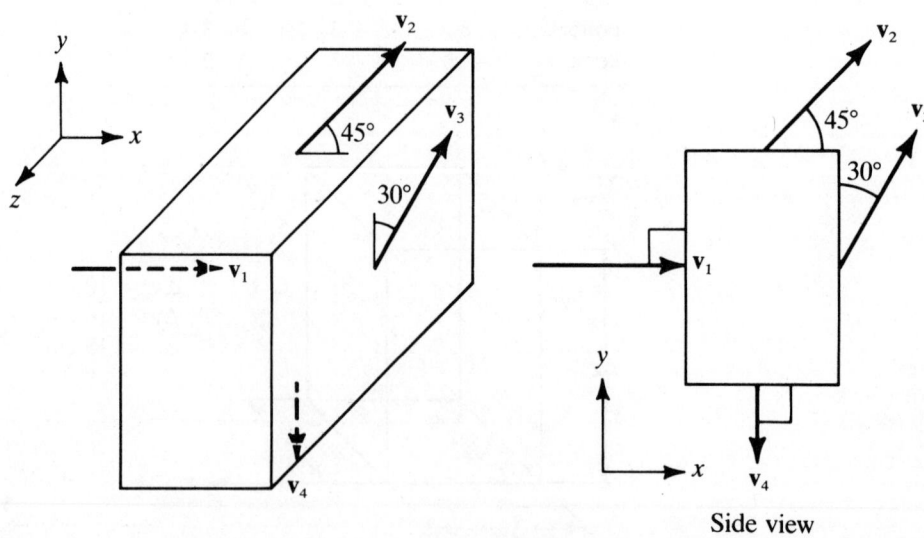

Side view

4.27 Water (with density of 1 g/cm³) is observed to be flowing with a velocity described by

$$\mathbf{v} = 2\hat{\mathbf{i}} + y\hat{\mathbf{j}} - z\mathbf{k}$$

Is there a source or sink at the origin? What is the value of the source or sink that lies within a cubic centimeter centered on the origin?

4.28 Water (with a density of 1 g/cm³) is observed to be flowing with a velocity described by $\mathbf{v} = x\hat{\mathbf{i}} + y\hat{\mathbf{j}} + 2\mathbf{k}$. Is there a sink or a source at the origin? What is the value of the sink or source that lies within a cubic centimeter centered on the origin?

4.29 Find the curl of the following vector fields:
(a) $\mathbf{v} = y\hat{\mathbf{i}} + x\hat{\mathbf{j}}$
(b) $\mathbf{v} = x\hat{\mathbf{i}} + y\hat{\mathbf{j}}$
(c) $\mathbf{v} = xz\hat{\mathbf{i}} - yz\hat{\mathbf{j}} + x^2\hat{\mathbf{k}}$
(d) $\mathbf{v} = y^2\hat{\mathbf{i}} + y^2\hat{\mathbf{j}} - 2yz\hat{\mathbf{k}}$
(e) $\mathbf{v} = xy^2z\hat{\mathbf{i}} + 2xyz\hat{\mathbf{j}} - 2xyz\hat{\mathbf{k}}$

4.30 Find $\mathbf{V} \times \mathbf{F}$ for the following forces:
(a) $\mathbf{F}_a = \hat{\mathbf{i}}(6abyz^3 - 20bx^3y^2) + \hat{\mathbf{j}}(6abxz^3 - 10bx^4y) + \hat{\mathbf{k}}18abxyz^2$
(b) $\mathbf{F}_b = \hat{\mathbf{i}}(18abyz^3 - 20bx^3y^2) + \hat{\mathbf{j}}(18abxz^3 - 10bx^4y) + \mathbf{k}6abxyz^2$
(c) $\mathbf{F}_c = \hat{\mathbf{i}}(6abx^2 + ab^2x^5) + \hat{\mathbf{j}}(7aby^2 + a^2b^3y) + \hat{\mathbf{k}}18abz^3$
(d) $\mathbf{F}_d = \hat{\mathbf{i}}F_x(x) + \hat{\mathbf{j}}F_y(y) + \hat{\mathbf{k}}F_z(z)$

4.31 Consider a square, 6 cm on a side, placed in a fluid with flow as sketched below. Use the edges of this square and Stokes's Theorem to describe the flow.

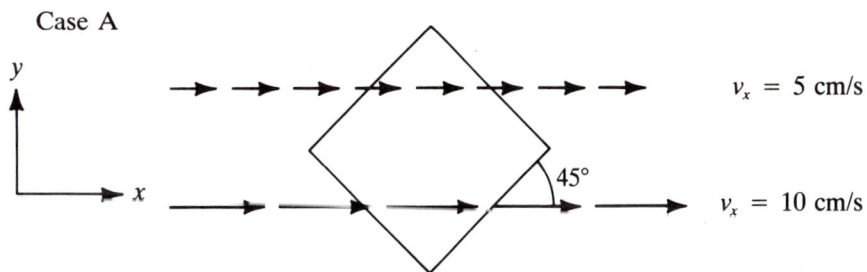

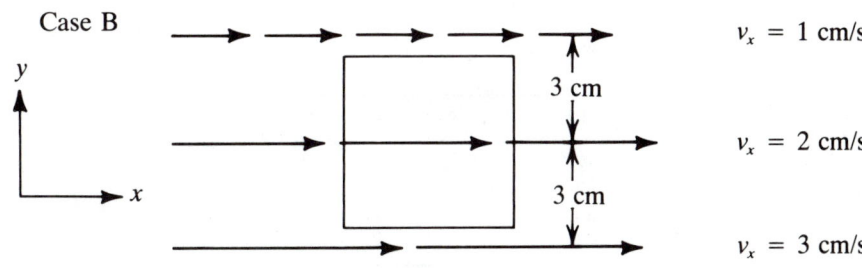

4.32 Using the numerical techniques of this chapter and Stokes's Theorem, find the average value of the curl of the fluid flow within the path sketched below if the flow has the average values given in the tables below.

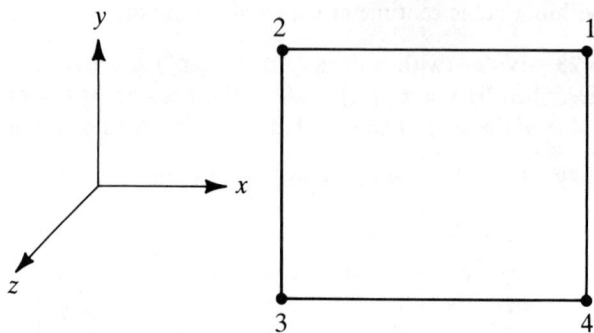

Path 1234 lies within the xy plane.

$$l_{12} = l_{34} = \Delta x = 10 \text{ cm}$$
$$l_{23} = l_{14} = \Delta y = 8 \text{ cm}$$

Case 1.

Side	v_x	v_y	v_z
12	1 m/s	0	0
23	2	0	0
34	3	0	0
41	2	0	0

Case 2.

Side	v_x	v_y	v_z
12	4 m/s	1	0
23	2	0	0
34	2	4	0
41	1	4	0

Case 3.

Side	v_x	v_y	v_z
12	1 m/s	1	0
23	2	1	0
34	3	2	0
41	4	2	0

4.33 Consider a square 6 cm on a side placed in a fluid with flow as sketched below. Use the edges of this square to demonstrate Stokes's Theorem.

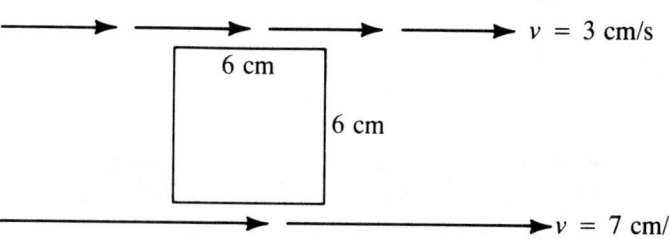

4.34 Quantum mechanical angular momentum operators are given by:

$$L_x = -i\left(y\frac{\partial}{\partial z} - z\frac{\partial}{\partial y}\right)$$

$$L_y = -i\left(z\frac{\partial}{\partial x} - x\frac{\partial}{\partial z}\right)$$

$$L_z = -i\left(x\frac{\partial}{\partial y} - y\frac{\partial}{\partial x}\right)$$

Show that $L_x L_y - L_y L_x = iL_z$ and that $\mathbf{L} \times \mathbf{L} = i\mathbf{L}$.

5

MOTION IN THREE DIMENSIONS

5.1 Introduction

Earlier we solved the equation $F = ma$ (or $F = m\ddot{x}$) for motion along a straight line. We considered a constant force F_0, a force that depended only on time $F(t)$, position $F(x)$, or velocity $F(v)$. We have even considered a force dependent on all three: $F(t, x, v)$, the forced harmonic oscillator. But even then the force—and the resulting motion—was restricted to a straight line.

We now want to consider the general case of $\mathbf{F} = m\mathbf{a}$ where the force $\mathbf{F}$ and the motion it causes can have components in all three dimensions. The most general form of the force will depend on time, position, and velocity—$\mathbf{F}(t, \mathbf{r}, \dot{\mathbf{r}})$. Of course $\mathbf{r}$ and $\dot{\mathbf{r}}$ are vectors, so this is really a form of shorthand. We could more explicitly write $\mathbf{F}(t, x, y, z, \dot{x}, \dot{y}, \dot{z})$. $\mathbf{F}$ is still a vector, so we still have a form of shorthand. We could very explicitly write:

$$m\ddot{x} = F_x(t, x, y, z, \dot{x}, \dot{y}, \dot{z}) \tag{5.1.1a}$$

$$m\ddot{y} = F_y(t, x, y, z, \dot{x}, \dot{y}, \dot{z}) \tag{5.1.1b}$$

$$m\ddot{z} = F_z(t, x, y, z, \dot{x}, \dot{y}, \dot{z}) \tag{5.1.1c}$$

We have three *coupled* second-order differential equations! We might only be able to solve this general case with a high-speed computer, but Nature is often not that difficult (for which we can be thankful). Many situations that are of considerable interest turn out to be *special* cases of the above equations, which *are* solvable in closed form.

5.2 Separable Forces

The easiest type of three-dimensional force to handle is a *separable* force where:

$$m\ddot{x} = F_x(t, x, \dot{x}) \tag{5.2.1a}$$

$$m\ddot{y} = F_y(t, y, \dot{y}) \tag{5.2.1b}$$

$$m\ddot{z} = F_z(t, z, \dot{z}) \tag{5.2.1c}$$

That is, the force in a particular direction depends on the component of position *in that direction*, the component of velocity *in that direction* and, perhaps, on time. Equations 5.2.1 are no longer coupled. Now we are simply solving three separate equations like those we have already handled. We have three separate one-dimensional problems to solve. This is no more difficult than the situation we faced in Chapter 2—for, indeed, it is exactly the same situation!

First consider a projectile moving in a uniform gravitational field in the absence of air resistance. Then Eqs. 5.2.1 become

$$m\ddot{x} = 0 \tag{5.2.2a}$$

$$m\ddot{y} = 0 \tag{5.2.2b}$$

$$m\ddot{z} = -mg \tag{5.2.2c}$$

All three of these are examples of constant acceleration, so we can write their solutions by inspection:

$$v_x = v_{0x} \tag{5.2.3a}$$

$$v_y = v_{0y} \tag{5.2.3b}$$

$$v_z = v_{0z} - gt \tag{5.2.3c}$$

and

$$x = x_0 + v_{0x}t \tag{5.2.4a}$$

$$y = y_0 + v_{0y}t \tag{5.2.4b}$$

$$z = z_0 + v_{0z}t - \tfrac{1}{2}gt^2 \tag{5.2.4c}$$

Or, we can write these as vectors:

$$\mathbf{v} = \mathbf{v}_0 - gt\hat{\mathbf{k}} \tag{5.2.5}$$

and

$$\mathbf{r} = \mathbf{r}_0 + \mathbf{v}_0 t - \tfrac{1}{2}gt^2\hat{\mathbf{k}} \tag{5.2.6}$$

Let us consider a projectile starting from the origin of a coordinate system so $\mathbf{r}_0 = 0$, oriented so $v_{0y} = 0$. Such orientation of the coordinate system does nothing to reduce the generality of this motion, but it makes y in Eq. (5.2.4b) equal to 0. We can immediately see that the motion of a ballistic trajectory is confined to the xz plane. Such careful choice of coordinate system greatly reduces the effort required for a solution.

We can solve for t in Eq. 5.2.4a:

$$t = \frac{x}{v_{0x}} \tag{5.2.7}$$

and substitute this into Eq. 5.2.4c:

$$z = v_{0z}\left(\frac{x}{v_{0x}}\right) - \frac{1}{2}g\left(\frac{x}{v_{0x}}\right)^2 \tag{5.2.8}$$

to obtain an equation of the path or *trajectory* taken in the xz plane. Inspection of Eq. 5.2.8 shows it to be some sort of parabola. We can rewrite this as:

$$\left(x - \frac{v_{0z}v_{0x}}{g}\right)^2 = -2\frac{v_{0x}^2}{g}\left(z - \frac{v_{0z}^2}{2g}\right) \tag{5.2.9}$$

from which we can more readily see that it is a parabola, concave downward, with a maximum altitude of

$$z_{max} = \frac{v_{0z}^2}{2g} \tag{5.2.10}$$

since

$$\left(z - \frac{v_{0z}^2}{2g}\right)$$

must remain negative because the left side of Eq. 5.2.9 is squared and, therefore, always positive. It crosses the horizontal axis (where z has the value of zero) at

$$x_{max} = 2\frac{v_{0z}v_{0x}}{g} \tag{5.2.11}$$

which is also called the *range* of the projectile.

In terms of an initial velocity $\mathbf{v}_0$, which is inclined an angle θ above the horizontal, the initial velocity components are

$$v_{0x} = v_0 \cos\theta \qquad v_{0z} = v_0 \sin\theta$$

so that the range may be written in terms of the initial speed v_0 and angle θ as

$$x_{max} = 2\frac{v_0^2 \sin\theta \cos\theta}{g} \tag{5.2.11a}$$

SECTION 5.2 / SEPARABLE FORCES

Since $\sin \theta = \cos(\pi/2 - \theta)$ or $\cos \theta = \sin(\pi/2 - \theta)$, it is clear that the ranges for a projectile fired at some angle θ and another fired at the complementary angle $(\pi/2 - \theta)$ will be identical (assuming they are fired with the same initial speed, of course!). That is, projectiles fired at 30° and 60° hit the ground at the same place; projectiles fired at 37° and 53° have the same range.

Since $\sin 2\theta = 2 \sin \theta \cos \theta$, we can rewrite the range equation as

$$x_{max} = \frac{v_0^2 \sin 2\theta}{g} \qquad (5.2.11b)$$

from which we can more readily see that the maximum range occurs for a firing angle of 45°.

We can approximate air resistance by including a *linear* velocity term in the force as

$$m\ddot{\mathbf{r}} = -mg\hat{\mathbf{k}} - c\mathbf{v} \qquad (5.2.12)$$

or

$$m\ddot{x} = -cv_x \qquad (5.2.13a)$$
$$m\ddot{y} = -cv_y$$
$$m\ddot{z} = -mg - cv_z \qquad (5.2.13b)$$

We shall still allow our projectile to start from the origin with no velocity in the y direction. Now we again have two uncoupled, separate, independent differential equations (we've already solved them in Chapter 2). We can simply write down our results from there:

$$v_x = v_{0x} e^{-(c/m)t} \qquad (5.2.14)$$

$$x = \frac{m}{c} v_{0x}(1 - e^{-(c/m)t}) \qquad (5.2.15)$$

and

$$v_z = \left(\frac{mg}{c} + v_{0z}\right)e^{-(c/m)t} - \frac{mg}{c} \qquad (5.2.16)$$

$$z = \left(\frac{m^2 g}{c^2} + \frac{m v_{0z}}{c}\right)(1 - e^{-(c/m)t}) - \frac{mg}{c}t \qquad (5.2.17)$$

Equations 5.2.15 and 5.2.17 give us the position as a function of time. Just as in the frictionless case, in order to get an equation in only x and z, we can solve for t from Eq. 5.2.15:

$$t = \frac{m}{c} \ln \frac{m v_{0x}}{m v_{0x} - cx} \qquad (5.2.18)$$

and substitute this into Eq. 5.2.17 to get

$$z = \left(\frac{m^2 g}{c^2} + \frac{mv_{0z}}{c}\right)\left(\frac{cx}{mv_{0x}}\right) - \frac{mg}{c}\frac{m}{c}\ln\frac{mv_{0x}}{mv_{0x} - cx}$$

$$z = \left(\frac{mg}{cv_{0x}} + \frac{v_{0z}}{v_{0x}}\right)x - \frac{m^2 g}{c^2}\ln\left(\frac{mv_{0x}}{mv_{0x} - cx}\right) \quad (5.2.19)$$

We can rewrite this and expand it as a power series to understand it better:

$$\begin{aligned}
z &= \left(\frac{mg}{cv_{0x}} + \frac{v_{0z}}{v_{0x}}\right)x + \frac{m^2 g}{c^2}\ln\left(1 - \frac{cx}{mv_{0x}}\right) \\
&= \left(\frac{mg}{cv_{0x}} + \frac{v_{0z}}{v_{0x}}\right)x + \frac{m^2 g}{c^2}\left[\left(\frac{-cx}{mv_{0x}}\right) - \frac{1}{2}\left(\frac{cx}{mv_{0x}}\right)^2 \right. \\
&\quad \left. + \frac{1}{3}\left(\frac{-cx}{mv_{0x}}\right)^3 - \frac{1}{4}\left(\frac{cx}{mv_{0x}}\right)^4 + \cdots\right] \\
&= \left(\frac{mg}{cv_{0x}} + \frac{v_{0z}}{v_{0x}}\right)x - \frac{mg}{cv_{0x}}x - \frac{1}{2}\frac{g}{v_{0x}^2}x^2 - \frac{1}{3}\frac{cg}{mv_{0x}^3}x^3 - \cdots
\end{aligned}$$

$$z = \frac{v_{0z}}{v_{0x}}x - \frac{1}{2}\frac{g}{v_{0x}^2}x^2 - \frac{1}{3}\frac{cg}{mv_{0x}^3}x^3 - \cdots \quad (5.2.20)$$

This looks like Eq. 5.2.8 as long as the position x or air resistance coefficient c is small enough that the higher-ordered terms may be neglected. But these higher-ordered terms, in general, cause the value of z (the altitude) to drop off faster as x or c increases. The trajectory starts off *looking* like a parabola but gets foreshortened as we look at larger and larger x. Figure 5.2.1 is a sketch of this trajectory.

We can see from Eq. 5.2.15 that x asymptotically approaches the maximum value of $x_{\max} = mv_{0x}/c$ if the time of flight is long enough. On the other hand, for a short time of flight (short compared to the characteristic time m/c), we expect the deviation from the frictionless case to be small. We

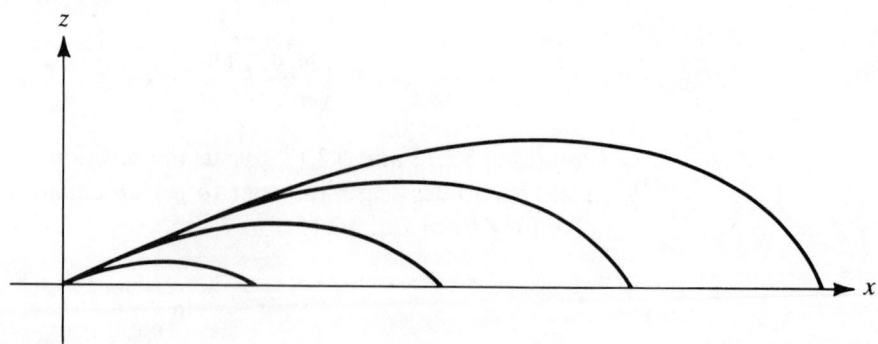

Figure 5.2.1 Ballistic trajectories with air resistance.

can set $z = 0$ in Eq. 5.2.20 and solve for x to get a first-order correction to x_{max}, the horizontal range. The result of that is

$$x_{max} = 2\frac{v_{0x}v_{0z}}{g} - \frac{8}{3}c\frac{cv_{0z}^2 v_{0z}}{mg^2} + \cdots \qquad (5.2.21)$$

You will probably be able to calculate x_{max}—or the entire trajectory for that matter—more easily using the numerical techniques you have already learned.

5.3 Three-Dimensional Harmonic Oscillator

A useful approximation of the motion of an atom in a crystal with cubic structure is a mass m held in place by three mutually perpendicular sets of springs, as sketched in Figure 5.3.1.

If there is no damping, the oscillations are given by

$$m\ddot{x} = -k_x x \qquad (5.3.1a)$$

$$m\ddot{y} = -k_y y \qquad (5.3.1b)$$

$$m\ddot{z} = -k_z z \qquad (5.3.1.c)$$

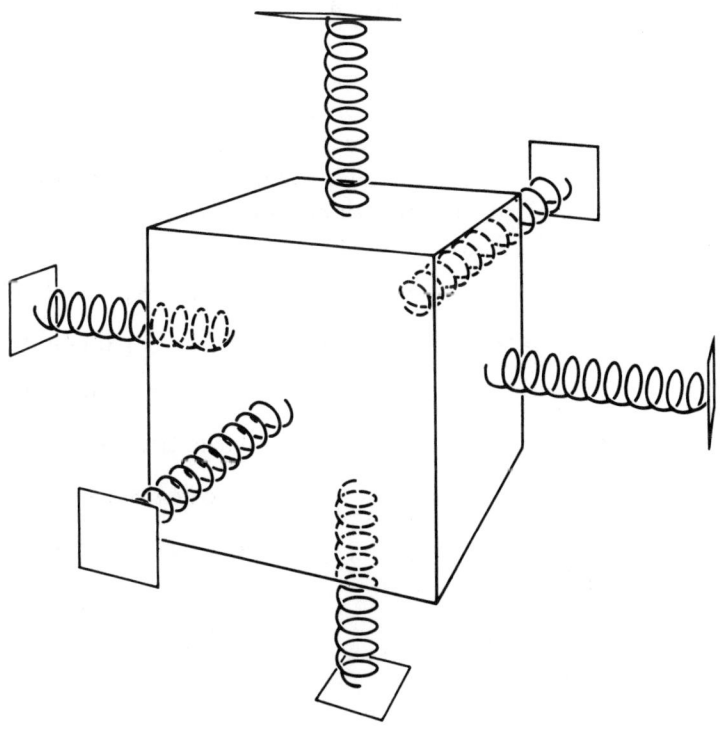

Figure 5.3.1 Three-dimensional harmonic oscillator.

and, in general, the three spring constants k_x, k_y, and k_z need not be equal. Each of these equations describes a simple harmonic oscillator so we can immediately write:

$$x = A \cos(\omega_x t + \alpha) \qquad (5.3.2a)$$

$$y = B \cos(\omega_y t + \beta) \qquad (5.3.2b)$$

$$z = C \cos(\omega_z t + \gamma) \qquad (5.3.2c)$$

where $\omega_x = \sqrt{k_x/m}$, $\omega_y = \sqrt{k_y/m}$, and $\omega_z = \sqrt{k_z/m}$. If there is no particular relationship between the ω's (or between the k's), then we can say only that the motion is confined to a rectangular solid of dimensions $2A \times 2B \times 2C$ and that the mass m will eventually come arbitrarily close to any chosen point in that box.

If, however, the ω's are related by integers through

$$\frac{\omega_x}{n_x} = \frac{\omega_y}{n_y} = \frac{\omega_z}{n_z} \qquad (5.3.3)$$

they are said to be *commensurable* and the motion of the mass either repeats itself or follows a closed path. If one of the amplitudes of Eq. 5.3.2 is also zero, then the path taken is a closed path in a plane. The path taken is then called a *Lissajou figure*. A few such paths are sketched in Figure 5.3.2 for the case of $z = 0$.

If the ω's (or k's) are identical, we say the system is isotropic—the mass feels the same force per displacement in any direction. We can write the force equation for this case as

$$\mathbf{F} = m\ddot{\mathbf{r}} = -k\mathbf{r} \qquad (5.3.4)$$

This is an example of an interesting and quite important class of forces, known as central forces, for which we can write $\mathbf{F} = F(\mathbf{r})\hat{\mathbf{r}}$. We shall study central forces in detail in Chapter 6. For the moment, though, it is instructive to look at the three-dimensional isotropic harmonic oscillator by rewriting Eqs. 5.3.2 as

$$x = A \cos(\omega t + \alpha) = A_1 \sin \omega t + A_2 \cos \omega t \qquad (5.3.5a)$$

$$y = B \cos(\omega t + \beta) = B_1 \sin \omega t + B_2 \cos \omega t \qquad (5.3.5b)$$

$$z = C \cos(\omega t + \gamma) = C_1 \sin \omega t + C_2 \cos \omega t \qquad (5.3.5c)$$

where we have made use of a basic trigonometric expansion in

$$A \cos(\omega t + \alpha) = A(\cos \omega t \cos \alpha - \sin \omega t \sin \alpha)$$

$$= (A \cos \alpha) \cos \omega t + (-A \sin \alpha) \sin \omega t$$

and then set

$$A_1 = A \cos \alpha \qquad (5.3.6a)$$

and

$$A_2 = -A \sin \alpha \qquad (5.3.6b)$$

$x = \sin 2t$
$y = \sin (3t + \frac{\pi}{4})$

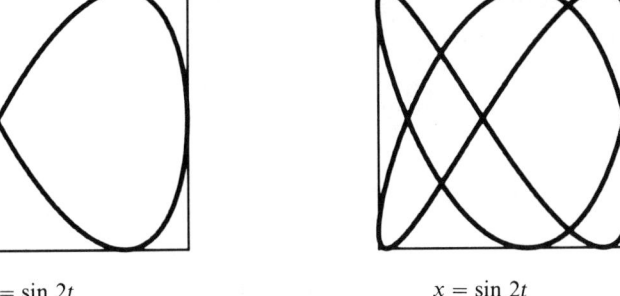
$x = \sin 2t$
$y = \sin (3t + \frac{\pi}{12})$

$x = \sin (2t + \frac{\pi}{6})$
$y = \sin (3t + \frac{\pi}{4})$

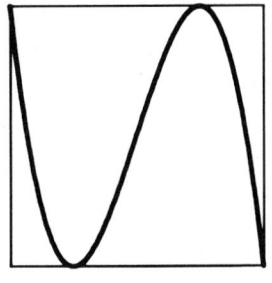
$x = \sin t$
$y = (\sin 3t)$

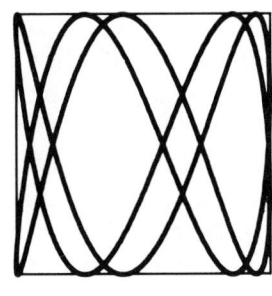
$x = (\sin t + \frac{\pi}{4})$
$y = \sin (2.5t + \frac{\pi}{4})$

$x = \sin (1.5t + \frac{\pi}{4})$
$y = \sin (2t + \frac{\pi}{4})$

Figure 5.3.2 Lissajou figures.

and similarly for the y and z equations. To determine what the *trajectory* or *path* of the motion really is—rather than simply being able to locate it from Eqs. 5.3.5 that use time as a parameter—we need to solve for, say, z in terms of x and y. We can multiply Eq. 5.3.5a by B_1, and Eq. 5.3.5b by A_1, and then subtract:

$$\begin{aligned} B_1 x &= A_1 B_1 \sin \omega t + A_2 B_1 \cos \omega t \\ A_1 y &= A_1 B_1 \sin \omega t + A_1 B_2 \cos \omega t \\ \hline B_1 x - A_1 y &= (A_2 B_1 - A_1 B_2) \cos \omega t \end{aligned} \quad (5.3.7)$$

$$\cos \omega t = \frac{B_1 x - A_1 y}{A_2 B_1 - A_1 B_2}$$

Likewise, we can solve for $\sin \omega t$ by multiplying Eq. 5.3.5a by B_2 and Eq. 5.3.5b by A_2 and then subtracting:

$$\begin{aligned} B_2 x &= A_1 B_2 \sin \omega t + A_2 B_2 \cos \omega t \\ A_2 y &= A_2 B_1 \sin \omega t + A_2 B_2 \cos \omega t \\ \hline B_2 x - A_2 y &= (A_1 B_2 - A_2 B_1) \sin \omega t \end{aligned} \quad (5.3.8)$$

$$\sin \omega t = \frac{B_2 x - A_2 y}{A_1 B_2 - A_2 B_1}$$

These can now be substituted into Eq. 5.3.5c for z:

$$z = C_1\left(\frac{B_2 x - A_2 y}{A_1 B_2 - A_2 B_1}\right) + C_2\left(\frac{B_1 x - A_1 y}{A_2 B_1 - A_1 B_2}\right)$$

$$z = \left(\frac{B_2 C_1}{A_1 B_2 - A_2 B_1} + \frac{B_1 C_2}{A_2 B_1 - A_1 B_2}\right) x$$

$$+ \left(\frac{-A_2 C_1}{A_1 B_2 - A_2 B_1} + \frac{-A_1 C_2}{A_2 B_1 - A_1 B_2}\right) y \quad (5.3.9)$$

$$z = (K_1) x + (K_2) y$$

where the values of K_1 and K_2 are the terms in parentheses in the equation above them. This is just the equation of a plane! No matter what the initial conditions, the motion of an isotropic harmonic oscillator is always confined to a plane that contains the origin. With a proper choice of coordinates, then, it behaves exactly as a two-dimensional harmonic oscillator. Since $\omega_x = \omega_y = \omega$, the path will be an ellipse—or a circle or straight line—both of which are special cases of an ellipse. In a later chapter we will discover that *all* central forces produce motion that is confined to a plane.

5.4 Potential Energy Function

In understanding motion in one dimension, we found it quite useful to define a potential energy function

$$V(x) = \int_x^{x_s} F(x)\, dx \quad (5.4.1)$$

from which we could readily obtain the force $F(x)$ as the negative of the derivative:

$$F(x) = -\frac{dV}{dx} \quad (5.4.2)$$

A potential energy function will prove similarly useful in our understanding of motion in three dimensions. But there are some important distinctions between one and three dimensions.

In one dimension, if the force depends only on the position, then Eq. 5.4.2 can be evaluated and $V(x)$ defined. The potential energy, you will recall, is the amount of *work* done by the force as it moves an object *from* position x *to* some reference position x_s. In three dimensions, we must specify the path

SECTION 5.4 / POTENTIAL ENERGY FUNCTION

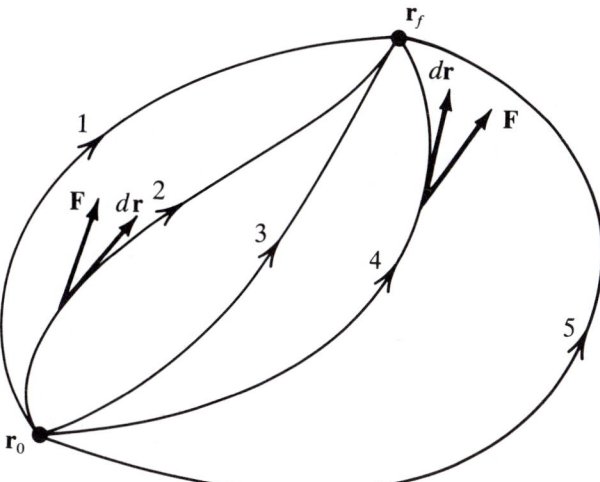

Figure 5.4.1 Work done along different paths.

along which the body moves. That is, the work done in moving from $\mathbf{r}_0$ to $\mathbf{r}_f$ is

$$W = \int_{\mathbf{r}_0}^{\mathbf{r}_f} \mathbf{F}(\mathbf{r}) \cdot d\mathbf{r} \qquad (5.4.3)$$

but we must specify the path to be taken in order to evaluate this *line integral*. It is conceivable that the work done along different paths from an initial position $\mathbf{r}_0$ to a final position $\mathbf{r}_f$ (or $\mathbf{r}_s$) could differ (see Figure 5.4.1). Then no potential energy function can be defined. Only when the work done depends solely on the initial and final positions, and is thus independent of the path taken, can a potential energy function be defined.

Before we pursue the details of such a potential energy function, let's digress and discuss the idea of line integrals in general.

EXAMPLE 5.4.1 Equation 5.4.3 is a line integral. The differential $d\mathbf{r}$ will always lie along a certain line—or path. In rectangular coordinates,

$$d\mathbf{r} = dx\hat{\mathbf{i}} + dy\hat{\mathbf{j}} + dz\hat{\mathbf{k}} \qquad (5.4.4)$$

but the values allowed for dx, dy, and dz will be only those corresponding to a certain line or path. The best way to explain this is by example.

Consider a mass m in a uniform gravitational field ($\mathbf{F} = -mg\hat{\mathbf{k}}$), initially located at $\mathbf{r}_0 = a\hat{\mathbf{k}}$, that is finally to be located at $\mathbf{r}_f = a\hat{\mathbf{i}}$. How much work is done by the force of gravity as the mass moves from $\mathbf{r}_0$ to $\mathbf{r}_f$ along the three paths indicated in Figure 5.4.2?

Verbally, we can describe path 1 as a vertical descent along the z-axis from $\mathbf{r}_0$ to $\mathbf{r}_1$ at the origin and then a horizontal path along the x-axis from $\mathbf{r}_1$ to $\mathbf{r}_f$. Along this first path, from $\mathbf{r}_0$ to $\mathbf{r}_1$, we hold x and y constant so $dx = 0$ and $dy = 0$.

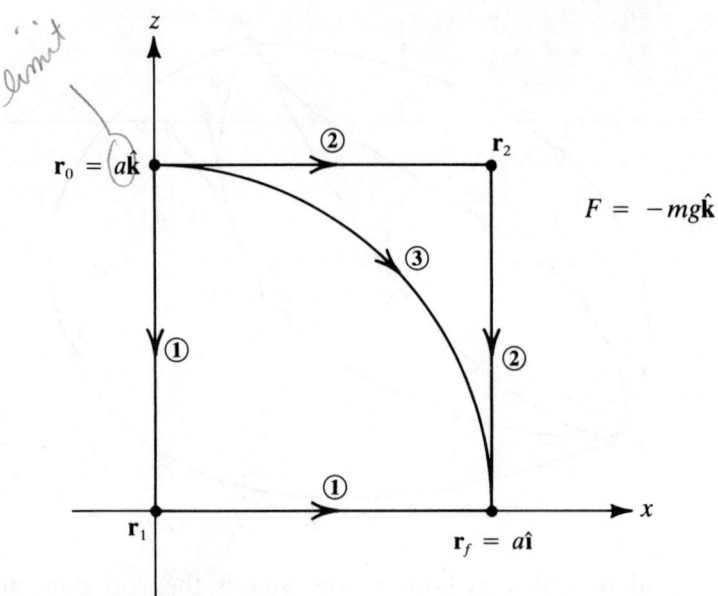

Figure 5.4.2 Work along a path.

We can write
$$d\mathbf{r} = dz\hat{\mathbf{k}} \tag{5.4.5}$$
and
$$W_{0\to 1} = \int_{\mathbf{r}_0}^{\mathbf{r}_1} \mathbf{F} \cdot d\mathbf{r}$$
$$= \int_{\mathbf{r}_0}^{\mathbf{r}_1} (-mg\hat{\mathbf{k}}) \cdot (dz\hat{\mathbf{k}})$$
$$= \int_a^0 (-mg)(dz)$$
$$= -mg \int_a^0 dz$$
$$= -mgz \Big|_a^0$$
$$W_{0\to 1} = mga \tag{5.4.6}$$

Along the second part of this path, y and z are held constant so $dy = 0$ and $dz = 0$. Therefore, we can write
$$d\mathbf{r} = dx\hat{\mathbf{i}}$$
and
$$W_{1\to f} = \int_{\mathbf{r}_1}^{\mathbf{r}_f} \mathbf{F} \cdot d\mathbf{r}$$
$$= \int_{\mathbf{r}_1}^{\mathbf{r}_f} (-mg\hat{\mathbf{k}}) \cdot (dx\hat{\mathbf{i}})$$
$$W_{1\to f} = 0 \tag{5.4.7}$$

Along path 1, the total work done is just the sum of Eqs. 5.4.6 and 5.4.7. Thus, the total work done by the force as the object moves along path 1 is

$$W_{\text{path 1}} = mga \tag{5.4.8}$$

Path 2 is quite similar. Along the first segment, from $\mathbf{r}_0$ to $\mathbf{r}_2$, we hold y and z at constant values so $dy = 0$ and $dz = 0$. Only x is free to change and we can write

$$d\mathbf{r} = dx\hat{\mathbf{i}} \tag{5.4.9}$$

As with the steps leading to Eq. 5.4.7, the force is perpendicular to this part of the path so it does no work along this segment.

$$W_{0 \to 2} = \int_{\mathbf{r}_2}^{\mathbf{r}_f} \mathbf{F} \cdot d\mathbf{r}$$

$$= \int_{\mathbf{r}_2}^{\mathbf{r}_f} (-mg\hat{\mathbf{k}}) \cdot (dx\hat{\mathbf{i}})$$

$$W_{0 \to 2} = 0 \tag{5.4.10}$$

From $\mathbf{r}_2$ we now move downward along the line $x = a$. That requires $dx = 0$ and $dy = 0$. We can write

$$d\mathbf{r} = dz\hat{\mathbf{k}} \tag{5.4.11}$$

and use this to evaluate the work done in going from $\mathbf{r}_2$ to $\mathbf{r}_f$:

$$W_{2 \to f} = \int_{\mathbf{r}_2}^{\mathbf{r}_f} \mathbf{F} \cdot d\mathbf{r}$$

$$= \int_{\mathbf{r}_2}^{\mathbf{r}_f} (-mg\hat{\mathbf{k}}) \cdot (dz\hat{\mathbf{k}})$$

$$= -mg \int_{a}^{0} dz$$

$$= -mgz \Big|_{a}^{0}$$

$$W_{2 \to f} = mga \tag{5.4.12}$$

As before, the total work done along path 2 is the sum of Eq. 5.4.10 and Eq. 5.4.12. Thus, the total work is

$$W_{\text{path 2}} = mga \tag{5.4.13}$$

Paths 1 and 2 may have been necessary (and, sufficient, I hope) to remind you of some details of line integrals that may have become rusty. But they're really fairly dull because the final result of Eqs. 5.4.8 and 5.10.13 may have been known from the beginning. Path number 3 is a little more interesting. Since the path is in the xz plane, we know y is constant (indeed, zero) so we have $dy = 0$. This path is along the quadrant of a circle. We might better see this in terms of polar coordinates as shown in Figure 5.4.3.

We can define this path by

$$r = a \tag{5.4.14}$$

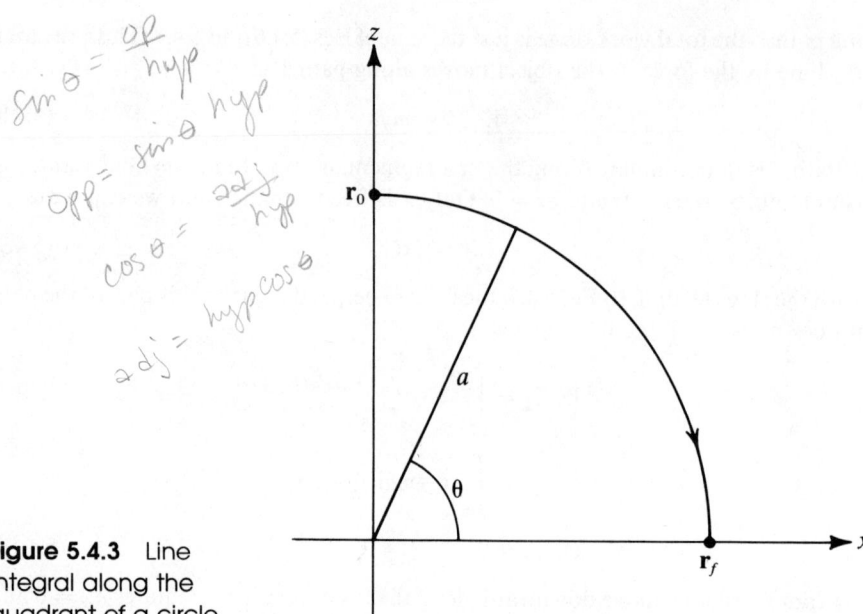

Figure 5.4.3 Line integral along the quadrant of a circle.

or, for rectangular coordinates x and z, this means

$$x = a \cos \theta \qquad z = a \sin \theta \qquad (5.4.15)$$

We want dx and dz, the changes in x and z, which are consistent with the constraints of Eq. 5.4.15. Taking differentials immediately gives

$$dx = -a \sin \theta \, d\theta \qquad dz = a \cos \theta \, d\theta \qquad (5.4.16)$$

which give an expression for $d\mathbf{r}$ as

$$d\mathbf{r} = (-a \sin \theta \, d\theta)\hat{\mathbf{i}} + (a \cos \theta \, d\theta)\hat{\mathbf{k}} \qquad (5.4.17)$$

or

$$d\mathbf{r} = a(-\hat{\mathbf{i}} \sin \theta + \hat{\mathbf{k}} \cos \theta) \, d\theta \qquad (5.4.17a)$$

The vector $\mathbf{r}_0$ corresponds to $\theta = \pi/2$, and $\mathbf{r}_f$ to $\theta = 0$. Now we can evaluate the work along this path:

$$W_3 = \int_{\mathbf{r}_0}^{\mathbf{r}_f} \mathbf{F} \cdot d\mathbf{r}$$

$$= \int_{\mathbf{r}_0}^{\mathbf{r}_f} (-mg\hat{\mathbf{k}}) \cdot (a)(-\hat{\mathbf{i}} \sin \theta + \hat{\mathbf{k}} \cos \theta \, d\theta)$$

$$= -mga \int_{\theta=\pi/2}^{\theta=0} \cos \theta \, d\theta$$

$$= [-mga]\left[\sin(0) - \sin\frac{\pi}{2}\right]$$

$$W_{\text{path 3}} = mga \qquad (5.4.18)$$

SECTION 5.4 / POTENTIAL ENERGY FUNCTION

So, we see from Eqs. 5.4.8, 5.4.13, and 5.4.18 that the work done by a uniform gravitational force gives every indication of being independent of the path. Such a force is called a *conservative* force. And this is comfortingly consistent with your earliest discussions of potential energy using $V = mgh$.

EXAMPLE 5.4.2 Lest you hastily jump to the conclusion that line integrals *don't* really depend on the path, let's make up a force, say:

$$\mathbf{F} = Az\hat{\mathbf{i}} + Bxz^2\hat{\mathbf{k}} \qquad (5.4.19)$$

and evaluate the work done as we move along the same three paths sketched in Figure 5.4.2 (note that A has units of N/m and B, N/m^3).

Along the first segment of path 1, from $\mathbf{r}_0$ to $\mathbf{r}_1$, x has a constant value of zero. So the force of Eq. 5.4.19 is simply:

$$\mathbf{F} = Az\hat{\mathbf{i}} \qquad (5.4.20)$$

As before, for this path, we have $dx = dy = 0$ or $d\mathbf{r} = dz\hat{\mathbf{k}}$. So, the work done is

$$W_{0\to 1} = \int_{\mathbf{r}_0}^{\mathbf{r}_1} \mathbf{F} \cdot d\mathbf{r} = \int_{\mathbf{r}_0}^{\mathbf{r}_1} (Az\hat{\mathbf{i}}) \cdot (dz\hat{\mathbf{k}}) = 0 \qquad (5.4.21)$$

We really could have gotten this "by inspection" since F has nonzero components only in the x direction and this segment of the path, along the z-axis, is perpendicular to the force.

For the second segment, from $\mathbf{r}_1$ to $\mathbf{r}_f$, the value of z is zero so the force is identically zero:

$$W_{1\to f} = 0 \qquad (5.4.22)$$

The force in Eq. 5.4.19 does *no* work as a body moves along path 1:

$$W_{\text{path 1}} = 0 \qquad (5.4.23)$$

Along the first segment of path 2, from $\mathbf{r}_0$ to $\mathbf{r}_2$, z is constant ($z = a$) but not zero. Thus, the force is now

$$\mathbf{F} = Aa\hat{\mathbf{i}} + Ba^2 x\hat{\mathbf{k}} \qquad (5.4.24)$$

As earlier, only x changes along this segment so $dy = dz = 0$ and $d\mathbf{r} = dx\hat{\mathbf{i}}$. Therefore,

$$W_{0\to 2} = \int_{\mathbf{r}_0}^{\mathbf{r}_2} \mathbf{F} \cdot d\mathbf{r}$$

$$= \int_{\mathbf{r}_0}^{\mathbf{r}_2} (Aa\hat{\mathbf{i}} + Ba^2 x\hat{\mathbf{k}}) \cdot (dx\hat{\mathbf{i}})$$

$$= \int_{x=0}^{x=a} Aa\, dx$$

$$= Aax \Big|_0^a$$

$$W_{0\to 2} = Aa^2 \qquad (5.4.25)$$

Along the second segment, from $\mathbf{r}_2$ to $\mathbf{r}_f$, only z changes so $dx = dy = 0$ and $d\mathbf{r} = dz\hat{\mathbf{k}}$. The variable x is a constant value ($x = a$), so the force is

$$\mathbf{F} = Az\hat{\mathbf{i}} + Baz^2\hat{\mathbf{k}} \quad (5.4.26)$$

and

$$W_{2 \to f} = \int_{\mathbf{r}_2}^{\mathbf{r}_f} \mathbf{F} \cdot d\mathbf{r}$$

$$= \int_{\mathbf{r}_2}^{\mathbf{r}_f} (Az\hat{\mathbf{i}} + Bax^2\hat{\mathbf{k}}) \cdot (dz\hat{\mathbf{k}})$$

$$= Ba \int_{z=a}^{z=0} z^2 \, dz$$

$$= \frac{B}{3} az^3 \Big|_a^0$$

$$W_{2 \to f} = -\frac{Ba^4}{3} \quad (5.4.27)$$

for

$$W_{\text{path 2}} = Aa^2 - \frac{Ba^4}{3} \quad (5.4.28)$$

That is, the work along path 2 is *considerably* different from that along path 1. We can immediately conclude that this particular force is *not* conservative. But let's continue with an evaluation of the work done along path 3, nonetheless.

Using polar coordinates as in the previous example is still the best way to evaluate the line integral for path 3. However, let's try a more cumbersome method by reducing this to an ordinary integral over, say, x (rather than over θ in the polar coordinate case). Along the quadrant of the circle that forms this path, we know:

$$x^2 + z^2 = a^2$$

or

$$z^2 = a^2 - x^2$$
$$z = \sqrt{a^2 - x^2} \quad (5.4.29)$$

Therefore, we can write the force as a function of x, only

$$\mathbf{F} = A\sqrt{a^2 - x^2}\,\hat{\mathbf{i}} + Bx(a^2 - x^2)\hat{\mathbf{k}} \quad (5.4.30)$$

But what of $d\mathbf{r}$ now? We still have $dy = 0$, of course. We need dz in terms of x and can get this by taking the differential of Eq. 5.4.29:

$$dz = \tfrac{1}{2}(a^2 - x^2)^{-1/2}(-2x\,dx)$$

$$dz = -\frac{x\,dx}{\sqrt{a^2 - x^2}} \quad (5.4.31)$$

SECTION 5.4 / POTENTIAL ENERGY FUNCTION

Now we can evaluate the work done along path 3 by

$$W_3 = \int_{r_0}^{r_f} \mathbf{F} \cdot d\mathbf{r}$$

$$= \int_{r_0(x=0)}^{r_f(x=a)} [(A\sqrt{a^2 - x^2}\hat{\mathbf{i}} + Bx(a^2 - x^2)\hat{\mathbf{k}})] \cdot \left[dx\hat{\mathbf{i}} - \frac{x\,dx\hat{\mathbf{k}}}{\sqrt{a^2 - x^2}} \right]$$

$$= \int_0^a (A\sqrt{a^2 - x^2} - Bx^2\sqrt{a^2 - x^2})\,dx$$

$$= \frac{A}{2}\left[x^2\sqrt{a^2 - x^2} + a^2 \sin^{-1}\left(\frac{x}{a}\right) \right]\Big|_0^a$$

$$+ B\left[\frac{x}{4}\sqrt{(a^2 - x^2)^3} + \frac{a^2}{8}\left(x\sqrt{a^2 - x^2} + a^2 \sin^{-1}\left(\frac{x}{a}\right) \right) \right]\Big|_0^a$$

$$W_{\text{path 3}} = \cdots = \frac{Aa^2\pi}{4} + \frac{Ba^4\pi}{16} = \frac{a^2\pi}{4}\left(A + \frac{Ba^2}{4} \right) \tag{5.4.32}$$

The work done as the body moves along path 3 is considerably different than our results for path 1 or path 2. This is simply further proof that the strange force given in Eq. 5.4.19 is, indeed, *not* conservative. Evaluation of the integrals for this case points out the great usefulness of a table of integrals. Using an integral table is *not* the mark of a poor mathematician. Rather, it indicates a physicist (or engineer or other scientist) who needs quick and ready solutions to complicated problems. You should expect to become very familiar with an integral table as this course progresses. Indeed, you may be familiar with one already!

Now that we're (again) familiar with line integrals, let's return to the question of finding a criterion to use in defining a potential energy function. Potential energy can be defined if and only if the work done in moving a body from initial position $\mathbf{r}_0$ to final position $\mathbf{r}_f$, such as

$$W = \int_{r_0}^{r_f} \mathbf{F} \cdot d\mathbf{r} \tag{5.4.3}$$

depends only on the endpoints $\mathbf{r}_0$ and $\mathbf{r}_f$; that is, if it is *independent* of the path.

We can make use of Stokes's Theorem in our criterion for conservative forces. If the work done in going from point A to point B in Figure 5.4.4 is independent of the path, then we know the net work done around a *closed* loop (from A to B and back again) must be zero. And this will be true for *any* closed path.

$$W_{\substack{\text{closed} \\ \text{path}}} = \oint_{\substack{\text{closed} \\ \text{path}}} \mathbf{F} \cdot d\mathbf{r} = 0 \tag{5.4.33}$$

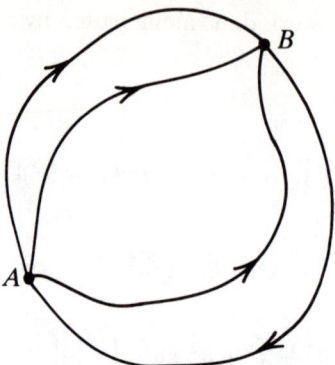

Figure 5.4.4 Work done along a *closed* path is zero for a conservative force.

From Stokes's Theorem we know

$$\oint_{\text{closed path}} \mathbf{F} \cdot d\mathbf{r} = \iint_{\text{surface}} (\mathbf{\nabla} \times \mathbf{F}) \cdot \hat{\mathbf{n}}\, dS = 0 \tag{5.4.34}$$

For this to be true for *any* surface, the integrand itself must be zero. That is,

$$\mathbf{\nabla} \times \mathbf{F} = 0 \tag{5.4.35}$$

for all conservative forces. This condition is both necessary and sufficient.

Whenever we encounter a force that satisfies this condition, $\mathbf{\nabla} \times \mathbf{F} = 0$, we can define a potential energy function as the work done by our force as it moves a test body *from* the position $\mathbf{r}$ *to* some standard reference position, $\mathbf{r}_s$.

$$V(\mathbf{r}) = \int_{\mathbf{r}}^{\mathbf{r}_s} \mathbf{F} \cdot d\mathbf{r} = -\int_{\mathbf{r}_s}^{\mathbf{r}} \mathbf{F} \cdot d\mathbf{r} \tag{5.4.36}$$

We can then define the components of the force in a manner similar to Eq. 5.4.2 for the one-dimensional case:

$$F_x = -\frac{\partial V}{\partial x} \tag{5.4.37a}$$

$$F_y = -\frac{\partial V}{\partial y} \tag{5.4.37b}$$

$$F_z = -\frac{\partial V}{\partial z} \tag{5.4.37c}$$

or

$$\mathbf{F} = -\mathbf{\nabla}V = -\frac{\partial V}{\partial x}\hat{\mathbf{i}} - \frac{\partial V}{\partial y}\hat{\mathbf{j}} - \frac{\partial V}{\partial z}\hat{\mathbf{k}} \tag{5.4.38}$$

SECTION 5.4 / POTENTIAL ENERGY FUNCTION

We can see that Eq. 5.4.35 is *necessary* by starting with Eqs. 5.4.37. Since the order of differentiation is irrelevant for any continuous function,

$$\frac{\partial F_x}{\partial y} = -\frac{\partial^2 V}{\partial y\, \partial x} = -\frac{\partial^2 V}{\partial x\, \partial y} = \frac{\partial F_y}{\partial x}$$

or

$$\frac{\partial F_x}{\partial y} - \frac{\partial F_y}{\partial x} = 0 \tag{5.4.39}$$

But

$$\frac{\partial F_x}{\partial y} - \frac{\partial F_y}{\partial x} = (\nabla \times \mathbf{F})_z = (\nabla \times \mathbf{F}) \cdot \hat{\mathbf{k}} \tag{5.4.40}$$

So Eq. 5.4.39 is simply a restatement of Eq. 5.4.35; it is a necessary condition for a potential energy function to exist.

But Eq. 5.4.35 is also *sufficient*. For *if* it holds, *then* the line integral of $\mathbf{F} \cdot d\mathbf{r}$ around any closed loop is zero. Thus, the work done in moving a body from point A to point B is path-independent and the potential energy function can be defined.

If the criterion of Eq. 5.4.35 is met; that is, if the curl of a force is zero, then a potential energy function exists from which the force can readily be derived using Eqs. 5.4.37 or Eq. 5.4.38. It is easy enough to determine if the curl of a force is zero. But how do we derive the potential energy function from the force?

We can begin with Eq. 5.4.36:

$$V(\mathbf{r}) = \int_{\mathbf{r}}^{\mathbf{r}_s} \mathbf{F} \cdot d\mathbf{r} \tag{5.4.36}$$

and *explicitly* carry out a line integral along any conveniently chosen path from a general position $\mathbf{r}$ or (x, y, z) to a conveniently chosen reference point $\mathbf{r}_s$ or (x_s, y_s, z_s).

Or we can begin with Eqs. 5.4.37 and do three separate ordinary integrals—being careful to watch what is variable and what is constant. In one dimension, we had

$$F = -\frac{dV}{dx}$$

from which we could define the potential energy as either

$$V = \int_x^{x_s} F\, dx = -\int_{x_s}^x F\, dx$$

in terms of a definite integral or

$$V = -\int F\, dx + C$$

in terms of an indefinite integral. The arbitrary constant C is a *constant of integration* and is quite necessary whenever an indefinite integral appears.

Likewise, we can write

$$F_x = -\frac{\partial V}{\partial x} \tag{5.4.41a}$$

which involves a *partial* derivative with respect to x in which y and z are treated as constants. If we now make this into an integral, we have

$$V = -\int F_x\, dx + C_1(y, z) \tag{5.4.42a}$$

The constant of integration is replaced by *any* function that does not depend on x; that is, it may very well be a function of y and z. Likewise,

$$F_y = -\frac{\partial V}{\partial y} \tag{5.4.41b}$$

yields

$$V = -\int F_y\, dy + C_2(x, z) \tag{5.4.42b}$$

and

$$F_z = -\frac{\partial V}{\partial z} \tag{5.4.41c}$$

yields

$$V = -\int F_z\, dz + C_3(x, y) \tag{5.4.42c}$$

But there is only a single potential energy function. The single potential energy function we seek is a potential energy function *consistent* with all three of the expressions above. Two examples may help make this clear.

EXAMPLE 5.4.3 Find the potential energy that corresponds to the force

$$\mathbf{F} = ax\hat{\mathbf{i}} + by\hat{\mathbf{j}} + cz\hat{\mathbf{k}}$$

Evaluation of the curl of this force shows that $\nabla \times \mathbf{F} = 0$, so a potential energy function does, indeed, exist.

We can, therefore, proceed to integrate each component of the force:

$$F_x = ax \qquad F_y = by \qquad F_z = cz$$

$$V = -\int ax\, dx + C_1(y, z) = -\frac{a}{2}x^2 + C_1(y, z)$$

$$V = -\int by\, dy + C_2(x, z) = -\frac{b}{2}y^2 + C_2(x, z)$$

$$V = -\int cz\, dz + C_3(x, y) = -\frac{c}{2}z^2 + C_3(x, y)$$

From these three expressions for V we find a single *consistent* equation of V:

$$V = -\frac{a}{2}x^2 - \frac{b}{2}y^2 - \frac{c}{2}z^2 + C$$

where C is just a constant, a number. You should check to see if this is correct by finding the force, starting from this potential using $\mathbf{F} = -\nabla V$.

EXAMPLE 5.4.4 Find the potential energy function associated with the force

$$\mathbf{F} = -yz\hat{\mathbf{i}} - xz\hat{\mathbf{j}} - xy\hat{\mathbf{k}}$$

As before, we first find that $\nabla \times \mathbf{F} = 0$ and, thus, assure ourselves that a potential energy function exists. Then we can proceed to integrate each component:

$$F_x = -yz \qquad F_y = -xz \qquad F_z = -xy$$

$$V = -\int(-yz)\,dx + C_1(y, z) = xyz + C_1(y, z)$$

$$V = -\int(-xz)\,dy + C_2(x, z) = xyz + C_2(x, z)$$

$$V = -\int(-xy)\,dz + C_3(x, y) = xyz + C_3(x, y)$$

From these expressions for V, we find a single *consistent* equation:

$$V = xyz + C$$

Check this by finding the force using the gradient.

Note that the potential energy is *not* given by $V = 3xyz$. This is *inconsistent* with the three separate expressions and it gives the wrong value for the force when the gradient is found.

5.5 Motion in Electromagnetic Fields

Electric charges at rest produce forces on a body of charge q that can be described in terms of an *electric field* $\mathbf{E}(\mathbf{r})$ by

$$\mathbf{F}_{el}(\mathbf{r}) = q\mathbf{E}(\mathbf{r}) \tag{5.5.1}$$

Moving charges or currents produce additional forces that we can describe in terms of another field, the magnetic induction $\mathbf{B}(\mathbf{r})$. These forces depend on the motion of our test body of charge q—they are velocity-dependent forces given by

$$\mathbf{F}_m = q\mathbf{v} \times \mathbf{B} \tag{5.5.2}$$

For convenience, we separate these two and say that Eq. 5.5.1 describes the *electric* force and Eq. 5.5.2 describes the *magnetic* force. This distinction is only for our convenience, as the two equations describe two aspects of the single *electromagnetic* force. We can also write the total force (sometimes called the Lorentz force) on a charged particle in the presence of both **E** and **B** fields as

$$\mathbf{F} = q\mathbf{E} + q\mathbf{v} \times \mathbf{B} \tag{5.5.3}$$

From electromagnetic theory, we find, for an electrostatic field, that

$$\mathbf{\nabla} \times \mathbf{E} = 0 \tag{5.5.4}$$

Therefore, we can immediately write

$$\mathbf{\nabla} \times \mathbf{F}_{el} = 0 \tag{5.5.5}$$

and conclude that the electrostatic force is conservative. Thus, we can define a potential energy $V(\mathbf{r})$ such that

$$\mathbf{F}_{el}(\mathbf{r}) = -\mathbf{\nabla} V(\mathbf{r}) \tag{5.4.38}$$

Parallel to Eq. 5.5.1, we can also define a *scalar electric potential* $\phi(\mathbf{r})$ in electromagnetic theory whereby

$$\mathbf{E}(\mathbf{r}) = -\mathbf{\nabla}\phi(\mathbf{r}) \tag{5.5.6}$$

Just as the electric field **E** is the force per unit charge, the scalar electric potential ϕ is the potential energy per unit charge (only electro*statics* is being considered).

Now, what of the magnetic force? Since the magnetic force is always *perpendicular* to the motion, it never does any work on our charged test body. Therefore, energy is conserved here as well. But since this magnetic force is velocity-dependent, we cannot define a potential energy function.

Thus, our ideas of conservation of energy can be applied to the entire electromagnetic force. We shall now look at a few very special cases of this electromagentic force.

EXAMPLE 5.5.1 *Uniform electric field; zero magnetic induction* **B**. We choose the z-axis to lie parallel to the direction of **E** so

$$\mathbf{E} = E_0 \hat{\mathbf{k}} \qquad \mathbf{B} = 0$$

Thus,

$$\mathbf{F} = qE_0 \hat{\mathbf{k}}$$

and with a constant force we can immediately write down the equations of motion

$$v_x = v_{x0} \qquad x = x_0 + v_{x0} t$$

$$v_y = v_{y0} \qquad y = y_0 + v_{y0} t$$

$$v_z = v_{z0} + \frac{qE_0}{m} t \qquad z = z_0 + v_{z0} t + \frac{1}{2} \frac{qE_0}{m} t^2$$

SECTION 5.5 / MOTION IN ELECTROMAGNETIC FIELDS

These have the same form as a projectile thrown in a uniform gravitational field because, of course, the problem really is the same.

EXAMPLE 5.5.2 *Uniform magnetic induction B; zero electric field.* We choose the z-axis parallel to **B**, so that

$$\mathbf{B} = B\hat{\mathbf{k}} \qquad \mathbf{E} = 0$$

Then,

$$\mathbf{F} = qB_0 \mathbf{v} \times \hat{\mathbf{k}} \tag{5.5.7}$$

There are several ways to solve for the motion produced by this force. We begin by writing the force explicitly:

$$\mathbf{F} = qB_0 (\dot{x}\hat{\mathbf{i}} + \dot{y}\hat{\mathbf{j}} + \dot{z}\hat{\mathbf{k}}) \times \hat{\mathbf{k}} = m\ddot{\mathbf{r}}$$

$$\mathbf{F} = qB_0 (\dot{y}\hat{\mathbf{i}} - \dot{x}\hat{\mathbf{j}}) = m(\ddot{x}\hat{\mathbf{i}} + \ddot{y}\hat{\mathbf{j}} + \ddot{z}\hat{\mathbf{k}})$$

As always, this single vector equation is really three equations:

$$\ddot{x} = \frac{qB_0}{m}\dot{y} \tag{5.5.8a}$$

$$\ddot{y} = \frac{-qB_0}{m}\dot{x} \tag{5.5.8b}$$

$$\ddot{z} = 0 \tag{5.5.8c}$$

This last equation says that there is no z-component of acceleration. Therefore,

$$v_z = v_{z0} \tag{5.5.9a}$$

$$z = z_0 + v_{z0} t \tag{5.5.9b}$$

The two remaining equations form a set of *coupled* second-order differential equations. There are several different methods we might choose to solve them. One very useful technique is to multiply one of them, say Eq. 5.5.8b, by $i = \sqrt{-1}$, add the two equations, and solve the resulting *complex variable* differential equation:

$$\ddot{x} + i\ddot{y} = \frac{qB_0}{m}\dot{y} - i\frac{qB_0}{m}\dot{x} = -i\frac{qB_0}{m}(\dot{x} + i\dot{y}) \tag{5.5.10}$$

In terms of complex variables, we can write

$$r = x + iy \tag{5.5.11a}$$

$$\dot{r} = \dot{x} + i\dot{y} \tag{5.5.11b}$$

$$\ddot{r} = \ddot{x} + i\ddot{y} \tag{5.5.11c}$$

Thus, Eq. 5.5.10 can be written as

$$\ddot{r} = -i\frac{qB_0}{m}\dot{r} \tag{5.5.12}$$

We can then immediately integrate this to find

$$\dot{r} = C_1 e^{-i(qB_0/m)t} \tag{5.5.13}$$

Integration of this immediately yields

$$r = C_2 e^{-i(qB_0/m)t} + C_3 \tag{5.5.14}$$

where

$$C_2 = -i \frac{C_1 m}{qB_0}$$

C_2 and C_3, of course, will eventually depend on the initial conditions. From our definition of r in Eq. 5.5.11, we can write

$$x = \text{Re}[r] = \text{Re}\left[C_2 e^{-i(qB_0/m)t} + C_3 \right] \tag{5.5.15a}$$

and

$$y = \text{Im}[r] = \text{Im}\left[C_2 e^{-i(qB_0/m)t} + C_3 \right] \tag{5.5.15b}$$

That is, x is the *real* part of r and y is the *imaginary* part of r. Since C_2 and C_3 are unknown constants at this point, we shall use a considerable amount of insight and write

$$C_2 = A e^{-i\phi} \quad \text{and} \quad C_3 = a + ib \tag{5.5.16}$$

Thus,

$$x = a + A\cos(\omega t + \phi) \tag{5.5.17a}$$

and

$$y = b - A\sin(\omega t + \phi) \tag{5.5.17b}$$

where $\omega = qB_0/m$. We see, then, that the xy-motion of our charged particle is simply circular motion of amplitude A centered about $x = a$, $y = b$. Of very special interest is the fact that the angular frequency ω or the period $T = 2\pi/\omega$ is independent of the velocity—it depends only on the strength of the field, B_0, and the charge-to-mass ratio, q/m. This unusual result has immense application for many particle accelerators, the cyclotron in particular—indeed, it is only because of this that they can operate at all.

The radius of this circular motion, A, depends on B, q, m, and the initial velocity or energy. If all but one of these variables is known, then a measurement of the radius A is enough to determine the remaining variable. In particular, a mass spectrometer often uses charged ions of known charge q and known energy E passing through a known uniform magnetic induction B. The particles then move in a circle of measurable radius A. From this we can determine their mass m with great precision.

SECTION 5.5 / MOTION IN ELECTROMAGNETIC FIELDS

Coupling this motion with constant velocity in the z direction, we see that the most general motion for a particle is traveling along a right circular helix. The position, pitch, and diameter of the helix depend on the initial conditions.

EXAMPLE 5.5.3 *Uniform electric field* **E** *and uniform magnetic induction* **B**. If **E** and **B** are parallel, the resulting motion is just a superposition of that from the previous two examples. This is because the new force is just a superposition of the earlier two and because Newton's Second Law is linear. Therefore, we shall consider **B** and **E** perpendicular to each other:

$$\mathbf{B} = B_0 \hat{\mathbf{k}} \qquad \mathbf{E} = E_0 \hat{\mathbf{i}} \qquad (5.5.18)$$

The force is now given by

$$\mathbf{F} = q\mathbf{E} + q\mathbf{v} \times \mathbf{B} \qquad (5.5.3)$$

or

$$\mathbf{F} = qE_0 \hat{\mathbf{i}} + q \begin{vmatrix} \hat{\mathbf{i}} & \hat{\mathbf{j}} & \hat{\mathbf{k}} \\ \dot{x} & \dot{y} & \dot{z} \\ 0 & 0 & B_0 \end{vmatrix} = qE_0 \hat{\mathbf{i}} + \hat{\mathbf{i}}(qB_0 \dot{y}) - \hat{\mathbf{j}}(qB_0 \dot{x}) \qquad (5.5.19)$$

Of course this vector equation really means

$$\ddot{x} = \frac{qE_0}{m} + \frac{qB_0}{m} \dot{y} \qquad (5.5.20a)$$

$$\ddot{y} = -\frac{qB_0}{m} \dot{x} \qquad (5.5.20b)$$

$$\ddot{z} = 0 \qquad (5.5.20c)$$

The motion in the z direction is straightforward if uninteresting—motion with constant speed:

$$v_z = v_{z0}$$
$$z = z_0 + v_{z0} t \qquad (5.5.21)$$

Equations 5.5.20a and 5.5.20b again present us with a set of *coupled* second-order differential equations. Our previous excursion into the realm of complex variables proved successful, so let's try it again:

$$\ddot{r} = \ddot{x} + i\ddot{y} = \frac{qE_0}{m} + \frac{qB_0}{m} \dot{y} + i\left(-\frac{qB_0}{m} \dot{x}\right)$$

$$= \frac{qE_0}{m} - i\frac{qB_0}{m} (\dot{x} + i\dot{y}) \qquad (5.5.22)$$

$$= \frac{qE_0}{m} - i\frac{qB_0}{m} \dot{r}$$

The solution to this *inhomogeneous* differential equation is

$$\dot{r} = C_1 e^{-i(qB_0/m)t} - i\frac{E_0}{B_0} \tag{5.5.23}$$

which should be checked by substitution. A further integration yields

$$r = C_2 e^{-i(qB_0/m)t} - i\frac{E_0}{B_0}t + C_3 \tag{5.5.24}$$

where, again,

$$C_2 = -i\frac{m}{qB_0}C_1 \quad \text{or} \quad C_1 = i\frac{qB_0}{m}C_2.$$

If we write $C_2 = Ae^{-i\phi}$ and $C_3 = a + bi$ and proceed as in the previous example, then

$$x = \text{Re}[r] = a + A\cos(\omega t + \phi) \tag{5.5.25a}$$

and

$$y = \text{Im}[r] = b - A\sin(\omega t + \phi) - \frac{E_0}{B_0}t \tag{5.5.25b}$$

For $E_0 = 0$, of course, the xy-motion is again circular. For $E_0 \neq 0$, the xy-motion is described as a *trochoid*—it is circular motion superposed on straight line motion in the negative y direction. (Remember, the electric field, E, was in the x direction!) This is the motion of a dot on a spinning disk as the axis of the disk moves with constant speed in the y direction.

These crossed magnetic and electric fields can be used as a velocity selector. Consider a charged particle initially moving in the y direction. From Eq. 5.5.20a you can see there will generally be a deflection in the x direction. But for one particular value of the velocity, $\dot{y}$, the magnetic and electric forces just cancel:

$$\ddot{x} = \frac{qE_0}{m} + \frac{qB_0}{m}\dot{y} = 0$$

$$\dot{y} = -\frac{E_0}{B_0} \tag{5.5.26}$$

This is illustrated in Figure 5.5.1, which shows a particle of positive charge q moving with a velocity $\mathbf{v} = -E_0/B_0\hat{\mathbf{j}}$ in a region with an electric field $\mathbf{E} = E_0\hat{\mathbf{i}}$ and a magnetic field $\mathbf{B} = B_0\hat{\mathbf{k}}$. Charged particles moving *slower* than E_0/B_0 "feel" a reduced magnetic force and, so, are pulled to their left. Particles moving *faster* than E_0/B_0 "feel" a greater magnetic force and are pulled to their right. Thus, only particles with a speed of exactly E_0/B_0 proceed through this region in a straight line. A beam of charged particles having various speeds (or energies) can be passed through this region with appropriate material placed as necessary to absorb those that deviate from a straight line. Only those with the desired velocity survive.

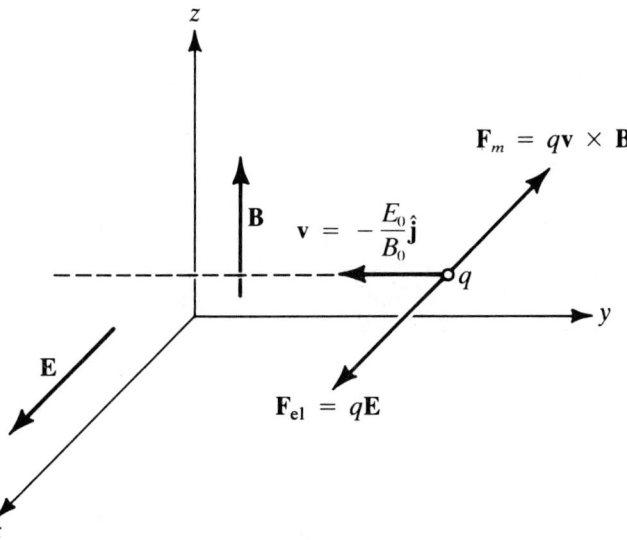

Figure 5.5.1 Crossed E- and B-fields used as a velocity selector.

It is easiest to understand this velocity selector in terms of the forces, as we did in arriving at Eq. 5.5.26. But we can now also understand it in terms of Eq. 5.5.25. Demanding no deviation in the x direction requires

$$A = 0$$

and that means

$$y = b - \frac{E_0}{B_0} t$$

from which we can immediately see

$$\dot{y} = -\frac{E_0}{B_0} \tag{5.5.27}$$

PROBLEMS

5.1 A particle at rest at the origin is suddenly affected at time $t = 0$ by a force given by $\mathbf{F} = b\hat{\mathbf{i}} + ct\hat{\mathbf{j}}$. Find its position and velocity at some time t later.

5.2 A particle is released from rest at the $\mathbf{r}_0 = x_0\hat{\mathbf{i}} + y_0\hat{\mathbf{j}}$ and experiences a force of $\mathbf{F} = bt\hat{\mathbf{i}} + ct^2\hat{\mathbf{j}}$. Find its position and velocity at some time t later.

5.3 A particle is released from $\mathbf{r}_0 = x_0\mathbf{i} + y_0\mathbf{j}$ with initial velocity $\mathbf{v}_0 = v_{0x}\mathbf{i} + v_{0y}\mathbf{j}$ at time $t = 0$, and experiences a force of $\mathbf{F} = be^{\beta t}\mathbf{i} + ce^{\varepsilon t}\mathbf{j}$. Find its position and velocity at some time t later.

5.4 An object is to be projected with initial velocity v_0 (initial speed v_0 at angle θ) to strike a target at $x = R$, $y = 0$. Neglecting air resistance, find the necessary firing angle θ.

*5.5 Due to a malfunction of a time machine, you find you have become a gunnery sergeant for a Roman Legion that is seiging a walled city. You are in charge of a catapult and must hurl a projectile to the *top* of a wall. The wall has a height of H, a width of W, and is located a distance L from your catapult.

What is the minimum initial speed necessary to hit the *top* of the wall? If the initial speed is actually *twice* this, what are the minimum and maximum angles θ that will allow you to hit the top of the wall? Neglect air resistance throughout.

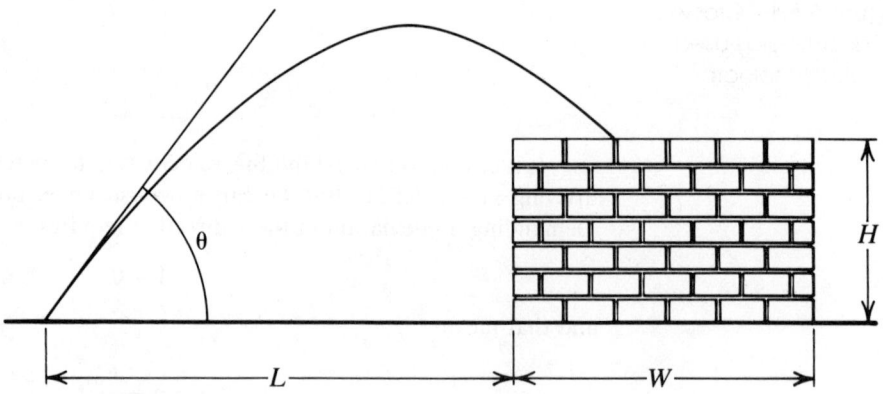

5.6 As a gunnery sergeant in charge of catapults for Hannibal, you are attempting to hurl a projectile to the *top* of a wall. The front of the wall is 50 m from your catapult. The wall is 10-m high and 20-m wide. If the catapult can only hurl objects at 53°, what range of initial speeds will allow the projectile to hit the top of the wall?

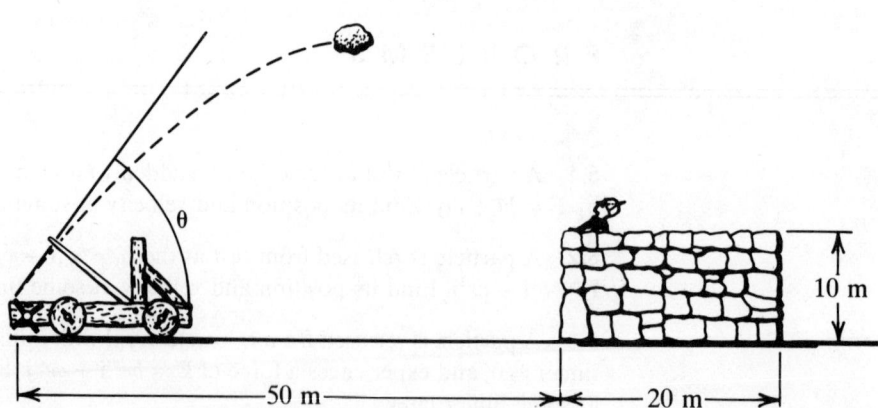

5.7

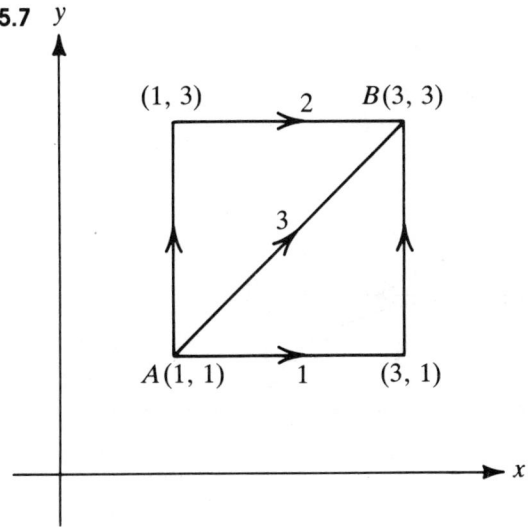

A body is to be moved from point A ($x = 1$, $y = 1$) to point B ($x = 3$, $y = 3$). Three paths are available as sketched:

Path 1: along the sides of a square, (1, 1) to (3, 1) to (3, 3)
Path 2: along the diagonal of a square, (1, 1) to (1, 3) to (3, 3)
Path 3: along the diagonal of a square, (1, 1) to (3, 3)

Calculate the work done in moving along each of the three paths for each of the following forces:

(a) $\mathbf{F} = xy\hat{\mathbf{i}} + y^2\hat{\mathbf{j}}$
(b) $\mathbf{F} = xy^2\hat{\mathbf{i}} + x^2y\hat{\mathbf{j}}$
(c) $\mathbf{F} = (x + 2)\hat{\mathbf{i}} + (y + 2)\hat{\mathbf{j}}$

5.8

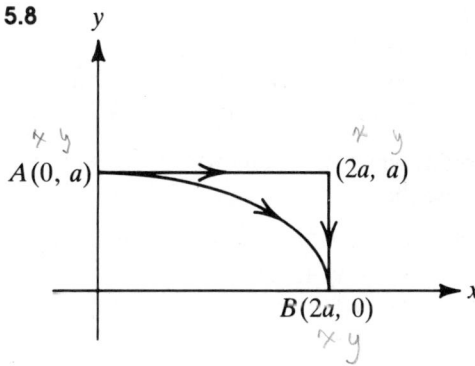

For the following forces, calculate the work done in moving from point $A(x = 0, y = a)$ to point $B(x = 2a, y = 0)$ along two different paths:

Path 1: straight lines from (0, a) to ($2a$, a) to ($2a$, 0)
Path 2: segment of an ellipse (use $x = 2a \sin \theta$, $y = a \cos \theta$).

Forces:
 (a) $\mathbf{F} = k\mathbf{r} = k(x\hat{\mathbf{i}} + y\hat{\mathbf{j}} + z\hat{\mathbf{k}})$
 (b) $\mathbf{F} = \dfrac{k}{x}\hat{\mathbf{i}} + \dfrac{k}{y}\hat{\mathbf{j}}$

5.9

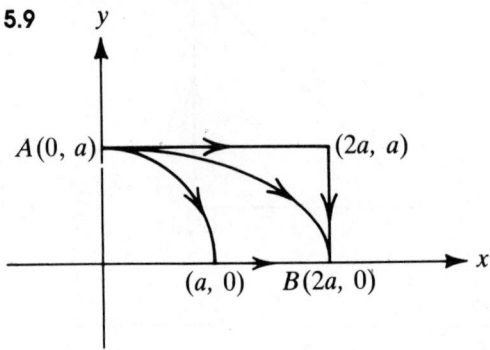

For the following forces, calculate the work done in moving a body from point $A(x = 0, y = a)$ to point $B(x = 2a, y = 0)$ along three different paths:

Path 1: straight lines from $(0, a)$ to $(2a, a)$ to $(2a, 0)$
Path 2: arc of circle from $(0, a)$ to $(a, 0)$ and then a straight line to $(2a, 0)$
Path 3: segment of an ellipse from $(0, a)$ to $(2a, 0)$. (Use $x = 2a \sin\theta$, $y = a\cos\theta$.)

Forces:
 (a) $\mathbf{F} = x\hat{\mathbf{i}} + y\hat{\mathbf{j}}$
 (b) $\mathbf{F} = y\hat{\mathbf{i}} - x\hat{\mathbf{j}}$
 (c) $\mathbf{F} = \dfrac{x}{y}(x\hat{\mathbf{i}} + y\hat{\mathbf{j}})$

5.10 Determine which of the following forces are conservative and find the potential energy function for those that are:
 (a) $\mathbf{F} = ax\hat{\mathbf{i}} + by\hat{\mathbf{j}} + cz\hat{\mathbf{k}}$
 (b) $\mathbf{F} = ax^2yz\hat{\mathbf{i}} + axy^2z\hat{\mathbf{j}} + bxyz^2\hat{\mathbf{k}}$
 (c) $\mathbf{F} = 5ay^2z^3\hat{\mathbf{i}} + 10axyz^3\hat{\mathbf{j}} + 15axy^2z^2\hat{\mathbf{k}}$
 (d) $\mathbf{F} = 6bxyz^2\hat{\mathbf{i}} + 3bx^2z^2\hat{\mathbf{j}} + 6bx^2yz\hat{\mathbf{k}}$

5.11 Determine which of the following forces are conservative and find the potential energy for those that are:
 (a) $\mathbf{F} = bxyz^2\hat{\mathbf{i}} + 2bx^2z^2\hat{\mathbf{j}} + 3bx^2yz\hat{\mathbf{k}}$
 (b) $\mathbf{F} = c(x\hat{\mathbf{i}} + y\hat{\mathbf{j}} + z\hat{\mathbf{k}})$
 (c) $\mathbf{F} = c(xy^2\hat{\mathbf{i}} + x^2y\hat{\mathbf{j}} + z^3\hat{\mathbf{k}})$
 (d) $\mathbf{F} = c(xy^2\hat{\mathbf{i}} + x^2y\hat{\mathbf{j}} + x^2z\hat{\mathbf{k}})$

5.12 Determine which of the following forces are conservative and find the potential energy for those that are:
 (a) $\mathbf{F} = (2axy + byz^2)\hat{\mathbf{i}} + (ax^2 + bxz^2)\hat{\mathbf{j}} + 2bxyz\hat{\mathbf{k}}$
 (b) $\mathbf{F} = (axy^2 + by^2z)\hat{\mathbf{i}} + (ax^2 + bxz^2)\hat{\mathbf{j}} + 2bxyz\hat{\mathbf{k}}$
 (c) $\mathbf{F} = (ay^2z^3 + 2bxy)\hat{\mathbf{i}} + (ay^2z^3 - 2bxy)\hat{\mathbf{j}} + axy^2z^2\hat{\mathbf{k}}$
 (d) $\mathbf{F} = (ay^2z^3 + 2bxy)\hat{\mathbf{i}} + 2(axyz^3 + bxy)\hat{\mathbf{j}} + 3axy^2z^2\hat{\mathbf{k}}$

5.13 A charged particle moves in a region of uniform electric and magnetic fields that are perpendicular. Let $\mathbf{E} = E\hat{\mathbf{j}}$ and $\mathbf{B} = B\hat{\mathbf{k}}$. The charged particle passes the origin with initial velocity $\mathbf{v}_0 = v_0\hat{\mathbf{i}}$. Solve for the motion of the particle. This motion traces out a path known as a cycloid. Show that it can be written in the form of

$$x = A \sin t + Bt$$
$$y = C(1 - \cos t)$$
$$z = 0$$

*5.14 A particle is released from rest at the origin at time $t = 0$. It then experiences a force given by

$$\mathbf{F} = \left(\frac{x}{1\,\text{m}} + \frac{2t^2}{\text{s}^2}\right)\hat{\mathbf{i}} + \left(\frac{y^2}{m^2} + \frac{3t}{\text{s}}\right)\hat{\mathbf{j}}$$

Find its position and velocity at the end of 10 seconds (this is best done by numerical methods).

*5.15 Air resistance is complex. For some shapes and higher speeds, a retarding force proportional to the *square* of the velocity is a better approximation. With such a retarding force, the total force is no longer separable. It is now easier to treat this problem numerically. Let the air resistance—assumed proportional to the *square* of the velocity—be such as to give a terminal velocity of 100 km/hr to a freely falling body. Plot the trajectory of an object fired with an initial speed of 50 km/hr at an angle of 37°. Compare this, by superimposing another graph on the same axes, with the trajectory expected if air resistance is neglected.

*5.16 Consider air resistance proportional to velocity acting on a particle fired with initial speed v_0 at angle θ_0. Plot the trajectories obtained for $v_0 = 20$ m/s and for both $\theta_0 = 30°$ and 60°. Use a linear air resistance giving a terminal velocity of 40 m/s for release from free fall. Compare these results with the case for zero air resistance by plotting all four trajectories on the same graph (this is best done by numerical methods).

6

COORDINATE SYSTEMS

6.1 Introduction

We have previously described a vector in terms of its components along the axes of a rectangular coordinate system. We have seen that the same vector can be described readily by two different sets of components along two different sets of rectangular coordinates. For many situations—for example, the motion of a scratch on a phonograph record or a satellite encircling Earth—there are easier ways of describing motion than using *rectangular* coordinates. We shall now extend our ideas of vectors to handle some of these cases.

6.2 Plane Polar Coordinates

Consider the point P in Figure 6.2.1. We know it is located by position vector $\mathbf{r}$, given by

$$\mathbf{r} = x\hat{\mathbf{i}} + y\hat{\mathbf{j}} \tag{6.2.1}$$

We can also locate it by giving its distance from the origin, r, and the polar angle, θ. Both r and θ are shown in Figure 6.2.1; the polar angle, θ, is measured counterclockwise from the x-axis. We can then define two new

SECTION 6.2 / PLANE POLAR COORDINATES

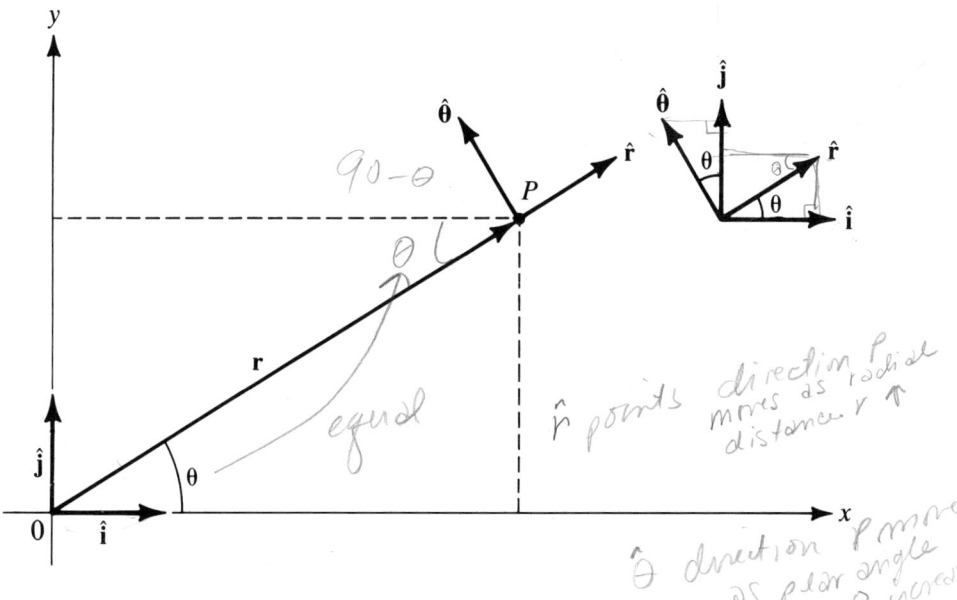

Figure 6.2.1 Plane polar coordinates.

perpendicular unit vectors—$\hat{\mathbf{r}}$, which points in the direction P would move as the radial distance r increases, and $\hat{\boldsymbol{\theta}}$, which points in the direction that P would move as the polar angle θ increases. The vector $\hat{\mathbf{r}}$ is sometimes written as $\hat{\mathbf{e}}_r$, and $\hat{\boldsymbol{\theta}}$ as $\hat{\mathbf{e}}_\theta$. In terms of *these* unit vectors, we can very simply write

$$\mathbf{r} = r\hat{\mathbf{r}} \tag{6.2.2}$$

Unit vector $\hat{\boldsymbol{\theta}}$ does not even appear. All the information about the polar angle θ is contained in the radial unit vector $\hat{\mathbf{r}}$ (i.e., $\hat{\mathbf{r}} = \hat{\mathbf{r}}(\theta)$).

The vectors $\hat{\mathbf{r}}$ and $\hat{\boldsymbol{\theta}}$ form a new coordinate system, the *plane polar coordinate system*, and are usually referred to simply as *polar coordinates*. If we need to switch back and forth from $\hat{\mathbf{i}}$ and $\hat{\mathbf{j}}$ to $\hat{\mathbf{r}}$ and $\hat{\boldsymbol{\theta}}$, Figure 6.2.1 shows this can be done by

$$\begin{aligned} \hat{\mathbf{r}} &= \hat{\mathbf{i}} \cos \theta + \hat{\mathbf{j}} \sin \theta & \hat{\mathbf{i}} &= \hat{\mathbf{r}} \cos \theta - \hat{\boldsymbol{\theta}} \sin \theta \\ \hat{\boldsymbol{\theta}} &= -\hat{\mathbf{i}} \sin \theta + \hat{\mathbf{j}} \cos \theta & \hat{\mathbf{j}} &= \hat{\mathbf{r}} \sin \theta + \hat{\boldsymbol{\theta}} \cos \theta \end{aligned} \tag{6.2.3}$$

which is just a direct application of the methods in Chapter 4.

To find the velocity expressed in polar coordinates, we simply obtain the time derivative of $\mathbf{r}$ from Eq. 6.2.2. Be careful, though, for *unlike* $\hat{\mathbf{i}}$, $\hat{\mathbf{j}}$, and $\hat{\mathbf{k}}$, the radial unit vector is *time-dependent* (therefore, the direction of $\hat{\mathbf{r}}$ changes).

$$\mathbf{v} = \dot{\mathbf{r}} = \frac{d\mathbf{r}}{dt} = \frac{dr}{dt} \hat{\mathbf{r}} + r \frac{d\hat{\mathbf{r}}}{dt} \tag{6.2.4}$$

In general, $d\hat{\mathbf{r}}/dt$ is not zero since $\hat{\mathbf{r}}$ certainly depends on the polar angle θ, which, in turn, depends on time. We can apply the usual chain rule and write

$$\frac{d\hat{\mathbf{r}}}{dt} = \frac{d\hat{\mathbf{r}}}{d\theta} \frac{d\theta}{dt} = \frac{d\hat{\mathbf{r}}}{d\theta} \dot{\theta} \tag{6.2.5}$$

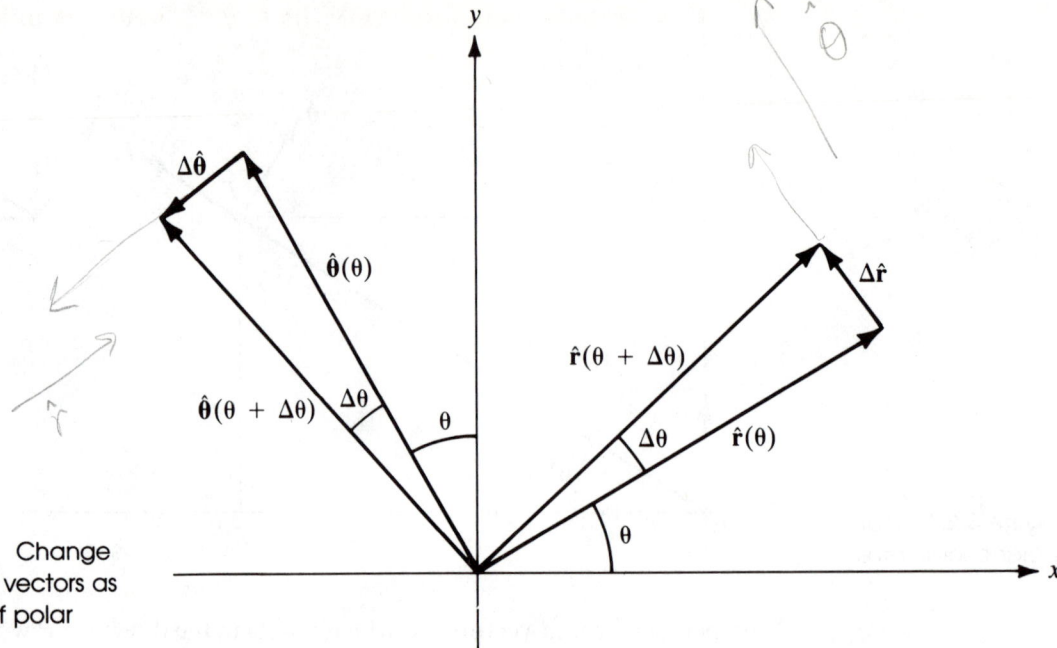

Figure 6.2.2 Change in polar unit vectors as a function of polar angle θ.

Figure 6.2.2 shows the relationships of unit vectors $\hat{\mathbf{r}}$ and $\hat{\boldsymbol{\theta}}$ for a particular angle θ and for $\theta + \Delta\theta$.

As the angle θ increases by $\Delta\theta$, the radial unit vector changes from $\hat{\mathbf{r}}(\theta)$ to $\hat{\mathbf{r}}(\theta + \Delta\theta)$ by an amount $\Delta\hat{\mathbf{r}}$. Likewise, the angular unit vector changes from $\hat{\boldsymbol{\theta}}(\theta)$ to $\hat{\boldsymbol{\theta}}(\theta + \Delta\theta)$ by an amount $\Delta\hat{\boldsymbol{\theta}}$. As $\Delta\theta$ approaches zero, of course, we can replace $\Delta\theta$ by $d\theta$ and $\Delta\hat{\mathbf{r}}$ becomes $d\hat{\mathbf{r}}$ while $\Delta\hat{\boldsymbol{\theta}}$ becomes $d\hat{\boldsymbol{\theta}}$. Notice that $\Delta\hat{\mathbf{r}}$ or $d\hat{\mathbf{r}}$ points in the direction of $\hat{\boldsymbol{\theta}}$ and, since the length of $\hat{\boldsymbol{\theta}}$ is unity, has a magnitude equal to the change in angle $\Delta\theta$ or $d\theta$. Likewise, $\Delta\hat{\boldsymbol{\theta}}$ or $d\hat{\boldsymbol{\theta}}$ points opposite to $\hat{\mathbf{r}}$ and in magnitude is equal to $\Delta\theta$ or $d\theta$. That is,

$$d\hat{\mathbf{r}} = \hat{\boldsymbol{\theta}}\, d\theta \quad \text{and} \quad d\hat{\boldsymbol{\theta}} = -\hat{\mathbf{r}}\, d\theta \qquad (6.2.6)$$

or

$$\frac{d\hat{\mathbf{r}}}{dt} = \hat{\boldsymbol{\theta}}\frac{d\theta}{dt} = \hat{\boldsymbol{\theta}}\dot{\theta} \quad \text{and} \quad \frac{d\hat{\boldsymbol{\theta}}}{dt} = -\hat{\mathbf{r}}\frac{d\theta}{dt} = -\hat{\mathbf{r}}\dot{\theta} \qquad (6.2.7)$$

We can use this to rewrite the velocity of Eq. 6.2.4 as

$$\mathbf{v} = \dot{\mathbf{r}} = \dot{r}\hat{\mathbf{r}} + r\dot{\theta}\hat{\boldsymbol{\theta}} \qquad (6.2.8)$$

Another time differentiation yields the acceleration in polar coordinates:

$$\mathbf{a} = \frac{d\mathbf{v}}{dt} = \ddot{\mathbf{r}} = \ddot{r}\hat{\mathbf{r}} + \dot{r}\frac{d\hat{\mathbf{r}}}{dt} + \dot{r}\dot{\theta}\hat{\boldsymbol{\theta}} + r\ddot{\theta}\hat{\boldsymbol{\theta}} + r\dot{\theta}\frac{d\hat{\boldsymbol{\theta}}}{dt}$$

$$= \ddot{r}\hat{\mathbf{r}} + \dot{r}\dot{\theta}\hat{\boldsymbol{\theta}} + \dot{r}\dot{\theta}\hat{\boldsymbol{\theta}} + r\ddot{\theta}\hat{\boldsymbol{\theta}} - r\dot{\theta}^2\hat{\mathbf{r}} \qquad (6.2.9)$$

$$\mathbf{a} = \ddot{\mathbf{r}} = (\ddot{r} - r\dot{\theta}^2)\hat{\mathbf{r}} + (r\ddot{\theta} + 2\dot{r}\dot{\theta})\hat{\boldsymbol{\theta}}$$

This all looks quite formidable. Its beauty, though, is in its simplicity *when used in situations of appropriate symmetry*. Consider an airplane moving in a circle of radius $r = b$ with constant speed $v = b\omega$. Expressions for **r**, **v**, and **a** in rectangular coordinates are involved and messy—even if straightforward. In polar coordinates they are simply

$$\mathbf{r} = b\hat{\mathbf{r}} \qquad (6.2.10)$$

$$r = b \qquad \theta = \omega t$$

$$\dot{r} = 0 \qquad \dot{\theta} = \omega \quad \text{(angular velocity)}$$

$$\ddot{r} = 0 \qquad \ddot{\theta} = 0$$

$$\mathbf{v} = \dot{\mathbf{r}} = b\omega\hat{\boldsymbol{\theta}} \qquad (6.2.11)$$

$$\mathbf{a} = \ddot{\mathbf{r}} = -b\omega^2\hat{\mathbf{r}} \qquad (6.2.12)$$

which come *directly* from Eqs. 6.2.8 and 6.2.9 upon substitution of r, $\dot{r}$, $\ddot{r}$, θ, $\dot{\theta}$, and $\ddot{\theta}$. It is also clear that the velocity is tangential with no radial component and that the acceleration experienced in such uniform circular motion is entirely radial, directed inward, with no tangential component.

6.3 Cylindrical Polar Coordinates

We can easily extend our notation to three dimensions by adding a z-component and creating *cylindrical* polar coordinates as sketched in Figure 6.3.1.

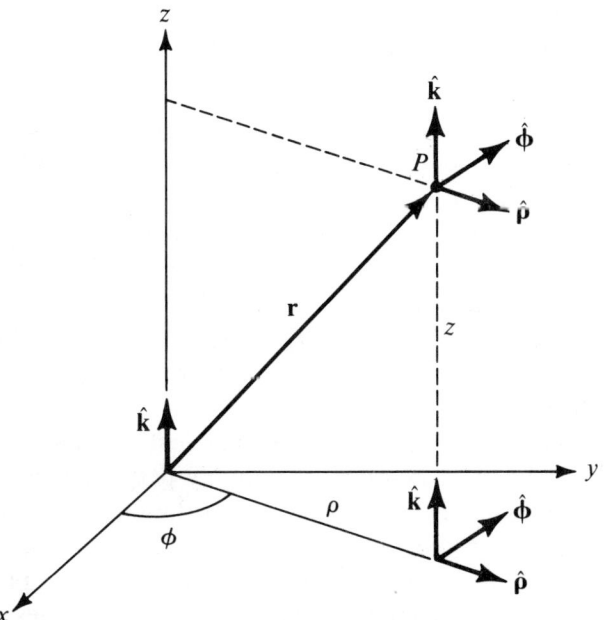

Figure 6.3.1
Cylindrical polar coordinates.

The location of a point P is given by its radial distance from the z-axis, ρ, its angular rotation from the x-axis, ϕ (just as in the previous case with r and θ), *and* its elevation above the xy plane, z. We can write the position vector **r** as

$$\mathbf{r} = \rho\hat{\boldsymbol{\rho}} + z\hat{\mathbf{k}} \tag{6.3.1}$$

Here ρ and ϕ for cylindrical polar coordinates are *exactly* analogous to r and θ for plane polar coordinates. Therefore,

$$\mathbf{v} = \dot{\mathbf{r}} = \dot{\rho}\hat{\boldsymbol{\rho}} + \rho\dot{\phi}\hat{\boldsymbol{\phi}} + \dot{z}\hat{\mathbf{k}} \tag{6.3.2}$$

$$\mathbf{a} = \ddot{\mathbf{r}} = (\ddot{\rho} - \rho\dot{\phi}^2)\hat{\boldsymbol{\rho}} + (2\dot{\rho}\dot{\phi} + \rho\ddot{\phi})\hat{\boldsymbol{\phi}} + \ddot{z}\hat{\mathbf{k}} \tag{6.3.3}$$

Cylindrical polar coordinates are important for the ease and beauty with which they describe motion *with applicable symmetry*. Consider an airplane flying as described before (see Eq. 6.2.10) while also *rising* at a constant rate of v_z. Then,

$$\mathbf{r} = b\hat{\boldsymbol{\rho}} + v_z t\hat{\mathbf{k}}$$

$$\mathbf{v} = b\omega\hat{\boldsymbol{\phi}} + v_z\hat{\mathbf{k}}$$

$$\mathbf{a} = -b\omega^2\hat{\boldsymbol{\rho}}$$

Again, we can see immediately that the acceleration is radial, directed inward. There is no vertical acceleration since the ascent is at constant vertical speed.

6.4 Spherical Polar Coordinates

One of the most useful coordinate systems is *spherical polar coordinates* (often simply called *spherical coordinates*), as shown in Figure 6.4.1. Such a system is useful whenever there is spherical symmetry—for example, describing the forces binding an electron to its nucleus, giving the location of the space shuttle in orbit around Earth, or charting the course of the *U.S.S. Enterprise* through the galaxy. In Figure 6.4.1, point P is located by giving its *radial distance* from the origin r, and *azimuthal angle* ϕ locating a plane whose angle of rotation is measured from the x-axis, and a *polar angle* θ, giving its angular location measured "down" from the z-axis. Note that θ can vary from 0 to π, and ϕ from 0 to 2π. This is sketched in Figure 6.4.1.

We can then define mutually perpendicular unit vectors $\hat{\mathbf{r}}$, $\hat{\boldsymbol{\theta}}$, and $\hat{\boldsymbol{\phi}}$ as shown in Figure 6.4.1 by giving the directions that P moves as r, θ, and ϕ increase, respectively. These unit vectors are sketched again in Figure 6.4.2.

Note that $\hat{\mathbf{r}}$ and $\hat{\boldsymbol{\theta}}$ both lie in a plane containing the z-axis and rotated an amount ϕ from the x-axis. The unit vector $\hat{\boldsymbol{\phi}}$ lies in the xy-plane. Figure 6.4.3

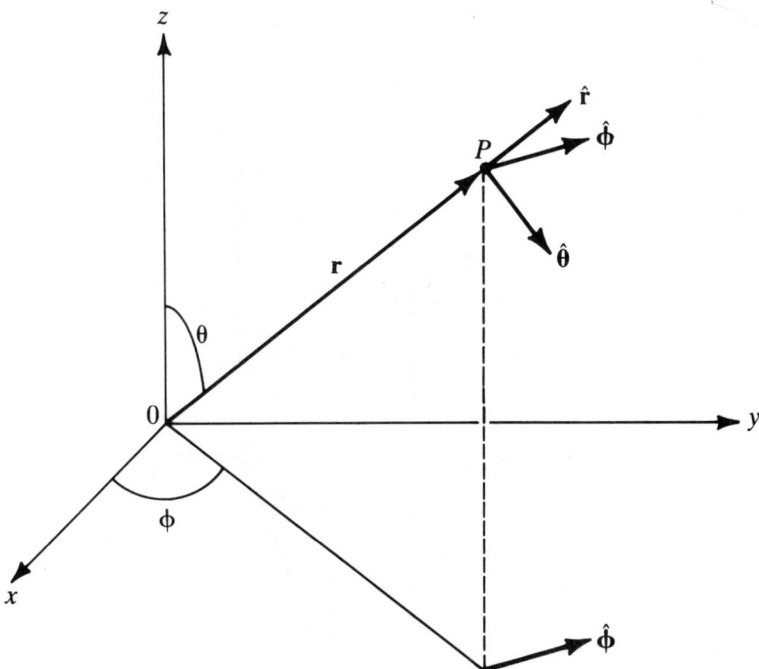

Figure 6.4.1 Spherical polar coordinates.

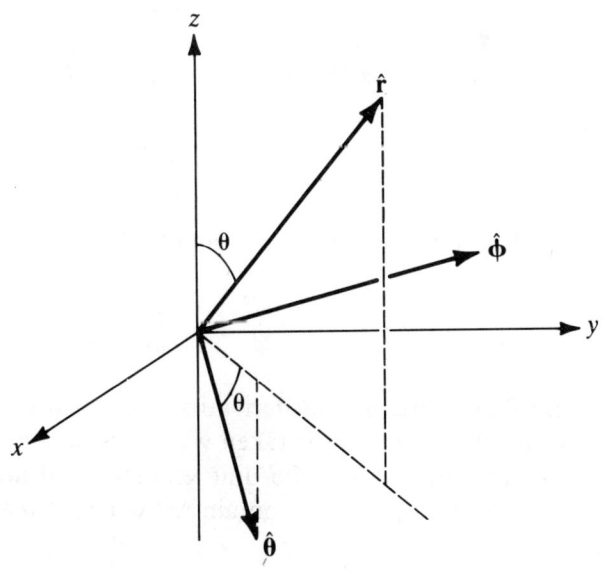

Figure 6.4.2 Unit vectors $\hat{r}$, $\hat{\theta}$, and $\hat{\phi}$ for spherical polar coordinates.

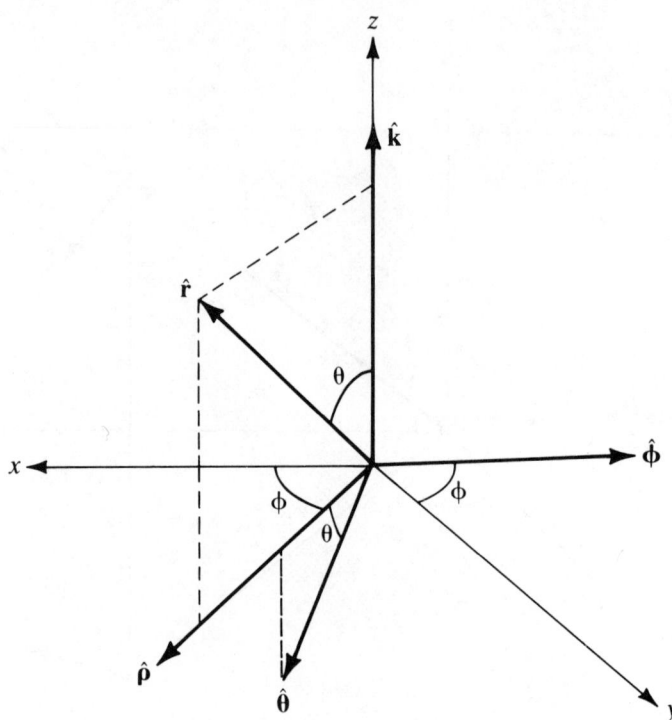

Figure 6.4.3 Unit vectors for spherical and cylindrical coordinates.

shows $\hat{\mathbf{r}}$, $\hat{\boldsymbol{\theta}}$, and $\hat{\boldsymbol{\phi}}$, the unit vectors for spherical coordinates, along with $\hat{\boldsymbol{\rho}}$, $\hat{\boldsymbol{\phi}}$, and $\hat{\mathbf{k}}$ for cylindrical coordinates ($\hat{\boldsymbol{\phi}}$ is the same for both).

As before, we can locate position P by a position vector $\mathbf{r}$ as shown in Figure 6.4.1:

$$\mathbf{r} = r\hat{\mathbf{r}} \tag{6.4.1}$$

We can now ask for expressions for $\mathbf{v} = \dot{\mathbf{r}}$ and $\mathbf{a} = \ddot{\mathbf{r}}$:

$$\mathbf{v} = \dot{\mathbf{r}} = \dot{r}\hat{\mathbf{r}} + r\frac{d\hat{\mathbf{r}}}{dt}$$

Now $d\hat{\mathbf{r}}/dt$ must be evaluated from the chain rule; since $\hat{\mathbf{r}} = \hat{\mathbf{r}}(\theta, \phi)$, we have

$$\frac{d\hat{\mathbf{r}}}{dt} = \frac{\partial \hat{\mathbf{r}}}{\partial \theta}\frac{d\theta}{dt} + \frac{\partial \hat{\mathbf{r}}}{\partial \phi}\frac{d\phi}{dt} \tag{6.4.2}$$

where $\partial/\partial\theta$ is the *partial* derivative with respect to θ—considering ϕ held constant; likewise, $\partial/\partial\theta$ is taken while θ is held constant.

Figure 6.4.4 shows the unit vectors for θ and ϕ and for $\theta + \Delta\theta$ and $\phi + \Delta\phi$. From that we can obtain $\Delta\hat{\mathbf{r}}/\Delta\theta$ or $\partial\hat{\mathbf{r}}/\partial\theta$ as $\Delta\theta \to 0$ and, likewise, for $\partial\hat{\mathbf{r}}/\partial\phi$.

SECTION 6.4 / SPHERICAL POLAR COORDINATES

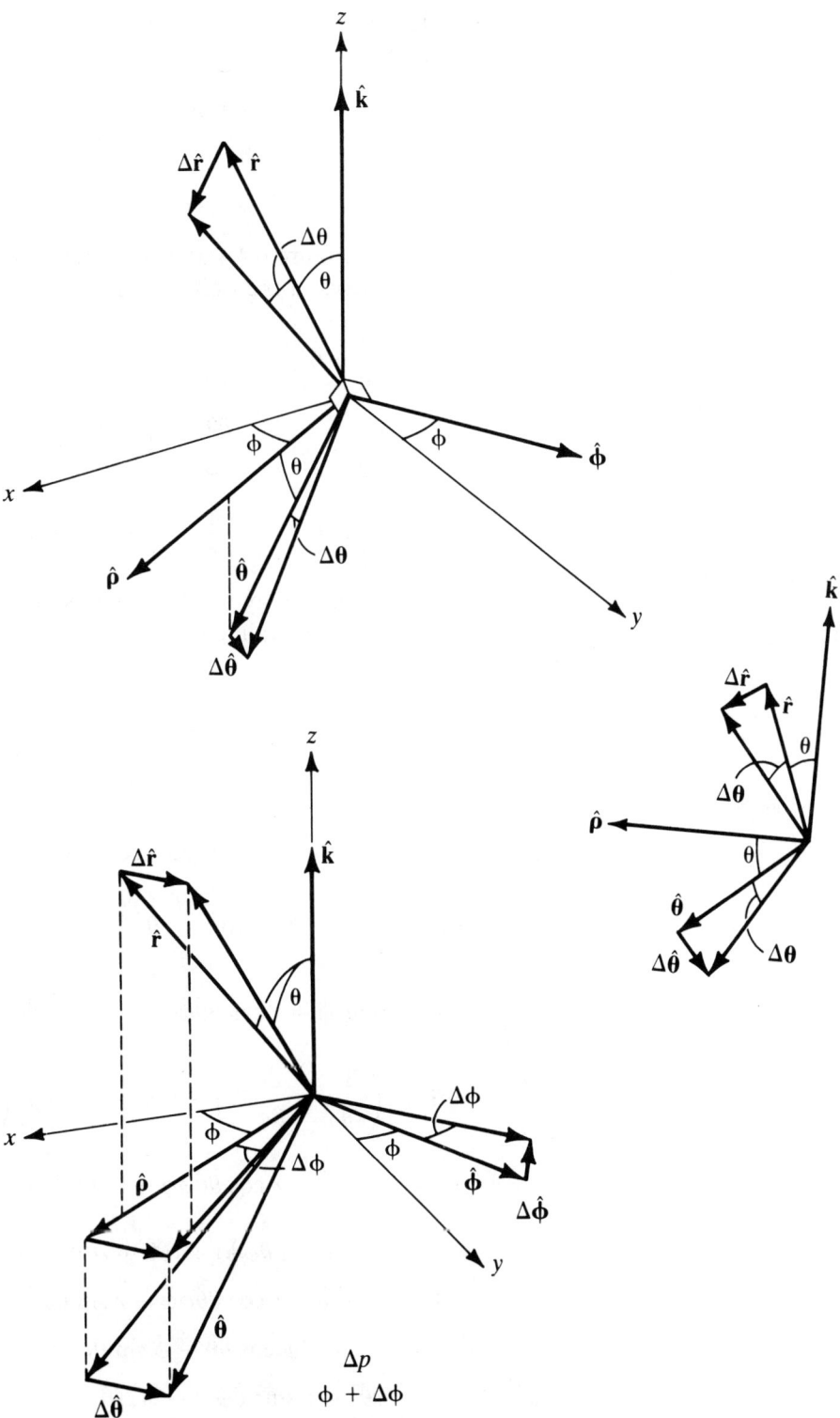

Figure 6.4.4 Unit vectors as functions of θ and ϕ.

From Figure 6.4.4 we can describe the spherical coordinate unit vectors $\hat{\mathbf{r}}$, $\hat{\boldsymbol{\theta}}$, and $\hat{\boldsymbol{\phi}}$ in terms of the cylindrical coordinate unit vectors $\hat{\boldsymbol{\rho}}$, $\hat{\boldsymbol{\phi}}$, and $\hat{\mathbf{k}}$:

$$\hat{\mathbf{r}} = \hat{\boldsymbol{\rho}} \sin \theta + \hat{\mathbf{k}} \cos \theta$$
$$\hat{\boldsymbol{\theta}} = \hat{\boldsymbol{\rho}} \cos \theta - \hat{\mathbf{k}} \sin \theta \qquad (6.4.3)$$
$$\hat{\boldsymbol{\phi}} = \hat{\boldsymbol{\phi}}$$

Careful use of Figure 6.4.4 and these relationships then lets us evaluate the partial derivatives of the unit vectors. That is,

$$\frac{\partial \hat{\mathbf{r}}}{\partial \theta} = \hat{\boldsymbol{\theta}} \qquad \frac{\partial \hat{\mathbf{r}}}{\partial \phi} = \frac{\partial \hat{\boldsymbol{\rho}} \sin \theta}{\partial \phi} = \hat{\boldsymbol{\phi}} \sin \theta$$

$$\frac{\partial \hat{\boldsymbol{\theta}}}{\partial \theta} = -\hat{\mathbf{r}} \qquad \frac{\partial \hat{\boldsymbol{\theta}}}{\partial \phi} = \frac{\partial \hat{\boldsymbol{\rho}} \cos \theta}{\partial \phi} = \hat{\boldsymbol{\phi}} \cos \theta \qquad (6.4.4)$$

$$\frac{\partial \hat{\boldsymbol{\phi}}}{\partial \theta} = 0 \qquad \frac{\partial \hat{\boldsymbol{\phi}}}{\partial \phi} = -\hat{\boldsymbol{\rho}} = -\hat{\boldsymbol{\theta}} \cos \theta - \hat{\mathbf{r}} \sin \theta$$

These relationships may not be obvious at first glance. But you should convince yourself of their validity. We can now finish our evaluation of **v** from Eq. 6.4.2:

$$\mathbf{v} = \dot{\mathbf{r}} = \dot{r}\hat{\mathbf{r}} + r\left(\frac{d\hat{\mathbf{r}}}{d\theta}\dot{\theta} + \frac{d\hat{\mathbf{r}}}{d\phi}\dot{\phi}\right)$$
$$\mathbf{v} = \dot{\mathbf{r}} = \dot{r}\hat{\mathbf{r}} + r\dot{\theta}\hat{\boldsymbol{\theta}} + r\sin\theta\dot{\phi}\hat{\boldsymbol{\phi}} \qquad (6.4.5)$$

Further differentiation to obtain the acceleration yields

$$\mathbf{a} = \ddot{\mathbf{r}} = \ddot{r}\hat{\mathbf{r}} + \dot{r}\frac{d\hat{\mathbf{r}}}{dt} + \dot{r}\dot{\theta}\hat{\boldsymbol{\theta}} + r\ddot{\theta}\hat{\boldsymbol{\theta}} + r\dot{\theta}\frac{d\hat{\boldsymbol{\theta}}}{dt}$$

$$+ \dot{r}\sin\theta\dot{\phi}\hat{\boldsymbol{\phi}} + r\cos\theta\dot{\theta}\dot{\phi}\hat{\boldsymbol{\phi}} + r\sin\theta\ddot{\phi}\hat{\boldsymbol{\phi}} + r\sin\theta\dot{\phi}\frac{d\hat{\boldsymbol{\phi}}}{dt}$$

$$= \ddot{r}\hat{\mathbf{r}} + \dot{r}\left(\frac{\partial \hat{\mathbf{r}}}{\partial \theta}\dot{\theta} + \frac{\partial \hat{\mathbf{r}}}{\partial \phi}\dot{\phi}\right) + \dot{r}\dot{\theta}\hat{\boldsymbol{\theta}} + r\ddot{\theta}\hat{\boldsymbol{\theta}} + r\dot{\theta}\left(\frac{\partial \hat{\boldsymbol{\theta}}}{\partial \theta}\dot{\theta} + \frac{\partial \hat{\boldsymbol{\theta}}}{\partial \phi}\dot{\phi}\right)$$

$$+ \dot{r}\sin\theta\dot{\phi}\hat{\boldsymbol{\phi}} + r\cos\theta\dot{\theta}\dot{\phi}\hat{\boldsymbol{\phi}} + r\sin\theta\ddot{\phi}\hat{\boldsymbol{\phi}} + r\sin\theta\dot{\phi}\left(\frac{\partial \hat{\boldsymbol{\phi}}}{\partial \theta}\dot{\theta} + \frac{\partial \hat{\boldsymbol{\phi}}}{\partial \phi}\dot{\phi}\right)$$

$$= \ddot{r}\hat{\mathbf{r}} + \dot{r}(\dot{\theta}\hat{\boldsymbol{\theta}} + \sin\theta\dot{\phi}\hat{\boldsymbol{\phi}}) + \dot{r}\dot{\theta}\hat{\boldsymbol{\theta}} + r\ddot{\theta}\hat{\boldsymbol{\theta}} + r\dot{\theta}(-\dot{\theta}\hat{\mathbf{r}} + \cos\theta\dot{\phi}\hat{\boldsymbol{\phi}})$$

$$+ \dot{r}\sin\theta\dot{\phi}\hat{\boldsymbol{\phi}} + r\cos\theta\dot{\theta}\dot{\phi}\hat{\boldsymbol{\phi}} + r\sin\theta\ddot{\phi}\hat{\boldsymbol{\phi}}$$

$$+ r\sin\theta\dot{\phi}(-\dot{\phi}\cos\theta\hat{\boldsymbol{\theta}} - \dot{\phi}\sin\theta\hat{\mathbf{r}})$$

$$\mathbf{a} = \ddot{\mathbf{r}} = \hat{\mathbf{r}}(\ddot{r} - r\dot{\theta}^2 - r\sin^2\theta\dot{\phi}^2) + \hat{\boldsymbol{\theta}}(2\dot{r}\dot{\theta} - r\sin\theta\cos\theta\dot{\phi}^2 + r\ddot{\theta})$$
$$+ \hat{\boldsymbol{\phi}}(2\dot{r}\dot{\phi}\sin\theta + 2r\dot{\phi}\dot{\theta}\cos\theta + r\sin\theta\ddot{\phi}) \qquad (6.4.6)$$

As an example for the use of spherical coordinates, consider a satellite in circular polar orbit about Earth with a constant radius $r = b$ and angular velocity ω. Then,

$$\mathbf{r} = b\hat{\mathbf{r}}$$

$r = b$	$\theta = \omega t$	$\phi = \text{constant}$
$\dot{r} = 0$	$\dot{\theta} = \omega$	$\dot{\phi} = 0$
$\ddot{r} = 0$	$\ddot{\theta} = 0$	$\ddot{\phi} = 0$

$$\mathbf{v} = b\omega\hat{\boldsymbol{\theta}}$$

$$\mathbf{a} = -b\omega^2\hat{\mathbf{r}}$$

We can clearly see that the acceleration is only radial and directed inward. The velocity is only tangential and lies in a plane defined by a constant value for ϕ.

Likewise, the motion of a satellite in a circular equatorial orbit can be described by

$$\mathbf{r} = b\hat{\mathbf{r}}$$

$r = b$	$\theta = \dfrac{\pi}{2}$	$\phi = \omega t$
$\dot{r} = 0$	$\dot{\theta} = 0$	$\dot{\phi} = \omega$
$\ddot{r} = 0$	$\ddot{\theta} = 0$	$\ddot{\phi} = 0$

$$\mathbf{v} = b\omega\hat{\boldsymbol{\phi}}$$

$$\mathbf{a} = -b\omega^2\hat{\mathbf{r}}$$

As before, it is very clear that the acceleration is only radial and directed inward. The velocity is only tangential and confined to a plane defined by $\theta = \pi/2$ (the equatorial plane).

Equations 6.4.5 and 6.4.6 are long and, indeed, somewhat messy in their derivation. Their use or their beauty is in the simplicity they lend to certain problems like the two simple examples just given. Expressing $\mathbf{r}$, $\mathbf{v}$, and $\mathbf{a}$ in more familiar rectangular coordinates would prove quite cumbersome. We shall recall this when we discuss central forces in Chapter 7.

6.5 Moving Coordinate Systems

We have already mentioned the idea of relative motion. Figure 6.5.1 shows a point P that can be described equally well by position vector $\mathbf{r}'$ measured in the $x'y'z'$ coordinate system (also called a *frame of reference*) or position

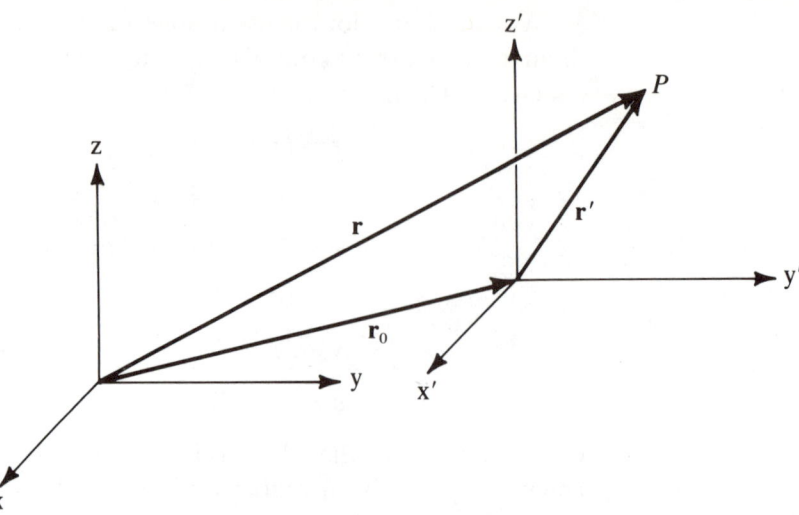

Figure 6.5.1 Point P located in two different reference frames.

vector $\mathbf{r}$ measured in the xyz coordinate system (or frame of reference). The vector $\mathbf{r}_0$ locates the origin of the $x'y'z'$ frame with respect to the origin of xyz.

From vector algebra we can see that

$$\mathbf{r} = \mathbf{r}_0 + \mathbf{r}' \tag{6.5.1}$$

The two coordinate systems need not remain fixed relative to each other. If they are not, then $\mathbf{r}_0$ will vary. Differentiation of Eq. 6.5.1 yields

$$\mathbf{v} = \mathbf{v}_0 + \mathbf{v}' \tag{6.5.2}$$

and

$$\mathbf{a} = \mathbf{a}_0 + \mathbf{a}' \tag{6.5.3}$$

where the unprimed letters refer to the velocity or acceleration of point P as seen by the xyz frame; the primed letters, $\mathbf{v}'$ and $\mathbf{a}'$, are the velocity and acceleration of the same point as seen by the $x'y'z'$ frame; and $\mathbf{v}_0$ and $\mathbf{a}_0$ are the velocity and acceleration of the origin of the $x'y'z'$ frame with respect to xyz.

We shall consider the xyz frame to be our laboratory frame of reference—a *fixed* inertial frame in which all the laws of physics hold (mainly, Newton's Laws of Motion will be of interest). For the present, we shall limit our discussion to relative motion wherein the axes of the "moving" $x'y'z'$ frame do not rotate with respect to our fixed xyz frame. We can then write

$$\mathbf{F} = m\mathbf{a} \quad \text{or} \quad m\mathbf{a} = \mathbf{F} \tag{6.5.4}$$

to explain the motion of an object as observed in the fixed or laboratory frame xyz. A "moving observer," that is, one located *in* the $x'y'z'$ frame, however, will see motion described by

$$m\mathbf{a}' = m(\mathbf{a} - \mathbf{a}_0) \quad \text{or} \quad m\mathbf{a}' = \mathbf{F} - m\mathbf{a}_0$$

or

$$\mathbf{F}' = \mathbf{F} + \mathbf{F}_{\text{fict}} = m\mathbf{a}' \tag{6.5.5}$$

(annotation: "real forces" pointing to $\mathbf{F}$)

where

$$\mathbf{F}_{\text{fict}} = -m\mathbf{a}_0$$

The vector $\mathbf{F}$ represents the resultant of all the "real" forces acting on the object of mass m—contact forces, electromagnetic forces, gravitational forces, and nuclear forces. The equation $\mathbf{F} = m\mathbf{a}$ is experimentally verifiable in the fixed inertial frame of xyz (indeed, an inertial frame *is* one in which Newton's laws are valid). But $\mathbf{F} = m\mathbf{a}'$ is *not* true in the moving $x'y'z'$ frame when there is an acceleration of this frame (i.e., when $\mathbf{a}_0 \neq 0$). That is, the net force—the sum of all the *real* forces—does not satisfy Newton's Second Law in this case. To make Newton's Second Law ($\mathbf{F} = m\mathbf{a}$) appear in the same *form* in this frame, we must add another term to the force—a *fictitious force*, if you will. This fictitious force is $-m\mathbf{a}_0$. Only then can we write an equation like Newton's Second Law that is valid: $\mathbf{F}' = m\mathbf{a}'$. Because we are convinced that Newton's Second Law correctly describes *all* motion, creating such a fictitious force is quite preferable to any alternative—like simply being unable to explain the motion. In the study of general relativity, these fictitious forces are specifically treated just like real forces caused by some new gravitational field. The result is the same. These fictitious forces—or *inertial forces* as they may also correctly be called—are indistinguishable from gravitational forces. Their effect is very real.

As an example, consider a railroad train travelling along a straight, smooth track. We can take an observer riding *on* the train as being in our $x'y'z'$ frame, and an observer sitting on the station platform, in our xyz frame. First, let the train travel at a constant velocity of $\mathbf{v}_0$ (that is, $\mathbf{a}_0 = 0$). If the observer on board the train releases a ball from rest, *he* will see it fall straight down. The equation $\mathbf{F}' = m\mathbf{a}'$ will explain his observations. The platform observer, however, sees the ball released with an initial horizontal velocity of $\mathbf{v}_0$. He sees the ball fall along a parabolic trajectory. Again, this corresponds to his calculations using $\mathbf{F} = m\mathbf{a}$. Although they observe different trajectories, each observer explains what he sees using the same law of physics.

Now let the train accelerate to the right with an acceleration $\mathbf{a}_0$. If the onboard observer releases a ball as before, the platform observer will still see it fall along a parabolic trajectory as before. $\mathbf{F} = m\mathbf{a}$ fully explains this just as before. But what does the onboard observer see this time? During the time it takes the ball to fall, the speed of the train (and onboard observer) has increased. The platform observer sees this manifested in the train moving *farther* to the right than the ball. The onboard observer, then, will see the ball fall to the left—toward the rear of the train. Some unexpected, unaccounted-for force is evidently present. This force turns out to be $-m\mathbf{a}_0$, the fictitious force mentioned earlier. Only with the inclusion of this term can our onboard observer determine that $\mathbf{F}' = m\mathbf{a}'$ is valid.

Another, and more common, example is that of a driver in an accelerating car. Think of yourself in a car as it accelerates from rest. *You* feel yourself pushed back into the seat; there is a *force* on you pushing you back into the seat. But your entire frame of reference is accelerating. An observer standing beside the road would simply see you caused to be accelerated by the car seat pushing on you. For you in the noninertial frame, there is an unexpected inertial force pushing you back into the seat. For the outside observer in the inertial frame, there is a very real force causing an acceleration.

6.6 Rotating Coordinate Systems

Now we shall consider relative motion such that the moving $x'y'z'$ frame may also be rotating with respect to the fixed xyz frame. In fact, we shall begin with the origins of the two located at the same place and the $x'y'z'$ frame *only* rotating about some vector $\boldsymbol{\omega}$ in xyz as shown in Figure 6.6.1. The direction of $\boldsymbol{\omega}$ establishes the axis of rotation and the magnitude ω is the angular speed of the rotation. The direction of $\boldsymbol{\omega}$ is chosen by another right-hand rule; that is, when the fingers of your right hand curl around in the general direction of the rotation, your thumb points in the direction of $\boldsymbol{\omega}$.

First consider some vector **B**. This can be any vector you care to make it. It can describe an electric field, a magnetic field, a position, a velocity, an acceleration, or anything else you wish. It is real and independent of the coordinate system we choose to use to describe it. We may describe **B** in either our fixed or rotating frames by

$$\mathbf{B} = B_x \hat{\mathbf{i}} + B_y \hat{\mathbf{j}} + B_z \hat{\mathbf{k}} \tag{6.6.1a}$$

$$\mathbf{B} = B_{x'} \hat{\mathbf{i}}' + B_{y'} \hat{\mathbf{j}}' + B_{z'} \hat{\mathbf{k}}' \tag{6.6.1b}$$

If we ask for the time derivative of this vector **B**, it makes a great deal of difference from which frame we are viewing **B**. Consider a vector **B** that is

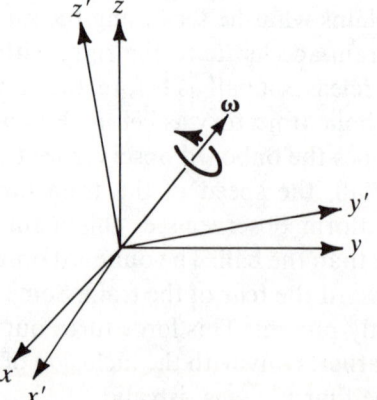 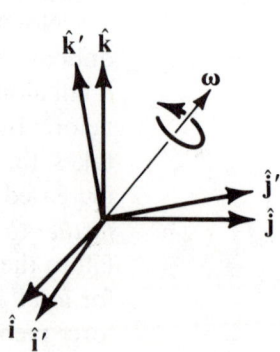

Figure 6.6.1 Rotating coordinates.

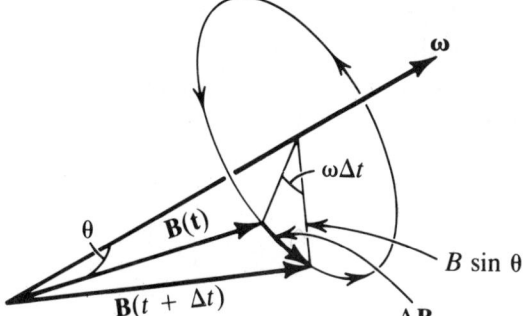

Figure 6.6.2 Motion in the fixed frame of a vector **B** at rest in a rotating frame.

constant with respect to $x'y'z'$. Then it is moving or changing along with the rotating frame as seen by the fixed frame xyz. Figure 6.6.2 shows such a vector **B** as it rotates about the axis $\boldsymbol{\omega}$.

There you can see that

$$\Delta B = (\omega \, \Delta t)(B \sin \theta)$$

$$\frac{\Delta B}{\Delta t} = \omega B \sin \theta$$

As $\Delta t \to 0$, we can write

$$\frac{\Delta B}{\Delta t} \to \frac{dB}{dt} = |\boldsymbol{\omega} \times \mathbf{B}|$$

In addition, we can see from Figure 6.6.2 that the *direction* of $\Delta \mathbf{B}$, or $d\mathbf{B}/dt$, is perpendicular to both $\boldsymbol{\omega}$ and **B**. Therefore,

$$\frac{d\mathbf{B}}{dt} = \boldsymbol{\omega} \times \mathbf{B} \tag{6.6.2}$$

This is the time derivative viewed from the unprimed, fixed frame, xyz. Remember that we arrived at Eq. 6.6.2 only for the special case of a vector **B** at rest in $x'y'z'$. Namely, Eq. 6.6.2 is true for **B** only when

$$\frac{d'\mathbf{B}}{dt} = 0 \tag{6.6.3}$$

where d'/dt means the time derivative as viewed from the primed coordinate system $x'y'z'$. Time derivatives of vectors differ in the two frames but time derivatives of scalars remain the same; they are merely time derivatives of numbers.

Let us take the time derivative in the fixed frame, d/dt, of equation 6.6.1a:

$$\frac{d\mathbf{B}}{dt} = \frac{dB_x}{dt}\hat{\mathbf{i}} + B_x\frac{d\hat{\mathbf{i}}}{dt} + \frac{dB_y}{dt}\hat{\mathbf{j}} + B_y\frac{d\hat{\mathbf{j}}}{dt} + \frac{dB_z}{dt}\hat{\mathbf{k}} + B_x\frac{d\hat{\mathbf{k}}}{dt}$$

$$\frac{d\mathbf{B}}{dt} = \dot{B}_x\hat{\mathbf{i}} + \dot{B}_y\hat{\mathbf{j}} + \dot{B}_z\hat{\mathbf{k}} \tag{6.6.4}$$

The dot indicates an *ordinary* time derivative of a scalar and the time derivatives of the unit vectors have all been set to zero since the unit vectors do not vary with time. This equation is exactly what we would expect.

Likewise, take the time derivative in the fixed reference frame, d/dt, of Eq. 6.6.1b:

$$\frac{d\mathbf{B}}{dt} = \frac{dB_{x'}}{dt}\hat{\mathbf{i}}' + B_{x'}\frac{d\hat{\mathbf{i}}'}{dt} + \frac{dB_{y'}}{dt}\hat{\mathbf{j}}' + B_{y'}\frac{d\hat{\mathbf{j}}'}{dt} + \frac{dB_{z'}}{dt}\hat{\mathbf{k}}' + B_{z'}\frac{d\hat{\mathbf{k}}'}{dt}$$

or (6.6.5)

$$\frac{d\mathbf{B}}{dt} = \dot{B}_{x'}\hat{\mathbf{i}}' + \dot{B}_{y'}\hat{\mathbf{j}}' + \dot{B}_{z'}\hat{\mathbf{k}}' + B_{x'}\frac{d\hat{\mathbf{i}}'}{dt} + B_{y'}\frac{d\hat{\mathbf{j}}'}{dt} + B_{z'}\frac{d\hat{\mathbf{k}}'}{dt}$$

The unit vectors $\hat{\mathbf{i}}'$, $\hat{\mathbf{j}}'$, and $\hat{\mathbf{k}}'$ are all fixed in frame $x'y'z'$; that is,

$$\frac{d'\hat{\mathbf{i}}'}{dt} = 0$$

Therefore, from Eqs. 6.6.2 and 6.6.3:

$$\frac{d\hat{\mathbf{i}}'}{dt} = \boldsymbol{\omega} \times \hat{\mathbf{i}}', \frac{d\hat{\mathbf{j}}'}{dt} = \boldsymbol{\omega} \times \hat{\mathbf{j}}', \frac{d\hat{\mathbf{k}}'}{dt} = \boldsymbol{\omega} \times \hat{\mathbf{k}}' \qquad (6.6.6)$$

Therefore, Eq. 6.6.5 can be rewritten as

$$\frac{d\mathbf{B}}{dt} = \dot{B}_{x'}\hat{\mathbf{i}}' + \dot{B}_{y'}\hat{\mathbf{j}}' + \dot{B}_{z'}\hat{\mathbf{k}}' + B_{x'}(\boldsymbol{\omega} \times \hat{\mathbf{i}}') + B_{y'}(\boldsymbol{\omega} \times \hat{\mathbf{j}}') + B_{z'}(\boldsymbol{\omega} \times \hat{\mathbf{k}}')$$

$$\frac{d\mathbf{B}}{dt} = \frac{d'\mathbf{B}}{dt} + \boldsymbol{\omega} \times B_{x'}\hat{\mathbf{i}}' + \boldsymbol{\omega} \times B_{y'}\hat{\mathbf{j}}' + \boldsymbol{\omega} \times B_{z'}\hat{\mathbf{k}}'$$

$$\frac{d\mathbf{B}}{dt} = \frac{d'\mathbf{B}}{dt} + \boldsymbol{\omega} \times \mathbf{B} \qquad (6.6.7)$$

This equation is now true for *any* vector **B**—the time derivative in the *fixed* frame is equal to what would be found by the *moving* frame plus an additional vector term, $\boldsymbol{\omega} \times \mathbf{B}$.

Since this applies to *any* vector **B**, it certainly applies to a position vector $\mathbf{r}'$. If we now apply Eq. 6.6.7 to a position vector $\mathbf{r}$ or $\mathbf{r}'$ locating some point P, we have

$$\mathbf{r} = \mathbf{r}' \qquad (6.6.8)$$

$$\mathbf{v} = \frac{d\mathbf{r}}{dt} = \frac{d'\mathbf{r}'}{dt} + \boldsymbol{\omega} \times \mathbf{r}'$$

$$\mathbf{v} = \dot{\mathbf{r}}' + \boldsymbol{\omega} \times \mathbf{r}' \qquad (6.6.9)$$

SECTION 6.6 / ROTATING COORDINATE SYSTEMS

where $\dot{\mathbf{r}}'$ is the time derivative seen by the moving frame, d'/dt. Application of Eq. 6.6.7 once more gives

$$\mathbf{a} = \frac{d\mathbf{v}}{dt} = \frac{d}{dt}(\dot{\mathbf{r}}' + \boldsymbol{\omega} \times \mathbf{r}')$$

$$\mathbf{a} = \ddot{\mathbf{r}}' + \dot{\boldsymbol{\omega}} \times \mathbf{r}' + 2\boldsymbol{\omega} \times \dot{\mathbf{r}}' + \boldsymbol{\omega} \times (\boldsymbol{\omega} \times \mathbf{r}') \quad (6.6.10)$$

where the dots indicate time differentiation as seen by the moving reference frame $x'y'z'$.

If we allow translation of the origins *and* rotation to occur simultaneously, then Eqs. 6.5.1, 6.5.2, and 6.5.3, combined with 6.6.8, 6.6.9, and 6.6.10 result in

$$\mathbf{r} = \mathbf{r}_0 + \mathbf{r}' \quad (6.6.11)$$

$$\mathbf{v} = \mathbf{v}_0 + \dot{\mathbf{r}}' + \boldsymbol{\omega} \times \mathbf{r}' \quad (6.6.12)$$

$$\mathbf{a} = \mathbf{a}_0 + \ddot{\mathbf{r}}' + \dot{\boldsymbol{\omega}} \times \mathbf{r}' + 2\boldsymbol{\omega} \times \dot{\mathbf{r}}' + \boldsymbol{\omega} \times (\boldsymbol{\omega} \times \mathbf{r}') \quad (6.6.13)$$

If we know the motion of an object as seen by the moving frame $x'y'z'$ (i.e., if we know $\mathbf{r}'$, $\dot{\mathbf{r}}'$, and $\ddot{\mathbf{r}}'$) and if we know how $x'y'z'$ is moving relative to our fixed frame xyz (i.e., if we know $\mathbf{r}_0$, $\mathbf{v}_0$, $\mathbf{a}_0$, and $\boldsymbol{\omega}$), then these equations allow us to completely describe the motion in our fixed frame xyz by giving us $\mathbf{r}$, $\mathbf{v}$, and $\mathbf{a}$.

The equation $\mathbf{F} = m\mathbf{a}$ holds in the fixed frame xyz but does not hold in the moving frame $x'y'z'$. Therefore, it will usually be $\mathbf{r}$, $\mathbf{v}$, and $\mathbf{a}$ that are understood or determined. From Eq. 6.6.13 we can then readily solve for $\ddot{\mathbf{r}}'$:

$$\ddot{\mathbf{r}}' = \mathbf{a} - \mathbf{a}_0 - 2\boldsymbol{\omega} \times \dot{\mathbf{r}}' - \boldsymbol{\omega} \times (\boldsymbol{\omega} \times \mathbf{r}') - \dot{\boldsymbol{\omega}} \times \mathbf{r}' \quad (6.6.14)$$

Multiplying through by the mass of the body, m, gives

$$\mathbf{F}' = m\ddot{\mathbf{r}}' = m\mathbf{a} - m\mathbf{a}_0 - 2m\boldsymbol{\omega} \times \dot{\mathbf{r}}' - m\boldsymbol{\omega} \times (\boldsymbol{\omega} \times \mathbf{r}') - m\dot{\boldsymbol{\omega}} \times \mathbf{r}' \quad (6.6.15)$$

Here $m\ddot{\mathbf{r}}'$ is the mass times the acceleration observed in the moving reference frame so it must be $\mathbf{F}'$, the *effective* force *observed* in that moving reference frame. On the other hand, $m\mathbf{a}$ is just $\mathbf{F}$, the *real* force as seen in the fixed frame. Recall that $\mathbf{F}$, or $m\mathbf{a}$, is due to contact forces—compressed springs, friction, tension in a rope, etc.—electromagnetic forces, gravitational forces, or nuclear forces. These are real forces in that their *cause* can be found, seen, or explained. The rest of the terms on the right-hand side are fictitious forces. $-m\mathbf{a}_0$ is the inertial force we've already discussed, due to relative acceleration of the origins of the two frames. It appears whether or not there is any relative rotation. $-2m\boldsymbol{\omega} \times \dot{\mathbf{r}}'$ is called the *Coriolis* force; $-m\boldsymbol{\omega} \times (\boldsymbol{\omega} \times \mathbf{r}')$ is called the *centrifugal* force. $-m\dot{\boldsymbol{\omega}} \times \mathbf{r}'$ is a fictitious force that exists (or seems to exist) whenever $\boldsymbol{\omega}$ changes. Since we will restrict ourselves to situations with constant $\boldsymbol{\omega}$, this term can be ignored.

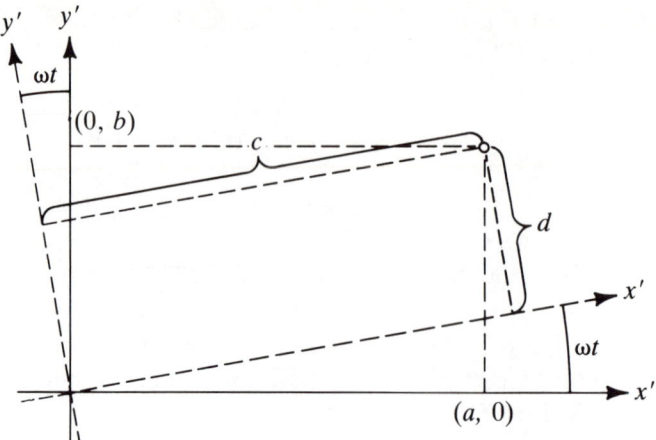

Figure 6.6.3 Coriolis force present in a rotating coordinate system.

The Coriolis force $-2m\boldsymbol{\omega} \times \dot{\mathbf{r}}'$ is a fictitious force present in a rotating coordinate system whenever an object moves. It is zero on a body at rest, but a body in motion experiences this fictitious force—always perpendicular to its velocity. As with all fictitious forces, this too comes about because the reference frame is moving—and with it, all the landmarks necessary to describe motion.

Consider, in Figure 6.6.3, an object (or person) located at position $(a, 0)$ moving with constant upward velocity **v**. Common sense, past experience, physical intuition, and Newton's laws lead us to *expect* the object to reach point (a, b) where $b = vt$ after some time t.

But in this same time t, the entire coordinate system—and with it all points of reference, or landmarks, if you will—has rotated through an angle ωt as shown. Instead of arriving at (a, b), we find the object at (c, d) where $c > a$ and $d < b$. This happens continuously and the straight path expected turns out instead to curve to the right. Such unusual behavior is explainable only by assuming there is an unexpected force acting

$$\mathbf{F}_{cor} = -2m\boldsymbol{\omega} \times \dot{\mathbf{r}}' \qquad (6.6.16)$$

You can imagine tossing a basketball back and forth while on a playground turntable or merry-go-round. When high speeds or large distances are involved, this Coriolis force must be taken into account. Airplane pilots must take account of it, as certainly must military pilots in firing bullets or rockets. The direction of rotation of cyclones—clockwise in the northern hemisphere and counter-clockwise in the southern hemisphere—is due to this Coriolis force.

Any particular motion will be described differently by observers in different reference frames. Equations 6.6.11, 6.6.12, 6.6.13, and the discussion following them enable us to take the observations of one frame and construct or predict what will be observed or described in another frame. Keep this in mind as you look at the following examples.

EXAMPLE 6.6.1

Consider a wheel moving along a straight path as sketched in Figure 6.6.4. If the wheel is of radius b and moving with linear speed v_0, then it has angular speed $\omega = v_0/b$ (that is, $v_0 = \omega b$). To an observer on the ground—in the xyz reference frame—what is the location of a point P on the wheel's rim and (more importantly) what acceleration does it have (or, equivalently, what net force is present) when it is at the top?

It is rather instructive to find the answer using several different methods. First describe the motion as seen by an observer in a frame *translating* along with the wheel, located at the axle of the wheel. This is the $x'y'z'$ frame sketched in Figure 6.6.4. In this frame, our moving observer would see

$$\mathbf{r}' = b\hat{\mathbf{j}}'$$
$$\dot{\mathbf{r}}' = \mathbf{v}' = b\omega\hat{\mathbf{i}}' = v_0\hat{\mathbf{i}}'$$
$$\ddot{\mathbf{r}}' = \mathbf{a}' = \frac{-v_0^2}{b}\hat{\mathbf{j}}' = -b\omega^2\hat{\mathbf{j}}'$$

The origin of the $x'y'z'$ frame is moving at velocity $\mathbf{v}_0 = v_0\hat{\mathbf{i}}$ with respect to xyz and is located at $\mathbf{r}_0 = v_0 t\hat{\mathbf{i}} + b\hat{\mathbf{j}}$. We have conveniently oriented our two frames so $\hat{\mathbf{i}} = \hat{\mathbf{i}}'$, $\hat{\mathbf{j}} = \hat{\mathbf{j}}'$, and $\hat{\mathbf{k}} = \hat{\mathbf{k}}'$. Therefore, from Eqs. 6.6.11, 6.6.12, and 6.6.13, we have

$$\mathbf{r} = v_0 t\hat{\mathbf{i}} + 2b\hat{\mathbf{j}}$$
$$\mathbf{v} = 2v_0\hat{\mathbf{i}}$$
$$\mathbf{a} = -\frac{v_0^2}{b}\hat{\mathbf{j}}$$

We might also consider this $x'y'z'$ frame fixed firmly to the wheel—"painted onto the wheel," if you like—and, thus, rotating with the wheel. It is of great convenience if we choose to look at it when the axes are once again aligned so that $\hat{\mathbf{i}} = \hat{\mathbf{i}}'$, $\hat{\mathbf{j}} = \hat{\mathbf{j}}'$, and

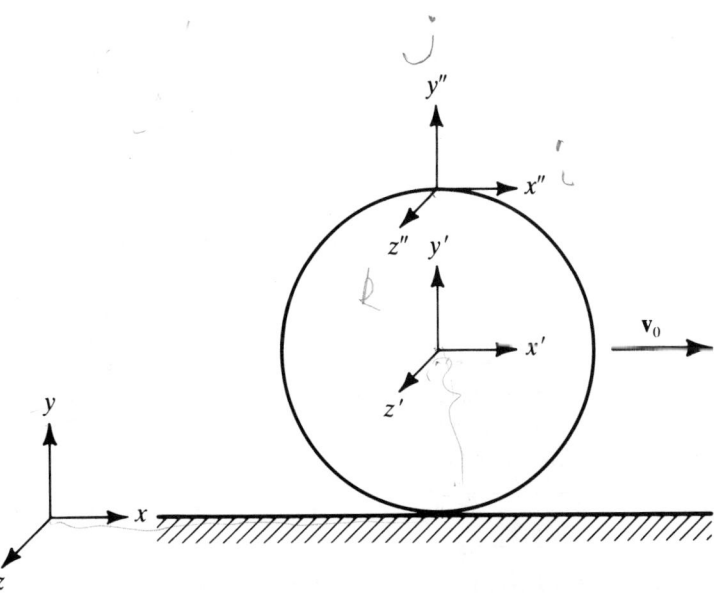

Figure 6.6.4 Motion of a wheel rolling along a straight line.

$\hat{\mathbf{k}} = \hat{\mathbf{k}}'$. No generality is lost in such a choice. *Now* an observer in $x'y'z'$ sees the following:

$$\mathbf{r}' = b\hat{\mathbf{j}}'$$

$$\dot{\mathbf{r}}' = 0$$

$$\ddot{\mathbf{r}}' = 0$$

We still have $\mathbf{r}_0 = v_0 t\hat{\mathbf{i}} + b\hat{\mathbf{j}}$, $\mathbf{v}_0 = v_0 \hat{\mathbf{i}}$, and $\mathbf{a}_0 = 0$. In order to use this rotating reference frame, we must use Eqs. 6.6.11, 6.6.12, and 6.6.13 and write for the angular rotation $\boldsymbol{\omega} = -\omega\hat{\mathbf{k}} = -(v_0/b)\hat{\mathbf{k}}$ (where the direction of $\boldsymbol{\omega}$ is chosen by the right-hand rule).

Then we shall have contributions from the following terms:

$$\boldsymbol{\omega} \times \mathbf{r}' = \left(-\frac{v_0}{b}\hat{\mathbf{k}}\right) \times (b\hat{\mathbf{j}}) = -v_0(\hat{\mathbf{k}} \times \hat{\mathbf{j}}) = +v_0\hat{\mathbf{i}}$$

$$\boldsymbol{\omega} \times (\boldsymbol{\omega} \times \mathbf{r}') = \left(-\frac{v_0}{b}\hat{\mathbf{k}}\right) \times (v_0\hat{\mathbf{i}}) = -\frac{v_0^2}{b}(\hat{\mathbf{k}} \times \hat{\mathbf{i}}) = -\frac{v_0^2}{b}\hat{\mathbf{j}}$$

where we have made frequent use of the equality of corresponding unit vectors in the two frames. The $\dot{\boldsymbol{\omega}} \times \mathbf{r}'$ and $\boldsymbol{\omega} \times \dot{\mathbf{r}}'$ terms contribute nothing since $\dot{\boldsymbol{\omega}} = \dot{\mathbf{r}}' = 0$. Finally, we get our already familiar results of

$$\mathbf{r} = (v_0 t\hat{\mathbf{i}} + b\hat{\mathbf{j}}) + (b\hat{\mathbf{j}}) = v_0 t\hat{\mathbf{i}} + 2b\hat{\mathbf{j}}$$

$$\mathbf{v} = (v_0\hat{\mathbf{i}}) + (v_0\hat{\mathbf{i}}) = 2v_0\hat{\mathbf{i}}$$

$$\mathbf{a} = -\frac{v_0^2}{b}\hat{\mathbf{j}}$$

Or we might begin by describing point P from a frame *attached to* the point P. This is shown in Figure 6.6.4 by the $x''y''z''$ frame. Now P is located at the origin, so it must remain at rest with no acceleration. Hence, we can very easily write

$$\mathbf{r}'' = 0$$

$$\dot{\mathbf{r}}'' = 0$$

$$\ddot{\mathbf{r}}'' = 0$$

So, once again, Eqs. 6.6.11, 6.6.12, and 6.6.13 give us

$$\mathbf{r} = v_0 t\hat{\mathbf{i}} + 2b\hat{\mathbf{j}}$$

$$\mathbf{v} = v_0\hat{\mathbf{i}}$$

$$\mathbf{a} = -\frac{v_0^2}{b}\hat{\mathbf{j}}$$

Individual terms, of course, are quite different than before but the final result is identical! Since $\dot{\mathbf{r}}''$ and $\ddot{\mathbf{r}}''$ were both zero, it makes no difference whether this $x''y''z''$ frame is considered to be "painted on the wheel" and rotating with it or merely translating.

EXAMPLE 6.6.2 Now consider a wheel of radius b that rolls along a circular path of radius ρ as sketched in Figure 6.6.5. Describe the motion of a point at the top of the wheel (like a

SECTION 6.6 / ROTATING COORDINATE SYSTEMS

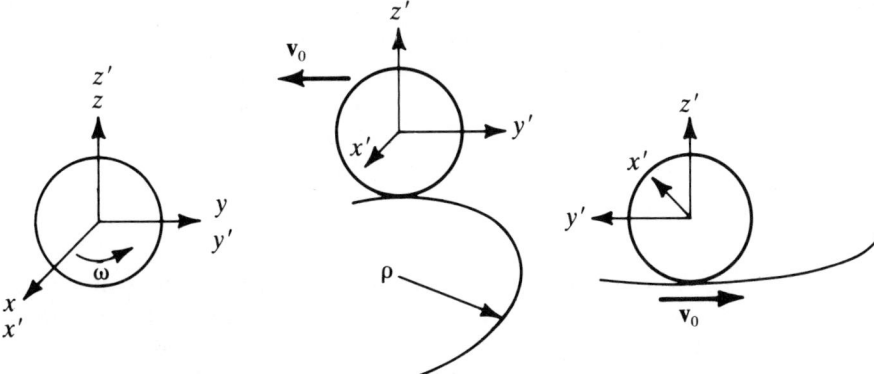

Figure 6.6.5 Wheel of radius b rolling along a circular path of radius ρ.

value stem) as seen by a *stationary observer* in the xyz frame. Begin by describing the point in the $x'y'z'$ frame, the *moving frame*. This frame moves with the axle of the wheel—think of it as carried *by the bicyclist* as he rides along the circle of radius ρ. The wheel rotates in this $x'y'z'$ reference frame. As before, we can readily write

$$\mathbf{r}' = b\hat{\mathbf{k}}'$$

$$\dot{\mathbf{r}}' = \mathbf{v}' = -v_0\hat{\mathbf{j}}'$$

$$\ddot{\mathbf{r}}' = \mathbf{a}' = -\frac{v_0^2}{b}\hat{\mathbf{k}}'$$

Again, we shall align the axes of $x'y'z'$ and xyz as shown in the left of Figure 6.6.5. The location and motion of the *origin* of the moving $x'y'z'$ frame (as it moves with the bicyclist along a circle of radius ρ) is given by

$$\mathbf{r}_0 = -\rho\hat{\mathbf{i}}$$

$$\mathbf{v}_0 = -v_0\hat{\mathbf{j}}$$

$$\mathbf{a}_0 = \frac{v_0^2}{\rho}\hat{\mathbf{i}} = \rho\omega^2\hat{\mathbf{i}}$$

where $v_0 = \rho\omega$ and $\boldsymbol{\omega} = \omega\hat{\mathbf{k}}$ give the angular velocity of the moving $x'y'z'$ frame. Equations 6.6.11, 6.6.12, and 6.6.13 now yield

$$\mathbf{r} = \mathbf{r}_0 + \mathbf{r}' = (-\rho\hat{\mathbf{i}}) + (b\hat{\mathbf{k}}) = -\rho\hat{\mathbf{i}} + b\hat{\mathbf{k}}$$

$$\mathbf{v} = \mathbf{v}_0 + \dot{\mathbf{r}}' + \boldsymbol{\omega} \times \mathbf{r}' = (-v_0\hat{\mathbf{j}}) + (-v_0\hat{\mathbf{j}}) + (\omega\hat{\mathbf{k}} \times b\hat{\mathbf{k}}) = -2v_0\hat{\mathbf{j}}$$

$$\mathbf{a} = \mathbf{a}_0 + \ddot{\mathbf{r}}' + \dot{\boldsymbol{\omega}} \times \mathbf{r}' + 2\boldsymbol{\omega} \times \dot{\mathbf{r}}' + \boldsymbol{\omega} \times (\boldsymbol{\omega} \times \mathbf{r}')$$

$$= \frac{v_0^2}{\rho}\hat{\mathbf{i}} + \left(-\frac{v_0^2}{b}\right)\hat{\mathbf{k}} + \mathbf{0} + 2\frac{v_0^2}{\rho}\hat{\mathbf{i}} + 0$$

$$= \left(\frac{v_0^2}{\rho} + 2\frac{v_0^2}{\rho}\right)\hat{\mathbf{i}} - \frac{v_0^2}{b}\hat{\mathbf{k}}$$

$$\mathbf{a} = 3\frac{v_0^2}{\rho}\hat{\mathbf{i}} - \frac{v_0^2}{b}\hat{\mathbf{k}}$$

for a description of $\mathbf{r}$, $\mathbf{v}$, and $\mathbf{a}$ for a point on top of the wheel as seen by the fixed observer in xyz.

EXAMPLE 6.6.3

Consider a bug on a rotating phonograph record. An observer sitting on the record sees this bug crawling radially outward with a constant speed and describes the bug's motion by

$$\mathbf{r}' = x'\hat{\mathbf{i}}' = v_0 t \hat{\mathbf{i}}'$$

$$\dot{\mathbf{r}}' = v_0 \hat{\mathbf{i}}'$$

$$\ddot{\mathbf{r}}' = 0$$

To predict what an outside, fixed observer sees, we need to evaluate the Coriolis term, $2\boldsymbol{\omega} \times \dot{\mathbf{r}}' = 2(\omega\hat{\mathbf{k}}) \times (v_0 \hat{\mathbf{i}}) = 2\omega v_0 \hat{\mathbf{j}}$, and the centripetal term, $\boldsymbol{\omega} \times (\boldsymbol{\omega} \times \mathbf{r}') = \omega\hat{\mathbf{k}} \times (\omega\hat{\mathbf{k}} \times v_0 t\hat{\mathbf{i}}) = -\omega^2 v_0 t\hat{\mathbf{i}}$. The term with $\dot{\boldsymbol{\omega}}$ is zero if we assume that the phonograph continues to turn at constant angular speed, and, of course, $\mathbf{a}_0$ is also zero. Thus, Eqs. 6.6.11, 6.6.12, and 6.6.13 give

$$\mathbf{r} = v_0 t \hat{\mathbf{i}}$$

$$\mathbf{v} = (v_0 \hat{\mathbf{i}}) + (\omega v_0 t \hat{\mathbf{j}}) = v_0 \hat{\mathbf{i}} + \omega v_0 t \hat{\mathbf{j}}$$

$$\mathbf{a} = -\omega^2 v_0 t \hat{\mathbf{i}} + 2\omega v_0 \hat{\mathbf{j}}$$

which may be easier to understand if we set $b = v_0 t$ for the radial distance the bug is from the center. Now the $\hat{\mathbf{i}}$-component of the acceleration is just the *centripetal* acceleration of an object moving in a circle. But this bug is moving in a spiral, so the $\hat{\mathbf{j}}$-component is also present.

We gain still more insight by looking at the forces as seen in the moving frame. Equation 6.6.15 gives us

$$\mathbf{F}' = m\ddot{\mathbf{r}}' = \mathbf{F}_{\text{real}} + m\omega^2 v_0 t \hat{\mathbf{i}} - 2m\omega v_0 \hat{\mathbf{j}}$$

If our bug suddenly steps onto an oily fingerprint where $\mathbf{F}_{\text{real}} = 0$ due to lack of friction, then the bug—as seen by our rotating observer—will appear to experience a force of

$$\mathbf{F}' = m\omega^2 v_0 t \hat{\mathbf{i}} - 2m\omega v_0 \hat{\mathbf{j}}$$

causing it to accelerate outward and to the right—the *centrifugal* force and the *Coriolis* force. The fixed observer, of course, simply sees the bug continuing in a straight line with constant velocity in the absence of an external force as another example of Newton's First Law.

6.7 Motion Observed on the Rotating Earth

Statics

Earth, since it rotates, establishes an accelerated reference frame and, therefore, a noninertial frame. That is, if we look closely we find that $\mathbf{F} = m\mathbf{a}$ must be modified (carefully). First, consider a body in equilibrium on the

SECTION 6.7 / MOTION OBSERVED ON THE ROTATING EARTH

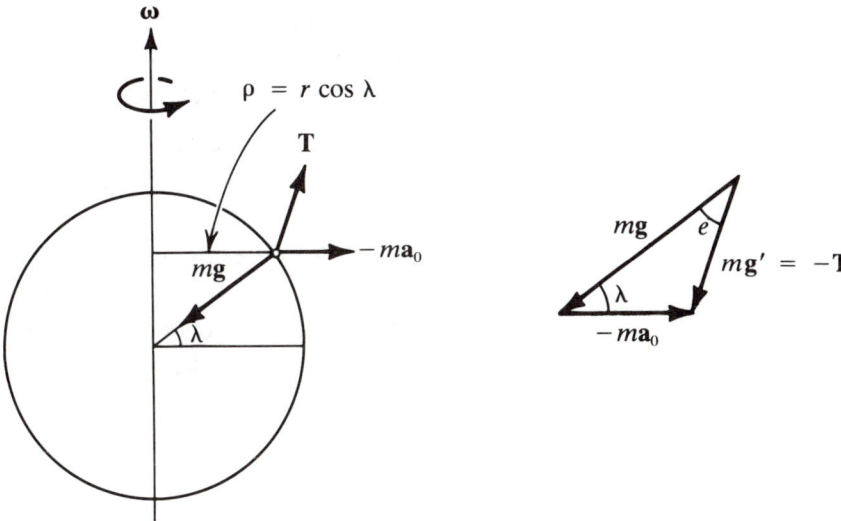

Figure 6.7.1 A plumb bob in static equilibrium on the rotating Earth.

rotating Earth as sketched in Figure 6.7.1. Locate the body—a plumb bob is an excellent example—at the origin of the $x'y'z'$ frame.

In our rotating $x'y'z'$ frame $\mathbf{r}' = 0$, $\dot{\mathbf{r}}' = 0$, and $\ddot{\mathbf{r}}' = 0$. Of course, $\dot{\boldsymbol{\omega}} = 0$. So only the acceleration $\mathbf{a}_0$ of the origin of the moving frame offers a contribution to Eq. 6.6.13 or 6.6.15. From the latter we have

$$m\ddot{\mathbf{r}}' = \mathbf{F}_{real} - m\mathbf{a}_0 = 0 \tag{6.7.1}$$

where

$$\mathbf{F}_{real} = \mathbf{T} + m\mathbf{g}$$

where $\mathbf{T}$ is the tension in a cord supporting the bob and $m\mathbf{g}$ is the real force of gravity directed toward the center of Earth and derivable from Newton's Law of Universal Gravitation. But our Earthbound observer *knows* that the tension $\mathbf{T}$ is simply

$$\mathbf{T} = -m\mathbf{g}' \tag{6.7.2}$$

the opposite of what he observes as the bob's weight or what he observes as the force of gravity pulling down on the bob. Therefore,

$$m\ddot{\mathbf{r}}' = -m\mathbf{g}' + m\mathbf{g} - m\mathbf{a}_0 = 0$$

or

$$\mathbf{g}' = \mathbf{g} - \mathbf{a}_0 \tag{6.7.3}$$

That is, the acceleration due to gravity *observed from* the moving Earth, $\mathbf{g}'$, is the vector difference between the acceleration we would expect (or would observe on a nonrotating Earth), $\mathbf{g}$, and the acceleration of the origin of the moving $x'y'z'$ frame, $\mathbf{a}_0$. This is equivalent to saying that the observed acceleration of gravity, $\mathbf{g}'$, is the vector sum of the real, expected acceleration, $\mathbf{g}$, and the centrifugal acceleration, $-\mathbf{a}_0$.

How large is $\mathbf{a}_0$ or how much do $\mathbf{g}$ and $\mathbf{g}'$ differ? On the right side of Figure 6.7.1 is a vector diagram corresponding to Eq. 6.7.3. From this diagram and the law of sines, we can write

$$\frac{\sin e}{ma_0} = \frac{\sin \lambda}{mg}$$

$$e \simeq \sin e = \frac{a_0}{g} \sin \lambda \tag{6.7.4}$$

where e is the angular difference between $\mathbf{g}$ and $\mathbf{g}'$, λ is the latitude, and a_0 is the centripetal acceleration given by

$$a_0 = \frac{v^2}{\rho} = \omega^2 \rho = \omega^2 r \cos \lambda \tag{6.7.5}$$

Therefore,

$$e = \frac{\omega^2 r \cos \lambda \sin \lambda}{g} = \frac{\omega^2 r \sin 2\lambda}{2g} \tag{6.7.6}$$

Cheat sheet

If we numerically evaluate e for $\lambda = 45°$ to give the maximum possible value of e (or the maximum deviation of g from g'), we find that

$$e_{\max} = 0.00174 \text{ radians} = 0.099981°$$

So, even though Earth rotates and, therefore, cannot be a completely perfect inertial frame, its deviation from such is very small if we consider only bodies at rest.

Dynamics

Bodies that *move* in the Earth frame of reference appear to have still other forces acting upon them. In an inertial (nonrotating Earth) frame, all motion is correctly described by $\mathbf{F} = m\ddot{\mathbf{r}}$ where, once again, $\mathbf{F}$ is the net real force. If we consider a body falling in a ballistic trajectory and neglect air resistance, then $\mathbf{F} = m\mathbf{g}$ and in the rotating Earth frame we observe motion given by

$$m\ddot{\mathbf{r}}' = m\mathbf{g} - m\mathbf{a}_0 - 2m\boldsymbol{\omega} \times \dot{\mathbf{r}}' - m\boldsymbol{\omega} \times (\boldsymbol{\omega} \times \mathbf{r}') \tag{6.7.7}$$

where we have already omitted the term in $\dot{\boldsymbol{\omega}}$ since we shall take Earth's rotation to be constant. The last term is very small, on the order of ω^2 (and $\omega = 0.26$ rad/hour), so this centrifugal term shall be omitted. We have seen that $\mathbf{g}' = \mathbf{g} - \mathbf{a}_0$ differs very little from $\mathbf{g}$ itself. Thus,

$$m\ddot{\mathbf{r}}' = m\mathbf{g}' - 2m\boldsymbol{\omega} \times \dot{\mathbf{r}}' \tag{6.7.8}$$

and this remaining Coriolis term will turn out to be important, as we said earlier, for bodies *moving* with respect to the Earth frame.

At a latitude of λ we can write

$$\boldsymbol{\omega} = \omega \cos \lambda \hat{\mathbf{j}}' + \omega \sin \lambda \hat{\mathbf{k}}' \tag{6.7.9}$$

SECTION 6.7 / MOTION OBSERVED ON THE ROTATING EARTH

so we can write the Coriolis term as

$$\boldsymbol{\omega} \times \dot{\mathbf{r}}' = \begin{vmatrix} \hat{\mathbf{i}}' & \hat{\mathbf{j}}' & \hat{\mathbf{k}}' \\ 0 & \omega \cos \lambda & \omega \sin \lambda \\ \dot{x}' & \dot{y}' & \dot{z}' \end{vmatrix}$$

$$= \hat{\mathbf{i}}'(\dot{z}'\omega \cos \lambda - \dot{y}'\omega \sin \lambda) + \hat{\mathbf{j}}'\dot{x}'\omega \sin \lambda - \hat{\mathbf{k}}'\dot{x}'\omega \cos \lambda \quad (6.7.10)$$

where x' is toward the east, y' is toward the north, and z' points up. Thus,

$$m\ddot{\mathbf{r}}' = mg'\hat{\mathbf{k}}' - 2m[\hat{\mathbf{i}}'(\dot{z}'\omega \cos \lambda - \dot{y}'\omega \sin \lambda) + \hat{\mathbf{j}}'\dot{x}'\omega \sin \lambda - \hat{\mathbf{k}}'\dot{x}'\omega \cos]$$

Since every position or component here carries a prime, indicating that *all* quantities are those measured in the Earth frame of reference, we may now conveniently omit all the primes and more simply write

$$m\ddot{\mathbf{r}} = mg\hat{\mathbf{k}} - 2m[\hat{\mathbf{i}}(\dot{z}\omega \cos \lambda - \dot{y}\omega \sin \lambda) + \hat{\mathbf{j}}\dot{x}\omega \sin \lambda - \hat{\mathbf{k}}\dot{x}\omega \cos \lambda] \quad (6.7.11)$$

which is equivalent to

$$\ddot{x} = -2\omega(\dot{z} \cos \lambda - \dot{y} \sin \lambda) \quad (6.7.12a)$$
$$\ddot{y} = -2\omega \dot{x} \sin \lambda \quad (6.7.12b)$$
$$\ddot{z} = -g + 2\omega \dot{x} \cos \lambda \quad (6.7.12c)$$

Such coupled differential equations may look formidable at first, but they are not really difficult to handle. Multiplying by dt and integrating immediately yields

$$\dot{x} = \dot{x}_0 + 2\omega(y \sin \lambda - z \cos \lambda) \quad (6.7.13a)$$
$$\dot{y} = \dot{y}_0 - 2\omega x \sin \lambda \quad (6.7.13b)$$
$$\dot{z} = \dot{z}_0 - gt + 2\omega x \cos \lambda \quad (6.7.13c)$$

where we have assumed $x_0 = y_0 = z_0 = 0$. (Remember, $\sin \lambda$ and $\cos \lambda$ are constants.) Now we can go back to Eq. 6.7.12a for $\ddot{x}$ and substitute values for $\dot{y}$ and $\dot{z}$ from Eqs. 6.7.13b and 6.7.13c:

$$\ddot{x} = -2\omega[(\dot{z}_0 - gt + 2\omega x \cos \lambda) \cos \lambda - (\dot{y}_0 - 2\omega x \sin \lambda) \sin \lambda]$$
$$= -2\omega \dot{z}_0 \cos \lambda + 2\omega gt \cos \lambda - 4\omega^2 x \cos^2 \lambda + 2\omega \dot{y}_0 \sin \lambda - 4\omega^2 x \sin^2 \lambda$$
$$= 2\omega gt \cos \lambda - 2\omega(\dot{z}_0 \cos \lambda - \dot{y}_0 \sin \lambda) - 4\omega^2 x$$
$$\ddot{x} = 2\omega gt \cos \lambda - 2\omega(\dot{z}_0 \cos \lambda - \dot{y}_0 \sin \lambda) \quad (6.7.14)$$

where we have neglected the last term since it involves the *square* of the angular velocity.

Now we have an ordinary, commonplace, uncoupled, second-order differential equation for x. Multiplying by dt and integrating gives

$$\int \ddot{x} \, dt = \int \frac{d\dot{x}}{dt} dt = \int_{\dot{x}_0}^{\dot{x}} d\dot{x} = 2\omega g \cos \lambda \int_0^t t \, dt - 2\omega(\dot{z}_0 \cos \lambda - \dot{y}_0 \sin \lambda) \int_0^t dt$$

$$\dot{x} = \dot{x}_0 + \omega g (\cos \lambda) t^2 - 2\omega(\dot{z}_0 \cos \lambda - \dot{y}_0 \sin \lambda) t \quad (6.7.15)$$

Further integration yields

$$x = \dot{x}_0 t + \tfrac{1}{3}\omega g(\cos \lambda)t^3 - \omega(\dot{z}_0 \cos \lambda - \dot{y}_0 \sin \lambda)t^2 \qquad (6.7.16)$$

This can now be used with Eqs. 6.7.13b and 6.7.13c to evaluate y and z:

$$\dot{y} = \dot{y}_0 - 2\omega(\sin \lambda)[\dot{x}_0 t + \tfrac{1}{3}\omega g \cos \lambda t^3 - \omega(\dot{z}_0 \cos \lambda - \dot{y}_0 \sin \lambda)t]$$
$$\dot{y} = \dot{y}_0 - 2\omega(\sin \lambda)(\dot{x}_0 t) + \omega^2(\cdots) \qquad (6.7.17)$$

Again, we shall neglect terms of second order in ω. Integration of this yields

$$y = \dot{y}_0 t - \omega \dot{x}_0 t^2 \sin \lambda \qquad (6.7.18)$$

Similarly, for z we have

$$\dot{z} = \dot{z}_0 - gt + 2\omega \cos \lambda[\dot{x}_0 t + \tfrac{1}{3}\omega g(\cos \lambda)t^3 - \omega(\dot{z}_0 \cos \lambda - \dot{y}_0 \sin \lambda)t]$$
$$\dot{z} = \dot{z}_0 - gt + 2\omega(\cos \lambda)\dot{x}_0 t + \omega^2(\cdots) \qquad (6.7.19)$$

$$z = \dot{z}_0 t - \tfrac{1}{2}gt^2 + \omega \dot{x}_0 t^2 \cos \lambda \qquad (6.7.20)$$

Consider dropping an object from rest, so that $\dot{x}_0 = \dot{y}_0 = \dot{z}_0 = 0$. When it has fallen a distance h ($z = -h$), where is it in terms of x and y? Equation 6.7.20 gives us the time it takes to fall:

$$-h = -\tfrac{1}{2}gt^2$$

or

$$t = \sqrt{\frac{2h}{g}}$$

Substituting this into Eqs. 6.7.16 and 6.7.18 gives the x- and y-coordinates of its position at that time:

$$x = \tfrac{1}{3}\omega g(\cos \lambda)\left(\frac{2h}{g}\right)^{3/2}$$
$$y = 0$$

That shows, as we expected (I hope!), that a ball dropped from rest (which would move straight downward in the absence of the Coriolis force) is deflected to the *east*. There is no deflection to the north.

6.8 Foucault Pendulum

Another interesting application of our understanding of motion in rotating reference frames is the Foucault pendulum. A heavy mass is attached to a long string, pulled away from equilibrium, and released. We might expect it to swing back and forth in a plane. And so it appears to do—at first. But closer observation shows that this plane of oscillation is slowly rotating about a vertical axis passing through the point of support.

It is really very easy to understand and explain this motion if the pendulum is suspended over the North Pole. Then we can say the plane of oscillation isn't rotating at all. It is really remaining fixed while we Earthbound observers actually rotate beneath it. Thus, we should expect the plane of oscillation of such a pendulum to make a complete rotation every 24 hours in a direction opposite to Earth's. And so it would—at the North Pole.

If the pendulum is *not* located at the North Pole, we find that the period of rotation of its plane of oscillation is increased (reaching infinity at the equator where no rotation of the plane of oscillation is observed at all!). We can no longer explain all the details so easily and quickly. Now we must look at all the forces on the pendulum as seen by us Earthbound observers. To do this explicitly, let us establish the origin of our $x'y'z'$ reference frame at the equilibrium position of the pendulum (and firmly fixed to Earth). Figure 6.8.1 shows the unit vector $\hat{\mathbf{k}}$ of such a frame. The vector $\boldsymbol{\omega}$ is parallel to the direction of Earth's rotation. The vector $\mathbf{T}$, the tension in the string, and $m\mathbf{g}$, the force of gravity, are the only two *real* forces acting on the pendulum. If we consider only these two forces, then we should get the ordinary, familiar simple harmonic motion of a pendulum swinging in a plane. But since we are observing this from a rotating frame, we know by now that the motion will be governed by

$$m\ddot{\mathbf{r}}' = \mathbf{T} + m\mathbf{g}' - 2m\boldsymbol{\omega} \times \dot{\mathbf{r}}'$$

where $\mathbf{g}' = \mathbf{g} - \mathbf{a}_0$ and terms in ω^2 and $\dot{\omega}$ have been neglected as before. Since all the quantities in this equation are those measured in the *same* frame, we may more conveniently write this as

$$m\ddot{\mathbf{r}} = \mathbf{T} + m\mathbf{g} - 2m\boldsymbol{\omega} \times \dot{\mathbf{r}} \qquad (6.8.1)$$

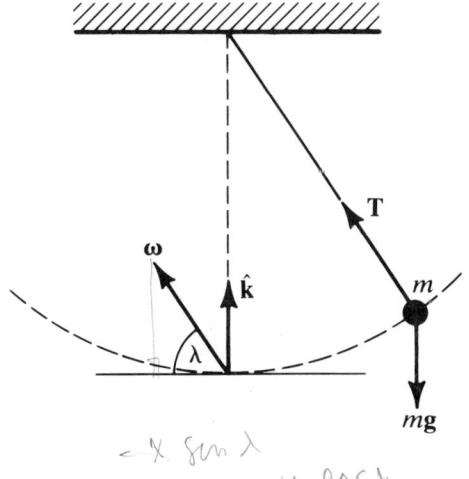

Figure 6.8.1 Real forces on a Foucault pendulum.

where **r** is *now* the position vector locating the pendulum in the Earth frame and **g** and **g**′ are essentially identical as we have shown earlier.

The Coriolis force, $-2m\boldsymbol{\omega} \times \dot{\mathbf{r}}$, is quite small. For any reasonably imaginable speeds of the pendulum its vertical component is entirely negligible when compared to mg. There are *no* horizontal forces perpendicular to $\dot{\mathbf{r}}$ other than the horizontal component of the Coriolis force. This horizontal component of the Coriolis force is responsible for the pendulum's unique behavior. To understand this motion, let us view it from yet *another* frame in which the plane of oscillation remains fixed. This new frame rotates about a vertical axis passing through the pendulum's support. It rotates with constant angular velocity $\boldsymbol{\Omega} = \Omega\hat{\mathbf{k}}$. In this new frame (call it $x''y''z''$), the motion is described by

$$\ddot{\mathbf{r}}'' = \ddot{\mathbf{r}} - 2\boldsymbol{\Omega} \times \dot{\mathbf{r}}'' - \boldsymbol{\Omega} \times (\boldsymbol{\Omega} \times \mathbf{r}'') - \dot{\boldsymbol{\Omega}} \times \mathbf{r}''$$

which we obtain from Eq. 6.6.13, but $\dot{\boldsymbol{\Omega}} = 0$, so

$$\ddot{\mathbf{r}}'' = \ddot{\mathbf{r}} - \Omega^2 \hat{\mathbf{k}} \times (\hat{\mathbf{k}} \times \mathbf{r}'') - 2\Omega\hat{\mathbf{k}} \times \dot{\mathbf{r}}'' \tag{6.8.2}$$

Recalling that

$$\dot{\mathbf{r}} = \dot{\mathbf{r}}'' + \Omega\hat{\mathbf{k}} \times \mathbf{r}'' \tag{6.8.3}$$

from Eq. 6.6.12, this now becomes

$$m\ddot{\mathbf{r}}'' = \mathbf{T} + m\mathbf{g} - 2m\boldsymbol{\omega} \times (\dot{\mathbf{r}}'' + \Omega\hat{\mathbf{k}} \times \mathbf{r}'') - m\Omega^2\hat{\mathbf{k}} \times (\hat{\mathbf{k}} \times \mathbf{r}'') - 2m\Omega\hat{\mathbf{k}} \times \dot{\mathbf{r}}''$$
$$= \mathbf{T} + m\mathbf{g} - 2m\boldsymbol{\omega} \times \dot{\mathbf{r}}'' - 2m\Omega\boldsymbol{\omega} \times (\mathbf{k} \times \mathbf{r}'')$$
$$\quad - m\Omega^2\hat{\mathbf{k}} \times (\hat{\mathbf{k}} \times \mathbf{r}'') - 2m\Omega\hat{\mathbf{k}} \times \dot{\mathbf{r}}''$$

$$m\ddot{\mathbf{r}}'' = \mathbf{T} + m\mathbf{g} - 2\Omega m\boldsymbol{\omega} \times (\hat{\mathbf{k}} \times \mathbf{r}'') - m\Omega^2\hat{\mathbf{k}} \times (\hat{\mathbf{k}} \times \mathbf{r}'') - 2m(\boldsymbol{\omega} + \Omega\hat{\mathbf{k}}) \times \dot{\mathbf{r}}'' \tag{6.8.4}$$

We can use the relation that

$$\mathbf{A} \times (\mathbf{B} \times \mathbf{C}) = \mathbf{B}(\mathbf{A} \cdot \mathbf{C}) - \mathbf{C}(\mathbf{A} \cdot \mathbf{B})$$

to evaluate

$$\boldsymbol{\omega} \times (\hat{\mathbf{k}} \times \mathbf{r}'') = \hat{\mathbf{k}}(\boldsymbol{\omega} \cdot \mathbf{r}'') - \mathbf{r}''(\boldsymbol{\omega} \cdot \hat{\mathbf{k}})$$

and

$$\hat{\mathbf{k}} \times (\hat{\mathbf{k}} \times \mathbf{r}'') = \hat{\mathbf{k}}(\hat{\mathbf{k}} \cdot \mathbf{r}'') - \mathbf{r}''(\hat{\mathbf{k}} \cdot \hat{\mathbf{k}})$$

(It may be worthwhile to remind ourselves that since **r** and **r**″ are measured in frames with a common origin, both observers see the same for their respective position vectors for any particular observation; i.e., $\mathbf{r} = \mathbf{r}''$. It's the way they *change* that differs—as given in Eq. 6.8.3.) Now we can write Eq. 6.8.4 for $m\ddot{\mathbf{r}}''$ as

$$m\ddot{\mathbf{r}}'' = \mathbf{T} + m\mathbf{g} - 2m\Omega[\hat{\mathbf{k}}(\boldsymbol{\omega} \cdot \mathbf{r}'') - \mathbf{r}''(\boldsymbol{\omega} \cdot \hat{\mathbf{k}})]$$
$$\quad - m\Omega^2[\hat{\mathbf{k}}(\hat{\mathbf{k}} \cdot \mathbf{r}'') - \mathbf{r}''] - 2m(\boldsymbol{\omega} + \Omega\hat{\mathbf{k}}) \times \dot{\mathbf{r}}'' \tag{6.8.5}$$

Every term except the last has components only along $\hat{\mathbf{k}}$ or $\mathbf{r}''$. So these terms could not possibly cause any rotation of the plane of oscillation. The last term could since it alone is perpendicular to the plane of oscillation defined by $\hat{\mathbf{k}}$ and $\mathbf{r}''$. But we have chosen to observe the pendulum from a frame in which the plane does *not* rotate. Hence, the last term must not cause a rotation in our new frame. To ensure this, we shall require that the horizontal component of this term vanishes:

$$\hat{\mathbf{k}} \cdot (\boldsymbol{\omega} + \Omega \hat{\mathbf{k}}) = 0 \tag{6.8.6}$$

or

$$\Omega = -\omega \sin \lambda \tag{6.8.7}$$

If our new *frame* rotates with this angular speed, then, of course, this is the angular speed of rotation of the plane of oscillation of the Foucault pendulum. At the equator, the latitude λ is zero, giving $\Omega = 0$; at the North Pole, $\lambda = 90°$, giving $\Omega = -\omega$ as we discussed earlier. It is interesting to note that we have gained a good understanding of the Foucault pendulum without ever explicitly and entirely solving the equation of motion for $\mathbf{r} = \mathbf{r}(t)$.

PROBLEMS

6.1 An object is located at

$$x = 3 \text{ m} \qquad y = 4 \text{ m} \qquad z = 5 \text{ m}$$

in a Cartesian coordinate system. Give its cylindrical polar coordinates and its spherical polar coordinates in frames whose origins coincide with the origin of the Cartesian frame.

6.2 A particle moves in a straight path along the line $x - 4$ m with a constant velocity $\mathbf{v} = (3 \text{ m/s})\hat{\mathbf{j}}$. Write its position and velocity in *polar* coordinates!

6.3 An object moves along a helical path so that its position is given by

$$\rho = R_0 \qquad \gamma = \omega t \qquad z = bt$$

in cylindrical polar coordinates. Find the speed as a function of time. Find the acceleration as a function of time.

6.4 An object moves on the surface of a sphere along a path given by

$$r = R_0 \qquad \gamma = \omega t \qquad \theta = bt$$

in spherical polar coordinates. Find the speed as a function of time. Find the acceleration as a function of time.

6.5 A particle moves in an elliptic orbit given by

$$\mathbf{r} = \hat{\imath} b \sin \omega t + \hat{\jmath} c \cos \omega t$$

Find its speed as a function of time. Find the acceleration as a function of time. If $b = c$, the ellipse becomes a circle. Show that its speed is then constant.

6.6 The location of a body in plane polar coordinates is given by

$$r = R_0 e^{kt} \qquad \theta = \omega t$$

Show that the angle between the velocity and acceleration vectors remains constant through time. Sketch the motion for $k = \omega$.

6.7 Derive an expression for the gradient in plane polar coordinates (*Hint*: use the chain rule for $\partial f / \partial x$ and $\partial f / \partial y$ and write $\hat{\imath}$ and $\hat{\jmath}$ in terms of $\hat{\mathbf{r}}$ and $\hat{\boldsymbol{\theta}}$). Show that $\nabla f(r) = \hat{\mathbf{r}}(\partial f / \partial r)$.

6.8 Derive an expression for the gradient in cylindrical polar coordinates.

6.9 Derive an expression for the gradient in spherical polar coordinates.

6.10 Demonstrate that the divergence in plane polar coordinates is given by

$$\nabla \cdot \mathbf{V} = \frac{1}{r} \frac{\partial}{\partial r}(rV_r) + \frac{1}{r} \frac{\partial V_\theta}{\partial \theta}$$

(That is, show that this reduces to $\partial V_x / \partial x + \partial V_y / \partial y$ when V_r and V_θ are expressed in terms of V_x and V_y and when the derivatives with respect to r and θ are replaced by derivatives with respect to x and y through the chain rule.)

6.11 Demonstrate that the divergence in cylindrical polar coordinates is given by

$$\nabla \cdot \mathbf{V} = \frac{1}{\rho} \frac{\partial}{\partial \rho}(\rho V_\rho) + \frac{1}{\rho} \frac{\partial V_\theta}{\partial \theta} + \frac{\partial V_z}{\partial z}$$

6.12 Demonstrate that the divergence of a vector function in spherical polar coordinates is given by

$$\nabla \cdot V = \frac{1}{r^2} \frac{\partial}{\partial r}(r^2 V_r) + \frac{1}{r \sin \theta} \frac{\partial}{\partial \theta}(V_\theta \sin \theta) + \frac{1}{r \sin \theta} \frac{\partial V_\phi}{\partial \phi}$$

6.13 Demonstrate that the curl in cylindrical polar coordinates is given by

$$\nabla \times \mathbf{V} = \hat{\boldsymbol{\rho}} \left(\frac{1}{\rho} \frac{\partial V_z}{\partial \theta} - \frac{\partial V_\theta}{\partial z} \right) + \hat{\boldsymbol{\theta}} \left(\frac{\partial V_\rho}{\partial z} - \frac{\partial V_z}{\partial \rho} \right) + \hat{\mathbf{k}} \left(\frac{1}{\rho} \frac{\partial}{\partial \rho}(\rho V_\theta) - \frac{1}{\rho} \frac{\partial V_\rho}{\partial \theta} \right)$$

6.14 Demonstrate that the curl in spherical polar coordinates is given by

$$\nabla \times \mathbf{V} = \hat{\mathbf{r}} \frac{1}{r \sin \theta} \left(\frac{\partial}{\partial \theta}(V_\phi \sin \theta) - \frac{\partial V_\theta}{\partial \phi} \right)$$

$$+ \hat{\boldsymbol{\theta}} \frac{1}{r \sin \theta} \left(\frac{\partial V_r}{\partial \phi} - \sin \theta \frac{\partial}{\partial r}(rV_\phi) \right)$$

$$+ \hat{\boldsymbol{\theta}} \frac{1}{r} \left(\frac{\partial}{\partial r}(rV_\theta) - \frac{\partial V_r}{\partial \theta} \right)$$

6.15 The gradient and curl in cylindrical coordinates are

$$\nabla V = \frac{\partial V}{\partial \rho}\hat{\boldsymbol{\rho}} + \frac{1}{\rho}\frac{\partial V}{\partial \theta}\hat{\boldsymbol{\theta}} + \frac{\partial V}{\partial z}\hat{\mathbf{k}}$$

$$\nabla \times \mathbf{F} = \left(\frac{1}{\rho}\frac{\partial F_z}{\partial \theta} - \frac{\partial F_\theta}{\partial z}\right)\hat{\boldsymbol{\rho}} + \left(\frac{\partial F_\rho}{\partial \theta} - \frac{\partial F_z}{\partial \rho}\right)\hat{\boldsymbol{\theta}} + \left(\frac{1}{\rho}\frac{\partial}{\partial \rho}(\rho F_\theta) - \frac{1}{\rho}\frac{\partial F_\rho}{\partial \theta}\right)\hat{\mathbf{k}}$$

Determine which of the following forces are conservative and find the potential energy function for those that are:
  (a) $\mathbf{F} = b(\rho\hat{\boldsymbol{\rho}} + z\hat{\mathbf{k}})$
  (b) $\mathbf{F} = b(\rho \cos\theta\hat{\boldsymbol{\rho}} + \rho \sin\theta\hat{\boldsymbol{\theta}} + z\hat{\mathbf{k}})$
  (c) $\mathbf{F} = b2\rho \cos\theta\hat{\boldsymbol{\rho}} - c\rho^2 \sin\theta\hat{\boldsymbol{\theta}} + 2dz\hat{\mathbf{k}}$
  (d) $\mathbf{F} = a\rho\hat{\boldsymbol{\rho}} + b \sin\theta\hat{\boldsymbol{\theta}} + c\rho z\hat{\mathbf{k}}$

6.16 The gradient and curl in spherical polar coordinates are

$$\nabla V = \frac{\partial V_r}{\partial r} + \frac{1}{r}\frac{\partial V}{\partial \theta}\hat{\boldsymbol{\theta}} + \frac{1}{r \sin\theta}\frac{\partial V}{\partial \phi}\hat{\boldsymbol{\phi}}$$

$$\nabla \times \mathbf{F} = \frac{1}{r \sin\theta}\left(\frac{\partial}{\partial \theta}(F_\phi \sin\theta) - \frac{\partial F_\theta}{\partial \phi}\right)\hat{\mathbf{r}}$$

$$+ \frac{1}{r \sin\theta}\left(\frac{\partial F_r}{\partial \phi} - \sin\theta\frac{\partial}{\partial r}(rF_\phi)\right)\hat{\boldsymbol{\theta}}$$

$$+ \frac{1}{r}\left(\frac{\partial}{\partial r}(rF_\theta) - \frac{\partial F_r}{\partial \theta}\right)\hat{\boldsymbol{\phi}}$$

Determine which of the following forces are conservative and find the potential energy function for those that are:
  (a) $\mathbf{F} = br \sin\theta \sin\phi\hat{\mathbf{r}} + br \cos\theta \sin\phi\hat{\boldsymbol{\theta}} + br \sin\theta \cos\phi\hat{\boldsymbol{\phi}}$
  (b) $\mathbf{F} = 2br \sin\theta \sin\phi\hat{\mathbf{r}} + br^2 \cos\theta \sin\phi\hat{\boldsymbol{\theta}} + br^2 \sin\theta \cos\phi\hat{\boldsymbol{\phi}}$
  (c) $\mathbf{F} = 2br \sin\theta \sin\phi\hat{\mathbf{r}} + br^2 \sin\theta \cos\phi\hat{\boldsymbol{\theta}} + br \sin^2\theta \cos\phi\hat{\boldsymbol{\phi}}$
  (d) $\mathbf{F} = b \sin^2\theta \sin\phi\hat{\mathbf{r}} + 2br \sin\theta \cos\theta \sin\phi\hat{\boldsymbol{\theta}} + br \sin^2\theta \cos\phi\hat{\boldsymbol{\phi}}$
  (e) $\mathbf{F} = br^2\hat{\mathbf{r}} + br \cos\theta\hat{\boldsymbol{\theta}} + br^2 \sin\phi\hat{\boldsymbol{\phi}}$
  (f) $\mathbf{F} = \frac{k}{r^2}\hat{\mathbf{r}}$

6.17 Derive an expression for $\dddot{\mathbf{r}} = d^3\mathbf{r}/dt = d\mathbf{a}/dt$—sometimes referred to as the *jerk*—in terms of the primed quantities measured in a moving reference frame as we have already done for $\mathbf{r}$, $\dot{\mathbf{r}}$, and $\ddot{\mathbf{r}}$.

6.18 A cup of coffee is dropped aboard an airplane accelerating along a smooth runway. If it falls a vertical distance of h, find its horizontal displacement when it hits the floor. Assume a constant horizontal acceleration a for the airplane.

6.19 A plumb bob of mass m is carried by a scientist aboard a moving train. Find the tension in the string and the deflection from vertical if
  (a) the train moves with constant acceleration $\mathbf{a}_0$ along a straight track.
  (b) the train rounds a curve of radius R at a constant speed v_0.

6.20 Consider a freely falling body. Write down and solve its equation of motion as seen by an observer in a noninertial reference frame that has an acceleration **g**. Then transform this motion back to the *fixed* inertial frame of a ground-based observer.

6.21 Consider a body that falls straight down subject to a frictional air drag proportional to its speed relative to the still air. Write down and solve the equation of motion as seen by an observer in a noninertial reference frame that has an acceleration **g**. Transform this back to the *fixed* inertial frame of a ground-based observer.

6.22 A bug crawls with speed v in a circular path of radius b atop a phonograph record that rotates with constant angular velocity ω. Describe this motion in a frame rotating with the record. Find the acceleration of the bug relative to a fixed, outside, inertial observer. To this inertial observer, what is the cause of this acceleration?

Two interesting cases are $v = b\omega$ and $v = -b\omega$. Find the acceleration seen by the inertial observer for both cases.

6.23 A bicycle wheel of radius b rolls along a straight path with speed v_0 as in the earlier example. What is the acceleration (with respect to the ground) of the valve stem when it is at the *bottom*?

6.24 As in an earlier example, a wheel of radius b rolls along a circular path of radius ρ. Find the acceleration (relative to the ground) of the valve stem when it is at the *bottom*.

*6.25 Consider a 1-g bug on a 30-cm (12-in.) phonograph record. The bug crawls outward along a radial scratch on the record at the rate of 3 cm/s. Initially, the phonograph is turning at 33 rev/min and the bug is at the center. Just as the bug starts to crawl, the switch is changed to 45 rev/min. Assume the rotational speed increases linearly and finally reaches 45 rev/min as the bug reaches the edge of the record. Find the force on the bug as a function of time. (While this is a straightforward *analytical* problem, it is also interesting to solve it by numerical techniques.)

6.26 Consider children playing on a playground turntable (or merry-go-round) 6 m in diameter, turning at 10 revolutions per minute. What will be the Coriolis force observed on a 600-g basketball thrown with a horizontal speed of 5 m/s? It is thrown between two children on opposite ends of a diameter. By how much does it miss its mark due to this Coriolis force?

6.27 An airplane is circling in a holding pattern with a radius of 5 km at a speed of 200 km/hr. A cup drops from rest a distance of 1 m on the airplane. Find the lateral displacement of its point of impact. (That is, where does it strike the floor?)

6.28 A satellite is in circular orbit around Earth with angular velocity ω. Define a coordinate system moving with the satellite at its origin and x-axis running from there to Earth and the y-axis in the direction of the satellite's motion. Show that a particle near the satellite is observed to experience the following inertial or fictitious force:

$$\mathbf{F}_{\text{inertial}} = (3\omega^2 x \hat{\mathbf{i}} - 2\boldsymbol{\omega} \times \mathbf{r})m$$

6.29 Find the magnitude and direction of the centripetal force on a 1-metric-ton airplane sitting on the ground at 40°N latitude.

6.30 Find the magnitude and direction of the Coriolis force on a 1 metric ton airplane travelling 600 km/hr due east at 40°N latitude.

6.31 A body is dropped from rest at a height h above Earth's surface at latitude λ. Calculate the lateral displacement of the point of impact due to the Coriolis force. For $h = 100\ m$ and $\lambda = 40°$, find the value of this displacement.

6.32 A body is thrown straight upward with initial velocity v_0 from a point on Earth's surface at latitude λ. Calculate the lateral displacement of its point of impact due to the Coriolis force. For $v = 10\ m/s$ and $\lambda = 40°$, find the value of this displacement.

6.33 A hockey puck is given an initial speed v on a frictionless surface at latitude λ. Find the magnitude and direction of the Coriolis force on the puck. Show that the puck will move in a circle. Find the radius of the circle.

6.34 Very large guns are capable of hurling projectiles several tens of kilometers (recall the shelling of Paris in both World Wars or the use of *U.S.S. New Jersey* during the Vietnam War). Neglect air resistance and rotational effects for the moment. Find the minimum initial velocity necessary to fire a shell 50 km north ($v = \sqrt{Rg}$, remember?). Now, considering the Coriolis force, how far will it be deflected if it is fired at a latitude of 40°?

6.35 Winds blow generally from west to east in the northern hemisphere. What pressure gradient is required to move this air of average density ρ with an average speed v? Make some reasonable estimates of ρ and v to provide an estimate of the pressure gradient in Pa/m.

6.36 It has been suggested that birds may be able to determine their latitude by sensing the Coriolis force. Consider a 500-gm bird in level flight at a latitude of 40°, flying with a speed of 40 km/hr. What force must be exerted sideways (to compensate for the Coriolis force) to enable the bird to fly in a straight line?

*6.37 Air resistance *with* Coriolis forces sounds like a formidable situation best handled with numerical techniques. Reconsider Problem 6-34. A shell is to be fired 50 km away. First order calculations, ignoring both air resistance and Coriolis forces, give a minimum initial speed of $v = \sqrt{Rg}$ where R is the range, and g the acceleration of gravity. Again, let $R = 50$ km. Use this as the initial speed and fire at 45°. Consider Coriolis forces at 40°N latitude and air resistance proportional to the velocity that would give a terminal velocity for free fall of 200 km/hr. Write a computer program and use it to find how far the projectile misses its target. Compare this with your analytical results if the Coriolis forces and air resistance are considered separately.

7
CENTRAL FORCES

7.1 Introduction

A central force is one whose direction is always along a radius; that is, either toward or away from a point we shall use as an origin (or force center), and whose magnitude depends solely upon the *distance* from that origin, r. We can write this as

$$\mathbf{F} = F(r)\hat{\mathbf{r}} \qquad (7.1.1)$$

where $\hat{\mathbf{r}}$ is a unit vector in the radial direction. Since $\hat{\mathbf{r}} = \mathbf{r}/r$, a central force can also be written as

$$\mathbf{F} = \frac{F(r)}{r}\mathbf{r} \qquad (7.1.1a)$$

Central forces are important because we encounter them so often in physics. The gravitational force is a central force. The electrostatic force between two charges is a central force. An isotropic three-dimensional harmonic oscillator, like an atom is a cubic crystal, is governed by a central force.

7.2 Potential Energy and Central Forces

The central force of Eq. 7.1.1 resembles the position-dependent forces $F(x)$ in Chapter 2. Then we found it very useful to talk of a potential energy $V(x)$. Can we construct a similar potential energy function for these central forces? That is, are central forces conservative? They are conservative if we can show that their curl vanishes; that is, if

$$\mathbf{\nabla} \times \mathbf{F} = 0 \qquad (7.2.1)$$

Using Eq. 7.1.1a, we write a central force as

$$\mathbf{F} = \frac{F(r)}{r}(x\hat{\mathbf{i}} + y\hat{\mathbf{j}} + z\hat{\mathbf{k}}) \qquad (7.2.2)$$

Thus, the three rectangular components are

$$F_x = \frac{x}{r}F(r) \qquad (7.2.3a)$$

$$F_y = \frac{y}{r}F(r) \qquad (7.2.3b)$$

and

$$F_z = \frac{z}{r}F(r) \qquad (7.2.3c)$$

Now use these to evaluate $\mathbf{\nabla} \times \mathbf{F}$:

$$\mathbf{\nabla} \times \mathbf{F} = \begin{vmatrix} \hat{\mathbf{i}} & \hat{\mathbf{j}} & \hat{\mathbf{k}} \\ \frac{\partial}{\partial x} & \frac{\partial}{\partial y} & \frac{\partial}{\partial z} \\ F_x & F_y & F_z \end{vmatrix}$$

$$= \hat{\mathbf{i}}\left(\frac{\partial F_z}{\partial y} - \frac{\partial F_y}{\partial z}\right) + \hat{\mathbf{j}}\left(\frac{\partial F_x}{\partial z} - \frac{\partial F_z}{\partial x}\right) + \hat{\mathbf{k}}\left(\frac{\partial F_y}{\partial x} - \frac{\partial F_x}{\partial y}\right) \qquad (7.2.4)$$

We can look at just the x-component for the present:

$$(\mathbf{\nabla} \times \mathbf{F})_x = \frac{\partial F_z}{\partial y} - \frac{\partial F_y}{\partial z} \qquad (7.2.5)$$

$$\frac{\partial F_z}{\partial y} = \frac{\partial}{\partial y}\left[\frac{z}{r}F(r)\right]$$

$$= z\frac{\partial}{\partial y}\left[\frac{F(r)}{r}\right] = z\frac{\partial}{\partial r}\left[\frac{F(r)}{r}\right]\frac{\partial r}{\partial y}$$

$$\frac{\partial F_z}{\partial y} = z\frac{\partial r}{\partial y}\frac{\partial}{\partial r}\left[\frac{F(r)}{r}\right] \qquad (7.2.6a)$$

Likewise,
$$\frac{\partial F_y}{\partial z} = y \frac{\partial r}{\partial z} \frac{\partial}{\partial r}\left[\frac{F(r)}{r}\right] \qquad (7.2.6b)$$

We can evaluate $\partial r/\partial y$ or $\partial r/\partial z$ by writing r in terms of x, y, and z:
$$r = (x^2 + y^2 + z^2)^{1/2}$$
$$\frac{\partial r}{\partial y} = \frac{1}{2}(x^2 + y^2 + z^2)^{-1/2}(2y) = \frac{y}{r} \qquad (7.2.7a)$$

Likewise,
$$\frac{\partial r}{\partial z} = \frac{z}{r} \qquad (7.2.7b)$$

We can now use this to evaluate the curl from Eq. 7.2.5:
$$(\nabla \times \mathbf{F})_x = \frac{\partial F_z}{\partial y} - \frac{\partial F_y}{\partial z}$$
$$= z \frac{\partial r}{\partial y} \frac{\partial}{\partial r}\left[\frac{F(r)}{r}\right] - y \frac{\partial r}{\partial z} \frac{\partial}{\partial r}\left[\frac{F(r)}{r}\right]$$
$$= z \frac{y}{r} \frac{\partial}{\partial r}\left[\frac{F(r)}{r}\right] - y \frac{z}{r} \frac{\partial}{\partial r}\left[\frac{F(r)}{r}\right]$$
$$= \frac{zy}{r} \frac{\partial}{\partial r}\left[\frac{F(r)}{r}\right] - \frac{yz}{r} \frac{\partial}{\partial r}\left[\frac{F(r)}{r}\right]$$
$$(\nabla \times \mathbf{F})_x = 0 \qquad (7.2.8)$$

We get similar results for the other two components, so
$$\nabla \times \mathbf{F} = 0 \qquad (7.2.9)$$

which means that a central force is conservative and is associated with a potential energy function
$$\mathbf{F} = \mathbf{F}(r) = -\nabla V(r) \qquad (7.2.10)$$

Now it is convenient to write the gradient operator ∇ in spherical coordinates. It can be shown that ∇ has the following form in spherical coordinates:
$$\nabla = \hat{\mathbf{r}} \frac{\partial}{\partial r} + \hat{\boldsymbol{\theta}} \frac{1}{r} \frac{\partial}{\partial \theta} + \hat{\boldsymbol{\phi}} \frac{1}{r \sin\theta} \frac{\partial}{\partial \phi} \qquad (7.2.11)$$

Since $V = V(r)$, Eq. 7.2.10 reduces to
$$\mathbf{F} = -\nabla V = -\frac{\partial V}{\partial r} \hat{\mathbf{r}}$$

or

$$F = -\frac{\partial V}{\partial r} \quad (7.2.12)$$

Again this resembles our earlier discussion of the one-dimensional force, $F(x)$. Just as then, we can calculate the potential energy by

$$V = V(r) = \int_r^{r_s} F(r)\,dr = -\int_{r_s}^r F(r)\,dr \quad (7.2.13)$$

7.3 Angular Momentum and Central Forces

From your background in physics you should recall that torque produces a change in angular momentum (just as force produces a change in linear momentum). In fact, the time rate of change of the angular momentum is equal to the torque:

$$\tau = \frac{d\mathbf{L}}{dt} \quad (7.3.1)$$

where τ is the torque (the rotational analogue of a force) given by

$$\tau = \mathbf{r} \times \mathbf{F} \quad (7.3.2)$$

and $\mathbf{L}$ is the angular momentum given by

$$\mathbf{L} = \mathbf{r} \times \mathbf{p} \quad (7.3.3)$$

where $\mathbf{p}$ is the linear momentum ($\mathbf{p} = m\mathbf{v}$). Central forces *exert no torques* about the origin. This is clear from Eq. 7.3.2, the definition of torque, since $\mathbf{r}$ and $\mathbf{F}$ are parallel. By the definition of a vector cross product, their cross product must vanish. If the torque is zero, then, from Eq. 7.3.1, the *angular momentum is constant*.

It may help to see Eq. 7.3.1 derived again. Most of this course—and, indeed, much of physics—can be viewed as a careful (my students often say "devious") application of Newton's Second Law. So we may as well begin with that.

$$\mathbf{F} = m\mathbf{a}$$

To find the torque, we can write

$$\tau = \mathbf{r} \times \mathbf{F} = \mathbf{r} \times (m\mathbf{a})$$

This is equivalent to Eq. 7.3.1 since, by Eq. 7.3.3,

$$\frac{d\mathbf{L}}{dt} = \frac{d}{dt}(\mathbf{r} \times \mathbf{p}) = \frac{d}{dt}(\mathbf{r} \times m\mathbf{v})$$

$$= \frac{d\mathbf{r}}{dt} \times m\mathbf{v} + \mathbf{r} \times \frac{d}{dt}(m\mathbf{v})$$

$$= \mathbf{v} \times m\mathbf{v} + \mathbf{r} \times m\mathbf{a}$$

$$\frac{d\mathbf{L}}{dt} = \mathbf{r} \times m\mathbf{a} = \boldsymbol{\tau}$$

since $\mathbf{v} \times \mathbf{v} = 0$.

Now, what does constant angular momentum tell about the motion? The angular momentum

$$\mathbf{L} = \mathbf{r} \times \mathbf{p}$$

is a vector perpendicular to the plane determined by the location vector $\mathbf{r}$ and the momentum $\mathbf{p}$ (or velocity $\mathbf{v}$). If $\mathbf{L}$ remains constant, this plane remains constant. That is, motion under a central force is confined to a plane. This is quite important. It is also very good news—the motion is describable in only two dimensions rather than the three that might well have been anticipated.

Constant angular momentum is all we need to have motion confined to a plane. But this important result is worth discussing a bit more. Figure 7.3.1 shows a body of mass m moving with velocity $\mathbf{v}$ located at position $\mathbf{r}$ from the origin and acted upon by a central force $\mathbf{F}$. The change in velocity $\Delta\mathbf{v}$ must be parallel to $\mathbf{F}$ from Newton's Second Law. Thus, the new velocity $\mathbf{v} + \Delta\mathbf{v}$ always lies in the plane defined by the present velocity $\mathbf{v}$ and the position $\mathbf{r}$. Likewise, the new position $\mathbf{r} + \Delta\mathbf{r}$ must lie in this same plane. All future motion is confined to this one plane.

Since we are dealing with a central force with a component in only the radial direction, any description of the motion will be far easier if we *begin* by using polar coordinates. Since we have already seen that the motion is

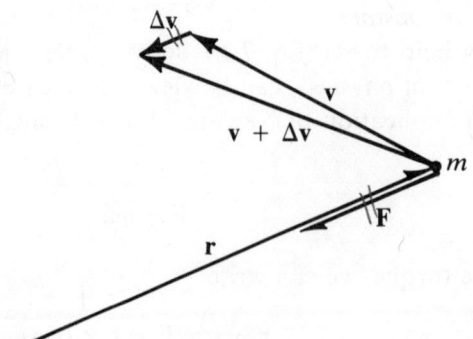

Figure 7.3.1 Planar motion results from central forces.

confined to a plane, it really does not matter whether we use spherical polar coordinates or plane polar coordinates. Using the results from Chapter 6 for expressing acceleration in polar coordinates, we can write Newton's Second Law as

$$\mathbf{F} = m\mathbf{a}$$

$$F(r)\hat{\mathbf{r}} = m[(\ddot{r} - r\dot{\theta}^2)\hat{\mathbf{r}} + (r\ddot{\theta} + 2\dot{r}\dot{\theta})\hat{\boldsymbol{\theta}}] \quad (7.3.4)$$

As always, this single vector equation implies two ordinary, scalar equations—one for each component:

$$F(r) = m\ddot{r} - mr\dot{\theta}^2 \quad (7.3.5a)$$

$$0 = mr\ddot{\theta} + 2m\dot{r}\dot{\theta} \quad (7.3.5b)$$

Let's look first at Eq. 7.3.5b, the θ-component of $\mathbf{F} = m\mathbf{a}$. With some insight and a little algebraic fidgeting, we can write this as

$$\frac{d}{dt}(mr^2\dot{\theta}) = mr^2\ddot{\theta} + 2mr\dot{r}\dot{\theta}$$

$$= r(mr\ddot{\theta} + 2m\dot{r}\dot{\theta})$$

$$= (r)(0)$$

$$\frac{d}{dt}(mr^2\dot{\theta}) = 0$$

$$mr^2\dot{\theta} = \text{constant}$$

But that is just the magnitude of the angular momentum L. Therefore,

$$mr^2\dot{\theta} = L = \text{constant} \quad (7.3.6)$$

Again, we see that the angular momentum remains constant for a central force. Sometimes physicists refer to a constant like this as being an *integral of the motion*. For central forces, angular momentum is an integral of the motion.

We can solve Eq. 7.3.5a using methods we developed in Chapter 2 for $F(x)$. To get something that resembles $m\ddot{x} = F(x)$, we can rewrite Eq. 7.3.5a as

$$m\ddot{r} = F(r) + mr\dot{\theta}^2 \quad (7.3.5a')$$

If we rename the right-hand side an *effective force*, F_{eff}, then this equation is identical to the one-dimensional case:

$$F_{\text{eff}}(r) = F(r) + mr\dot{\theta}^2 \quad (7.3.7)$$

$$m\ddot{r} = F_{\text{eff}}(r) \quad (7.3.8)$$

The $mr\dot{\theta}^2$ term (which is *really* part of the acceleration) looks like another term in the force. We have seen this idea before—in Chapter 6. We can treat $mr\dot{\theta}^2$ just like another force—it's a fictitious force. In this case, we can call it the *centrifugal force*.

Just as for the one-dimensional case, we can use the effective force in Eq. 7.3.7 to define an *effective potential*:

$$V_{\text{eff}}(r) = \int_r^{r_s} F_{\text{eff}}(r)\, dr$$

$$= \int_r^{r_s} [F(r) + mr\dot{\theta}^2]\, dr$$

$$= \int_r^{r_s} \left[F(r) + mr\left(\frac{L}{mr^2}\right)^2\right] dr$$

$$= \int_r^{r_s} F(r)\, dr + \frac{L^2}{m}\int_r^{r_s} r^{-3}\, dr$$

$$V_{\text{eff}}(r) = V(r) + \frac{L^2}{2mr^2} \tag{7.3.9}$$

where we have used $\dot{\theta} = L/mr^2$ from Eq. 7.3.6 and set the upper limit, r_s, to infinity. Thus, our *effective potential energy* is the real potential energy, $V(r)$, plus an additional term $L^2/2mr^2$. This can be called the centrifugal potential energy. It is also often called the centrifugal barrier.

Stop for a moment and reflect on what we've done. We have taken a two-dimensional force and acceleration problem and written it in the *form* of a one-dimensional problem that we already understand (and can solve). In doing this, we had to introduce fictitious forces as we've done before. We're really looking at the radial motion as viewed from a reference frame rotating along with our body of mass m. In this rotating frame, the force looks different. This ability to redefine things, to rewrite things so they *look* like something we already understand, is very important for a physicist.

Total energy is another *integral of the motion* (i.e., $E = $ constant) since central forces are conservative. The velocity has a radial component $\dot{r}$ and a tangential component $r\dot{\theta}$ so we can write the total energy as

$$T + V = \frac{1}{2}m\dot{r}^2 + \frac{1}{2}mr^2\dot{\theta}^2 + V(r) = E \tag{7.3.10}$$

Again, using $\dot{\theta} = L/mr^2$ from Eq. 7.3.6, we have

$$\frac{1}{2}m\dot{r}^2 + \frac{1}{2}mr^2\left(\frac{L}{mr^2}\right)^2 + V(r) = E$$

SECTION 7.3 / ANGULAR MOMENTUM AND CENTRAL FORCES

Just as for one-dimensional $F(x)$, we can solve for $\dot{r}$ and then integrate to find $r = r(t)$:

$$\dot{r} = \frac{dr}{dt} = \sqrt{\frac{2}{m}\left[E - V(r) - \frac{L^2}{2mr^2}\right]}$$

$$t = \int_{r_0}^{r} \frac{dr}{\sqrt{\frac{2}{m}\left[E - V(r) - \frac{L^2}{2mr^2}\right]}} \quad (7.3.11)$$

This would result in $t = t(r)$, from which we could then get $r = r(t)$. Notice that $L^2/(2mr^2)$ again plays the same role as a potential energy.

If we expended the necessary effort to solve Eqs. 7.3.5 explicitly as indicated in Eq. 7.3.11, we would still be faced with the question "what does the orbit *look* like?" So, let us instead look for a solution to r in terms of θ—or, at least, an equation involving only r and θ. We can do this if we begin with Eq. 7.3.5a and then make a change of variable, substituting $r = 1/u$ (u is a function of r; $u = u(r)$). We will also need Eq. 7.3.6.

$$\dot{\theta} = \frac{L}{m}u^2$$

$$r = \frac{1}{u} = u^{-1}$$

$$\dot{r} = -u^{-2}\frac{du}{dt} = -u^{-2}\frac{du}{d\theta}\frac{d\theta}{dt} = -u^{-2}\dot{\theta}\frac{du}{d\theta}$$

$$\dot{r} = -u^{-2}\left(\frac{L}{m}u^2\right)\frac{du}{d\theta} = -\frac{L}{m}\frac{du}{d\theta}$$

$$\ddot{r} = -\frac{L}{m}\frac{d}{dt}\frac{du}{d\theta} = -\frac{L}{m}\frac{d^2u}{d\theta^2}\frac{d\theta}{dt} = -\frac{L}{m}\dot{\theta}\frac{d^2u}{d\theta^2}$$

$$\ddot{r} = -\frac{L^2}{m^2}u^2\frac{d^2u}{d\theta^2}$$

$F(r) = m\ddot{r} - mr\dot{\theta}^2$ becomes

$$F\left(\frac{1}{u}\right) = m\left(-\frac{L^2}{m^2}u^2\frac{d^2u}{d\theta^2}\right) - m\frac{1}{u}\left(\frac{L^2}{m^2}u^4\right)$$

$$F\left(\frac{1}{u}\right) = -\frac{L^2}{m}u^2\frac{d^2u}{d\theta^2} - \frac{L^2}{m}u^3$$

We can rearrange this to read

$$\frac{d^2u}{d\theta^2} + u = -\frac{m}{L^2}\frac{1}{u^2}F\left(\frac{1}{u}\right) \quad (7.3.12)$$

Equation 7.3.12 is a second-order differential involving only the radial variable u and the angular variable θ. We must, of course, know the form of $F(1/u)$ or $F(r)$ before we can solve it. There is one particular form of this force that is more important than all others because of the frequency with which we encounter it in our attempts to understand the Universe. This is the inverse-square force, which we will investigate in the next section.

7.4 Inverse-Square Force

If we investigate the force between two charges, such as the force binding an electron to its nucleus, or if we investigate the force between two masses, such as the force binding the moon in its orbit around Earth or Earth in her orbit around the Sun, we find that in both these cases the force varies *inversely* as the *square* of the distance. As a specific example, let us take two electric charges some distance apart. We measure and record the force between them (i.e., the force one exerts on the other). If we then move them so the distance between them is *twice* what we measured initially, we will find the force reduced to *one-fourth* its initial value. Moving the charges to *three* times the distance reduces the force to *one-ninth* the initial value. The same thing would occur if we looked at the gravitational force between two masses. We can represent this behavior by writing

$$F(r) = \frac{K}{r^2} \qquad (7.4.1)$$

where $K < 0$ for an attractive force and $K > 0$ for a repulsive force. In particular, for gravitational forces K equals $-GMm$, where G is a universal constant. In SI units, $G = 6.67 \times 10^{-11}$ Nm^2/kg^2 and M and m are the two masses involved. For electrical forces, $K = (1/4\pi\varepsilon_0) Qq$ where ε_0 is called the *permittivity of free space* (it's just another universal constant like G). It has the value, in SI units, of $\varepsilon_0 = 8.85 \times 10^{-12}$ C^2/Nm^2, and Q and q are the electric charges involved. Since Q and q can each be positive or negative, K can be positive or negative. Hence, the electrical force can be either attractive or repulsive. The gravitational force, in contrast, is always attractive.

To find out what sort of orbit a body follows under the influence of this inverse-square force, we can use Eq. 7.3.12 with the force now given explicitly by Eq. 7.4.1:

$$\frac{d^2u}{d\theta^2} + u = -\frac{m}{L^2}\frac{1}{u^2}(Ku^2)$$

$$\frac{d^2u}{d\theta^2} + u = -\frac{mK}{L^2} \qquad (7.4.2)$$

SECTION 7.4 / INVERSE-SQUARE FORCE

This is a second-order *inhomogeneous* differential equation, not unlike the ones we studied earlier in our discussion of the *forced* harmonic oscillator. As then, we must first solve the corresponding *homogeneous* equation:

$$\frac{d^2 u}{d\theta^2} + u = 0 \tag{7.4.3}$$

This equation should be *quite* familiar to you by now. It's the same one we have solved for the *simple* harmonic oscillator (with $\omega = 1$), so we can immediately write the solution to this:

$$u = A \cos(\theta - \theta_0) \tag{7.4.4}$$

Now we must find a particular solution u_p to satisfy Eq. 7.4.2. A little thought (and substitution into Eq. 7.4.2) will show that

$$u_p = -\frac{mK}{L^2} \tag{7.4.5}$$

satisfies this nicely. Now our complete solution is obtained by adding Eqs. 7.4.4 and 7.4.5 to get

$$u = A \cos(\theta - \theta_0) - \frac{mK}{L^2} \tag{7.4.6}$$

Recalling our definition of u as $1/r$, we have

$$\frac{1}{r} = A \cos(\theta - \theta_0) - \frac{mK}{L^2} \tag{7.4.6a}$$

or

$$r = \frac{1}{A \cos(\theta - \theta_0) - (mK/L^2)} \tag{7.4.7}$$

Now we must ask "what in the world is this?" We can choose our coordinate axes so that $\theta_0 = 0$ to simplify things just a little:

$$r = \frac{1}{A \cos\theta - (mK/L^2)} \tag{7.4.8}$$

This is a *conic section*. From plane geometry, any conic section can be written as

$$r = r_0 \frac{1 + e}{1 + e \cos\theta} \tag{7.4.9}$$

$e = 1 \Rightarrow$ parabola
$e > 1 \Rightarrow$ hyperbola
$e < 1 \Rightarrow$ ellipse

where e is called the *eccentricity* of the orbit. For $e < 1$, the orbit is an ellipse. For the special case of $e = 0$, this ellipse becomes a circle. For $e = 1$, the orbit is a parabola; for $e > 1$, a hyperbola. The geometry of these different conic sections is shown in Figure 7.4.1.

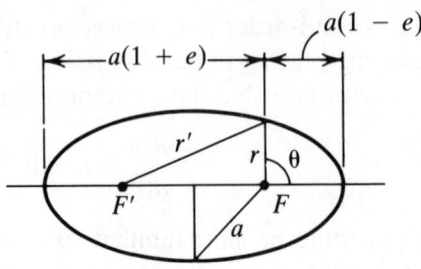

Ellipse $(e < 1)$
$r + r' = 2a$

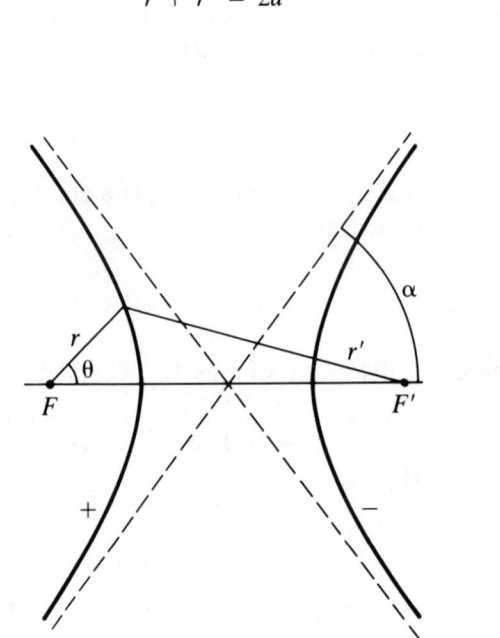

Parabola $(e = 1)$
$r = r'$

$r' - r = 2a$ $r' - r = -2a$

Figure 7.4.1 Geometry of conic sections.

Hyperbola $(e > 1)$

Therefore, the orbit of a body under the influence of an inverse-square force must be a conic section—an ellipse, parabola, or hyperbola. To force Eq. 7.4.8 into the *form* of Eq. 7.4.9, we may write it as

$$r = \frac{1}{-mK/L^2} \cdot \frac{1}{1 + (-AL^2/mK)\cos\theta}$$

$$r = -\frac{L^2}{mK} \cdot \frac{1}{1 + (-AL^2/mK)\cos\theta} \qquad (7.4.10)$$

SECTION 7.4 / INVERSE-SQUARE FORCE

Thus,
$$e = -\frac{AL^2}{mK} \tag{7.4.11}$$

(remember that for the gravitational force, $K = -GMm$) and

$$r_o = -\frac{L^2}{mK}\frac{1}{1+e} \tag{7.4.12}$$

As we can see from Eq. 7.4.9 or Figure 7.4.1, r_0 is the minimum value of the radius. We can also define the maximum radius from Eq. 7.4.9; we shall label that r_1:

$$r_1 = r_0 \frac{1+e}{1-e} \tag{7.4.13}$$

or

$$r_1 = -\frac{L^2}{mK}\frac{1}{1-e} \tag{7.4.14}$$

The smallest radius, r_0, is called the *perihelion* (for an object in solar orbit) or *perigee* (for an Earth orbit) while the largest radius, r_1, is called the *aphelion* or *apogee*. These two radii are turning points if we look at the one-dimensional motion of a body influenced by the effective potential energy, V_{eff}, of Eq. 7.3.9. As always, these turning points depend on the total energy E. We may identify them as those radii for which V_{eff} equals E.

For the inverse-square force, $F = K/r^2$, and the real potential $V(r)$ is given by

$$V(r) = \frac{K}{r} \tag{7.4.15}$$

This, with Eq. 7.3.9, defines the *effective potential*:

$$V_{eff} = \frac{K}{r} + \frac{L^2}{2mr^2} \tag{7.4.16}$$

This is equal to the total energy, E, for the turning points $r = r_0$ and $r = r_1$. This can be seen in Figure 7.4.2 (remember that $\dot{r} = 0$ for the turning points).

We can write this as a quadratic in $1/r$:

$$\frac{L^2}{2m}\left(\frac{1}{r}\right)^2 + K\left(\frac{1}{r}\right) - E = 0$$

and readily solve for $1/r$:

$$\frac{1}{r} = \frac{-K \pm [K^2 - 4(L^2/2m)(-E)]^{1/2}}{2(L^2/2m)}$$

or

$$\frac{1}{r_0} = -\frac{Km}{L^2} + \left[\left(\frac{mK}{L^2}\right)^2 + \frac{2mE}{L^2}\right]^{1/2} \tag{7.4.17}$$

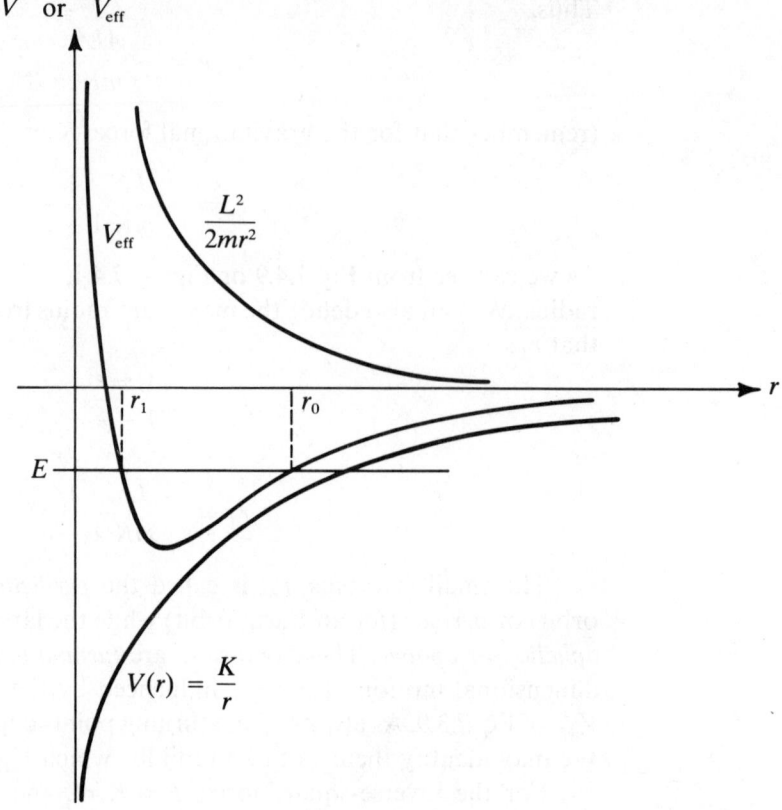

Figure 7.4.2 Effective potential energy for an attractive inverse-square force.

and

$$\frac{1}{r_1} = -\frac{Km}{L^2} - \left[\left(\frac{mK}{L^2}\right)^2 + \frac{2mE}{L^2}\right]^{1/2} \quad (7.4.18)$$

To compare these with our earlier expressions, it is easier if we invert Eqs. 7.4.12 and 7.4.14 so that

$$\frac{1}{r_0} = -\frac{mK}{L^2}(1 + e) = -\frac{mK}{L^2}\left(1 - \frac{AL^2}{mK}\right)$$

$$\frac{1}{r_0} = -\frac{mK}{L^2} + A \quad (7.4.19)$$

and

$$\frac{1}{r_1} = -\frac{mK}{L^2}(1 - e) = -\frac{mK}{L^2}\left(1 + \frac{AL^2}{mK}\right)$$

$$\frac{1}{r_1} = -\frac{mK}{L^2} - A \quad (7.4.20)$$

Comparing Eq. 7.4.17 with 7.4.19, and Eq. 7.4.18 with 7.4.20, we see that

$$A = \left[\frac{m^2 K^2}{L^4} + \frac{2mE}{L^2} \right]^{1/2} \quad (7.4.21)$$

We can use this in Eq. 7.4.11 to solve for the eccentricity of the orbit:

$$e = \sqrt{1 + \frac{2EL^2}{mK^2}} \quad (7.4.22)$$

For $K < 0$, we have an *attractive* force, so we can have a *bound orbit* or a closed orbit with reasonable values for r_0 and r_1 (real and positive). For $0 \leq e < 1$, the orbit is an ellipse (a bound orbit has $E < 0$). That is the case for Earth in orbit round the Sun or the moon (or an artificial satellite) in orbit around Earth. Let us restrict ourselves to this important case for the moment.

7.5 Kepler's Laws

Based on detailed astronomical data taken by Tycho Brahe, early in the seventeenth century Johannes Kepler announced three general laws that described the motion of the planets around the Sun. Newton's Law of Universal Gravitation, given shortly after Kepler's laws, was readily accepted because it provided a description of the planets' motion entirely consistent with Kepler's laws.

Kepler's laws can be stated as follows:

1. Planets move in orbits that are ellipses with the Sun at one focus (elliptical orbits).
2. Areas swept out by the radius vector from the Sun to a planet in equal times are equal (equal areas in equal times).
3. The *square* of a planet's period is proportional to the *cube* of the semimajor axis of its orbit ($T^2 \propto r^3$).

We have seen in the previous section that the first of these laws (that of elliptical orbits) follows directly from Newton's Law of Universal Gravitation (i.e., from the inverse-square nature of the force of gravity).

Equal areas being swept in equal times is a consequence of the angular momentum's being constant. This can be seen if we start with Figure 7.5.1, which shows a body in an elliptical orbit about an origin O.

For a small change in angle $d\theta$ as shown, the area swept out as a body moves from $\mathbf{r}$ to $\mathbf{r} + d\mathbf{r}$ is

$$dA = \tfrac{1}{2} r \, dh$$

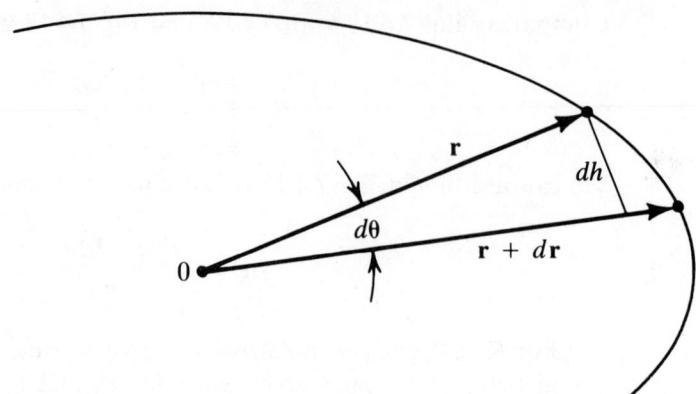

Figure 7.5.1 Equal areas are swept by a radius vector in equal times.

where dh is the perpendicular distance between the two radius vectors; this is just the area of a right triangle. But

$$dh = r\, d\theta$$

by the definition of angular measurement. Thus,

$$dA = \tfrac{1}{2} r\,(r\,d\theta) = \tfrac{1}{2} r^2\, d\theta \tag{7.5.1}$$

or

$$\frac{dA}{dt} = \frac{1}{2} r^2 \frac{d\theta}{dt} = \frac{1}{2} r^2 \dot\theta \tag{7.5.2}$$

But $mr^2\dot\theta = L$, the angular momentum, which we know to be constant for any central force. Therefore,

$$\frac{dA}{dt} = \frac{1}{2}\frac{L}{m} = \text{constant} \tag{7.5.3}$$

again agreeing with Kepler's second law.

We have discussed two of Kepler's laws. Kepler's Third Law is a consequence of the inverse-square nature of the gravitational force. We can readily integrate Eq. 7.5.3 to find

$$A = \frac{LT}{2m} \tag{7.5.4}$$

or

$$T = \frac{2mA}{L} \tag{7.5.5}$$

where A is the area of the orbit and T is the period, the time necessary to complete one entire orbit. The area of an ellipse is

$$A = \pi a^2 \sqrt{1 - e^2} \tag{7.5.6}$$

where a is the semimajor axis; that is, half of the maximum diameter:

$$a = \tfrac{1}{2}(r_0 + r_1) \qquad (7.5.7)$$

From Eqs. 7.4.12 and 7.4.13 we can find that

$$a = -\frac{L^2}{mK}\frac{1}{1-e^2}$$

Putting these together, we have

$$T = \frac{2m\pi}{\sqrt{m(-K)}}\, a^{3/2}$$

or

$$\frac{T^2}{a^3} = \left[\frac{2m\pi}{\sqrt{m(-K)}}\right]^2 = \text{constant} \qquad (7.5.8)$$

which, indeed, is just Kepler's Third Law.

So, you can see, Kepler's Second Law is true because gravity is a central force. And the remaining two of Kepler's laws are the result of the inverse-square nature of the gravitational force as postulated by Newton in his Law of Universal Gravitation. This agreement of Newton's *theory* with Kepler's Laws based upon careful *observation* was of primary importance in the initial acceptance of Newton's theory. Such agreement of theory and observation is at the very heart of scientific method. What does a theory predict? What are the observed results? Do theory and observation agree?

7.6 Orbital Transfers and "Gravitational Boosts"

Circular orbits are quite interesting for a variety of reasons. Most of the planetary orbits in our solar system are very nearly circular. For a circular orbit, the force remains constant (in magnitude) since the radial distance r is constant. In this section we shall investigate some of the methods we might use to send an interplanetary probe from one orbit to another.

What is the speed, v_c, of a body in a circular orbit of radius r_0? Any body moving in any circle must have a *centripetal* force on it to keep it moving in a circle:

$$F_c = -\frac{mv_c^2}{r_0} \qquad (7.6.1)$$

CHAPTER 7 / CENTRAL FORCES

This centripetal force points toward the center (hence, the negative sign). For a circular orbit, our central inverse-square force must supply this centripetal force. That is,

$$\frac{K}{r_0^2} = -m\frac{v_c^2}{r_0}$$

or

$$v_c^2 = -\frac{K}{mr_0} \tag{7.6.2}$$

Now consider a body a distance r_0 from the origin, moving with a velocity v_0 perpendicular to r_0, as sketched in Figure 7.6.1. From Eq. 7.4.12, the eccentricity of an orbit is given by

$$e = -\frac{L^2}{mKr_0} - 1$$

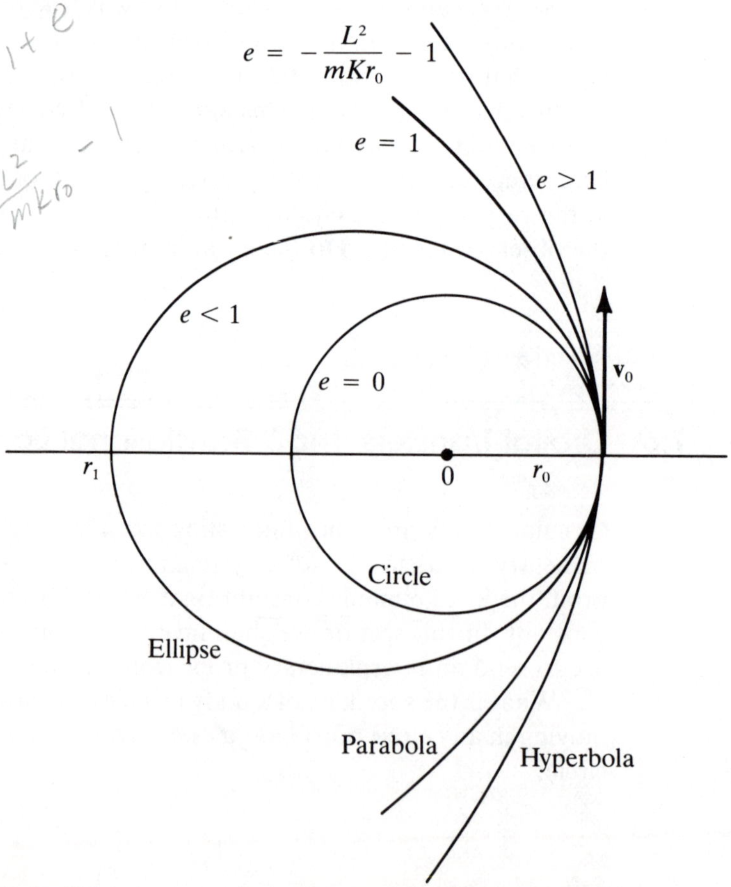

Figure 7.6.1 Initial conditions determine the orbit.

SECTION 7.6 / ORBITAL TRANSFERS AND "GRAVITATIONAL BOOSTS"

But the angular momentum is simply $L = m r_0 v_0$ since the velocity is perpendicular to the radius vector in this case. Thus,

$$e = -\frac{m r_0 v_0^2}{K} - 1 \tag{7.6.3}$$

As a check on Eq. 7.6.1, we can set $e = 0$ for a circular orbit and find (setting $v_0 = v_c$ for this circular orbit)

$$v_c^2 = -\frac{K}{m r_0} \tag{7.6.2}$$

once more.

We can substitute Eq. 7.6.2 into Eq. 7.6.3 and find

$$e = \left(\frac{v_0}{v_c}\right)^2 - 1 \tag{7.6.4}$$

For $v_0 > v_c$, then, r_0 is the perihelion of the orbit and we can use Eqs. 7.4.14 and 7.6.4 to find r_1, the aphelion or maximum distance from the center:

$$r_1 = -\frac{L^2}{mK} \cdot \frac{1}{1 - e}$$

$$= -\frac{(m r_0 v_0)^2}{m(-m r_0 v_c^2)} \cdot \frac{1}{1 - [(v_0/v_c)^2 - 1]}$$

$$r_1 = r_0 \left(\frac{v_0}{v_c}\right)^2 \frac{1}{2 - (v_0/v_c)^2} \tag{7.6.5}$$

As an application of this, consider the circular orbits of planets E (say, Earth) and M (perhaps Mars) as sketched in Figure 7.6.2. At an initial

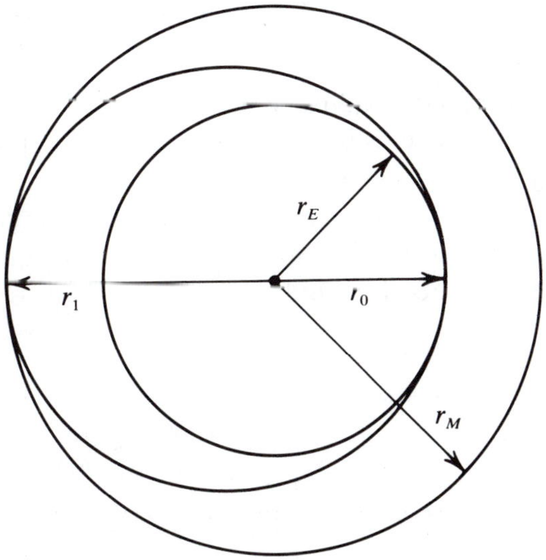

Figure 7.6.2 Orbital transfers.

minimum radius of Earth's orbit ($r_0 = r_E$), what speed v_0 would an interplanetary probe need so that its maximum radius would equal that of Mars' orbit ($r_1 = r_M$)?

Solve Eq. 7.6.6 with the conditions that $r_0 = r_E$ and $r_1 = r_M$.

$$r_M = r_E \frac{(v_0/v_c)^2}{2 - (v_0/v_c)^2}$$

$$2\frac{r_M}{r_E} - \frac{r_M}{r_E}\left(\frac{v_0}{v_c}\right)^2 = \left(\frac{v_0}{v_c}\right)^2$$

$$\left(\frac{v_0}{v_c}\right)^2\left(1 + \frac{r_M}{r_E}\right) = 2\frac{r_M}{r_E}$$

$$\left(\frac{v_0}{v_c}\right)^2 = \frac{2(r_M/r_E)}{1 + (r_M/r_E)}$$

$$v_0 = v_c \sqrt{\frac{2(r_M/r_E)}{1 + (r_M/r_E)}} \tag{7.6.6}$$

Here v_c is the speed required for a *circular* orbit of radius r_E, so we may as well relabel this by setting $v_c = v_E$, the orbital speed of Earth in her (nearly) circular orbit. Now v_0 is the speed a planetary probe (or *any* body) must have at Earth's orbit r_E to carry it into an orbit with a maximum radius of r_M which, for our example, is the radius of Mars' orbit.

Before proceeding, look at Eq. 7.6.6 again and take the limit as r_M approaches infinity. That is, what speed would carry a probe from Earth's orbit to some distance infinitely far away? Since both numerator and denominator tend to infinity as the limit is taken, l'Hôpital's Rule* must be applied. The result is

$$\lim_{r_M \to \infty} v_0 = v_c \sqrt{2} = v_E \sqrt{2} \tag{7.6.7}$$

what is, from some radius r_E the *escape velocity* (the speed needed to move infinitely far away) is just equal to the speed required for a circular orbit at that radius multiplied by $\sqrt{2}$.

Now back to the example at hand with a finite value of r_M. How do the orbital velocities of Earth and Mars compare? Equivalently, we may ask how the speed required to maintain a circular orbit depends upon the radius.

---

* L'Hôpital's rule is

$$\lim_{x \to x_0} \frac{f(x)}{g(x)} = \lim_{x \to x_0} \frac{f'(x)}{g'(x)}$$

and is very useful when the initial expression, $f(x)/g(x)$, is undefined for $x = x_0$; $f'(x) = df/dx$ and $g'(x) = dg/dx$.

We have already seen that for a body to move in a circle, there must be a centripetal force on it of

$$F_c = -\frac{mv_c^2}{r_0}$$

Only the gravitational force

$$F_g = -\frac{K}{r_0^2}$$

can supply this force. So these two forces must be equal. Therefore,

$$v_c^2 = -\frac{K}{mr_0} \qquad (7.6.2)$$

192 to maintain earth orbit

once more. Now let's examine this a little more closely.

This *looks* like the circular orbital speed has an unexpected dependence upon the *mass* of the orbiting body. This is not the case, though, because

$$K = -GMm \qquad (7.6.8)$$

is dependent upon this same mass in just the right way. So,

$$v_c^2 = \frac{GM}{r}$$

For the two particular circular orbits in our example, then,

$$v_E^2 = \frac{GM}{r_E} \qquad (7.6.9a)$$

and

$$v_M^2 = \frac{GM}{r_M} \qquad (7.6.9b)$$

or

$$\frac{v_M}{v_E} = \sqrt{\frac{r_E}{r_M}} \qquad (7.6.10)$$

In words, this means that the ratio of the (circular) orbital speeds of *any* two bodies varies inversely as the square root of the ratio of the radii of their (circular) orbits. Planets farther from the Sun move slower—with smaller tangential velocities—than planets closer to the Sun.

What of the speed of our planetary probe that leaves the orbit of Earth and arrives at the orbit of Mars? How does its speed vary? How does it compare to either Earth's speed or to Mars'?

v_0 is the velocity our probe would need while located at Earth's orbit ($r_E = r_0$) in order to reach Mars' orbit. Therefore, let us call this velocity v_{EM}, the *transfer velocity* (from r_E to r_M). From Eq. 7.6.6, we know

$$v_{EM} = v_E \sqrt{\frac{2(r_M/r_E)}{1 + (r_M/r_E)}} \tag{7.6.11}$$

Now, to determine the probe's speed once it gets to Mars—its *arrival velocity* A_{ME}—look at the energy of a body in an inverse-square, gravitational field:

$$E = T + V(r)$$

$$E = \tfrac{1}{2}mv^2 + \frac{K}{r} \tag{7.6.12}$$

This equation holds true, and E remains constant, throughout the orbit. Specifically, for the minimum and maximum radii:

$$E = \tfrac{1}{2}mv_0^2 + \frac{K}{r_0} = \tfrac{1}{2}mv_1^2 + \frac{K}{r_1}$$

In terms of our explicit example, we can write this as

$$\tfrac{1}{2}mv_{EM}^2 + \frac{K}{r_E} = \tfrac{1}{2}mA_{ME}^2 + \frac{K}{r_M} \tag{7.6.13}$$

Solving for A_{ME}, the arrival velocity, gives

$$A_{ME}^2 = v_{EM}^2 - 2\frac{GM}{r_E} + 2\frac{GM}{r_M} \tag{7.6.14}$$

where we have once again used $K = -GMm$ to eliminate the misleading appearances of a mass-dependent speed. Using Eqs. 7.6.9a and 7.6.9b, we can simplify this to

$$A_{ME}^2 = v_{EM}^2 - 2v_E^2 + 2v_M^2 \tag{7.6.15}$$

Simplifying further by use of Eq. 7.6.10 gives

$$A_{ME}^2 = v_E^2 \frac{2(r_M/r_E)}{1 + (r_M/r_E)} - 2v_E^2 + 2v_M^2$$

$$A_{ME}^2 = 2\left[v_M^2 - v_E^2 \frac{1}{1 + (r_M/r_E)}\right]$$

SECTION 7.6 / ORBITAL TRANSFERS AND "GRAVITATIONAL BOOSTS"

Equation 7.6.9 simplifies this still more to

$$A_{ME}^2 = 2v_M^2 \left[1 - \left(\frac{r_M}{r_E}\right)\frac{1}{1 + (r_M/r_E)}\right]$$

$$A_{ME}^2 = 2v_M^2 \left[1 - \left(\frac{r_M}{r_E}\right)\frac{1}{(r_E + r_M)/r_E}\right]$$

$$= 2v_M^2 \left[1 - \frac{r_M}{r_E + r_M}\right]$$

$$= 2v_M^2 \left(\frac{(r_E + r_M) - r_M}{r_E + r_M}\right)$$

$$A_{ME}^2 = 2v_M^2 \left[\frac{r_E}{r_E + r_M}\right] \tag{7.6.16}$$

Look at the term in brackets, $r_E/(r_E + r_M)$. For Mars, $r_M > r_E$, so

$$\frac{r_E}{r_E + r_M} < \frac{1}{2}$$

Therefore,

$$A_{ME}^2 < v_M^2$$

or

$$A_{ME} < v_M \tag{7.6.17}$$

Arrival V at Mars from E

Therefore, the arrival velocity, A_{ME}, is less than the orbital velocity of Mars, v_M. This means, in our example, that the planetary probe will be moving slower than Mars when it arrives at the Martian orbit. Mars will approach it from *behind* and overtake it. An observer *on Mars* would see the probe approaching it *from* the direction opposite to Mars' own motion. This will prove to be quite useful.

Consider, first, though, another planetary probe fired from Earth to an inner planet like Venus or Mercury. To reach such an inner planet, we must have $v_0 < v_c$, of course. That would seem to cause a problem with Eq. 7.6.4 since the eccentricity is now negative. Also, our initial position is now the maximum value of r instead of the minimum it was before. The real eccentricity of the orbit is now just the negative of our value from Eq. 7.6.4 or 7.6.5. But inspection of Eq. 7.4.13 shows that the roles of r_0 and r_1 are reversed if e is replaced by $-e$. So we may still use all of our previous results for this situation as well if we let r_0 represent the maximum radius (aphelion) and r_1, the minimum radius (perihelion). If we let r_M now be the radius of Mercury's orbit, then with $r_M < r_E$, Eq. 7.6.15 now yields

$$A_{ME} > v_M \tag{7.6.18}$$

for Mercury. The probe now comes speeding up on the (unsuspecting?) planet and overtakes it. Quite the opposite approach for an inner or an outer planet.

Let's return to Mars and observe the interaction of Mars with our planetary probe. Our observer (a foreign correspondent?) there sees the probe approach—*from* the direction Mars is travelling *toward*—with a speed of

$$\Delta v = v_M - A_{ME} \tag{7.6.19}$$

From the viewpoint of our Martian observer, this probe is seen approaching Mars from far, far away (infinity, for practical Martian purposes). It can then loop around Mars and go off in nearly the same direction from which it came—leaving with the same *speed* as that with which it approached. That is what our Martian observer sees. To an *outside* observer, then, our probe must leave with a final speed v_f given by

$$v_f = v_M + \Delta v \tag{7.6.20}$$

or

$$v_f = v_M \left[2 - \sqrt{\frac{2r_E}{r_E + r_M}} \right] \tag{7.6.21}$$

That is, after approaching Mars and looping around it (as seen from Mars), our probe leaves with an *increased* speed. This is sketched in Figure 7.6.3. Our probe now leaves Mars, going *outside* Mars' orbit. Applying Eq. 7.6.7, with appropriate changes, will give the maximum radius of the new orbit.

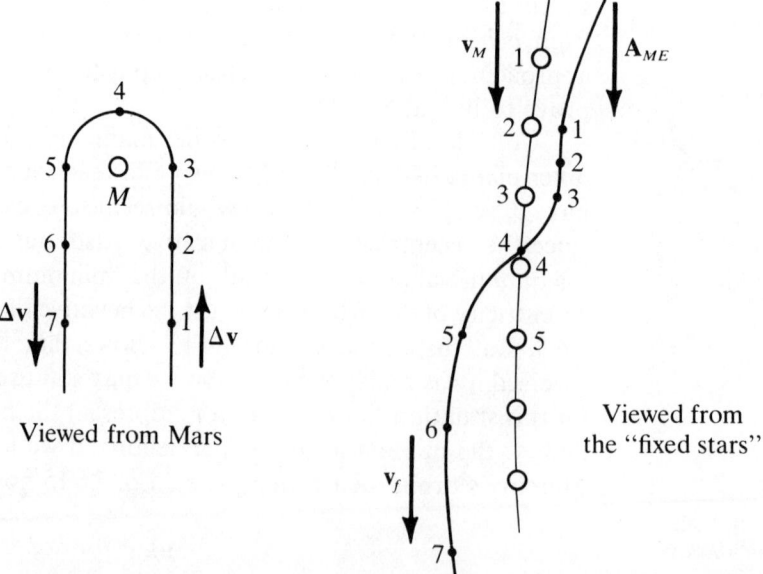

Figure 7.6.3
Gravitational boost from an outer planet.

A similar maneuver around an inner planet like Mercury can be used to brake the probe's speed and allow it to move to a new orbit with r_M (say, Mercury's radius) as the *maximum* radius and a new smaller minimum radius. Such a technique would be useful (or necessary) to get a probe *close* to the Sun. Without such "gravitational braking" it's far more difficult (i.e., requires higher relative rocket speeds) to actually hit the Sun than to reach escape velocity and escape the solar system entirely. Such gravitational braking is sketched in Figure 7.6.4.

It is interesting to ask what is required in order to achieve escape velocity for our probe using a gravitational *boost*. From Eq. 7.6.2 and the discussion that followed, the escape velocity from the orbit of Mars is just $\sqrt{2}$ multiplied by the orbital speed of Mars. Namely,

$$v_{esc} = \sqrt{2}\, v_M \tag{7.6.22}$$

So we now set this equal to the final velocity v_f of Eq. 7.6.20:

$$v_f = v_M \left[2 - \sqrt{\frac{2r_E}{r_E + r_M}} \right] = \sqrt{2}\, v_M = v_{esc} \tag{7.6.23}$$

Solving this yields

$$r_M = r_E \left[\frac{2}{6 - 4\sqrt{2}} - 1 \right]$$

$$r_M = 4.81\, r_E \tag{7.6.24}$$

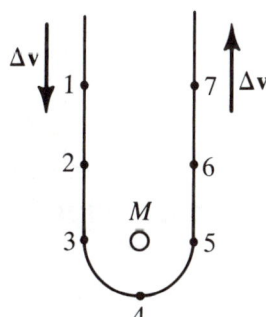

Viewed from Mercury

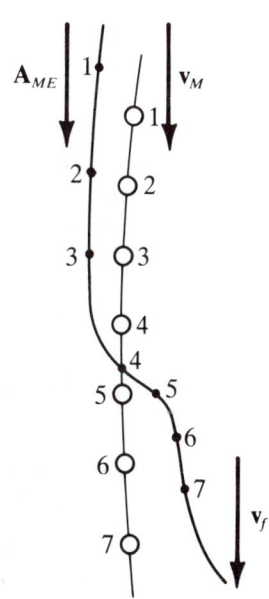

Viewed from the "fixed stars"

Figure 7.6.4
Gravitational braking from an inner planet.

That means the following: Suppose a planetary probe leaves a circular orbit of radius r_E and travels to a maximum distance from the Sun of $4.81 \, r_E$. If it then gets a maximum gravitational boost from a planet in orbit *there*, our probe will leave *that planet* with escape velocity. That is, it will leave that planet on an orbit that will carry it infinitely far away. The requirement to escape the solar system is just that a gravitational boost be done by a planet whose orbit is (at least) 4.81 times as great as the circular orbit from which the probe originates (namely, Earth's orbit). Mars doesn't satisfy this requirement. But Jupiter or Saturn certainly would since

$$r_{\text{Jupiter}} = 5.19 \, r_E \tag{7.6.25}$$

The "Grand Tours of the Planets" past Jupiter and Saturn in 1979 and 1981 by Pioneer 10 and Pioneer 11 made use of this effect.

We now return to Eq. 7.3.8 to determine the maximum radius of the probe's new orbit if escape velocity is not obtained—as would be the case for gravitational boost off of Mars. r_M is now the *minimum* radius of the new orbit. From Eq. 7.6.5, the *new* maximum radius r_1 is given by

$$r_1 = r_M \frac{\left(\dfrac{v_f}{v_M}\right)^2}{2 - \left(\dfrac{v_f}{v_M}\right)^2}$$

Solving, using Eq. 7.6.21, gives:

$$r_1 = r_M \frac{[2 - \sqrt{2r_E/(r_E + r_M)}]^2}{2 - [2 - \sqrt{2r_E/(r_E + r_M)}]^2} \tag{7.6.26}$$

These results hold true for $r_M < r_E$, as in the case of gravitational braking using an inner planet. But for that case, r_1 is the new *minimum* radius (r_M is the new maximum radius).

7.7 Radial Oscillations about a Circular Orbit

Radial motion, $r = r(t)$, is entirely describable in terms of $V_{\text{eff}}(r)$, as we saw earlier in Eqs. 7.3.9 and 7.3.11. For an effective potential energy like that sketched in Figure 7.7.1, there may be a radius $r = a$ that locates a minimum in V_{eff} and, thus, a stable equilibrium. Such an equilibrium value for the radius means that a *circular* orbit of radius $r = a$ is possible with $r(t) = a =$ constant.

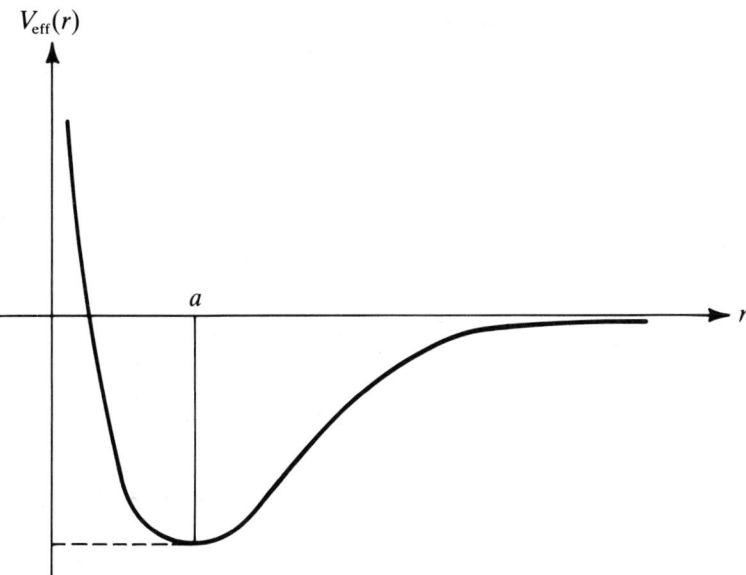

Figure 7.7.1 Stable equilibrium for a stable circular orbit.

As always, we can find such a minimum by setting the derivative (or slope) of the function equal to zero.

$$V_{\text{eff}} = V(r) + \frac{L^2}{2mr^2} \tag{7.3.9}$$

$$\frac{dV_{\text{eff}}}{dr} = \frac{dV}{dr} + \frac{L^2}{2m}(-2r^{-3})$$

$$\frac{dV_{\text{eff}}}{dr} = -F(r) - \frac{L^2}{mr^3} \tag{7.7.1}$$

Setting $dV_{\text{eff}}/dr = 0$ then gives

$$F(r) = -\frac{L^2}{mr^3}$$

or, since $r = a$,

$$F(a) = -\frac{L^2}{ma^3} \tag{7.7.2}$$

for a circular orbit of radius a.

Perturbation of the orbit so that the radius is not quite equal to a will cause oscillations of the radius about $r = a$. As we discussed earlier, as long as these (or any other) oscillations about an equilibrium position are small, they look like simple harmonic oscillations of a mass on a spring with an effective spring constant of

$$k_{\text{eff}} = \left.\frac{d^2 V_{\text{eff}}}{dr^2}\right|_{r=a} \tag{3.3.7}$$

We can differentiate Eq. 7.7.1 once more to obtain

$$\frac{d^2 V_{eff}}{dr^2} = -\frac{dF}{dr} + \frac{3L^2}{mr^4} \qquad (7.7.3)$$

If we use $F' = dF/dr$ and Eq. 7.7.2, then we can write this as

$$k_{eff} = \frac{d^2 V_{eff}}{dr^2}\bigg|_{r=a} = -F'(a) + \frac{3L^2}{ma^4}$$

$$k_{eff} = -F'(a) - \frac{3}{a}\left(-\frac{L^2}{ma^3}\right)$$

$$k_{eff} = -F'(a) - \frac{3}{a} F(a) \qquad (7.7.4)$$

Radial oscillations, then, will occur with a frequency ω_r given by

$$\omega_r = \sqrt{\frac{k_{eff}}{m}} \qquad (7.7.5)$$

or

$$\omega_r = \sqrt{\frac{-F'(a) - \frac{3}{a} F(a)}{m}} \qquad (7.7.5a)$$

Of course this is equivalent to saying that the radius increases and decreases with a period given by

$$T_r = \frac{2\pi}{\omega_r} = 2\pi \sqrt{\frac{m}{-F'(a) - \frac{3}{a} F(a)}} \qquad (7.7.6)$$

Oscillations occur for $k_{eff} > 0$. But for $k_{eff} < 0$, there is no restoring force. There will be no oscillation. Equation 7.7.2 will describe a circular orbit with $r = a$ but it will not be stable. Any perturbation from it will cause the radius to continue to deviate from $r = a$.

What conditions determine a *stable* orbit? We require $k_{eff} > 0$.

$$k_{eff} = -F'(a) - \frac{3}{a} F(a) > 0$$

$$F'(a) + \frac{3}{a} F(a) < 0 \qquad (7.7.7)$$

for a stable orbit. If we look at a force expressible as a power of r, like $F = -cr^n$, what restriction does Eq. 7.7.7 place on n for a stable orbit?

$$F(r) = -cr^n \quad \text{or} \quad F(a) = -ca^n \qquad (7.7.8)$$

$$F'(r) = -cnr^{n-1} \quad \text{or} \quad F'(a) = -cna^{n-1} \qquad (7.7.9)$$

For $F = -cr^n$, then, Eq. 7.7.7 becomes

$$-cna^{n-1} - \frac{3}{a}ca^n < 0$$

$$-n - 3 < 0$$

or

$$n > -3 \tag{7.7.10}$$

That is, for a radial power force, $F = -cr^n$, stable orbits are possible *only* for n greater than -3.

While the radius is oscillating about the equilibrium orbit of radius $r = a$, what is happening to the angular location of our body? We can solve for $\dot{\theta}$ (we could define $\dot{\theta} = \omega_\theta$ but that seems redundant) from Eq. 7.3.6.

$$\dot{\theta} = \frac{L}{mr^2} = \frac{L}{ma^2}$$

where we have used $r = a$ since we're concerned with *small* oscillations about an equilibrium (or *small* perturbations of a circular orbit).

We can solve for L using Eq. 7.7.2:

$$F(a) = -\frac{L^2}{ma^3}$$

$$L^2 = -ma^3 F(a)$$

$$L = \sqrt{-ma^3 F(a)}$$

$$\dot{\theta} = \frac{L}{ma^2} = \sqrt{\frac{-ma^3 F(a)}{(ma^2)^2}}$$

$$\dot{\theta} = \sqrt{\frac{-F(a)}{ma}} \tag{7.7.11}$$

We now define ψ, an *apsidal angle*, as the angle moved through as r goes from maximum to minimum. This is sketched in Figure 7.7.2. Using Eq. 7.7.6, we can write ψ as

$$\psi = (\tfrac{1}{2}T_r)\dot{\theta}$$

$$= \tfrac{1}{2}2\pi \sqrt{\frac{m}{-F'(a) - \frac{3}{a}F(a)}} \sqrt{\frac{-F(a)}{ma}}$$

$$\psi = \pi \left[3 + a\frac{F'(a)}{F(a)}\right]^{-1/2} \tag{7.7.12}$$

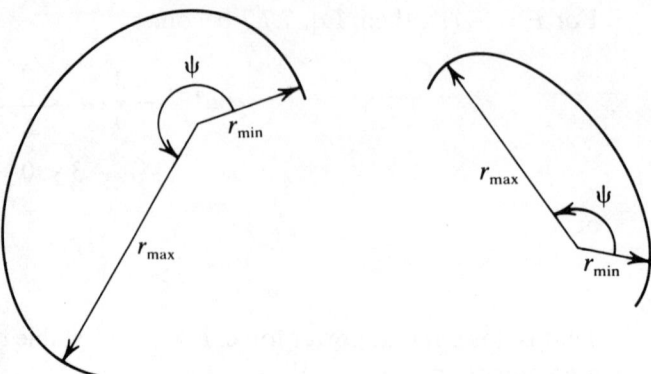

Figure 7.7.2 Apsidal angles.

This is true for any $F(r)$. But it's interesting to look at this apsidal angle for a radial power force, $F = -cr^n$. Then, using Eqs. 7.7.8, 7.7.9, and 7.7.12.

$$\psi = \pi\left[3 + a\frac{-cna^{n-1}}{-ca^n}\right]^{-1/2}$$

$$\psi = \pi[3 + n]^{-1/2} = \frac{\pi}{\sqrt{3+n}} \quad (7.7.13)$$

In order to have a *closed orbit*, that is, one that repeats itself, we must require

$$\psi = \frac{1}{m}\pi, \quad m = \text{an integer} \quad (7.7.14)$$

With the further restriction that $n > -3$, this leaves only limited possibilities for integer values of n that lead to closed, stable orbits; namely,

$n = -2 \quad F = \dfrac{K}{r^2}, \quad$ inverse square force

$n = 1 \quad F = -kr, \quad$ simple harmonic oscillator

$n = 6 \quad F = kr^6$

For other values of n, the orbits are not closed. That is, a body does not retrace its previous positions. However, for $n > -3$, oscillations about a stable circular orbit of radius $r = a$ can occur. Such orbits will be confined between a minimum radius and a maximum radius.

7.8 Gravity

If we look at the motion of only a single planet orbiting a far-distant Sun, then we can readily treat this motion as due to a simple, central, gravitational force,

SECTION 7.8 / GRAVITY

$$F = -G\frac{Mm}{r^2}$$

or

$$\mathbf{F} = -G\frac{Mm}{r^3}\mathbf{r} \tag{7.8.1}$$

Either of these equations assumes that the Sun is at the origin of the coordinate system. We can make this more general by replacing $\mathbf{r}$ by $\mathbf{r} - \mathbf{r}_s$ where $\mathbf{r}$ locates the planet of mass m and $\mathbf{r}_s$ locates the Sun of mass M. Now $\mathbf{r} - \mathbf{r}_s$ is the location vector drawn from Sun to planet. Then Eq. 7.8.1 becomes

$$\mathbf{F} = -G\frac{Mm}{|\mathbf{r} - \mathbf{r}_s|^3}(\mathbf{r} - \mathbf{r}_s) \tag{7.8.2}$$

The *net* gravitational force on a particle of mass m due to several other particles, each of mass M_i, is but the vector sum of all the gravitational forces acting on mass m. That is,

$$\mathbf{F} = -G\sum \frac{M_i m}{|\mathbf{r} - \mathbf{r}_i|^3}(\mathbf{r} - \mathbf{r}_i) \tag{7.8.3}$$

where $\mathbf{r}$ is still the position vector of the particle with mass m and $\mathbf{r}_i$ locates mass M_i.

If, instead of discrete particles of mass M_i, we want to determine the gravitational force due to some extended mass with density ρ, then the sum becomes an integral. That is,

$$\mathbf{F} = -G\iiint \frac{\rho(\mathbf{r}')m}{|\mathbf{r} - \mathbf{r}'|^3}(\mathbf{r} - \mathbf{r}')\, dv' \tag{7.8.4}$$

Here, as always, the gravitational force on mass m is proportional to that mass. So it will be useful to define a *gravitational field* as the ratio of this force to the mass. Labeling the gravitational field $\mathbf{g}$, we have

$$\mathbf{g}(\mathbf{r}) = -G\sum \frac{M_i}{|\mathbf{r} - \mathbf{r}_i|^3}(\mathbf{r} - \mathbf{r}_i) \tag{7.8.5}$$

or

$$\mathbf{g}(\mathbf{r}) = -G\iiint_{v'} \frac{\rho(\mathbf{r}')}{|\mathbf{r} - \mathbf{r}'|^3}(\mathbf{r} - \mathbf{r}')\, dv' \tag{7.8.6}$$

Notice that this is completely analogous to the electric field, $\mathbf{E}$, that is defined as the force per charge,

$$\mathbf{E} = \frac{\mathbf{F}}{q}$$

that you have already seen and used in electromagnetism. There it turned out very useful to define an *electric potential*, the potential energy of a system divided by the test charge q used to find the potential energy. This electric potential—the voltage—varied from place to place as the test charge q was moved from place to place. The same idea will prove just as useful here.

Using Eq. 7.6.8 in Eq. 7.2.13, we can readily see that the potential energy of two particles with masses M and m located a distance r apart is given by

$$V = -G\frac{Mm}{r} \tag{7.8.7}$$

If, instead, mass m is located at position $\mathbf{r}$, and M at $\mathbf{r}_s$, then this becomes

$$F = -G\frac{Mm}{|\mathbf{r} - \mathbf{r}_s|} \tag{7.8.7a}$$

If we want to know the potential energy of a test particle of mass m located at $\mathbf{r}$ due to the gravitational interaction with several other particles of mass M_i located at positions $\mathbf{r}_i$, just take the sum of the various potential energies due to mass m interacting with each mass M_i to get

$$V(r) = -G\sum \frac{M_i m}{|\mathbf{r} - \mathbf{r}_i|} \tag{7.8.8}$$

Or, if there is some distribution of mass with density ρ that interacts with our test particle of mass m, then its potential energy is

$$V(\mathbf{r}) = -G \iiint_{v'} \frac{\rho(\mathbf{r}')m}{|\mathbf{r} - \mathbf{r}'|} \, dv' \tag{7.8.9}$$

Following the example of electrostatics, we shall define a *gravitational potential* as the potential energy V of a test mass m divided by that mass. Labeling this as Φ, we have

$$\Phi \equiv \frac{V}{m} \tag{7.8.10}$$

Thus,

$$\Phi(\mathbf{r}) = -G\sum \frac{M_i}{|\mathbf{r} - \mathbf{r}_i|} \tag{7.8.11}$$

or

$$\Phi(\mathbf{r}) = -G \iiint_{v'} \frac{\rho(\mathbf{r}')}{|\mathbf{r} - \mathbf{r}'|} \, dv' \tag{7.8.12}$$

The gravitational field and gravitational potential are clearly related just like the force and potential energy. That is,

$$\mathbf{g}(\mathbf{r}) = -\nabla \Phi(r) \tag{7.8.13}$$

and

$$\Phi(\mathbf{r}) = -\int_{\mathbf{r}_{\text{ref}}}^{\mathbf{r}} \mathbf{g} \cdot d\mathbf{r} \tag{7.8.14}$$

These results are best understood through application.

EXAMPLE 7.8.1 Consider a particle of mass m at point P, located by vector $\mathbf{r}$ measured from an origin 0 as shown in Figure 7.8.1. There is a uniform spherical *shell* of mass M and radius R concentric with the origin. What force is experienced by mass m at position $\mathbf{r}$? This calls for an application of Eq. 7.8.4.

Since $\rho(\mathbf{r}')$ is nonzero only for $\mathbf{r}'$, locating a point on this spherical shell, the indicated triple integral is reduced to a double integral. The representative mass M_i of Eq. 7.8.3 became $\rho(\mathbf{r}') \, dv'$; now it becomes $\rho(\mathbf{r}') \, dS'$. A small, representative mass dM located at $\mathbf{r}'$ is given by

$$dM = \rho \, dS' = \rho(R \, d\theta)(R \sin\theta \, d\phi)$$

$$dM = \rho R^2 \sin\theta \, d\theta \, d\phi \tag{7.8.15}$$

where R and θ are defined in Figure 7.8.1 and ϕ is the azimuthal angle measuring rotation *about* the vector $\mathbf{r}$. Then Eq. 7.8.4 becomes

$$\mathbf{F} = -G \int_{\theta=0}^{\pi} \int_{\phi=0}^{2\pi} \frac{\imath}{\imath^3} \rho m R^2 \sin\theta \, d\theta \, d\phi \tag{7.8.16}$$

The density is just the mass M divided by the area

$$\rho = \frac{M}{S} = \frac{M}{4\pi R^2} \tag{7.8.17}$$

and readily passes through the integral signs.

While the integrand remains a *vector*, the spherical symmetry involved will greatly simplify the actual evaluation. Consider the force on m due to a small mass

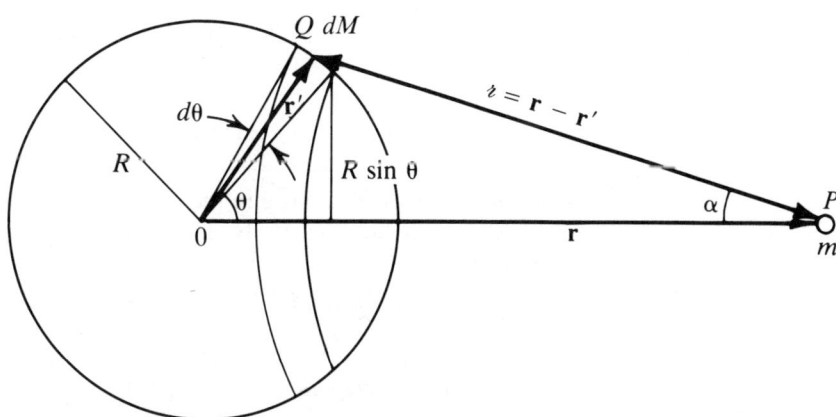

Figure 7.8.1 Gravitation due to a uniform spherical shell.

element dM as shown in Figure 7.8.1. Resolving this force $d\mathbf{F}$ into components parallel and perpendicular to $\mathbf{r}$,

$$dF_\perp = dF \sin \alpha \tag{7.8.18}$$

and

$$dF_\parallel = dF \cos \alpha \tag{7.8.19}$$

where angle α is shown in Figure 7.8.1.

As the integration over azimuthal angle ϕ is carried out, the small mass element dM sweeps out a *ring* on the spherical shell. Thus, it is easy to see that the vector sum of all the perpendicular force components must be zero—for a small mass element dM at some angle ϕ there is an identical one just opposite it at $\phi + \pi$. And the perpendicular components of the forces from these mass element pairs precisely cancel.

The net force, then, is only from the parallel components of Eq. 7.8.19. Integration over ϕ results in a constant factor of 2π from

$$\int_{\phi=0}^{2\pi} d\phi = \phi \Big|_{\phi=0}^{2\pi} = 2\pi$$

Thus, the magnitude of the force is given by

$$F = -2\pi G \rho m R^2 \int_{\theta=0}^{\pi} \frac{1}{\imath^2} \cos \alpha \sin \theta \, d\theta \tag{7.8.20}$$

and we must express r and α in terms of θ before this can be evaluated.

The triangle OPQ and the Law of Cosines can be used to write

$$\imath^2 = R^2 + r^2 - 2Rr \cos \theta \tag{7.8.21}$$

Differentiation yields

$$2\imath \, d\imath = 2Rr \sin \theta \, d\theta$$

Thus,

$$\sin \theta \, d\theta = \frac{\imath \, d\imath}{Rr} \tag{7.8.22}$$

Applying the Law of Cosines, using angle α, we have

$$R^2 = r^2 + \imath^2 - 2r\imath \cos \alpha$$

or

$$\cos \alpha = \frac{r^2 + \imath^2 - R^2}{2r\imath} \tag{7.8.23}$$

Substituting these expressions into the integral changes it into an integral over r—so we must watch the limits carefully. The result is

$$F = -2\pi G \rho m R^2 \int_{(\theta=0),\, \imath=r-R}^{\imath=r+R,\,(\theta=\pi)} \left(\frac{1}{r^2}\right)\left(\frac{r^2 + \imath^2 - R^2}{2rr}\right)\left(\frac{d\imath}{R\imath}\right)$$

$$F = -\pi G \rho m R \frac{1}{r^2} \int_{\imath=r-R}^{\imath=r+R} \left(1 + \frac{r^2 - R^2}{\imath^2}\right) d\imath$$

Or, putting in the density explicitly from Eq. 7.8.17,

$$F = \frac{-GmM}{4Rr^2} \int_{\imath=r-R}^{\imath=r+R} \left(1 + \frac{r^2 - R^2}{\imath^2}\right) d\imath$$

$$= \frac{-GmM}{4Rr^2} \left[\imath - (r^2 - R^2)\left(\frac{1}{\imath}\right)\right]_{\imath=r-R}^{\imath=r+R}$$

$$= \frac{-GmM}{4Rr^2} \left[(r+R) - (r-R) - (r^2 - R^2)\left(\frac{1}{r+R} - \frac{1}{r-R}\right)\right]$$

$$= \frac{-GmM}{4Rr^2} \left[2R - (r^2 - R^2)\frac{(r-R)-(r+R)}{(r+R)(r-R)}\right]$$

$$= \frac{-GmM}{4Rr^2} [2R + 2R]$$

$$F = \frac{-GmM}{r^2} \quad (7.8.24)$$

And this is *exactly* the same force as if the entire mass M were concentrated at the center of the sphere!

It is rather interesting to explicitly calculate the force for $r < R$; that is, for a point *inside* the spherical shell. But, alas, this is left as a homework problem.

We have just calculated the force. But force and potential energy are directly related. Further, potential energy is a scalar and scalars are generally easier to work with than vectors.

EXAMPLE 7.8.2 So, again referring to Figure 7.8.1, let us now evaluate the gravitational potential energy of a mass m located at a distance r from the center of a spherical shell of mass M and radius R.

From Eq. 7.8.9, we can write the gravitational potential energy as

$$V = -\frac{GmM}{4\pi} \int_{\theta=0}^{\pi} \int_{\phi=0}^{2\pi} \frac{1}{\imath} \sin\theta \, d\theta \, d\phi \quad (7.8.25)$$

We can immediately integrate over ϕ. Using Eq. 7.8.22, we can write this as

$$V = -\frac{GmM}{2} \int_{\theta=0}^{\theta=\pi} \frac{1}{\imath} \left(\frac{\imath \, d\imath}{Rr}\right)$$

$$= -\frac{GmM}{2Rr} \int_{\imath=r-R}^{\imath=r+R} d\imath$$

$$= -\frac{GmM}{2Rr} [(r+R) - (r-R)]$$

$$V = -\frac{GmM}{r} \quad (7.8.26)$$

This, too, is precisely the gravitational potential energy we would have if the total mass M were concentrated at the center of the sphere.

Of course we could have carried out these same operations *without* a test mass present at point P. The equivalent results would have been the gravitational field $\mathbf{g}$ and the gravitational potential Φ.

7.9 Rutherford Scattering

Thus far we have considered—or implied—attractive forces. An interesting and rather important exception to this is the force of repulsion between two electric charges of the same sign. This, too, is an inverse-square law, $F = K/r^2$, where $K = Qq/4\pi\varepsilon_0$.

That is,

$$F = \frac{1}{4\pi\varepsilon_0} \frac{Qq}{r^2} \tag{7.9.1}$$

where

$$\varepsilon_0 = 8.85 \times 10^{-12} \frac{C^2}{N-m^2}$$

or

$$\frac{1}{4\pi\varepsilon_0} = 8.99 \times 10^9 \frac{Nm^2}{C^2}$$

F is the force between two electric charges, Q and q, separated by distance r.

In particular, consider a massive atomic nucleus with a charge of $Q = Ze$ (where Z is the atomic number), which we may consider fixed used as a target for incoming alpha particles with charge $q = 2e$. $e = 1.6 \times 10^{-19}$ C is the elementary electric charge. For such a case,

$$K = \frac{Qq}{4\pi\varepsilon_0} = \frac{2Ze^2}{4\pi\varepsilon_0} = Z(4.6 \times 10^{-28} \, N \cdot m^2) \tag{7.9.2}$$

Since K is positive for a repulsive force, Eq. 7.4.22 tells us that $e > 1$ or that the incoming alpha-particle will travel along a *hyperbolic* trajectory. Referring back to Figure 7.4.1, the trajectory will be the negative branch of the hyperbola. This is drawn again in Figure 7.9.1.

We are interested in the scattering angle θ that describes the difference between the alpha particle's initial direction and the final direction. We can easily relate this to the angle α of Figures 7.4.1 and 7.9.1 by

$$\theta = \pi - 2\alpha \tag{7.9.3}$$

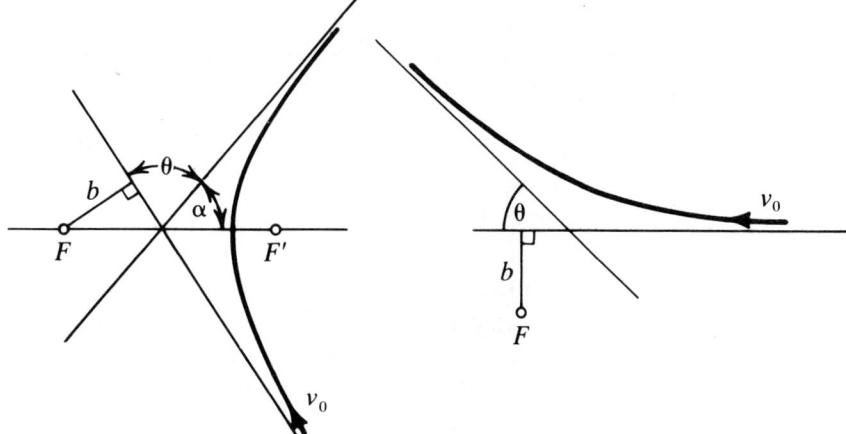

Figure 7.9.1 Scattering of alpha-particles from a heavy nucleus. Impact parameter = b, scattering angle = θ, incoming velocity = v_0, target location = F.

or

$$\frac{\theta}{2} = \frac{\pi}{2} - \alpha$$

and

$$\tan \frac{\theta}{2} = \cot \alpha \tag{7.9.4}$$

which will prove useful shortly. From analytic geometry it can be shown that the eccentricity of the orbit is related to the angle α by

$$\cos \alpha = \frac{1}{e} \tag{7.9.5}$$

Therefore,

$$\tan \frac{\theta}{2} = \cot \alpha = \frac{\cos \alpha}{\sin \alpha}$$

$$\tan \frac{\theta}{2} = \frac{1/e}{\sqrt{1 - (1/e^2)}}$$

$$\tan \frac{\theta}{2} = \cdots = \frac{1}{\sqrt{e^2 - 1}} \tag{7.9.6}$$

Using Eq. 7.4.22, we have

$$\tan \frac{\theta}{2} = \frac{1}{\sqrt{[1 + (2EL^2/mK^2)] - 1}}$$

$$\tan \frac{\theta}{2} = \sqrt{\frac{mK^2}{2EL^2}} \tag{7.9.7}$$

For a repulsive force, the potential energy, $V(r) = K/r$, is always positive. So the total energy E is always positive. When the alpha is far away ($r \approx \infty$), the potential energy is zero and the total energy is just

$$E = \tfrac{1}{2}mv_0^2 \tag{7.9.8}$$

where v_0 is the initial or incoming velocity measured far from the scattering center. Extend a straight line in the direction of v_0, past the nucleus at the origin as shown in Figure 7.9.1. This is, of course, the path the alpha would take in the *absence* of the force. A perpendicular drawn to this line from the scattering center is called the *impact parameter*. It is labeled b in the Figure. This impact parameter measures how close the alpha would pass to the target—if there were no force. Hence, the name.

The angular momentum is constant. So let's calculate the angular momentum at the very beginning, when the alpha is far away. It's easy then; the angular momentum is simply

$$L = mv_0 b \tag{7.9.9}$$

Now we can readily solve for the scattering angle θ:

$$\tan\frac{\theta}{2} = \sqrt{\frac{mK^2}{2(\tfrac{1}{2}mv_0^2)(mv_0^b)^2}} \quad (mv_0 \cdot b)^2$$

$$\tan\frac{\theta}{2} = \frac{K}{bmv_0^2} \tag{7.9.10}$$

θ can be measured experimentally. We can determine the impact parameter b by measuring the scattering angle θ since

$$b = \frac{1}{\tan(\theta/2)} \frac{K}{mv_0^2} \tag{7.9.11}$$

A particle scattering through an angle θ initially had an impact parameter b—and vice versa. Consider particles coming in toward the target with impact parameters between b and $b + db$. They will be scattered through angles lying between θ and $\theta + d\theta$ as sketched in Figure 7.9.2.

We can experimentally measure the number of alphas scattered through angles between θ and $\theta + d\theta$. We'll write that as $dN/d\theta$. That clearly depends upon—in fact, is proportional to—the number of incoming alpha particles, N. And it's also proportional to the number of nuclei—the number of "scattering centers"—that are present in front of the beam, n. Thus,

$$\frac{dN}{d\theta} = Nn \text{ (something else)}$$

That "something else" we call the *differential scattering cross section* and write as $d\sigma/d\theta$. Then we can write

$$\frac{dN}{d\theta} = Nn\frac{d\sigma}{d\theta} \tag{7.9.12}$$

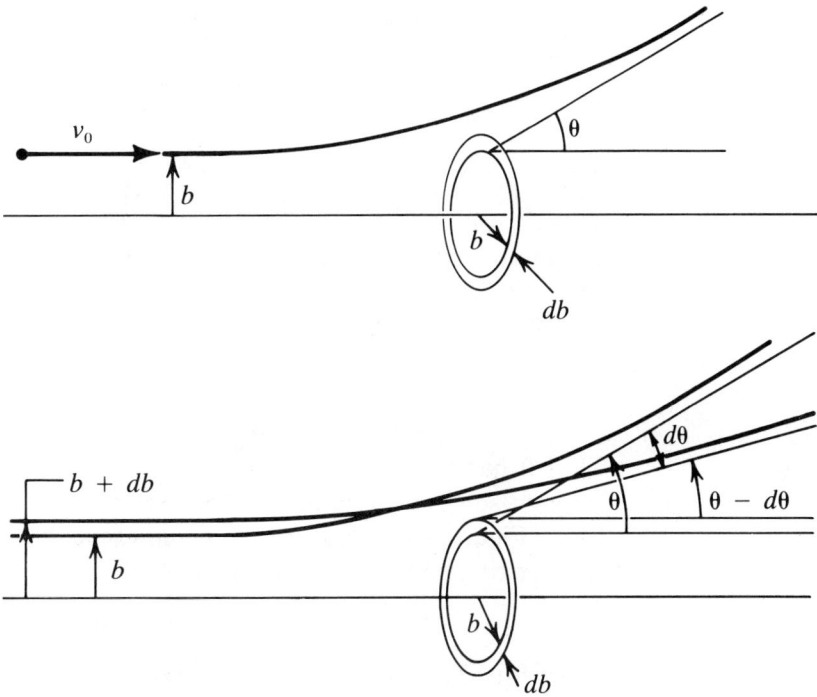

Figure 7.9.2
Scattering
cross section.

or

$$\frac{d\sigma}{d\theta} = \frac{1}{Nn}\frac{dN}{d\theta} \qquad (7.9.13)$$

Measurements involving θ, such as $dN/d\theta$, are readily made. But it is more useful to make our *calculations* in terms of b and d/db. Equation 7.9.10 allows us to change variables by using the chain rule:

$$\tan\frac{\theta}{2} = \frac{K}{mv_0^2}\frac{1}{b}$$

$$d\left(\tan\frac{\theta}{2}\right) = \frac{K}{mv_0^2}d\left(\frac{1}{b}\right)$$

$$\left(\sec^2\frac{\theta}{2}\right)\frac{1}{2}d\theta = -\frac{K}{mv_0^2}\frac{db}{b^2}$$

$$\frac{d\theta}{2\cos(2\theta/2)} = -\frac{K}{mv_0^2}\frac{db}{b^2}$$

$$db = -\frac{mv_0^2 b^2}{K}\frac{d\theta}{2\cos(2\theta/2)}$$

Using Eq. 7.9.11, we can write db as

$$db = -\frac{mv_0^2}{K}\left[\frac{K}{mv_0^2 \tan(\theta/2)}\right]^2 \frac{d\theta}{2\cos(2\theta/2)}$$

$$db = (-)\frac{K d\theta}{2mv_0^2 \sin(2\theta/2)} \tag{7.9.14}$$

The minus sign indicates that as b increases (so $db > 0$), the scattering angle θ decreases (so $d\theta < 0$).

We can rewrite Eq. 7.9.12 as

$$\frac{dN}{N} = n\, d\sigma \tag{7.9.15}$$

The left-hand side is just the proportion of incoming alphas that are scattered through some angle θ, divided by the total number of incoming particles. From Figure 7.9.2, $d\sigma$ is just the *area* (hence the term *cross section*) of the circular ring between b and $b + db$.

$$d\sigma = 2\pi b\, db \tag{7.9.16}$$

[*need minus sign*]

From Eqs. 7.9.11 and 7.9.15, we have

$$d\sigma = 2\pi\left(\frac{1}{\tan(\theta/2)}\frac{K}{mv_0^2}\right)\left(\frac{K\, d\theta}{2mv_0^2 \sin(2\theta/2)}\right)$$

$$= \pi\left(\frac{K}{mv_0^2}\right)^2 \frac{\cos(\theta/2)}{\sin(\theta/2)}\frac{d\theta}{\sin^2(\theta/2)}$$

$$= \pi\left(\frac{K}{mv_0^2}\right)^2 \frac{\sin(\theta/2)\cos(\theta/2)}{\sin^4(\theta/2)} d\theta$$

$$= \pi\left(\frac{K}{mv_0^2}\right)^2 \frac{\tfrac{1}{2}\sin\theta}{\sin^4(\theta/2)} d\theta$$

$$d\sigma = 2\pi\left(\frac{K}{2mv_0^2}\right)^2 \frac{\sin\theta}{\sin^4(\theta/2)} d\theta$$

Therefore,

$$\frac{dN}{d\theta} = Nn\frac{d\sigma}{d\theta} = 2\pi Nn\left[\frac{K}{2mv_0^2}\right]^2 \frac{\sin\theta}{\sin^4(\theta/2)} \tag{7.9.17}$$

Beginning with a repulsive, inverse-square force, we have predicted that a graph of the number of particles scattered through some angle *versus* that scattering angle will be governed or explained by Eq. 7.9.17. In 1911, Ernest Lord Rutherford (with able assistance from his students Geiger and Marsden) did an alpha scattering experiment much like the one we have discussed. He did this to determine the interior structure of an atom. The electron with its negative charge had been identified as a separate entity or particle. But no one knew the structure of the positive charge. The total atom was known to be neutral.

Sir J. J. Thompson proposed a model in which an atom's positive charge was evenly distributed throughout its entire volume. The electrons, too, were expected to be sprinkled uniformly throughout this volume—like "plums in a pudding." Hence, this is usually referred to as Thompson's plum pudding model. If we conducted scattering experiments on such an assemblage of positive and negative charges, there would be very little scattering since the entirety of such an atom is nearly uniformly neutral throughout.

But Rutherford's experimental results were startling. Very large scattering angles—even back scattering of nearly 180°—were observed. This led Rutherford to propose a "nuclear model" in which the structure of the atom resembled our own solar system. All of the positive charge resided in an incredibly dense nucleus at the very center of the atom. Most of the volume of the atom was simply empty. At some distance, the electrons (with very little mass) orbited the nucleus, being held in orbit by an electrostatic force between the opposite charges.

Scattering from such an atom as this presents an entirely different set of expectations than from Thompson's model. The alpha is essentially unaffected by the orbiting electrons due to its large mass (about 8000 times that of an electron). Thus the electrons affect the alpha only slightly more than a bumblebee running into the windshield of your automobile. But there is a large effect from the positive nucleus. Indeed, the alpha would be scattered just as we have described—scattered due to the influence of an inverse-square force. If Rutherford's model were correct, then he would expect the angular distribution of scattered particles detected to be given by Eq. 7.9.17. And, indeed, that is just what he found. This established the nuclear model of the atom. It also caused scientists to begin to wonder how all those positive charges—the protons in the nucleus—could be bound together so closely, so tightly. The answer to that question involves a bagpipe of a different tartan and is far beyond the reach of classical mechanics.

PROBLEMS

7.1 Show that *any* central force—one that may be written as $\mathbf{F} = F(r)\hat{\mathbf{r}}$—is a conservative force by *directly* evaluating the integral

$$\int_{\mathbf{r}_1}^{\mathbf{r}_2} \mathbf{F} \cdot d\mathbf{r}$$

and showing that for any path between $\mathbf{r}_1$ and $\mathbf{r}_2$, the integral depends *only* upon the scalars r_1 and r_2. Express F and $d\mathbf{r}$ in spherical coordinates.

7.2 An object of mass m moves so that its spherical coordinates are given by $r(t)$, $\theta(t)$, and $\phi(t)$. It is acted upon by some noncentral force with components of F_r, F_θ, and $\mathbf{F}_\phi$. Calculate the angular momentum and torque about the origin and show—for each component—that

$$\frac{d\mathbf{L}}{dt} = \mathbf{N}$$

7.3 An object of mass m moves so that its cylindrical coordinates are given by $\rho(t)$, $\theta(t)$, and $z(t)$. It is acted upon by some noncentral force with components F_ρ, F_θ, and F_z. Calculate the angular momentum and torque about the origin and show—for each component—that

$$\frac{d\mathbf{L}}{dt} = \mathbf{N}$$

7.4 A comet is observed a distance of 2.0×10^8 km from the Sun, traveling toward the Sun with a velocity of 27.3 km/sec, with an angle of 30° with its radius from the Sun (the Sun's mass is $2.0 = 10^{30}$ kg). Is the orbit hyperbolic, parabolic, or elliptic? How close to the Sun will it come?

7.5 The orbits of planets in our solar system are nearly circular. Orbits of comets, however, are markedly noncircular. Halley's comet has an orbital eccentricity of 0.967 and a perihelion of 89×10^6 km. Calculate its period, aphelion, and speed at perihelion and aphelion.

7.6 A comet is observed at a distance of D astronomical units from the Sun, traveling with a speed of Q times Earth's orbital speed. Show that the orbit is hyperbolic, parabolic, or elliptic as $Q^2 D$ is greater than, equal to, or less than 2, respectively.

7.7 Show that the motion of an object *repelled* by a central force of the form

$$\mathbf{F}(r) = Kr\hat{\mathbf{r}}, \qquad K > 0$$

must be hyperbolic.

7.8 Place a dry-ice puck of mass m on a smooth table and connect it by a string through a frictionless hole in the table to another object of mass M suspended below. Thus, the puck experiences a constant central force,

$$\mathbf{F}(r) = -Mg\hat{\mathbf{r}}$$

Fully describe the types of motion possible.

7.9 A body of mass m experiences an attractive inverse-cubed force,

$$\mathbf{F}(r) = \frac{K}{r^3}\hat{\mathbf{r}}, \qquad K < 0$$

Find its total energy and angular momentum when it moves in a circular orbit of radius r_c. What is its period then? If it is perturbed slightly from this orbit, what will be its period for small radial oscillations about r_c?

7.10 A body experiences an attractive inverse-cubed force,

$$\mathbf{F}(r) = \frac{K}{r^3}\hat{\mathbf{r}}, \qquad K < 0$$

Solve the orbital equation for $r = r(\theta)$ and show that the solutions can be written in each of the following forms:

$$\frac{1}{r} = A \cos[\beta(\theta - \theta_0)]$$

$$\frac{1}{r} = A \cosh[\beta(\theta - \theta_0)]$$

$$\frac{1}{r} = A \sinh[\beta(\theta - \theta_0)]$$

$$\frac{1}{r} = A (\theta - \theta_0)$$

$$\frac{1}{r} = \frac{1}{r_0} e^{\pm \beta \theta}$$

For what values of total energy and angular momentum will each type of orbit be found? Express constants A and β in terms of total energy and angular momentum for each case. Sketch a typical orbit for each case.

7.11 Show that for a repulsive inverse-cube force of the form

$$\mathbf{F}(r) = \frac{K}{r^3} \hat{\mathbf{r}}, \qquad K > 0$$

motion will take the form of

$$\frac{1}{r} = A \cos[\beta(\theta - \theta_0)]$$

Find values for the constants A and β in terms of total energy, angular momentum, mass, and initial position and velocity.

7.12 The true potential energy for an isotropic harmonic oscillator is given by

$$V = \tfrac{1}{2} k r^2$$

Draw a graph of the *effective* potential energy for radial motion for a body of mass m and angular momentum L subjected to this potential. Discuss the types of motion possible. Determine the frequency of revolution, $\dot\theta$, for circular motion and the frequency of small radial oscillations, ω_r, about such a circular orbit. Describe the types of orbits to be expected for small disturbances from a circular orbit.

7.13 A body of mass m moves in a region where its potential energy is given by

$$V = Cr^4, \qquad C > 0$$

Find its total energy and angular momentum when it moves in a circular orbit of radius r_c. What is its period then? If it is perturbed slightly from this orbit, what will be its period for small radial oscillations?

7.14 Motion of particles inside a nucleus can often be described well in terms of the Yukawa potential,

$$V = \frac{ke^{-ar}}{r}, \quad k < 0, a > 0$$

Calculate the force. How does it compare with (or differ from) an inverse-square force? Graph this Yukawa potential and discuss the types of motion that are available for various energies and angular momenta. Find the energy and angular momentum necessary for a circular orbit of radius r_c.

7.15 In a remote section of another galaxy, the science officer aboard the *U.S.S. Enterprise* observes a planetoid moving in a spiral orbit given by $r = r_0 e^{a\theta}$. Show that the central force causing this is inverse-cube and that θ varies as the logarithm of time.

7.16 In another remote section of yet another galaxy, observations from the interstellar trading ship, the *Millennium Hawk*, show a planetoid in a circular orbit that passes through the *origin* of the central field. Such an orbit can be written as $r = r_0 \cos \theta$. Show that the central force causing that motion must be an inverse-fifth radial power force.

7.17 A Voyager space probe discovers an object moving in a spiral orbit about a force center given by $r = a\theta$. Can θ increase linearly with t if this is a central force? If not, determine $\theta(t)$ for this being a central force.

7.18 Find an expression for the central force that will provide radial-quadratic spiral orbits given by

$$r = b\theta^2$$

7.19 Prove Kepler's Third Law for circular orbits. That is, show that the square of the period divided by the cube of the radius is constant for all circular orbits in an inverse-square, gravitational field.

7.20 Calculate the height above Earth's surface for a geosynchronous satellite. Such a satellite goes in a circular, equatorial orbit with a period of exactly 24 hours. Such a satellite is quite useful for communications. Why? Would it be as useful in a polar orbit? Take Earth's radius to be 6200 km.

7.21 Calculate the Sun's mass from the data of Earth's period and a mean radius of 1.46×10^8 km (assume a circular orbit).

7.22 A space probe to Mars would probably take an elliptical orbit with Earth at its perigee and Mars at its apogee. Find the length of time a probe in such an orbit would require to reach Mars. The radius of Mars' orbit is 1.53 times that of Earth's (consider both Earth and Mars to be in circular orbits).

7.23 A space probe to Venus would probably follow an elliptical path with Earth at its apogee and Venus at its perigee. Find the amount of time required for a probe in such an orbit to reach Venus. The radius of Venus' orbit is 0.72 times that of Earth's (consider both Earth and Venus to be in circular orbits).

PROBLEMS

7.24 To what maximum distance could a space probe from Earth go after undergoing a maximum "gravitational boost" off of Mars? (The radius of Mars' orbit is 1.53 that of Earth's.)

7.25 What would be the eccentricity of the orbit of a space probe that leaves Earth and experiences maximal "gravitational braking" from the planet Mercury (whose orbit is 0.39 times that of Earth)? How much energy would a 100-kg probe lose in such a maneuver?

7.26 Consider an interplanetary probe sent to Jupiter by means of a "gravitational boost" from some intermediate planet (between the orbits of Earth and Jupiter). What is the minimum radius required of the orbit of this intermediate planet? If an *additional* gravitational boost were to be accomplished from Jupiter, what would be the eccentricity of the final orbit?

7.27 A satellite is to be launched from Earth's surface. Assume Earth is a uniform sphere of radius R. Neglect atmospheric friction. The satellite will be launched at an angle θ with the vertical, with an initial speed v_0, so that it *coasts* without power until its velocity is horizontal at a radius of R_1 from the center of Earth. At that point a horizontal thrust is applied to add an additional velocity Δv_1 to the satellite's velocity. The final orbit is to be an ellipse with perigee R_1 and apogee R_2 (measured from the center of Earth). Find the required initial speed v_0 and additional speed Δv_1 in terms of R, θ, R_1, R_2, and g (the acceleration due to gravity at the surface of Earth).

7.28 Two manned satellites travel in the same direction along a circular orbit of radius a. Initially the two are some distance apart. In an attempt to rendezvous, the one behind increases its speed by Δv (without changing its direction). Discuss what happens to its orbit and its position relative to the other ship. How can it eventually dock with the other ship?

7.29 Find the apsidal angle for nearly circular orbits due to a central force only slightly different from an inverse-square force; namely,

$$F(r) = \frac{K}{r^{2.1}}, \quad K < 0$$

What would such motion look like?

7.30 Find the apsidal angle for nearly circular orbits due to a central force given by

$$\mathbf{F}(r) = K \frac{e^{-br}}{r^2} \hat{\mathbf{r}}, \quad K < 0$$

What would such motion look like?

7.31 For a central force given by

$$\mathbf{F}(r) = K \frac{e^{-br}}{r^2} \hat{\mathbf{r}}, \quad K < 0$$

what values of r are necessary for stable circular orbits?

7.32 Calculate the frequency of small radial oscillations, ω_r, about a circular orbit for an inverse-square force like gravity. Show that this is the same as the frequency of revolution, ω. What does this equality mean for orbits due to an inverse-square force?

7.33 Show that the gravitational field inside a spherical shell vanishes by calculating the field directly.

7.34 Show that the gravitational field inside a spherical shell vanishes by calculating the potential.

7.35 Find the gravitational force that would be experienced by a body of mass m dropped into a well or tunnel drilled entirely through Earth along a diameter, passing through the center. Neglecting friction, what kind of motion would this force predict (or cause)? If the motion is periodic, calculate its period.

7.36 An object slides along a straight, frictionless tunnel between two points on Earth's surface (like Toronto and Atlanta). The tunnel does *not* pass through the center of Earth. Show that the object should undergo simple harmonic motion. How long would such a trip from Vancouver to Orlando require? From Auckland to São Paulo? From Los Angeles to London?

7.37 An initially parallel beam of very small, smooth, spherical ball bearings (bee-bees) collides with a larger, fixed, smooth *cylinder* of radius R (this is now a two-dimensional problem). The incoming "particles" scatter elastically—that is, with angle of incidence equal to angle of reflection. Find the scattering angle in terms of impact parameter. Find an expression for scattering cross section, $d\sigma/d\theta$.

7.38 A galactic probe has a velocity v_0 as it approaches a far-distant start of mass M and radius R. Find the cross section σ for striking the star.

7.39 Project particles with mass m and velocity v at a far-distant target. The fixed target repels the incoming particles by a central force given by

$$\mathbf{F}(r) = Kr\hat{\mathbf{r}}, \quad K > 0$$

Determine the scattering angle θ in terms of impact parameter b, mass m, force constant k, and initial velocity v.

7.40 A uniform spherical star of mass m_s and radius r_s moves at constant velocity v through a region of interstellar debris. Show that all debris ahead of it inside a cylinder of radius

$$R = r_s \sqrt{1 + \frac{2Gm_s}{r_s v^2}}$$

will collide with the star.

*7.41 Write a computer program that calculates the position of a body orbiting due to a central force. Check it out first with an inverse-square force like

$$F = -\frac{k}{r^2}$$

Set k equal to 1. Then an initial position of $r_0 = 1$ and tangential speed of $v_0 = 1$ (with zero radial speed) should produce a circular orbit. Plot the orbits for $v_0 = 1.1, 1.3$, and 1.5. Once you are convinced that your program runs correctly, change the form of the force to

$$F = -r^6$$

and use $r_0 = 1$, $v_0 = 1.5$. Also try

$$F = -\frac{1}{r^3}$$

with $r_0 = 1$ and $v_0 = 1.2$.

*7.42 Atmospheric friction causes a satellite in a low orbit to slowly spiral inward (do you remember *Skylab I*?). Consider a 1000-kg satellite in a circular orbit 200 km above Earth's surface (use $r_E = 6200$ km for Earth's radius). Let the atmospheric friction be constant (a dubious assumption!) and proportional to the velocity; namely,

$$\mathbf{F}_{af} = -\left(10^{-4}\frac{kg}{s}\right)\mathbf{v}$$

Write a computer program to handle the calculations and find the number of revolutions it makes as the height decreases from 200 km to 20 km.

*7.43 A better approximation to atmospheric frictional drag on our 1000-kg satellite of the previous problem might be

$$\mathbf{F}_{af} = -Ce^{-b(r-R)}\mathbf{v}$$

with $C = 250$ kg/s, $b = 1/100$ km, and $R = 6200$ km. Write a computer program to handle the calculations with this new force. Begin with a circular orbit 200 km above Earth's surface. Make a graph of height as a function of number of revolutions.

8

SYSTEMS OF PARTICLES

So far we have considered the motion of a single body caused by the external forces acting upon it—for example, the motion of a rocket, an oscillating mass, or a planet in orbit around the Sun. Now we shall inquire into the motion of several bodies, several particles, that may be acted upon by both external and internal forces—for example, forces between the particles or forces of one particle upon another. Earlier we introduced and used the idea of conservation of energy. Now, with a system of particles, conservation principles take on a new and increased importance.

8.1 *N* Particles, the General Case

For the general case, consider N particles labeled $1, 2, 3, \ldots, N$. They are located at positions $\mathbf{r}_1, \mathbf{r}_2, \ldots, \mathbf{r}_N$, respectively, and move with velocities $\mathbf{v}_1, \mathbf{v}_2, \ldots, \mathbf{v}_N$ or $\dot{\mathbf{r}}_1, \dot{\mathbf{r}}_2, \ldots, \dot{\mathbf{r}}_N$, all measured relative to some origin 0. The particles have masses $m_1, m_2, \ldots, m_N$. This general situation is illustrated in Figure 8.1.1. Any particular particle, say, the ith particle, experiences a *net* force that is the result of an external force on it, $\mathbf{F}_i$, and all the internal forces from all the other particles. If we write the internal force on particle i exerted by particle k as $\mathbf{F}_{ik}$, then

$$\mathbf{F}(\text{net})_i = \mathbf{F}_i + \sum_{\substack{k=1 \\ k \neq i}}^{N} \mathbf{F}_{ik} \qquad (8.1.1)$$

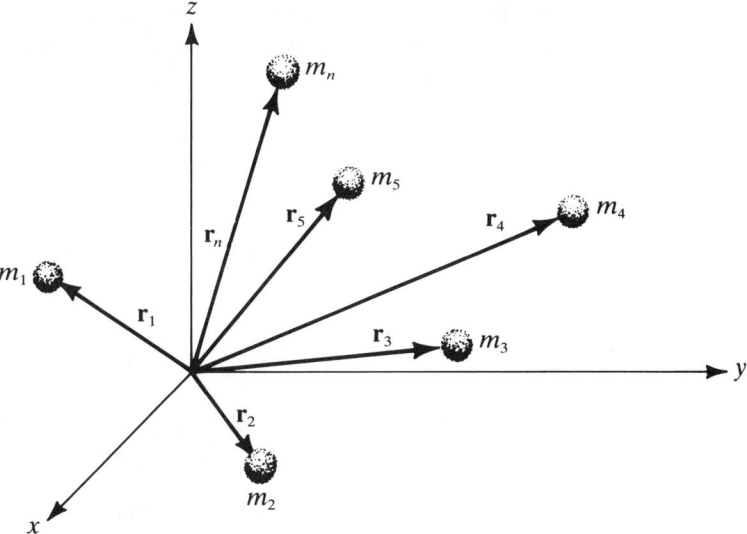

Figure 8.1.1 A system of *N* particles.

This is the net force acting upon mass m_i, so we can attempt to solve for the motion of this particle in a straightforward manner by applying Newton's Second Law, $\mathbf{F}(\text{net}) = m\ddot{\mathbf{r}}$. In this situation we arrive at

$$m_i \ddot{\mathbf{r}}_i = \mathbf{F}(\text{net})_i = \mathbf{F}_i + \sum_{\substack{k=1 \\ k \neq i}}^{N} \mathbf{F}_{ik} \tag{8.1.2}$$

It is straightforward, but also deceiving to write down this equation. Force $\mathbf{F}_{ik}$ is a function of $\mathbf{r}_i$ and $\mathbf{r}_k$, so this equation couples the solution of the ith particle to the solutions of the motions of all the other particles as well. Since these are vector equations, that means that this direct approach leads to $3N$ coupled, second-order differential equations! Direct solution is out of the question for all but very special cases.

The *center of mass* has aptly been called a "uniquely interesting point." And so it is. While determination of the detailed motion of every particle in a large system is out of the question, determination of the motion of the center of mass is far easier. The center of mass, $\mathbf{R}$, is just the average position of the mass and is given by

$$\mathbf{R} = \frac{\sum m_i \mathbf{r}_i}{\sum m_i} \tag{8.1.3}$$

Or, the three coordinates of the center of mass can be written as

$$X = \frac{1}{M} \sum m_i x_i$$

$$Y = \frac{1}{M} \sum m_i y_i \tag{8.1.4}$$

$$Z = \frac{1}{M} \sum m_i z_i$$

where $M = \sum m_i$ is the total mass of the system. Differentiation of the center of mass yields the velocity of the center of mass:

$$\mathbf{V} = \dot{\mathbf{R}} = \frac{1}{M} \sum m_i \dot{\mathbf{r}}_i \qquad (8.1.5)$$

or

$$V_x = \dot{X} = \frac{1}{M} \sum m_i \dot{x}_i$$

$$V_y = \dot{Y} = \frac{1}{M} \sum m_i \dot{y}_i \qquad (8.1.6)$$

$$V_z = \dot{Z} = \frac{1}{M} \sum m_i \dot{z}_i$$

Further differentiation yields the acceleration of the center of mass:

$$\mathbf{A} = \ddot{\mathbf{R}} = \frac{1}{M} \sum m_i \ddot{\mathbf{r}}_i \qquad (8.1.7)$$

or

$$A_x = \ddot{X} = \frac{1}{M} \sum m_i \ddot{x}_i$$

$$A_y = \ddot{Y} = \frac{1}{M} \sum m_i \ddot{y}_i \qquad (8.1.8)$$

$$A_z = \ddot{Z} = \frac{1}{M} \sum m_i \ddot{z}_i$$

We now return to Eq. 8.1.2 and sum up all of the net forces acting on all N particles:

$$\sum_{i=1}^{N} m_i \ddot{\mathbf{r}}_i = \sum_{i=1}^{N} \mathbf{F}_i + \sum_{i=1}^{N} \sum_{\substack{k=1 \\ k \neq i}}^{N} \mathbf{F}_{ik} \qquad (8.1.9)$$

The last term is the summation of every internal force in this system. This term is zero because every force $\mathbf{F}_{ji}$ will be canceled by its counterpart $\mathbf{F}_{ij}$ and

$$\mathbf{F}_{ji} = -\mathbf{F}_{ij} \qquad (8.1.10)$$

by Newton's Third Law. This means that it is impossible to change the motion of the center of mass by internal forces alone. The remaining force term is the resultant of all the *external* forces exerted on all the individual particles. The left-hand side is the total mass multiplied by the acceleration of the center of mass. That is,

$$M\ddot{\mathbf{R}} = \mathbf{F} \qquad (8.1.11)$$

where

$$\mathbf{F} = \sum_{i=1}^{N} \mathbf{F}_i \qquad (8.1.12)$$

Thus, the motion of the center of mass of any system of particles can be described by Eq. 8.1.11, which is identical to an equation describing the net force on a single object. This is a most effective technique, which we shall use over and over again. Once a new or complicated problem can be written in the *form* of a familiar problem, the solution is assured. The importance of this should not be overlooked.

8.2 Momentum

The momentum **p** of a particle with mass m and velocity **v** is defined by

$$\mathbf{p} = m\mathbf{v} \qquad (8.2.1)$$

Newton's Second Law can then be written in terms of momentum as

$$\mathbf{F}_{net} = \frac{d\mathbf{p}}{dt} \qquad (8.2.2)$$

This may well appear as mere relabeling of quantities since

$$\frac{d\mathbf{p}}{dt} = \frac{d}{dt}(m\mathbf{v}) = m\frac{d\mathbf{v}}{dt} = m\mathbf{a} \qquad (8.2.3)$$

as long as the mass remains constant. But if the mass varies—at high velocities with Relativity, of course, but also at moderate speeds with rockets that lose mass or conveyor belts that gain mass—the *momentum form* of Newton's Second Law is still correct and more useful than $\mathbf{F} = m\mathbf{a}$. Equation 8.2.2 is always correct.

If we apply this to a system of particles as before, we have the change in momentum of particle i given by

$$\frac{d\mathbf{p}_i}{dt} = \mathbf{F}_i(net) = \mathbf{F}_i + \sum_k \mathbf{F}_{ik} \qquad (8.2.4)$$

Summing this over all N particles, we have

$$\frac{d\mathbf{P}}{dt} = \frac{d}{dt}\sum_i \mathbf{p}_i = \sum_i \frac{d\mathbf{p}_i}{dt} = \sum_i \mathbf{F}_i(net) = \sum_i \mathbf{F}_i + \sum_i \sum_k \mathbf{F}_{ik} \qquad (8.2.5)$$

The *total momentum* of the system is the vector sum of the momenta of all the bodies in the system

$$\mathbf{P} \equiv \sum_i \mathbf{p}_i = \sum m_i \mathbf{v}_i = M\mathbf{V} \tag{8.2.6}$$

That is, this vector sum of the momenta of the individual particles of the system is the same as the momentum of a single particle whose mass is the total mass of the system and whose velocity is that of the center of mass of the system.

Just as before, we can easily argue that the last term in Eq. 8.2.5 must be zero. But rather than use the previous argument, we shall now use the *method of virtual work*, or the *method of virtual displacement*. Suppose that the entire system of particles is displaced some small distance $\delta \mathbf{r}$. Every particle in the system is displaced this same amount $\delta \mathbf{r}$; there is no *relative* movement of the particle. Thus, the internal forces will do no work. The work done by the internal forces on the ith particle is given by

$$\delta W_i = \delta \mathbf{r} \cdot \sum_k \mathbf{F}_{ik} \tag{8.2.7}$$

Summed over all particles, the net work done by internal forces for such a virtual displacement $\delta \mathbf{r}$ is

$$\delta W = \sum_i \delta W_i = \delta \mathbf{r} \cdot \left(\sum_i \sum_k \mathbf{F}_{ik} \right) \tag{8.2.8}$$

But the homogeneity or uniformity of space requires that the work done by these internal forces be identically zero for *any* displacement $\delta \mathbf{r}$. That is,

$$\delta \mathbf{r} \cdot \left(\sum_i \sum_k \mathbf{F}_{ik} \right) = 0 \tag{8.2.9}$$

Therefore, the term in parentheses must itself be zero, just as we found earlier.

Equation 8.2.5 can now be written as

$$\frac{d\mathbf{P}}{dt} = \mathbf{F} \tag{8.2.10}$$

where, again, $\mathbf{P}$ is the total momentum and $\mathbf{F}$ is the resultant of all the external forces.

Again, we have reduced the motion of the center of mass of a system of particles to that of a single particle.

EXAMPLE 8.2.1 Consider an explosive projectile fired as shown in Figure 8.2.1. If it remains intact, the projectile will follow a familiar parabolic trajectory as shown. Suppose, instead, that it explodes. Where the individual pieces then go clearly depends on the details of the explosion. But the forces of the explosion are internal forces, so they don't affect the total momentum or the center of mass motion. The momentum is affected by the force

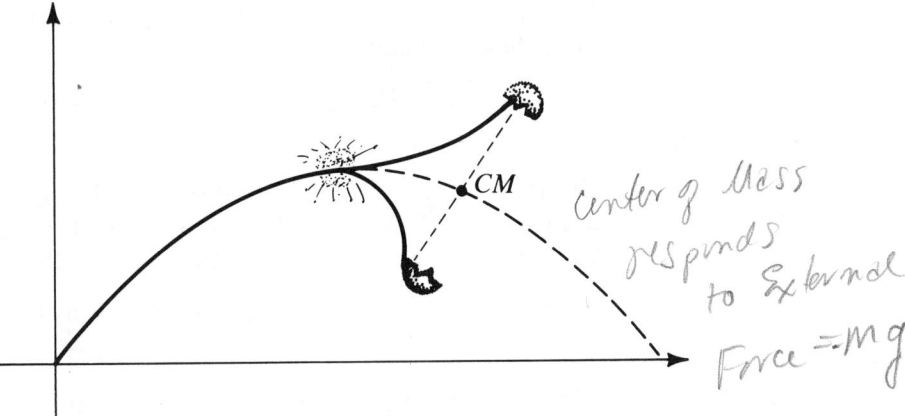

Figure 8.2.1 Trajectory of an explosive charge.

of gravity, $\mathbf{F} = M\mathbf{g}$. This is true before the explosion and remains true afterwards. The center of mass continues along its parabolic trajectory as shown by the dotted line. Only when the first fragment strikes the ground is this no longer true. But that changes the external force, too.

8.3 Motion with a Variable Mass—Rockets

Figure 8.3.1 shows a rocket of mass m moving with velocity $\mathbf{v}$. Fuel is burned and exhausted from the rocket engine at a rate of $|dm/dt|$ with a velocity of $\mathbf{u}$ *relative to the rocket*. Note that we can easily get into trouble with the sign of dm/dt; since m is the mass of the rocket and that mass is decreasing, dm/dt must be negative.

At time t, the mass of the rocket is m and its velocity is $\mathbf{v}$, so the momentum is

$$\mathbf{P}(t) = m\mathbf{v} \tag{8.3.1}$$

Consider the future momentum of this "system," the rocket and fuel. At time $t + dt$, the mass of the rocket has changed to $m + dm$ (not $m - dm$; dm is

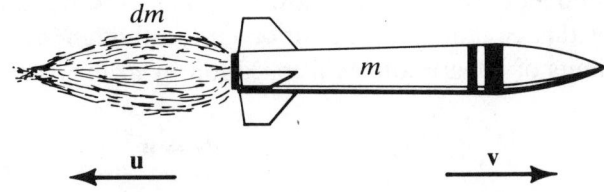

Figure 8.3.1 Rocket with mass m and velocity $\mathbf{v}$.

intrinsically negative!) and its velocity has changed to $\mathbf{v} + d\mathbf{v}$, so the rocket alone has a momentum of

$$\mathbf{P}_{\text{rocket}}(t + dt) = (m + dm)(\mathbf{v} + d\mathbf{v})$$
$$= m\mathbf{v} + m\, d\mathbf{v} + dm\, \mathbf{v} + dm\, d\mathbf{v} \qquad (8.3.2)$$

But the expended fuel must also be considered. At this same time, $t + dt$, an amount of fuel $|dm| = -dm$ is moving with velocity $\mathbf{v} + \mathbf{u}$ *relative to some fixed coordinate frame*, so the momentum of the fuel is

$$\mathbf{P}_{\text{fuel}}(t + dt) = (-dm)(\mathbf{v} + \mathbf{u})$$
$$= -dm\, \mathbf{v} - dm\, \mathbf{u} \qquad (8.3.3)$$

Therefore, the total momentum of the system, the sum of momenta for rocket and for fuel, is just

$$\mathbf{P}(t + dt) = \mathbf{P}_{\text{rocket}}(t + dt) + \mathbf{P}_{\text{fuel}}(t + dt)$$
$$= m\mathbf{v} + m\, d\mathbf{v} - dm\, \mathbf{u} \qquad (8.3.4)$$

where the two $dm\mathbf{v}$ terms canceled and the second-order term, $dm\, d\mathbf{v}$, has been neglected.

Thus, the *change* in momentum is just

$$d\mathbf{P} = m\, d\mathbf{v} - dm\, \mathbf{u} \qquad (8.3.5)$$

The time *rate* of change, $d\mathbf{P}/dt$, is equal to the *external* force, $\mathbf{F}$, so

$$\mathbf{F} = \frac{d\mathbf{P}}{dt} = m\frac{d\mathbf{v}}{dt} - \frac{dm}{dt}\mathbf{u} \qquad (8.3.6)$$

A very interesting case occurs when $\mathbf{F} = 0$. This only happens when a rocket is in deep space, far from any other body that might exert a gravitational force on it. But it is a useful case to consider in developing an understanding of rocket propulsion. For this case,

$$m\frac{d\mathbf{v}}{dt} - \frac{dm}{dt}\mathbf{u} = 0$$

or

$$m\frac{d\mathbf{v}}{dt} = \frac{dm}{dt}\mathbf{u} \qquad (8.3.7)$$

The left side of this equation is the *thrust*, the force felt *by the rocket* of mass m causing it to accelerate at $d\mathbf{v}/dt$. Thus, the thrust of a rocket engine is just the product of the fuel mass flow rate dm/dt and the exhaust velocity $\mathbf{u}$. Solution of this equation is straightforward, although we must be careful with the limits of integration. Multiplication by dt/m gives

$$d\mathbf{v} = \mathbf{u}\frac{dm}{m} \qquad (8.3.8)$$

which can then be integrated:

$$\int_{v_0}^{v} d\mathbf{v} = \mathbf{u} \int_{m_0}^{m} \frac{dm}{m}$$

to get

$$\mathbf{v} - \mathbf{v}_0 = \mathbf{u} \ln m \Big|_{m_0}^{m} = \mathbf{u} \ln \frac{m}{m_0} \quad (8.3.9)$$

Since $m_0 > m$, it is sometimes more convenient to write this as

$$\mathbf{v} = \mathbf{v}_0 - \mathbf{u} \ln \frac{m_0}{m} \quad (8.3.10)$$

Remember that $\mathbf{v}$ and $\mathbf{u}$ are in opposite directions. High final velocities require large values for the exhaust velocity $\mathbf{u}$ and/or large values for m_0/m, where m is the final mass, the mass of the rocket structure (frame, engine, tanks, and so on), and the payload, and m_0 is all of that plus the fuel. Requiring a large value for m_0/m is often referred to as requiring a large "fuel to payload ratio." But the *structure* plays a vital role in this ratio.

To reduce the mass of the structure that is being accelerated, and thus increase this important ratio of m_0/m, rocket designers use *staged rockets*. The structure of the first stage of such a rocket will be very massive in order to contain all the necessary fuel, a large engine, and so on. When its fuel is expended, all of this structure is jettisoned from the rest of the rocket so that all future forces are applied to a much smaller mass to produce much greater accelerations. All rockets used for satellite launches and the like are staged.

Another special and interesting case of Eq. 8.3.6 is that of a rocket on or near Earth's surface. Then the external force on the rocket is its weight, $\mathbf{F} = m\mathbf{g}$:

$$m \frac{d\mathbf{v}}{dt} - \frac{dm}{dt} \mathbf{u} = m\mathbf{g} \quad (8.3.11)$$

This can be solved in the same manner as before:

$$d\mathbf{v} = \mathbf{u} \frac{dm}{m} + \mathbf{g} \, dt$$

$$\int_{v_0}^{v} d\mathbf{v} = \mathbf{u} \int_{m_0}^{m} \frac{dm}{m} + \mathbf{g} \int_{0}^{t} dt$$

$$\mathbf{v} = \mathbf{v}_0 - \mathbf{u} \ln \frac{m_0}{m} + \mathbf{g}t \quad (8.3.12)$$

Remembering that $\mathbf{v}$ and $\mathbf{u}$ go in opposite directions, the corresponding scalar equation for a rocket fired vertically upward from rest is

$$v = u \ln \frac{m_0}{m} - gt \quad (8.3.13)$$

For a rocket to rise at all, its thrust must exceed its weight:

$$\left|\frac{dm}{dt}\right| u > m_0 g \qquad (8.3.14)$$

For some rather large present-day rockets, this condition is barely satisfied. Most large rockets move very slowly at first. Their high final speed is a result of continued acceleration—and the value of that acceleration increases as the remaining mass of the rocket decreases.

8.4 Motion with a Variable Mass—Conveyor Belts

Figure 8.4.1 is a sketch of a conveyor belt. At the left, mass falls from a hopper at a rate of dm/dt, landing on a conveyor belt that moves at a constant speed v. The mass initially has no horizontal component of velocity and soon after landing on the conveyor belt its horizontal velocity is v. Therefore, the momentum has changed. What force F must be exerted by the belt in order to keep the system moving at a constant speed? Alternately, it may be asked, what power is necessary to keep such a system operating under this constant, steady-state condition?

All of the mass on the conveyor belt moves with horizontal velocity v. Even as the mass falls off the end of the conveyor, it still carries a horizontal velocity component of v. But for every increment in time dt, an amount of mass dm falls onto the conveyor with zero horizontal velocity. This mass is then accelerated to the conveyor's velocity v. Thus, the force F that must be exerted is

$$F = \frac{dp}{dt} = \frac{dm}{dt} v \qquad (8.4.1)$$

or, the power required is

$$\text{Power} = Fv = \frac{dm}{dt} v^2 \qquad (8.4.2)$$

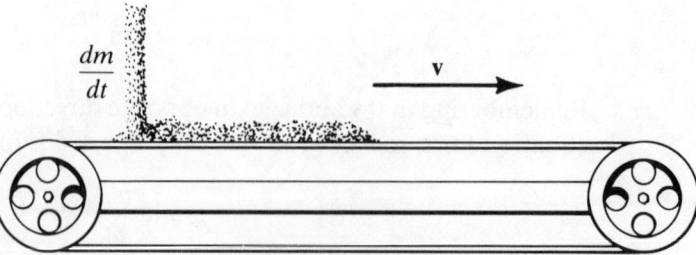

Figure 8.4.1 Conveyor belt moving with velocity **v**.

8.5 Collisions

Collisions provide us with an interesting example of the power of conservation. A collision usually involves two objects that are initially far from each other. This initial configuration can be characterized by the masses and initial velocities of both bodies. Then there is a collision—the word itself conjures up two billiard balls crashing into momentary contact or two automobiles with bumpers in contact. Of course, these are legitimate collisions. But a collision can also be any interaction of internal forces between the two bodies—for example, an incoming alpha particle experiencing a repulsive Coulomb force from a target nucleus or a marauding comet being deflected from its path by the gravitational attraction of our Sun. Sometime after the collision—due to "contact forces" or to "forces at a distance"—the two bodies are again far removed from each other. This final configuration can also be described by the masses and final velocities of both bodies.

To fully predict the final state from the initial state, it would seem that we would have to know all of the details of the force of interaction between the two colliding bodies. But the details of such a force may be quite complicated. For example, the magnitude and duration of the force between two colliding billiard balls is neither obvious nor trivial!

As it turns out, the ideas of conservation allow us to predict many and occasionally all of the details of the final state with very little information about the force of interaction itself. For example, consider an object with mass m_1 and initial velocity v_{10} approaching a stationary object with mass m_2, as shown in Figure 8.5.1(a). When the two masses are far apart, v_{10} is the velocity of m_1 (that is, before there is any appreciable interaction, such as

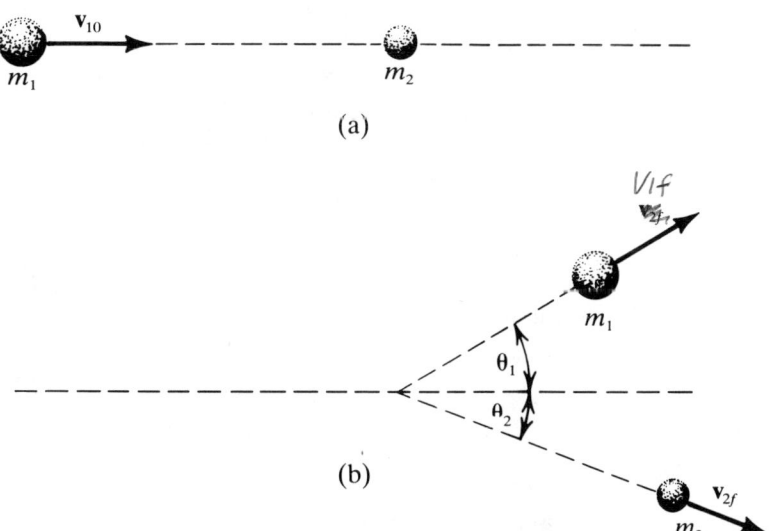

Figure 8.5.1 Collisions. (a) Initial configuration before the collision; (b) Final configuration afterward.

occurs when the "collision" is a long-ranged interaction like the $1/r^2$ dependence of Coulomb or gravitational forces).

It might seem that we should also consider the case with both masses moving toward each other. But if both masses are moving, the collision may then be viewed from a reference frame moving along with the initial velocity of m_2. In that frame, m_2 is initially at rest. Therefore, there is no loss of generality to consider the situation shown in Figure 8.5.1 with an incoming mass m_1 approaching a stationary target of mass m_2.

The initial momentum $\mathbf{P}_0$ of this system is given by

$$\mathbf{P}_0 = m_1 \mathbf{v}_{10} \tag{8.5.1}$$

The horizontal or x-component of this is merely

$$P_{0x} = m_1 v_{10} \tag{8.5.2}$$

while

$$P_{0y} = 0 \tag{8.5.3}$$

for the vertical or y-component. After the collision, object 1 moves off with final velocity $\mathbf{v}_{1f}$ and object 2 moves off with final velocity $\mathbf{v}_{2f}$, as shown in Figure 8.5.1(b), and the final momentum of the system is given by

$$\mathbf{P}_f = m_1 \mathbf{v}_{1f} + m_2 \mathbf{v}_{2f} \tag{8.5.4}$$

Thus, the components of the final momentum are

$$P_{fx} = m_1 v_{1f} \cos \theta_1 + m_2 v_{2f} \cos \theta_2 \tag{8.5.5}$$

and

$$P_{fy} = m_1 v_{1f} \sin \theta_1 - m_2 v_{2f} \sin \theta_2 \tag{8.5.6}$$

where θ_1 and θ_2 are the angles between the final velocities $\mathbf{v}_{1f}$ and $\mathbf{v}_{2f}$ and the initial velocity $\mathbf{v}_{10}$, respectively, as shown in Figure 8.5.1(b). Note that the initial velocity and the two final velocities lie in the same plane. With our present choice of coordinates, the initial momentum only has an x-component. If the y-axis is chosen so that $\mathbf{v}_{1f}$ lies in the xy plane, then $\mathbf{v}_{2f}$ must also lie in this plane. Otherwise, it would carry a z-component of momentum that was not present initially.

Whatever the manner of the collision, it is the result of *internal* forces. Thus, in the absence of external forces, there will be no change in the total momentum of the system. Therefore, we can set the initial momentum equal to the final momentum:

$$m_1 v_{10} = m_1 v_{1f} \cos \theta_1 + m_2 v_{2f} \cos \theta_2 \tag{8.5.7}$$

$$0 = m_1 v_{1f} \sin \theta_1 - m_2 v_{2f} \sin \theta_2 \tag{8.5.8}$$

This provides two equations, but there are seven variables. Usually, m_1 and m_2 and the initial velocity v_{10} are known, but that still leaves us with four variables to find: v_{1f}, v_{2f}, θ_1, and θ_2. Obviously, we need more information.

Collisions in which the total *kinetic* energy is conserved (called *totally elastic collisions*) are especially interesting. The *total* energy is always conserved. But total energy must include the internal energy of the masses. This internal energy can change if the temperature increases due to the collision or if work is done in deforming the shape of one or both of the masses. In relativity, we must consider the energy associated with or stored in the masses themselves. In that case, the masses may even change in the collision, which would effect our expressions for the total energy. This is quite common in nuclear reactions.

In Chapter 2, kinetic energy was defined for a particle. The initial kinetic energy of this system is

$$T_0 = \tfrac{1}{2} m_1 v_{10}^2 \tag{8.5.9}$$

The final kinetic energy of the system is simply the sum of the kinetic energies of the two masses:

$$T_f = \tfrac{1}{2} m_1 v_{1f}^2 + \tfrac{1}{2} m_2 v_{2f}^2 \tag{8.5.10}$$

For a totally elastic collision, the final kinetic energy is the same as the initial kinetic energy. That is,

$$\tfrac{1}{2} m_1 v_{10}^2 = \tfrac{1}{2} m_1 v_{1f}^2 + \tfrac{1}{2} m_2 v_{2f}^2 \tag{8.5.11}$$

Collisions between billiard balls or hard plastic pucks on an air hockey board are very good approximations of totally elastic collisions. All collisions between nuclei and particles that leave the masses unchanged are truly totally elastic.

Equation 8.5.11, along with Eqs. 8.5.7 and 8.5.8, now give us three equations with four unknowns. Thus, the ideas of conservation of energy and momentum do not allow us to completely solve for the final state without knowing the internal force itself. But if one of the four variables v_{1f}, v_{2f}, θ_1, or θ_2 is known, these three equations do allow us to finish the solution of the final state. Therefore, we shall spend some time discussing various special cases.

Head-on collision: For a head-on collision or a "direct hit," $\theta_2 = 0$. From Eq. 8.5.8, this immediately requires that $\theta_1 = 0$ as well (unless $v_{1f} = 0$, in which case θ_1 cannot be reasonably defined). All the motion is now confined to one dimension and our three equations are reduced to two:

$$m_1 v_{10} = m_1 v_{1f} + m_2 v_{2f} \tag{8.5.12}$$

$$m_1 v_{10}^2 = m_1 v_{1f}^2 + m_2 v_{2f}^2 \tag{8.5.13}$$

These equations can be conveniently rewritten as

$$v_{10} - v_{1f} = \frac{m_2}{m_1} v_{2f} \tag{8.5.14}$$

$$v_{10}^2 - v_{1f}^2 = \frac{m_2}{m_1} v_{2f}^2 \tag{8.5.15}$$

Dividing Eq. 8.5.15 by Eq. 8.5.14 yields

$$v_{10} + v_{1f} = v_{2f} \qquad (8.5.16)$$

Substituting this into Eq. 8.5.14 yields

$$v_{1f} = \frac{m_1 - m_2}{m_1 + m_2} v_{10} \qquad (8.5.17)$$

which readily provides v_{2f} upon substitution into Eq. 8.5.16:

$$v_{2f} = \frac{2m_1}{m_1 + m_2} v_{10} \qquad (8.5.18)$$

Thus, v_{2f} is always positive; mass m_2 always moves forward in any head-on collision. But interesting things happen to mass m_2 for various choices of masses:

1. For equal masses, $v_{1f} = 0$; the incoming mass stops and the target mass moves out with a velocity identical to the initial velocity of the incoming mass.
2. For a small incoming mass colliding with a heavy mass ($m_1 < m_2$), $v_{1f} < 0$; the light incoming mass recoils or bounces off the heavy mass in the backward direction. It is often useful in physics to look at or think of extreme examples. A collision with $m_1 < m_2$ might be an MGB sports car running into a Mack truck—the MGB bounces backward and the Mack truck moves forward just a little. Or consider a bumblebee flying into the back of a parked car.
3. For a large incoming mass colliding with a small target mass ($m_1 > m_2$), $v_{1f} > 0$; the incoming mass continues on in the forward direction with reduced speed. Mass m_2 moves with a greater velocity. An extreme example might be a Mack truck running into the back of a parked MGB. The sports car is jolted forward with high speed while the truck continues on with a somewhat reduced speed.

Equation 8.5.16 readily gives the *relative* speed of the two masses after the collision

$$v_{21} = v_{2f} - v_{1f} = v_{10} \qquad (8.5.19)$$

For any choice of masses, the relative speed of approach before the collision is always equal to the relative speed of departure following the collision. And this will prove to be true in cases more general than just head-on collisions.

General case: Moving the v_{1f} terms to the left side of momentum equations 8.5.7 and 8.5.8, squaring the resulting equations, and adding them and dividing by m_1^2, yields

$$v_{10}^2 - 2v_{10}v_{1f} \cos \theta_1 + v_{1f}^2 = \left(\frac{m_2}{m_1}\right)^2 v_{2f}^2 \qquad (8.5.20)$$

which is independent of θ_2. The kinetic energy equation 8.5.11 yields an expression for v_{2f}^2:

$$v_{2f}^2 = \frac{m_1}{m_2}(v_{10}^2 - v_{1f}^2) \tag{8.5.21}$$

Combining these equations gives a quadratic equation for v_{1f} in terms of the masses, the initial velocity, and the scattering angle θ_1:

$$v_{1f} = v_{10}\left[\frac{m_1}{m_1 + m_2}\right]\left[\cos\theta_1 \pm \sqrt{\cos^2\theta_1 - \frac{m_1^2 - m_2^2}{m_1^2}}\right] \tag{8.5.22}$$

Immediately we can see restrictions on the possible values for θ_1 when the incoming mass is large. For $m_1 > m_2$, the final term in the radical is intrinsically positive. Since a complex value for v_{1f} is physically meaningless, the speed v_{1f} must remain real. So the expression within the radical must remain nonnegative, or

$$\cos^2\theta_1 \geq \frac{m_1^2 - m_2^2}{m_1^2} \tag{8.5.23}$$

or

$$\cos^2\theta_{1\text{ max}} = 1 - \frac{m_2^2}{m_1^2} \tag{8.5.24}$$

Since $\cos\theta$ decreases as θ increases, that means

$$\theta_1 \leq \theta_{1\text{ max}} \tag{8.5.25}$$

Any value of the scattering angle θ_1 is allowed up to the maximum scattering angle given by Eq. 8.5.24. Figure 8.5.2 is a graph of maximum scattering angle as a function of the ratio of the masses.

For $m_1 < m_2$, Eq. 8.5.23 provides no restriction on scattering angle, and the incoming mass can scatter through any angle—including backscatter up

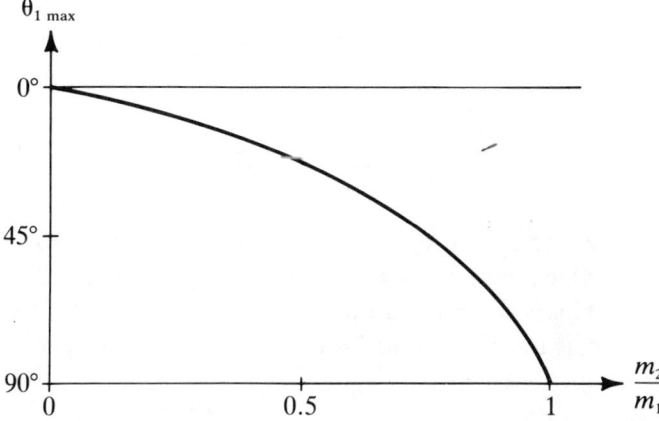

Figure 8.5.2
Maximum scattering angle as a function of the ratio of the masses for $m_1 > m_2$.

to $\theta = \pi$. We have already seen this result in our discussion of head-on collisions.

One further item of interest is the angle between the outgoing particles, $\theta_1 + \theta_2$. This is particularly interesting for the case of equal masses. If we multiply the x-momentum equation, Eq. 8.5.7, by $\cos\theta_1$, and the y-momentum equation, Eq. 8.5.8, by $\sin\theta_1$, the sum of these two is

$$v_{10}\cos\theta_1 = v_{1f} + v_{2f}(\cos\theta_1\cos\theta_2 - \sin\theta_1\sin\theta_2)$$

or

$$v_{10}\cos\theta_1 = v_{1f} + v_{2f}\cos(\theta_1 + \theta_2) \qquad (8.5.26)$$

But for $m_1 = m_2$, Eq. 8.5.22 yields

$$v_{1f} = v_{10}\cos\theta_1 \qquad (8.5.27)$$

Therefore,

$$\cos(\theta_1 + \theta_2) = 0$$

or

$$\theta_1 + \theta_2 = \frac{\pi}{2} \qquad (8.5.28)$$

The angle between exiting particles of equal mass is a right angle. This is often useful on a pool table because the cue ball always leaves at 90° to the direction of the struck ball. (This is only true if there is no spin or "english" on the balls involving additional rotational effects. Then the balls can accurately be treated as point particles.)

8.6 Center of Mass Frame

In Figure 8.1.1, and with Eq. 8.1.3, we have defined the center of mass. For a two-body system, the center of mass is given by the simple expression

$$\mathbf{R} = \frac{m_1\mathbf{r}_1 + m_2\mathbf{r}_2}{m_1 + m_2} \qquad (8.6.1)$$

where $\mathbf{r}_1$ and $\mathbf{r}_2$ are the location vectors or coordinates of bodies 1 and 2, which have masses m_1 and m_2, respectively, and $\mathbf{R}$ locates the center of mass. This is sketched in Figure 8.6.1. Another vector that appears in that diagram, $\mathbf{r}$, is the relative position of object 1 with respect to object 2. That is,

$$\mathbf{r} = \mathbf{r}_1 - \mathbf{r}_2 \qquad (8.6.2)$$

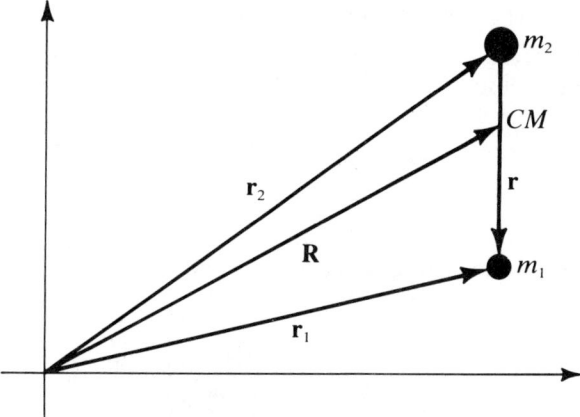

Figure 8.6.1 The two-body problem.

Both the center of mass coordinate **R** and this new relative coordinate **r** will prove useful. Equations 8.6.1 and 8.6.2 give the transformation *from* $\mathbf{r}_1$ and $\mathbf{r}_2$ *to* **R** and **r**. The inverse transformations are

$$\mathbf{r}_1 = \mathbf{R} + \frac{m_2}{m_1 + m_2}\mathbf{r} \tag{8.6.3}$$

and

$$\mathbf{r}_2 = \mathbf{R} - \frac{m_1}{m_1 + m_2}\mathbf{r} \tag{8.6.4}$$

Applying Newton's Second Law to each object in the normal manner provides

$$m_1\ddot{\mathbf{r}}_1 = \mathbf{F}_1^{\text{Total}} = \mathbf{F}_1^{\text{ex}} + \mathbf{F}_{12} \tag{8.6.5}$$

and

$$m_2\ddot{\mathbf{r}}_2 = \mathbf{F}_2^{\text{Total}} = \mathbf{F}_2^{\text{ex}} + \mathbf{F}_{21} \tag{8.6.6}$$

where $\mathbf{F}_1^{\text{ex}}$ and $\mathbf{F}_2^{\text{ex}}$ are the *external* forces on objects 1 and 2, $\mathbf{F}_{12}$ is the force exerted *on* object 1 *by* 2, and $\mathbf{F}_{21}$ is the force exerted *on* object 2 *by* object 1. $\mathbf{F}_{12}$ and $\mathbf{F}_{21}$, then, are *internal* forces. By Newton's Third Law, these are equal in magnitude and opposite in direction.

To reduce subscripts, let us define **f** by

$$\mathbf{f} \equiv \mathbf{F}_{12} = -\mathbf{F}_{21} \tag{8.6.7}$$

Just as we did in Section 8.1, we can add Eqs. 8.6.5 and 8.6.6, using Eq. 8.6.7, to obtain

$$m_1\ddot{\mathbf{r}}_1 + m_2\ddot{\mathbf{r}}_2 = \mathbf{F}_1^{\text{ex}} + \mathbf{F}_2^{\text{ex}} = \mathbf{F} \tag{8.6.8}$$

where $\mathbf{F} = \mathbf{F}_1^{\text{ex}} + \mathbf{F}_2^{\text{ex}}$, just as in Eq. 8.1.11. The left side, of course, is $M\ddot{\mathbf{R}}$, so

$$\mathbf{F} = M\ddot{\mathbf{R}} \tag{8.6.9}$$

Just as we saw earlier, the center of mass moves as if all of the external forces, **F**, acted upon a single mass M located at the center of mass **R**. Describing the motion of the center of mass is a simple, ordinary, *one*-body problem.

Now consider the special cases where either

$$\mathbf{F}_1^{ex} = \mathbf{F}_2^{ex} = 0 \tag{8.6.10}$$

when no external forces are present, or

$$\frac{\mathbf{F}_1^{ex}}{m_1} = \frac{\mathbf{F}_2^{ex}}{m_2} \tag{8.6.11}$$

where the forces on the objects are proportional to the masses, as we find in a *uniform* gravitational field.

For either of these special cases, we multiply Eq. 8.6.5 by m_2 and Eq. 8.6.6 by m_1 and then subtract. The result is

$$m_1 m_2 (\ddot{\mathbf{r}}_1 - \ddot{\mathbf{r}}_2) = (m_1 + m_2)\mathbf{f} \tag{8.6.12}$$

We may simplify this by introducing a quantity called the *reduced mass*:

$$\mu = \frac{m_1 m_2}{m_1 + m_2} \tag{8.6.13}$$

Now we may write

$$\mathbf{f} = \mu \ddot{\mathbf{r}} \tag{8.6.14}$$

So the *relative* motion of the two bodies may now be written as another single-body problem.

The relative motion is the same as the motion of an object with mass μ under the influence of a force **f**, which is the internal force between the two objects. Thus, we have reduced the two-body problem down to two separate single-body problems, which may be solved separately using all the techniques developed earlier.

While discussing central forces in the previous chapter, we looked at a satellite orbiting a *fixed* Earth, a planet orbiting a *fixed* Sun, or alpha scattering from a *fixed* nucleus. Due to the masses involved, these are reasonable approximations. But they remain approximations. As the moon orbits Earth, *both* the moon and Earth move about the center of mass of the Earth-moon system. To find that motion exactly—rather than approximately—we must solve for the *relative* motion using Eq. 8.6.14 and the reduced mass μ. Then Eqs. 8.6.3 and 8.6.4 immediately provide the exact motion of both the moon and Earth. Note that the reduced mass can be used in writing these two transformations as

$$\mathbf{r}_1 = \mathbf{R} + \frac{\mu}{m_1}\mathbf{r} \tag{8.6.15}$$

and

$$\mathbf{r}_2 = \mathbf{R} - \frac{\mu}{m_2}\mathbf{r} \qquad (8.6.16)$$

Figure 8.5.1 has already described a collision as viewed from the usual reference frame at rest with respect to the laboratory; we shall call this the LAB frame. Collisions take on an elegant and useful symmetry when viewed from a reference frame moving along with the center of mass; we shall call this the CM frame. As viewed from this CM frame, the total momentum of the two bodies is zero both before and after the collision. This is shown in Figure 8.6.2, where the values in pointed brackets $\langle\ \rangle$ indicate "internal" quantities, or that they are measured in the CM frame.

Objects 1 and 2 are located in the CM frame by

$$\langle \mathbf{r}_1 \rangle = \mathbf{r}_1 - \mathbf{R} \qquad (8.6.17)$$

and

$$\langle \mathbf{r}_2 \rangle = \mathbf{r}_2 - \mathbf{R} \qquad (8.6.18)$$

Expressed in terms of the relative position $\mathbf{r}$ using Eqs. 8.6.3 and 8.6.4, and the reduced mass of Eq. 8.6.13, these become

$$\langle \mathbf{r}_1 \rangle = \frac{\mu}{m_1}\mathbf{r} \qquad (8.6.19)$$

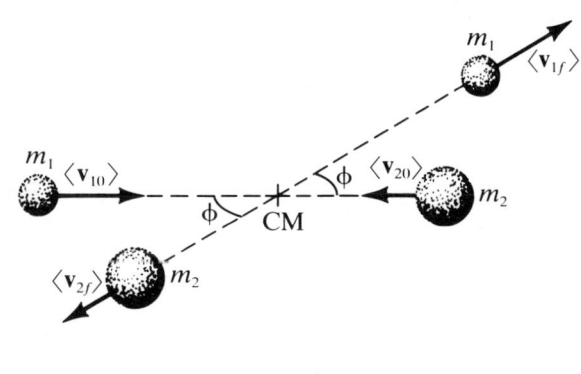
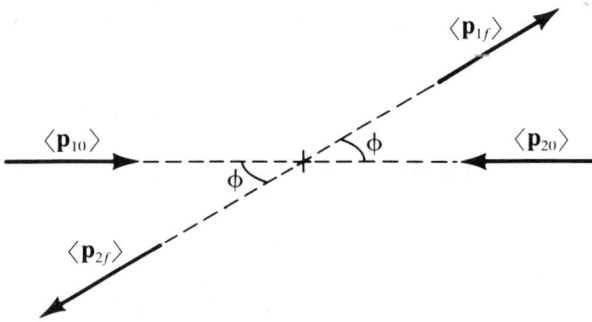

Figure 8.6.2 A collision viewed from the CM frame.

and
$$\langle \mathbf{r}_2 \rangle = -\frac{\mu}{m_2}\mathbf{r} \tag{8.6.20}$$

Differentiation yields
$$\langle \mathbf{v}_1 \rangle = \frac{\mu}{m_1}\mathbf{v} \tag{8.6.21}$$

and
$$\langle \mathbf{v}_2 \rangle = -\frac{\mu}{m_2}\mathbf{v} \tag{8.6.22}$$

from which we readily get
$$\langle \mathbf{p}_1 \rangle = m_1 \langle \mathbf{v}_1 \rangle = \mu \mathbf{v} \tag{8.6.23}$$

and
$$\langle \mathbf{p}_2 \rangle = m_2 \langle \mathbf{v}_2 \rangle = -\mu \mathbf{v} \tag{8.6.24}$$

which clearly justifies the earlier statement that the momentum in the CM frame is always zero.

Thus, the two particles approach with momenta equal in magnitude and opposite in direction. After the collision they leave each other with momenta that are again equal in magnitude, but opposite in direction. Each object, then, has "scattered" through the same angle ϕ. Now we must relate this ϕ of the CM frame back to θ_1 and θ_2 of the LAB frame. To do this we must look more closely at the "internal" coordinates measured from the center of mass.

We differentiate Eq. 8.6.17 to get
$$\langle \mathbf{v}_1 \rangle = \mathbf{v}_1 - \mathbf{V} \tag{8.6.25}$$

This vector addition is shown in Figure 8.6.3. From that diagram we can find that
$$\tan \theta_1 = \frac{\langle v_1 \rangle \sin \phi}{\langle v_1 \rangle \cos \phi + V} \tag{8.6.26}$$

For object 1, moving with initial velocity v_{10} approaching object 2 at rest in the LAB frame, we find that the center of mass is moving with velocity
$$\mathbf{V} = \frac{m_1}{m_1 + m_2}\mathbf{v}_1 = \frac{\mu}{m_2}\mathbf{v}_{10} \tag{8.6.27}$$

Or, combining this with Eq. 8.6.21,
$$\tan \theta_1 = \frac{\sin \phi}{\cos \phi + \dfrac{V}{v_{1f}}} = \frac{\sin \phi}{\cos \phi + \dfrac{m_1 v_{10}}{m_2 v_f}} \tag{8.6.28}$$

Figure 8.6.3 Relation between velocities and scattering angles in the CM and LAB frames.

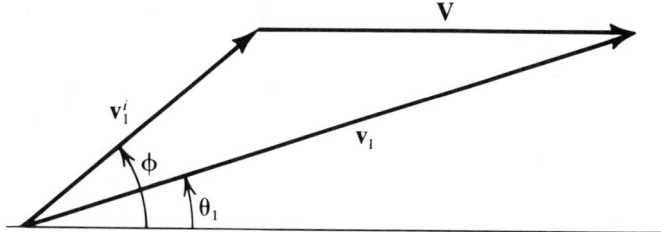

Since $v_{20} = 0$, $v_{10} = v_0$; the initial speed of object 1 is also the initial *relative* speed. v_f is, of course, the final *relative* speed. Thus,

$$\tan\theta_1 = \frac{\sin\phi}{\cos\phi + \dfrac{m_1 v_0}{m_2 v_f}} \tag{8.6.29}$$

An elastic collision requires that the *relative* speed remain unchanged, or that $v_0 = v_f$. Thus, for an elastic collision,

$$\tan\theta_1 = \frac{\sin\phi}{\cos\phi + \dfrac{m_1}{m_2}} \tag{8.6.30}$$

For equal masses a few choice trig substitutions allow this to be reduced to the very simple form of

$$\theta_1 = \tfrac{1}{2}\phi \tag{8.6.31}$$

In Chapter 7, we solved for the motion of objects acted upon by forces due to a *fixed* force center. The results are *exactly* true for an infinite m_2. The results are a *close approximation* for $(m_1/m_2) \ll 1$. But what of other situations? The developments of this section provide a mechanism to take care of that. Namely, replace m by μ and all the equations of Chapter 7 describe the relative motion.

At the end of Chapter 7 we discussed Rutherford scattering. The results obtained there are exact if the scattering nucleus is somehow bound into a crystal lattice structure so it really is fixed. But the results are still reasonably accurate for alphas colliding with, say, gold nuclei for $m_2/m_1 = 50$, which is clearly much larger than unity even if not quite infinite. But what is the cross section for Rutherford scattering when the incoming and target masses are the same? Equation 7.9.1 gives the result with a fixed scattering center. Changing m to μ provides the results measured in *relative* coordinates or in angles measured in the CM frame. Thus,

$$d\sigma = 2\pi\left(\frac{K}{2\mu v_0^2}\right)^2 \frac{\sin\phi}{\sin^4\dfrac{\phi}{2}}\, d\phi \tag{8.6.32}$$

where $K = (Qq/4\pi)\varepsilon_0$, as defined by Eq. 7.9.2. But for $m_1 = m_2$, we've just found $\phi = 2\theta$, so

$$d\sigma = 2\pi \left(\frac{K}{2\mu v_0^2}\right)^2 \frac{\sin 2\theta_1}{\sin^4 \theta_1} (2d\theta_1) \qquad (8.6.33)$$

or

$$\frac{d\sigma}{d\theta_1} = \left(\frac{K}{2\mu v_0^2}\right)^2 \frac{4\cos\theta_1}{\sin^4\theta_1} 2\pi \sin\theta_1 \qquad (8.6.34)$$

PROBLEMS

8.1 Find the center of mass, the velocity of the center of mass, total linear momentum, and total kinetic energy of the following systems of particles:
  (a) System A
  $m_1 = 1$ kg, $\mathbf{r}_1 = (\hat{\mathbf{i}} + 2\hat{\mathbf{j}} + 3\hat{\mathbf{k}})$ m, $\mathbf{v}_1 = 2\hat{\mathbf{i}}$ m/s
  $m_2 = 1$ kg, $\mathbf{r}_2 = (-\hat{\mathbf{i}} + \hat{\mathbf{j}} + \hat{\mathbf{k}})$ m, $\mathbf{v}_2 = (2\hat{\mathbf{i}} + 3\hat{\mathbf{j}})$ m/s
  $m_3 = 2$ kg, $\mathbf{r}_3 = (-\hat{\mathbf{i}} + 2\hat{\mathbf{j}} - \hat{\mathbf{k}})$ m, $\mathbf{v}_3 = (\hat{\mathbf{j}} + 2\hat{\mathbf{k}})$ m/s
  (b) System B
  $m_1 = 1$ kg, $\mathbf{r}_1 = (3\hat{\mathbf{i}} + 2\hat{\mathbf{j}} + \hat{\mathbf{k}})$ m, $\mathbf{v}_1 = (\hat{\mathbf{i}} + \hat{\mathbf{j}})$ m/s
  $m_2 = 2$ kg, $\mathbf{r}_2 = (-2\hat{\mathbf{i}} + \hat{\mathbf{j}})$ m, $\mathbf{v}_2 = (\hat{\mathbf{i}} + \hat{\mathbf{j}} + \hat{\mathbf{k}})$ m/s
  $m_3 = 3$ kg, $\mathbf{r}_3 = (\hat{\mathbf{i}} + \hat{\mathbf{j}} + \hat{\mathbf{k}})$ m, $\mathbf{v}_3 = (2\hat{\mathbf{i}} + \hat{\mathbf{j}})$ m/s
  (c) System C
  $m_1 = 2$ kg, $\mathbf{r}_1 = 2\hat{\mathbf{i}}$ m, $\mathbf{v}_1 = (\hat{\mathbf{i}} + \hat{\mathbf{j}})$ m/s
  $m_2 = 1$ kg, $\mathbf{r}_2 = 3\hat{\mathbf{j}}$ m, $\mathbf{v}_2 = (\hat{\mathbf{j}} + \mathbf{k})$ m/s
  $m_3 = 1$ kg, $\mathbf{r}_3 = 1\hat{\mathbf{k}}$ m, $\mathbf{v}_3 = (2\hat{\mathbf{i}} + 3\mathbf{k})$ m/s

8.2 Three particles, each of unit mass, have the following positions:

$$\mathbf{r}_1 = 3\hat{\mathbf{i}} + 4t\hat{\mathbf{j}} + 5t^2\mathbf{k}$$
$$\mathbf{r}_2 = (3 + t)\hat{\mathbf{i}} + (1 + t^2)\hat{\mathbf{j}}$$
$$\mathbf{r}_3 = (1 + t^2 + 2t^3)\hat{\mathbf{k}}$$

For $t = 0$, calculate the following:
  (a) Position of the center of mass.
  (b) Velocity of the center of mass.
  (c) Linear momentum.
  (d) Kinetic energy of the system.

8.3 Objects of mass m, $2m$, $3m$, and $4m$ are located (in that order) at the corners of a massless square plate with side length l. Find the center of mass of this system.

8.4 Objects of mass m, $2m$, and $3m$ are located at the corners of a massless equilateral triangular plate with side length l. Find the center of mass of this system. Show your choice of coordinate system clearly in a sketch.

8.5 Body 1 with mass $m_1 = m$ moves with position given by

$$\mathbf{r}_1 = 3\hat{\mathbf{i}} + 2t\hat{\mathbf{j}} + 3t^2\hat{\mathbf{k}}$$

while body 2 with mass $m_2 = 2m$ moves according to

$$\mathbf{r}_2 = 3t\hat{\mathbf{i}} + 2t^2\hat{\mathbf{j}} + 3\hat{\mathbf{k}}$$

Find the acceleration of the center of mass of the system of these two bodies.

8.6 Water is poured into a large flask at the rate of 240 ml/min from a height of 2 m. The flask has a mass of 200 gm when empty and rests on a spring scale. Find the scale reading after water has poured into the flask for 1 minute.

8.7 A rocket with an initial mass of 50,000 kg is to be fired vertically from Earth's surface. The engine's exhaust burned gasses with a speed of 4,000 m/s. What is the minimum mass flow rate to the engines necessary to ensure lift-off?

*8.8 A rocket with initial mass of 50,000 kg burns fuel at the rate of 150 kg/sec and exhausts it at 6000 m/s. If the rocket is fired vertically from Earth's surface, find its height and velocity when all 45,000 kg of fuel is expended. Consider gravity as constant. (This may be done using numerical methods.)

8.9 Consider a two-stage rocket built to give a 100-kg payload a final velocity of 10,000 km/hr. The exhaust velocity of both stages' rockets is 1400 m/s. The rocket is initially at rest. The design criteria requires that the structure of each stage has a mass that is 20% of the fuel of that stage. Show that a single-stage rocket cannot be used. Find the minimum mass necessary for a two-stage rocket. Neglect gravitational effects.

8.10 Consider a conveyor belt inclined an angle θ from the horizontal. The axes of the rollers at each end are horizontal. The belt forms an inclined plane, moving along the incline. Material falls from a hopper at a rate of $\dot{m} = dm/dt$, sticks to the belt, and travels a distance l before it falls off the lower end of the incline. There is a constant friction force F on the system. Find the steady state speed of the conveyor belt.

8.11 A conveyor belt is inclined an angle θ with the horizontal, moving along the incline. Mass is deposited on the belt at a rate of $\dot{m} = dm/dt$ and travels a distance l before leaving the belt at the upper end. Find the power required to keep the belt moving at a constant speed v.

8.12 A raindrop falls vertically from rest through a fog or mist. It accumulates additional mass by condensation as it falls. Its mass increases linearly so its mass is given by $m = m_0 + \lambda t$. There is a frictional drag equal to $-kmv$. Find the velocity of the raindrop as a function of time (in terms of g, m_0, λ, and k).

*8.13 A spherical raindrop falls from rest through fog. Additional mass—proportional to its cross-section—accumulates on the drop as it falls through the fog. That is,

$$\frac{dm}{dt} = \lambda(\pi r^2)v$$

In addition to the force of gravity, there is a frictional force due to air resistance that is linear in the velocity and the cross-section:

$$F_f = -kv(\pi r^2)$$

Draw a graph of velocity versus time and position time. This can also be handled numerically. Use $\lambda = 0.1$ g/sec/cm^2 and $k = 4$ g/cm^2/sec.

8.14 Consider a spherical raindrop falling through a fog or mist. Assume that the raindrop accumulates mass at a rate proportional to its cross-section and to its speed. The raindrop starts from rest while infinitely small. Find the acceleration in terms of velocity and radius of the drop.

8.15 Three particles of mass m, $2m$, and $4m$ lie on a straight line along which the particles move. Initially the particles are located at points 0, 2, and 4 and move with velocities $4v_0$, $2v_0$, and v_0, respectively. Find the final velocities of the particles if all collisions are perfectly elastic.

8.16 Three gliders of mass m, $2m$, and $4m$ sit at rest on a frictionless air track at positions 0, 50 cm, and 100 cm, respectively. The first glider, of mass m, is given an initial speed of 20 cm/sec toward the next glider. What is the speed of the third glider if all collisions are elastic?

8.17 Two gliders approach each other on a frictionless air track. Their speeds (relative to the air track) are 10 and 20 cm/sec. Find the ratio of their masses if the one initially traveling 20 cm/s remains at rest after they collide elastically.

8.18 Two frictionless air track gliders approach each other with speeds v_1 and v_2 (relative to the air track). Find the ratio of their masses necessary that one remain at rest after they collide elastically.

8.19 A block of wood of mass M rests on a horizontal surface. A bullet of mass m is fired horizontally with speed v_0 into the block. What fraction of the initial kinetic energy of the bullet is lost to heat?

8.20 A shell breaks into two fragments with masses m_1 and m_2 due to an explosion that adds kinetic energy T to the system. Find the relative speed of the two fragments after the explosion.

8.21 An incident particle scatters through angle θ_1 from a target initially at rest. The target leaves, making an angle θ_2 with the initial direction of the incident particle. Use θ_1 and θ_2 to find the ratio of the two masses involved, m_1/m_2.

8.22 A neutron in a nuclear reactor scatters from a deuteron (an atom of "heavy hydrogen" in a molecule of "heavy water" made up of a proton and a neutron; thus, its mass is essentially twice that of a neutron). Consider the deuteron initially at rest. The neutron has an initial speed of 100 m/s and leaves after scattering elastically through an angle of 20°.

Find the scattering angle of the deuteron and final speeds of the deuteron and neutron. What are the scattering angles in the center of mass (CM) frame?

8.23 An alpha particle with initial speed v_0 collides with an atom of Carbon-12, initially at rest, and is scattered through an angle of 30°. Take the mass of the Carbon-12 atom to be three times the mass of the alpha. Find the scattering angle for the

Carbon-12 atom. Find the final velocities of both particles. Assume a perfectly elastic collision. Find both angles in the center of mass (CM) frame.

*8.24 Consider a two-stage rocket built to give a 100-kg payload a final velocity of 30,000 km/hr. The exhaust velocity of both stages' rockets is 2300 m/s. The design criteria require that the structure of each stage has a mass of 15% of the mass of the fuel *plus payload* (the second stage is the payload of the first stage). The rocket is initially at rest, far from Earth (i.e., neglect gravity). Show that a single stage rocket can not be used. Find the minimum mass necessary for a two-stage rocket. The slight change in design criteria (compared to Problem 8.10) greatly complicates the algebra so that you should solve it by numerical means.

*8.25 The rocket of Problem 8.10 reaches a height greater than one Earth radius. The effect of gravity there is less than one-fourth its value at Earth's surface. Therefore, gravity should not be considered constant. Use numerical techniques to solve for position and velocity using the actual, inverse-square gravitational force.

9
RIGID BODIES

In Chapter 8 we studied the behavior of a system in which the particles were free to move about. Indeed, it was the particles' relative motion that proved interesting. Now we shall consider the motion of *rigid bodies* or systems of particles in which there is no relative motion. The distances between and the orientations of the various particles remain constant over time.

Most rigid bodies that are of practical interest are extended bodies made up of physical matter. Even though they are actually composed of discrete atoms, we consider them to be continuous (as they appear to be until we use a microscope on a molecular scale).

9.1 Center of Mass

We have already defined the center of mass for a system of discrete, separate, individual particles in Eq. 8.1.3. Instead of a discrete, finite summation over masses, m_i, we must now carry out an integral over infinitesimal masses, dm, or, equivalently, over a volume of space, dV. That is,

$$\sum_{i=1}^{N} m_i \rightarrow \iiint_V dm = \iiint_V \rho \, dV \qquad (9.1.1)$$

where

$$\rho = \frac{dm}{dV} \qquad (9.1.2)$$

is the density of the rigid body in question. The density may, in general, be a function of position. However, in many interesting cases, the density is uniform.

Using Eqs. 9.2.1 and 8.1.3, the center of mass of an extended, rigid body then can be written as

$$\mathbf{R} = \frac{1}{M} \iiint_V \mathbf{r}\rho \, dV \qquad (9.1.3)$$

or, in component form, as

$$X = \frac{1}{M} \iiint_V x\rho \, dV$$

$$Y = \frac{1}{M} \iiint_V y\rho \, dV \qquad (9.1.4)$$

$$Z = \frac{1}{M} \iiint_V z\rho \, dV$$

where

$$M = \iiint_V \rho \, dV \qquad (9.1.5)$$

is the total mass. The form to be taken by dV and the limits of integration depend quite strongly on the geometry of the body in question. (Such multiple integrals are crucial in advanced physics or engineering so using them must become automatic and second nature. See Appendix C for a review and discussion.)

9.2 Angular Momentum

In Section 7.3, we defined the angular momentum, $\mathbf{L}$, of a small object relative to some point 0 by

$$\mathbf{L} = \mathbf{r} \times \mathbf{P} \qquad (7.3.3)$$

and found that this angular momentum changes due to the application of torque, τ, by

$$\tau = \frac{d\mathbf{L}}{dt} \tag{7.3.1}$$

where the torque is defined by

$$\tau = \mathbf{r} \times \mathbf{F} \tag{7.3.2}$$

The angular momentum of a system of N particles, then, is given by

$$\mathbf{L} = \sum_{\alpha=1}^{N} \mathbf{r}_\alpha \times \mathbf{p}_\alpha \tag{9.2.1}$$

The subscript has been changed from the more usual i or j to α for reasons that will become clear shortly. For a continuous, extended body, this summation will be replaced by an integral. But for the present, we will use the summation to complete the development in a clear manner.

For a rigid body rotating with angular velocity $\boldsymbol{\omega}$ about an origin 0, the velocity or momentum of particle α, measured about the origin 0, is given by

$$\mathbf{p}_\alpha = m_\alpha \mathbf{v}_\alpha = m_\alpha (\boldsymbol{\omega} \times \mathbf{r}_\alpha) \tag{9.2.2}$$

Thus,

$$\mathbf{L} = \sum_{\alpha=1}^{N} m_\alpha \mathbf{r}_\alpha \times (\boldsymbol{\omega} \times \mathbf{r}_\alpha)$$

Therefore,

$$\mathbf{L} = \sum_\alpha m_\alpha \mathbf{r}_\alpha (\boldsymbol{\omega} \times \mathbf{r}_\alpha) \tag{9.2.3}$$

This can be expanded directly by

$$\boldsymbol{\omega} \times \mathbf{r} = \begin{vmatrix} \hat{\mathbf{i}} & \hat{\mathbf{j}} & \hat{\mathbf{k}} \\ \omega_x & \omega_y & \omega_z \\ x & y & z \end{vmatrix}$$

$$= \hat{\mathbf{i}}(\omega_y z - \omega_z y) + \hat{\mathbf{j}}(\omega_z x - \omega_x z) + \hat{\mathbf{k}}(\omega_x y - \omega_y x)$$

Then

$$\mathbf{r} \times (\boldsymbol{\omega} \times \mathbf{r}) = \begin{vmatrix} \hat{\mathbf{i}} & \hat{\mathbf{j}} & \hat{\mathbf{k}} \\ x & y & z \\ (\omega_y z - \omega_z y) & (\omega_z x - \omega_x z) & (\omega_x y - \omega_y x) \end{vmatrix}$$

$$= \hat{\mathbf{i}}[y(\omega_x y - \omega_y x) - z(\omega_z x - \omega_x z)] +$$
$$\hat{\mathbf{j}}[z(\omega_y z - \omega_z y) - x(\omega_x y - \omega_y x)] +$$
$$\hat{\mathbf{k}}[x(\omega_z x - \omega_x z) - y(\omega_y z - \omega_z y)]$$

Thus, the x-component of the angular momentum **L** is given by

$$L_x = \sum_\alpha m_\alpha(y_\alpha^2 \omega_x + z_\alpha^2 \omega_x - x_\alpha y_\alpha \omega_y - x_\alpha z_\alpha \omega_z) \tag{9.2.4}$$

For symmetry, we can add and subtract $x_\alpha^2 \omega_x$ to have

$$L_x = \sum_\alpha m_\alpha(x_\alpha^2 \omega_x + y_\alpha^2 \omega_x + z_\alpha^2 \omega_x - x_\alpha x_\alpha \omega_x - x_\alpha y_\alpha \omega_y - x_\alpha z_\alpha \omega_z)$$

$$L_x = \sum_\alpha m_\alpha[(x_\alpha^2 + y_\alpha^2 + z_\alpha^2)\omega_x - (x_\alpha \omega_x + y_\alpha \omega_y + z_\alpha \omega_z)x_\alpha]$$

$$L_x = \sum_\alpha m_\alpha[r_\alpha^2 \omega_x - (\mathbf{r}_\alpha \cdot \boldsymbol{\omega})x_\alpha] \tag{9.2.5a}$$

Likewise,

$$L_y = \sum_\alpha m_\alpha[r_\alpha^2 \omega_y - (\mathbf{r}_\alpha \cdot \boldsymbol{\omega})y_\alpha] \tag{9.2.5b}$$

$$L_z = \sum_\alpha m_\alpha[r_\alpha^2 \omega_z - (\mathbf{r}_\alpha \cdot \boldsymbol{\omega})z_\alpha]$$

These three may be combined to form the vector equivalent:

$$\mathbf{L} = \sum_\alpha m_\alpha[r_\alpha^2 \boldsymbol{\omega} - (\mathbf{r}_\alpha \cdot \boldsymbol{\omega})\vec{r}_\alpha] \tag{9.2.6}$$

This means that **L** is *not* parallel to $\boldsymbol{\omega}$ in general. This result may well be surprising. We shall shortly look at examples where that is the case. It is only for special cases that we can write the familar expression

$$\mathbf{L} = I\boldsymbol{\omega} \tag{9.2.7}$$

where the angular momentum **L** is just proportional to—and, therefore, parallel to—the angular velocity $\boldsymbol{\omega}$. We will define I, the moment of inertia for rotation about a particular axis, shortly.

A ready example of the general case where angular momentum and angular velocity are *not* parallel is sketched in Figure 9.2.1. There a dumbbell—two bodies each of mass m, connected by a lightweight rod of length $2a$—rotates about the z-axis with angular velocity ω. The axis of the dumbbell makes an angle ψ with the z-axis. The angular momentum for each body of mass m is

$$\mathbf{L} = \mathbf{r} \times \mathbf{p} = \mathbf{r} \times m\mathbf{v}$$

$$\mathbf{L} = m\mathbf{r} \times (\boldsymbol{\omega} \times \mathbf{r})$$

which is just

$$\mathbf{L} = m\mathbf{r} \times \mathbf{v}$$

The result is the same for both masses. The angular momentum, **L**, is a vector as shown in the diagram. It is *not* parallel to the angular velocity $\boldsymbol{\omega}$.

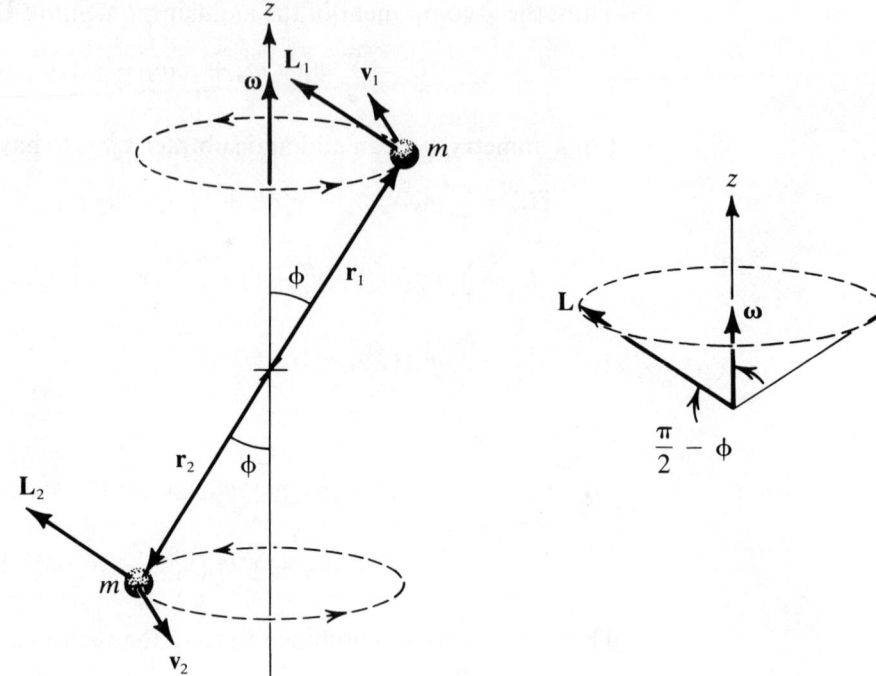

Figure 9.2.1 A rotating dumbbell. Note that **L** = **r** × **p** is *not* parallel to **ω**.

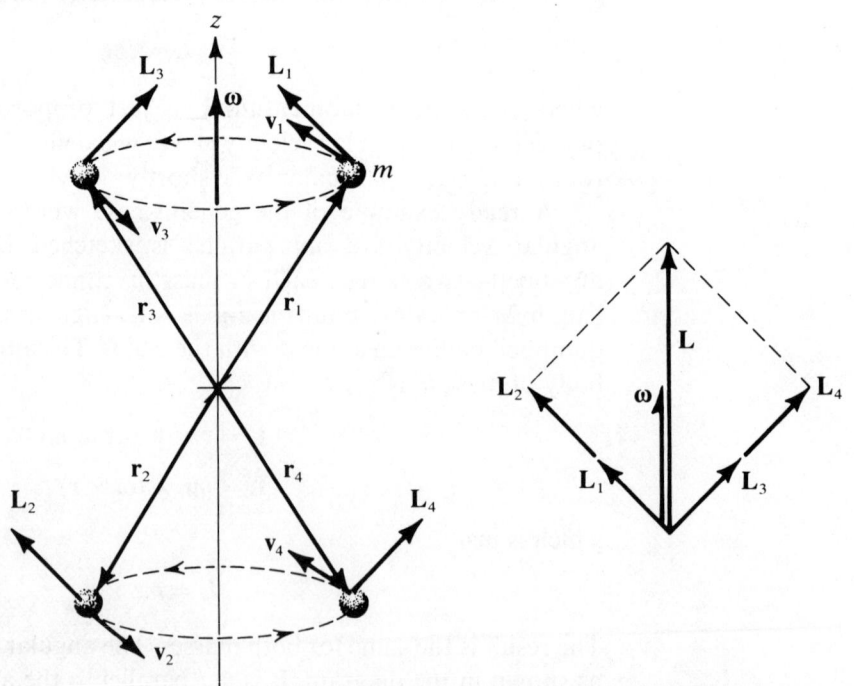

Figure 9.2.2 Two rotating dumbbells. With added symmetry, **L** and **ω** are parallel.

Why, then, is this result so startling? Figure 9.2.2 shows *two* dumbbells rotating about the z-axis as before. The four masses lie in a common plane. Now the angular momenta for the individual particles cancel so that only their z-components remain. For this case—with additional symmetry—**L** and **ω** are, indeed, parallel. Such situations are much easier to visualize. You can fall into the trap of thinking that must always be the case, but now you know better.

9.3 Rotation about an Axis

Let us restrict our discussion for a while to situations for which Eq. 9.2.7 is true. This will include reasonably "regular" bodies rotating about an axis of symmetry. To determine the moment of inertia for rotation about a fixed axis, we may look at the kinetic energy of all the objects comprising a rigid body:

$$T = \sum_\alpha \frac{1}{2} m_\alpha v_\alpha^2 \tag{9.3.1}$$

While $\mathbf{v}_\alpha = \boldsymbol{\omega} \times \mathbf{r}_\alpha$ as before, we may write the scalar equation

$$v_\alpha = \omega R_\alpha \tag{9.3.2}$$

where R_α is the *perpendicular* distance from the axis of rotation to object α. R_α is sometimes called the *moment arm* of object α relative to the axis of rotation. Then

$$T = \sum_\alpha \frac{1}{2} m_\alpha R_\alpha^2 \omega^2$$

or

$$T = \frac{1}{2} \left(\sum_\alpha m_\alpha R_\alpha^2 \right) \omega^2 \tag{9.3.3}$$

So

$$T = \tfrac{1}{2} I \omega^2 \tag{9.3.4}$$

where

$$I = \sum_\alpha m_\alpha R_\alpha^2 \tag{9.3.5}$$

is the *moment of inertia*. For a continuous, extended body this becomes

$$I = \iiint_V \rho R_\perp^2 \, dV \tag{9.3.6}$$

CHAPTER 9 / RIGID BODIES

Again, $R_\perp$ is the *perpendicular distance* from the axis of rotation; it is *not* an actual distance from an origin. For example, if an object is rotating about the z-axis, then

$$R_\perp^2 = x^2 + y^2 \tag{9.3.7}$$

so that

$$I_z = \iiint_V \rho(x^2 + y^2)\, dV \tag{9.3.8}$$

Integration is carried out in just the same manner as when we found the center of mass earlier in this chapter.

Notice that the moment of inertia plays exactly the same role for rotation as the mass does for translation. It could readily be called the *rotational mass*, just as torque is a *rotational force*. For rotation about an axis, the equations of motion are in the same *form* as the equations of motion for straight line motion. Therefore, all the techniques for solving translational motion may be immediately applied. This similarity—indeed, identity—is shown in Table 9.3.1.

We have solved the motion of a simple pendulum earlier, in Section 3.7. But it is worthwhile to look at it once more, using the ideas of rotation about a fixed axis. Figure 9.3.1(a) is a schematic of a simple pendulum. A small body

TABLE 9.3.1 Comparison of rectilinear motion and rotation about fixed axis

Rectilinear	Rotation
Position, x	Angular displacement, θ
Speed, $v = \dfrac{dx}{dt}$	Angular velocity, $\omega = \dfrac{d\theta}{dt}$
Acceleration, $a = \dfrac{dv}{dt}$	Angular acceleration, $\alpha = \dfrac{d\omega}{dt}$
Mass, m	Moment of inertia, $I = \sum mR^2$
Momentum, $p = mv$	Angular momentum, $L = I\omega$
Force, F	Torque, $\tau = rF\sin\theta$
$F = \dfrac{dp}{dt}$	$\tau = \dfrac{dL}{dt}$
$F = ma$	$\tau = I\alpha$
$T = \tfrac{1}{2}mv^2$	$T = \tfrac{1}{2}I\omega^2$
$V(x) = \displaystyle\int_x^{x_r} F(x)\, dx$	$V(\theta) = \displaystyle\int_\theta^{\theta_r} \tau(\theta)\, d\theta$
$F = -\dfrac{dV}{dx}$	$\tau = -\dfrac{dV}{d\theta}$
$x = x_0 + v_0 t + \tfrac{1}{2}at^2$	$\theta = \theta_0 + \omega_0 t + \tfrac{1}{2}\alpha t^2$
$v = v_0 + at$	$\omega = \omega_0 + \alpha t$

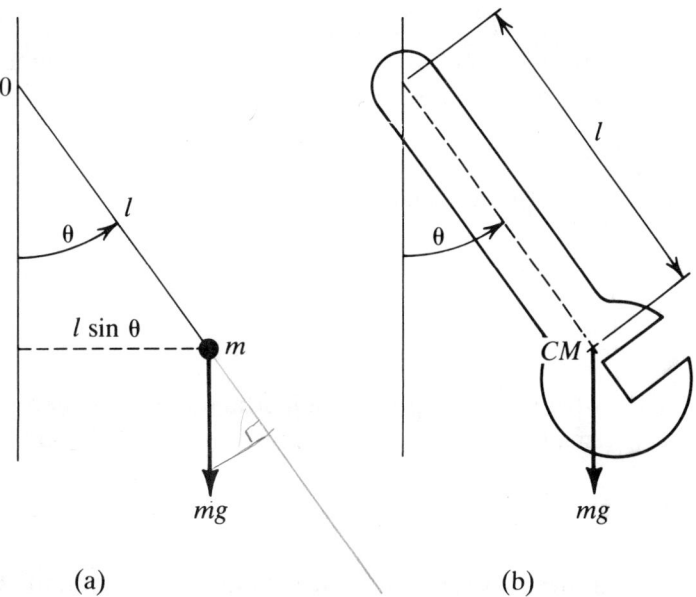

Figure 9.3.1 (a) A simple pendulum; (b) a physical pendulum.

of mass m swings about the pivot 0, always swinging in the plane of the paper, attached to a string or massless rod of length l. The axis of rotation is perpendicular to the page. From the figure, we have

$$I = ml^2 \qquad (9.3.9)$$

and

$$\tau = -mgl \sin\theta \qquad (9.3.10)$$

Thus,

$$\tau = I\alpha = I\ddot{\theta}$$

$$-mgl \sin\theta = ml^2\ddot{\theta}$$

$$\ddot{\theta} = -\frac{g}{l}\sin\theta \qquad (9.3.11)$$

Of course this is just Eq. 3.7.2, which we encountered earlier. The exact solution remains just as messy, requiring "elliptic integrals." A numerical solution is far easier. For small angles, the approximation

$$\sin\theta \simeq \theta \qquad (9.3.12)$$

can again be used giving

$$\ddot{\theta} = -\frac{g}{l}\theta \qquad (9.3.13)$$

which is, of course, the equation of a harmonic oscillator.

Or, consider the *physical pendulum* shown in Figure 9.3.1(b). There some extended body is suspended at an origin 0. The center of mass is located a distance l from the origin or axis of rotation. The body is moved an angle θ away from equilibrium. Thus, there is a restoring torque that will try to rotate this back to equilibrium;

$$\tau = -mgl \sin \theta \tag{9.3.14}$$

This torque then causes an angular acceleration given by

$$\tau = I\ddot{\theta}$$

or

$$-mgl \sin \theta = I\ddot{\theta} \tag{9.3.15}$$

where I is the moment of inertia for rotation about the origin 0. Staying with small angles so Eq. 9.3.12 remains valid, this gives

$$\ddot{\theta} = -\frac{mgl}{I}\theta \tag{9.3.16}$$

which is, once more, the equation of a harmonic oscillator. You should see that Eq. 9.3.13 is just a special case of this more general form.

Additional Applications

In the absence of a torque, the angular momentum remains constant, according to Eq. 7.3.1. For many interesting situations the moment of inertia is variable. As the moment of inertia changes, the angular velocity must change inversely to maintain a constant value for the angular momentum. For example, as a twirling ice skater or ballet dancer pulls her arms in close to the axis of rotation, the value of her moment of inertia decreases. And so her angular velocity increases to keep the angular momentum constant. Or consider a diver as he executes a somersault. He leaves the board with some angular momentum, but with arms and legs outstretched, his moment of inertia is large. But by pulling himself into a "tuck position," he greatly reduces his moment of inertia, causing his angular velocity to greatly increase. Just before entering the water, he again stretches out his arms and legs. This increases his moment of inertia and he enters the water with very little angular velocity. All the while his angular momentum has remained constant.

A rapidly spinning top or a gyroscope is rather interesting to watch and to understand. Of course, it does *not* rotate about a fixed axis, so Eq. 9.2.7 is not valid. But if the gyroscope is spinning *rapidly* about its axis, then Eq. 9.2.7 ($\mathbf{L} = I\boldsymbol{\omega}$) is *approximately* valid. We shall investigate the behavior of a gyroscope, assuming that it is spinning rapidly enough about its axis that its angular momentum vector always lies along its axis of rotation. That means the corrections due to wobbles or any other movements are small. Of course, as the gyroscope slows, this approximation becomes worse.

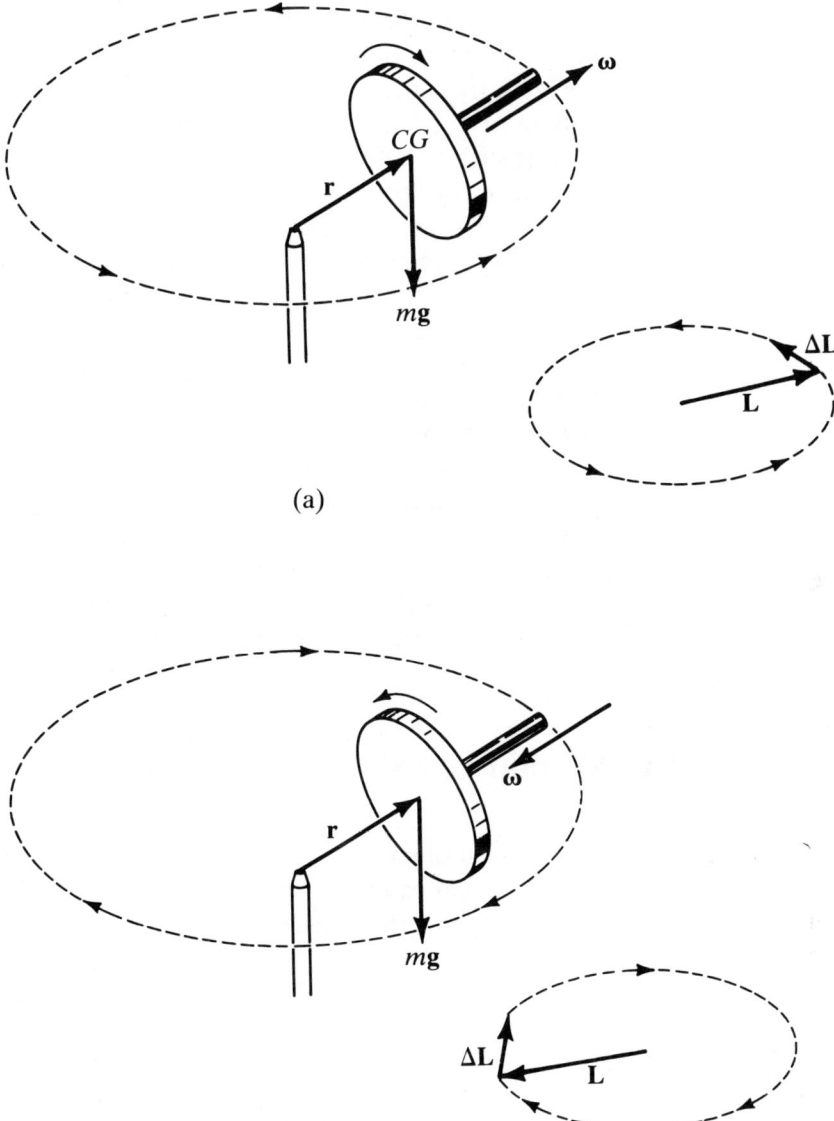

Figure 9.3.2
Gyroscopic motion.

Consider a gyroscope rotating as shown in Figure 9.3.2. It is rotating rapidly enough that the angular momentum vector is aligned with the axis of rotation. The gyroscope is free to rotate about the point of support as shown. Due to its weight, there is a *torque* exerted about that point equal to

$$\tau = r \times (m\mathbf{g}) \tag{9.3.17}$$

which points *into* the page for this figure. From Eq. 7.3.1 we know that the *change* in the angular momentum, $d\mathbf{L}$, must be in the same direction,

$$d\mathbf{L} = \tau \, dt \tag{9.3.18}$$

Thus, the tip of the angular momentum vector traces out a circle, going into the paper on the right and coming out on the left.

How, then, must the gyroscope move in order to be consistent with this motion of the angular momentum? Figure 9.3.2(a) shows a gyroscope rotating such that the angular velocity **ω** or angular momentum **L** is a vector pointing *away* from the point of support. For this case, the tip of the gyroscope traces out a circle going into the paper on the right and coming out on the left—counterclockwise if viewed from above. Such motion is called *precession*.

Figure 9.3.2(b) shows a gyroscope rotating in the opposite direction. The torque is the same, so the *change* in the angular momentum is the same. But now the gyroscope must precess in the *opposite* direction—clockwise as viewed from above.

Such a gyroscope obviously behaves very differently, depending on whether it is stopped or spinning rapidly. Spinning rapidly, it does not "defy gravity" or "violate the laws of physics," as is sometimes stated by toy manufacturers. The precession is caused by a torque due to the force of gravity. The net force is zero since the support pushes up with a force equal to its weight. Its precession is, indeed, *predicted* by the laws of physics.

9.4 Moment of Inertia Theorems

The moment of inertia is given by Eq. 9.3.6. We now pursue this further and look at special cases and some theorems that will be useful in calculating a variety of moments of inertia.

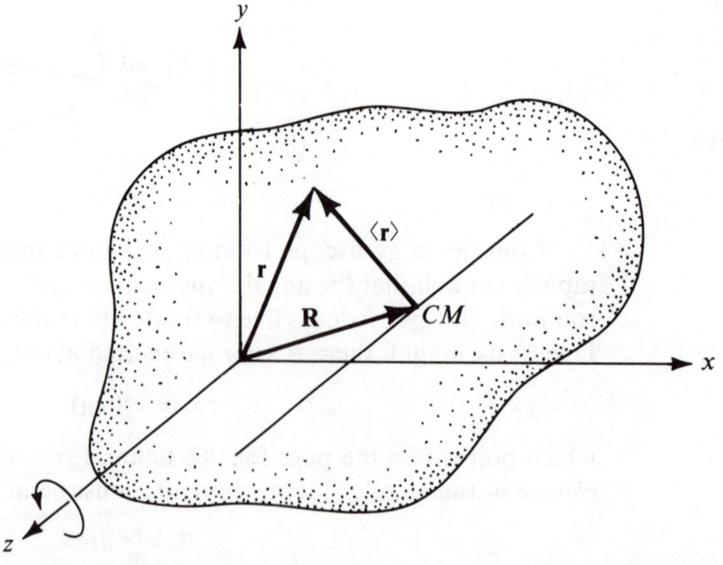

Figure 9.4.1 Parallel-axis theorem.

SECTION 9.4 / MOMENT OF INERTIA THEOREMS

Parallel-Axis Theorem: Consider a body rotating about some axis. Without loss of generality, we may define a coordinate system such that this axis of rotation is the z-axis as shown in Figure 9.4.1. The vector $\mathbf{R}$ locates the center of mass and the position $\mathbf{r}$ of any mass dm or $\rho\, dV$ is given by

$$\mathbf{r} = \langle \mathbf{r} \rangle + \mathbf{R} \tag{9.4.1}$$

where $\langle \mathbf{r} \rangle$ is the "relative coordinate" as measured from the center of mass.

The moment of inertia for rotation about the z-axis as shown is given by Eq. 9.3.8. Let us express this in terms of the coordinates of the center of mass and the relative coordinates:

$$I = \iiint_V \rho(x^2 + y^2)\, dV$$

$$= \iiint_V [(\langle x \rangle + X)^2 + (\langle y \rangle + Y)^2] \rho\, dV$$

$$= \iiint_V [(\langle x \rangle^2 + \langle y \rangle^2) + (X^2 + Y^2) + 2(X\langle x \rangle + Y\langle y \rangle)] \rho\, dV$$

$$I = \iiint_V (\langle x \rangle^2 + \langle y \rangle^2)\rho\, dV + R_\perp^2 M \tag{9.4.2}$$

where

$$R_\perp^2 = X^2 + Y^2 \tag{9.4.3}$$

and we have made use of

$$\iiint_V \langle x \rangle \rho\, dV = 0 \tag{9.4.4}$$

and

$$\iiint_V \langle y \rangle \rho\, dV = 0 \tag{9.4.5}$$

since these merely locate the center of mass *relative to itself*. We can simplify Eq. 9.4.2 by writing I:

$$I = I_{\text{CM}} + MR_\perp^2 \tag{9.4.6}$$

where

$$I_{\text{CM}} = \iiint_V (\langle x \rangle^2 + \langle y \rangle^2) \rho\, dV \tag{9.4.7}$$

is the moment of inertia for rotation about an axis passing through center of mass and parallel to the z-axis.

> *That is, the moment of inertia for rotation about any axis is equal to the sum of the moment of inertia for rotation about a <u>parallel</u> axis passing through the center of mass plus the moment of inertia about the given axis for the total mass concentrated at the center of mass.*

Perpendicular-Axis Theorem: Consider a thin, flat plate or lamina lying in the xy plane as shown in Figure 9.4.2. For rotation about the z-axis, its moment of inertia is

$$I_z = \iiint_V (x^2 + y^2)\rho \, dV \tag{9.4.8}$$

For rotation about the x-axis, its moment of inertia is

$$I_x = \iiint_V y^2 \rho \, dV \tag{9.4.9}$$

And for rotation about the y-axis, its moment of inertia is

$$I_y = \iiint_V x^2 \rho \, dV \tag{9.4.10}$$

Since this is a lamina, $z = 0$, there is no z^2-term in either of these expressions. Therefore,

$$I_z = I_x + I_y \tag{9.4.11}$$

> *That is, the sum of the moments of inertia for a thin flat plate or lamina for rotation about two <u>perpendicular</u> axes lying in the plane of the plate*

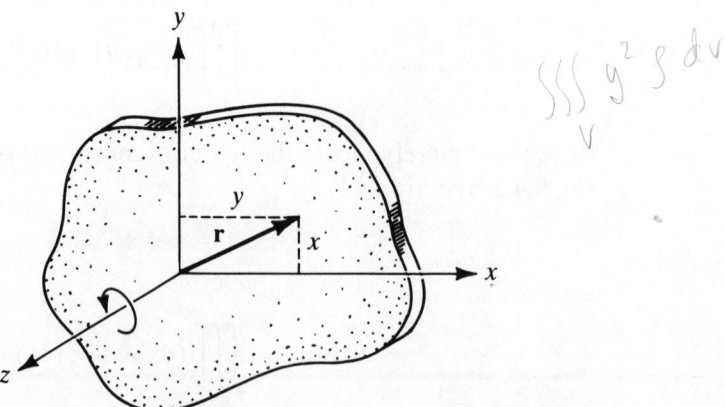

Figure 9.4.2
Perpendicular-axis theorem.

SECTION 9.4 / MOMENT OF INERTIA THEOREMS

or lamina is equal to the moment of inertia for rotation about a perpendicular axis passing through their point of intersection.

These two theorems are quite helpful. Consider a hoop or ring, of mass M and radius a, as sketched in Figure 9.4.3. The calculation of the moment of inertia about the z-axis is trivial. Since all the mass is at a distance of a from the axis, the moment of inertia is

$$I_z = Ma^2 \qquad (9.4.12)$$

But much of the symmetry is lost if we require the moment of inertia about some other axis. Consider axis AA', parallel to the z-axis, perpendicular to the plane of the ring and passing through the edge of it. Direct solution of Eq. 9.3.6 to determine the moment of inertia about this axis is no longer trivial. But the parallel-axis theorem immediately provides

$$I_{AA'} = I_z + Ma^2$$

or

$$I_{AA'} = 2Ma^2 \qquad (9.4.13)$$

Likewise, consider a rotation about an axis lying in the plane of the ring—like the x- or y-axis. Again, direct solution of Eq. 9.3.6 to find the moment of inertia is not trivial. But symmetry requires

$$I_x = I_y \qquad (9.4.14)$$

so the perpendicular-axis theorem immediately gives

$$I_z = I_x + I_x = I_y + I_y$$

or

$$I_x = I_y = \tfrac{1}{2}Ma^2 \qquad (9.4.15)$$

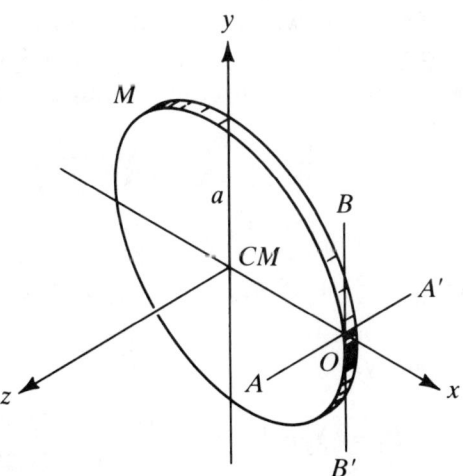

Figure 9.4.3 Moments of inertia of a hoop about various axes.

Further application of the parallel-axis theorem allows us to easily find the moment of inertia about axis BB', in the plane of the ring and tangent to its edge:

$$I_{BB'} = I_y + Ma^2$$
$$I_{BB'} = \tfrac{3}{2}Ma^2 \tag{9.4.16}$$

For one more application, we shall check these results by applying the perpendicular-axis theorem at point 0:

$$I_{AA'} = I_{BB'} + I_x$$
$$2Ma^2 = \tfrac{3}{2}Ma^2 + \tfrac{1}{2}Ma^2$$

And, of course, this result checks. Consistency is a useful check to be used whenever possible.

9.5 The Inertia Tensor

Let us return now to the general definition of angular momentum:

$$\mathbf{L} = \sum_{\alpha=1}^{N} m_\alpha [r_\alpha^2 \boldsymbol{\omega} - (\mathbf{r}_\alpha \cdot \boldsymbol{\omega})\mathbf{r}_\alpha] \tag{9.2.6}$$

The angular momentum $\mathbf{L}$ clearly depends on the angular velocity $\boldsymbol{\omega}$. But the two are not necessarily parallel. This seems surprising, but remember that rotation about a fixed direction—all of Sections 9.3 and 9.4—is only a *special* case of rotation. The right-hand side is *not* just a vector multiplied by a scalar.

Part of the right-hand side will turn out to be a *tensor*—we shall call it the *inertia tensor*. Therefore, we shall adopt tensor notation and write (x, y, z) as (x_1, x_2, x_3) and $\mathbf{L} = (L_x, L_y, L_z)$ as (L_1, L_2, L_3). That is, subscripts 1, 2, and 3 shall indicate, respectively, the x-, y-, and z-components of a vector. In particular, then,

$$r_\alpha^2 = x_{\alpha 1}^2 + x_{\alpha 2}^2 + x_{\alpha 3}^2 \tag{9.5.1}$$

Thus,

$$L_1 = \sum_\alpha m_\alpha[(x_{\alpha 1}^2 + x_{\alpha 2}^2 + x_{\alpha 3}^2)\omega_1 - x_{\alpha 1}(x_{\alpha 1}\omega_1 + x_{\alpha 2}\omega_2 + x_{\alpha 3}\omega_3)] \tag{9.5.2}$$

Or, more compactly,

$$L_1 = \sum_{\alpha=1}^{N} m_\alpha \left[\left(\sum_{i=1}^{3} x_{\alpha i}^2 \right)\omega_1 - x_{\alpha 1} \left(\sum_{i=1}^{3} x_{\alpha i}\omega_i \right) \right] \tag{9.5.3}$$

SECTION 9.5 / THE INERTIA TENSOR

Similar expressions hold for L_2 and L_3, or we may write the general form as

$$L_j = \sum_{\alpha=1}^{N} m_\alpha \sum_{i=1}^{3} [x_{\alpha i}^2 \omega_j - x_{\alpha j} x_{\alpha i} \omega_i] \tag{9.5.4}$$

This may be further simplified by a notation called the Krönicker delta, δ_{ij} where

$$\delta_{ij} = 1 \text{ for } i = j \tag{9.5.5}$$

$$\delta_{ij} = 0 \text{ for } i \neq j \tag{9.5.6}$$

Thus,

$$L_j = \sum_{\alpha=1}^{N} m_\alpha \sum_{i=1}^{3} \left[\delta_{ij} \sum_{k} x_{\alpha k}^2 - x_{\alpha i} x_{\alpha j} \right] \omega_i \tag{9.5.7}$$

Equation 9.5.7 is perhaps best written in matrix form as

$$\begin{pmatrix} L_1 \\ L_2 \\ L_3 \end{pmatrix} = \begin{pmatrix} \sum m_\alpha (x_{\alpha 2}^2 + x_{\alpha 3}^2) & -\sum m_\alpha x_{\alpha 1} x_{\alpha 2} & -\sum m_\alpha x_{\alpha 1} x_{\alpha 3} \\ -\sum m_\alpha x_{\alpha 1} x_{\alpha 2} & \sum m_\alpha (x_{\alpha 1}^2 + x_{\alpha 3}^2) & -\sum m_\alpha x_{\alpha 2} x_{\alpha 3} \\ -\sum m_\alpha x_{\alpha 1} x_{\alpha 3} & -\sum m_\alpha x_{\alpha 2} x_{\alpha 3} & \sum m_\alpha (x_{\alpha 1}^2 + x_{\alpha 2}^2) \end{pmatrix} \begin{pmatrix} \omega_1 \\ \omega_2 \\ \omega_3 \end{pmatrix} \tag{9.5.8}$$

Or, as

$$\begin{pmatrix} L_1 \\ L_2 \\ L_3 \end{pmatrix} = \begin{pmatrix} I_{11} & I_{12} & I_{13} \\ I_{21} & I_{22} & I_{23} \\ I_{31} & I_{32} & I_{33} \end{pmatrix} \begin{pmatrix} \omega_1 \\ \omega_2 \\ \omega_3 \end{pmatrix} \tag{9.5.9}$$

This may look less menacing if it is written in a more familiar style as

$$\begin{pmatrix} L_x \\ L_y \\ L_z \end{pmatrix} = \begin{pmatrix} \sum m_\alpha (y_\alpha^2 + z_\alpha^2) & -\sum m_\alpha x_\alpha y_\alpha & -\sum m_\alpha x_\alpha z_\alpha \\ -\sum m_\alpha x_\alpha y_\alpha & \sum m_\alpha (x_\alpha^2 + z_\alpha^2) & -\sum m_\alpha y_\alpha z_\alpha \\ -\sum m_\alpha x_\alpha z_\alpha & -\sum m_\alpha y_\alpha z_\alpha & \sum m_\alpha (x_\alpha^2 + y_\alpha^2) \end{pmatrix} \begin{pmatrix} \omega_x \\ \omega_y \\ \omega_z \end{pmatrix} \tag{9.5.10}$$

or, as

$$\begin{pmatrix} L_x \\ L_y \\ L_z \end{pmatrix} = \begin{pmatrix} I_{xx} & I_{xy} & I_{xz} \\ I_{yx} & I_{yy} & I_{yz} \\ I_{zx} & I_{zy} & I_{zz} \end{pmatrix} \begin{pmatrix} \omega_x \\ \omega_y \\ \omega_z \end{pmatrix} \tag{9.5.11}$$

Thus, we can write

$$\mathbf{L} = \{I\} \boldsymbol{\omega} \tag{9.5.12}$$

where $\{I\}$ is, in fact, a second-rank *tensor* and the operation of a second-rank tensor on a vector (which is a first-rank tensor) results in a first-rank tensor, as shown explicitly by Eqs. 9.5.8 through 9.5.11. $\{I\}$ is known as the *inertia tensor*. For a continuous, extended body, the elements of $\{I\}$ are given by

$$I_{ij} = \iiint_V \rho \left[\left(\delta_{ij} \sum_{k=1}^{3} x_k^2 \right) - x_i x_j \right] dv \tag{9.5.13}$$

As you can see from Eqs. 9.5.8 or 9.5.10, the *diagonal elements* are the ordinary *moments of inertia.* The off-diagonal elements are known as the *products of inertia.* Notice that the inertia tensor is symmetric; that is

$$I_{ij} = I_{ji} \tag{9.5.14}$$

so there are only six independent terms.

9.6 Principal Axes

We can simplify our mathematical calculations considerably if we choose the coordinate axes so that the off-diagonal elements, the products of inertia, all vanish. The axes of this coordinate system are known as the *principal axes* of the body. For such a case, the inertia tensor becomes

$$\{I\} = \begin{pmatrix} I_{11} & 0 & 0 \\ 0 & I_{22} & 0 \\ 0 & 0 & I_{33} \end{pmatrix} \tag{9.6.1}$$

and the angular momentum of Eq. 9.5.12 takes the form of

$$\begin{aligned} L_1 &= I_{11}\omega_1 \\ L_2 &= I_{22}\omega_2 \\ L_3 &= I_{33}\omega_3 \end{aligned} \tag{9.6.2}$$

Remember, this is equivalent to

$$\begin{aligned} L_x &= I_{xx}\omega_x \\ L_y &= I_{yy}\omega_y \\ L_z &= I_{zz}\omega_z \end{aligned} \tag{9.6.2a}$$

where ω_x, ω_y, and ω_z are the *components* of the angular velocity $\boldsymbol{\omega}$. Note that *expressing* quantities in terms of these principal axes does *not*, somehow, cause **L** and $\boldsymbol{\omega}$ to be parallel. But *rotation* about one of these principal axes does exactly that. For example, rotation about the 1-axis, (the x-axis, if you prefer) means

$$\begin{aligned} \omega_1 &= \omega \\ \omega_2 &= 0 \\ \omega_3 &= 0 \end{aligned} \tag{9.6.3}$$

and Eq. 9.6.2 then gives

$$L_1 = I_{11}\omega$$
$$L_2 = 0 \quad (9.6.4)$$
$$L_3 = 0$$

or

$$\mathbf{L} = I\boldsymbol{\omega} \quad (9.6.5)$$

where

$$I = I_{11} \quad (9.6.6)$$

That is, for rotation about a principal axis, the angular momentum is directly proportional to (and, therefore, *parallel* to) the angular velocity. The constant of proportionality is the moment of inertia about the axis of rotation.

This has an important application in the manufacture, use, and maintenance of rotating equipment—automobile tires, airplane propellers, fan blades, steam turbines, and crank shafts in engines are ready examples. Any rotating body is *statically balanced* if its center of mass lies on the axis of rotation. Such a body will be in neutral equilibrium if stopped in any orientation—once stopped, it exhibits no tendency to rotate back to some particular orientation. A rotating body is *dynamically balanced* if the axis of rotation is also a principal axis. In that case, the angular momentum vector lies along the axis of rotation and remains constant.

If a body is not dynamically balanced, the angular momentum vector will vary; it will describe a cone about the axis of rotation as shown in Figure 9.6.1. But any change in angular momentum must have an associated torque (by Eq. 7.3.1). This torque is perpendicular to the axis of rotation and must be

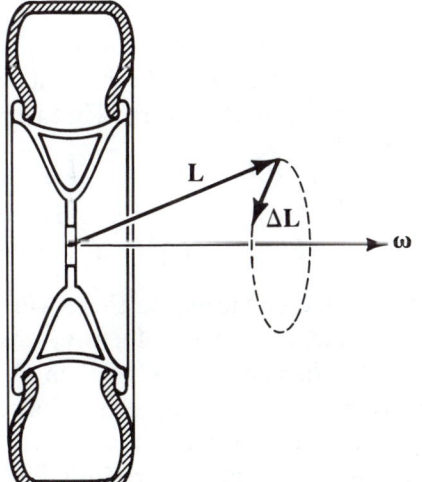

Figure 9.6.1 An automobile wheel *not* balanced dynamically.

applied by the bearings. Thus, an automobile wheel that is not dynamically balanced (even though it is statically balanced) may cause noticeable or violent vibration. The last decade has seen rise in the availability of machines (usually controlled by a microprocessor) that allow quick and easy dynamic balancing of automobile tires. Thus, principal axes are important not only because they simplify the mathematics, but because they generate real physical consequences.

So, how do we find principal axes? Often the fruitful approach is by inspection! If a body has some symmetry, it is often possible to choose principal axes by inspection so that the products of inertia vanish. For instance, the symmetry axis for a right circular cylinder (or cone or top) is a principal axis. The other two principal axes are any two axes perpendicular to this one and to each other. The principal axes for a rectangular solid are axes perpendicular to the faces if the origin is chosen to coincide with the center of mass. If that is not possible, we may require Eq. 9.6.5 to be true and solve for the orientations of the angular velocity ω that allow this. That is, we may require

$$\mathbf{L} = I\boldsymbol{\omega} \tag{9.6.5}$$

We can write this and the general case of Eq. 9.5.9 in component form as

$$\begin{aligned} L_1 &= I\omega_1 = I_{11}\omega_1 + I_{12}\omega_2 + I_{13}\omega_3 \\ L_2 &= I\omega_2 = I_{21}\omega_1 + I_{22}\omega_2 + I_{23}\omega_3 \\ L_3 &= I\omega_3 = I_{31}\omega_1 + I_{32}\omega_3 + I_{33}\omega_3 \end{aligned} \tag{9.6.7}$$

Rearranging terms, this becomes

$$\begin{aligned} (I_{11} - I)\omega_1 + I_{12}\omega_2 + I_{13}\omega_3 &= 0 \\ I_{21}\omega_1 + (I_{22} - I)\omega_2 + I_{23}\omega_3 &= 0 \\ I_{31}\omega_1 + I_{32}\omega_2 + (I_{33} - I)\omega_3 &= 0 \end{aligned} \tag{9.6.8}$$

which is a set of three simultaneous equations.

For a solution to exist, the determinant of the coefficients of the unknowns ω_1, ω_2, and ω_3 must be zero. That is,

$$\begin{vmatrix} I_{11} - I & I_{12} & I_{13} \\ I_{21} & I_{22} - I & I_{23} \\ I_{31} & I_{32} & I_{33} - I \end{vmatrix} = 0 \tag{9.6.9}$$

This is, of course, a cubic equation. So there will be three roots for I. Each root represents the moment of inertia for rotation about one principal axis. The direction of the principal axis is then found by substituting the corresponding value of I back into Eq. 9.6.8 and solving for ω_1, ω_2, and ω_3. Since Eq. 9.6.8 is a set of homogeneous simultaneous equations, only the *ratio* of $\omega_1 : \omega_2 : \omega_3$ can be found. But that ratio is just the ratio of the direction

SECTION 9.6 / PRINCIPAL AXES

cosines relative to the original axes that specify the direction of the particular principal axis. Perhaps this is best illustrated by example.

EXAMPLE 9.6.1 Find the principal axes and their associated moments of inertia for a cube of mass M and sides a. For an origin at one corner, with axes directed along the edges as shown in Figure 9.6.2, we can evaluate the elements of the inertia tensor as follows:

$$I_{xx} = \iiint_V \rho(y^2 + z^2)\, dv$$

$$= \rho \int_0^a dx \int_0^a \left[\int_0^a (y^2 + z^2)\, dy \right] dz$$

$$= \rho a \int_0^a \left[\frac{1}{3} y^3 \Big|_0^a + z^2 y \Big|_0^a \right] dz$$

$$= \rho a \int_0^a \left(\frac{1}{3} a^3 + a z^2 \right) dz$$

$$= \rho \left(\frac{1}{3} a^4 \int_0^a dz + a^2 \int_0^a z^2\, dz \right)$$

$$= \rho \left(\frac{1}{3} a^5 + \frac{1}{3} a^5 \right)$$

$$= \frac{M}{a^3} \left(\frac{2}{3} a^5 \right)$$

$$I_{xx} = \frac{2}{3} M a^2 \qquad (9.6.10)$$

By symmetry, the three moments of inertia are identical for this choice of axes so that

$$I_{xx} = I_{yy} = I_{zz} = \tfrac{2}{3} M a^2 \qquad (9.6.11)$$

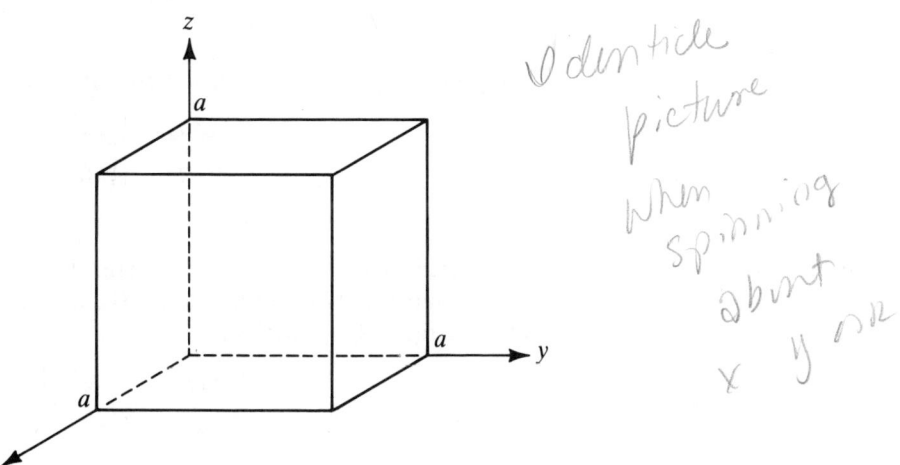

Figure 9.6.2 Cube with origin at one corner.

Likewise, we can evaluate the products of inertia by

$$I_{xy} = -\iiint_V \rho xy \, dv$$

$$= -\rho \int_0^a dz \int_0^a \left[\int_0^a xy \, dx\right] dy$$

$$= -\rho a \int_0^a \left(\frac{1}{2}a^2\right) y \, dy$$

$$= -\frac{1}{2}\rho a^3 \int_0^a y \, dy$$

$$= -\frac{1}{2}\rho a^3 \left(\frac{1}{2}a^2\right)$$

$$= -\frac{1}{4}\rho a^5$$

$$= -\frac{1}{4}\left(\frac{M}{a^3}\right)a^5$$

$$I_{xy} = -\frac{1}{4}Ma^2 \tag{9.6.12}$$

By symmetry, all of the products of inertia are equal so the inertia tensor may now be written as

$$\{I\} = \begin{pmatrix} \frac{2}{3}Ma^2 & -\frac{1}{4}Ma^2 & -\frac{1}{4}Ma^2 \\ -\frac{1}{4}Ma^2 & \frac{2}{3}Ma^2 & -\frac{1}{4}Ma^2 \\ -\frac{1}{4}Ma^2 & -\frac{1}{4}Ma^2 & \frac{2}{3}Ma^2 \end{pmatrix} \tag{9.6.13}$$

Clearly, the axes specified along the edge are *not* principal axes. To find the principal axes (still retaining the same origin at a corner), we must set

$$\begin{vmatrix} \frac{2}{3}Ma^2 - I & -\frac{1}{4}Ma^2 & -\frac{1}{4}Ma^2 \\ -\frac{1}{4}Ma^2 & \frac{2}{3}Ma^2 - I & -\frac{1}{4}Ma^2 \\ -\frac{1}{4}Ma^2 & -\frac{1}{4}Ma^2 & \frac{2}{3}Ma^2 - I \end{vmatrix} = 0 \tag{9.6.14}$$

Roots of this, which may be verified by substitution, are

$$I_1 = \frac{1}{6}Ma^2$$
$$I_2 = \frac{11}{12}Ma^2 \tag{9.6.15}$$
$$I_3 = \frac{11}{12}Ma^2$$

This degeneracy of roots, that $I_2 = I_3$, means that the remaining root corresponds to an axis of symmetry. Now, to find that axis, we substitute $I = I_1$ into Eq. 9.6.8 and solve for the ratio of $\omega_1 : \omega_2 : \omega_3$:

$$(\tfrac{2}{3} - \tfrac{1}{6})\omega_1 - \tfrac{1}{4}\omega_2 - \tfrac{1}{4}\omega_3 = 0$$
$$-\tfrac{1}{4}\omega_1 + (\tfrac{2}{3} - \tfrac{1}{6})\omega_2 - \tfrac{1}{4}\omega_3 = 0 \tag{9.6.16}$$
$$-\tfrac{1}{4}\omega_1 - \tfrac{1}{4}\omega_2 + (\tfrac{2}{3} - \tfrac{1}{6})\omega_3 = 0$$

SECTION 9.6 / PRINCIPAL AXES

Multiplying every term by 4 yields

$$2\omega_1 - \omega_2 - \omega_3 = 0$$
$$-\omega_1 + 2\omega_2 - \omega_3 = 0 \qquad (9.6.17)$$
$$-\omega_1 - \omega_2 + 2\omega_3 = 0$$

which looks more tractable. We now multiply the first by 2 and add it to the second one:

$$4\omega_1 - 2\omega_2 - 2\omega_3 = 0$$
$$-\omega_1 + 2\omega_2 - \omega_3 = 0$$
$$3\omega_1 \qquad - 3\omega_3 = 0$$
$$\omega_1 = \omega_3$$

Substituting this into any of the equations then gives

$$\boxed{\omega_1 = \omega_2 = \omega_3} \qquad (9.6.18)$$

Hence the principal axis corresponding to $I_1 = \frac{1}{6}Ma^2$ must lie along the diagonal of the cube. The degeneracy of $I_2 = I_3$ means that any axis in the plane perpendicular to the diagonal is also a principal axis. Once an axis is chosen in this plane, the remaining axis is the mutually perpendicular one. The principal axes are always mutually perpendicular.

These principal moments of inertia are *eigenvalues* or *characteristic values* of the inertia tensor. The corresponding directions are *eigenvectors* or *characteristic vectors*. These ideas and mathematical formulation will be encountered in various topics throughout physics.

Note that the choice of *origin* is crucial in determining the principal axes. Had we chosen an origin at the center of the cube and axes parallel to the edges as shown in Figure 9.6.3, these axes would have turned out to be principal axes. In most cases, such a choice of origin at the center of mass will be more useful.

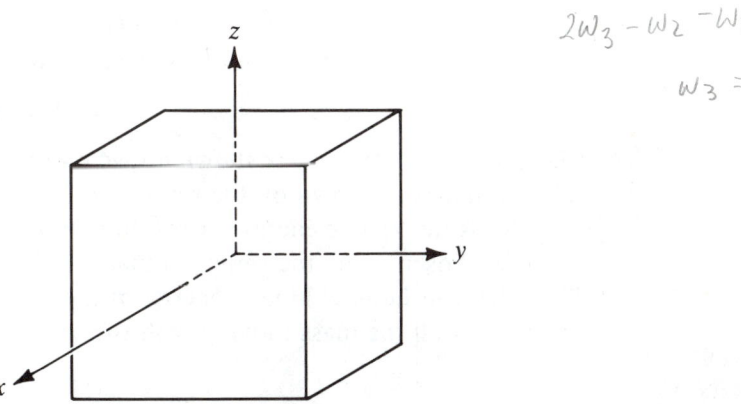

Figure 9.6.3 Cube with origin at the center of mass.

9.7 Kinetic Energy

Consider once more a rigid body composed of N particles rotating about an origin 0. The kinetic energy due to this rotation is

$$T = \sum_\alpha \frac{1}{2} m_\alpha v_\alpha^2 = \sum \frac{1}{2} m_\alpha \mathbf{v}_\alpha \cdot \mathbf{v}_\alpha \tag{9.7.1}$$

As with Eq. 9.2.2, we can write

$$\mathbf{v}_\alpha = \boldsymbol{\omega} \times \mathbf{r}_\alpha \tag{9.7.2}$$

so that

$$T = \frac{1}{2} \sum [(\boldsymbol{\omega} \times \mathbf{r}_\alpha) \cdot m_\alpha \mathbf{v}_\alpha] \tag{9.7.3}$$

Using the vector identity

$$(\mathbf{A} \times \mathbf{B}) \cdot \mathbf{C} = \mathbf{A} \cdot (\mathbf{B} \times \mathbf{C}) \tag{9.7.4}$$

this becomes

$$T = \frac{1}{2} \boldsymbol{\omega} \cdot \sum \mathbf{r}_\alpha \times (m_\alpha \mathbf{v}_\alpha) \tag{9.7.5}$$

This summation is, by definition, the angular momentum of the rotating body. Hence,

$$T = \frac{1}{2} \boldsymbol{\omega} \cdot \mathbf{L} \tag{9.7.6}$$

In terms of the inertia tensor, this can be written as

$$T = \tfrac{1}{2} \boldsymbol{\omega}^T \{I\} \boldsymbol{\omega} \tag{9.7.7}$$

or

$$T = \tfrac{1}{2} (\omega_1 \ \ \omega_2 \ \ \omega_3) \begin{pmatrix} I_{11} & I_{12} & I_{13} \\ I_{21} & I_{22} & I_{23} \\ I_{31} & I_{32} & I_{33} \end{pmatrix} \begin{pmatrix} \omega_1 \\ \omega_2 \\ \omega_3 \end{pmatrix} \tag{9.7.8}$$

where $\boldsymbol{\omega}^T$ is the row matrix shown above on the left and the *transpose* of the column matrix $\boldsymbol{\omega}$ shown on the right.

This is the kinetic energy for rotation about an origin 0. Quite often the origin is chosen to be the center of mass, which may itself also be moving. Then the *translational* kinetic energy of the body must also be included by considering all the mass moving with the center of mass so that

$$T_{\text{trans}} = \tfrac{1}{2} M V^2 \tag{9.7.9}$$

where V is the speed of the center of mass. The total kinetic energy is just the sum of Eqs. 9.7.7 and 9.7.9. In order to place these in the *same* form, it is sometimes written as

$$T = \tfrac{1}{2}\boldsymbol{\omega} \cdot \mathbf{L} + \tfrac{1}{2}\mathbf{V} \cdot \mathbf{P} \tag{9.7.10}$$

9.8 Euler's Equations

As we have seen, the motion of a rigid body is governed by the following equations:

$$\mathbf{F} = \frac{d\mathbf{P}}{dt} \tag{9.8.1}$$

$$\boldsymbol{\tau} = \frac{d\mathbf{L}}{dt} \tag{9.8.2}$$

where $\mathbf{F}$ is the resultant external force and

$$\mathbf{P} = M\mathbf{V} \tag{9.8.3}$$

In words, the total linear momentum is equal to the total mass M multiplied by $\mathbf{V}$, the velocity of the center of mass. Likewise, $\boldsymbol{\tau}$ is the resultant external torque and the angular momentum $\mathbf{L}$ is given by

$$\mathbf{L} = \{I\}\boldsymbol{\omega} \tag{9.8.4}$$

where $\{I\}$ is the inertia tensor and $\boldsymbol{\omega}$ is the angular velocity. As discussed earlier, for rotation about a *fixed* axis the resemblance between Eqs. 9.8.1 and 9.8.2 and Eqs. 9.8.3 and 9.8.4 is strong. Indeed, all of the earlier methods of solving for rectilinear motion then become directly applicable.

But for the general case, the rotational equations become considerably more complicated. First, $\mathbf{P}$ and $\mathbf{V}$ are always parallel since they are related by scalar M, whereas $\mathbf{L}$ and $\boldsymbol{\omega}$ need not be parallel since $\{I\}$ is a tensor. This is further complicated since, while the value of M remains constant, the value of $\{I\}$ with respect to a fixed laboratory reference frame does not remain constant. The value of $\{I\}$ with respect to the LAB frame changes as the body rotates. We shall handle this difficulty by using a reference frame rotating with the body (thus, all the earlier work of Chapter 6 will prove useful). A third complication of rotational motion occurs because no symmetrical set of three coordinates (analogous to x, y, and z) exists with which to uniquely describe the orientation of such a rigid body. To see this, consider a rotation about the x-axis of $\pi/2$ followed by a rotation about the y-axis of $\pi/2$. (Actually do this using a book or box.) Then consider the same two rotations *in the opposite order*. The orientation is quite different for the two cases. We shall defer handling this problem until after we introduce Lagrangian mechanics in the next chapter.

Equation 9.8.2 holds for an inertial frame, the LAB frame. To have a constant value for the inertia tensor, it must be expressed in a rotating frame attached to the body, the body frame. From Eq. 6.2.11 we can write

$$\tau = \frac{d\mathbf{L}}{dt} = \frac{d'\mathbf{L}}{dt} + \boldsymbol{\omega} \times \mathbf{L} \tag{9.8.5}$$

where d'/dt is the time derivative *as seen in the rotating, body-fixed frame*. Now, with $\{I\}$ remaining constant, we can write

$$\tau = \frac{d'}{dt}[\{I\} \cdot \boldsymbol{\omega}] + \boldsymbol{\omega} \times \{I\}\boldsymbol{\omega}$$

$$= \frac{d'\{I\}}{dt}\boldsymbol{\omega} + \{I\}\frac{d'\boldsymbol{\omega}}{dt} + \boldsymbol{\omega} \times \{I\}\boldsymbol{\omega}$$

$$\tau = \{I\}\frac{d\boldsymbol{\omega}}{dt} + \boldsymbol{\omega} \times \{I\}\boldsymbol{\omega} \tag{9.8.6}$$

where

$$\frac{d'\{I\}}{dt} = 0 \tag{9.8.7}$$

since we choose to evaluate $\{I\}$ in the body frame. By Eq. 6.2.11,

$$\frac{d\boldsymbol{\omega}}{dt} = \frac{d'\boldsymbol{\omega}}{dt} + \boldsymbol{\omega} \times \boldsymbol{\omega}$$

$$\frac{d\boldsymbol{\omega}}{dt} = \frac{d'\boldsymbol{\omega}}{dt} \tag{9.8.8}$$

Choosing principal axes provides the most convenient form for Eq. 9.8.6. Written in component form, this becomes

$$\begin{aligned}\tau_1 &= I_1\dot{\omega}_1 + (I_3 - I_2)\omega_2\omega_3 \\ \tau_2 &= I_2\dot{\omega}_2 + (I_1 - I_3)\omega_1\omega_3 \\ \tau_3 &= I_3\dot{\omega}_3 + (I_2 - I_1)\omega_2\omega_1\end{aligned} \tag{9.8.9}$$

These are known as *Euler's equations* for the motion of a rigid body. It is worth reminding ourselves that these equations are nothing but the components of Eq. 9.8.2,

$$\tau = \frac{d\mathbf{L}}{dt} \tag{9.8.2}$$

Now the angular momentum **L** is defined by quantities measured in the BODY frame. In this BODY frame, the angular velocity **ω** is given by

$$\boldsymbol{\omega} = \omega_1 \hat{\mathbf{e}}_1 + \omega_2 \hat{\mathbf{e}}_2 + \omega_3 \hat{\mathbf{e}}_3$$

where $\hat{\mathbf{e}}_1$, $\hat{\mathbf{e}}_2$, and $\hat{\mathbf{e}}_3$ are unit vectors along the principal axes, the axes of the BODY frame.

In the absence of a torque, how can a body rotate so that its angular velocity—and axis of rotation—does not change? Look at the first Euler equation, which then becomes

$$0 = 0 + (I_3 - I_2)\omega_2 \omega_3$$

This and the other two require that if $\omega_1 \neq 0$, then $\omega_2 = \omega_3 = 0$. That is a rotation about a principal axis. Or, if $\omega_2 \neq 0$, then $\omega_1 = \omega_3 = 0$. Or, if $\omega_3 \neq 0$, then $\omega_1 = \omega_2 = 0$. ==Only rotations about principal axes keep the angular velocity—and axis of rotation—unchanged==. For the general case, to ensure a constant angular velocity, $d\boldsymbol{\omega}/dt = 0$, there must be a torque present according to Eq. 9.8.6,

$$\boldsymbol{\tau} = \boldsymbol{\omega} \times \{I\}\boldsymbol{\omega} \tag{9.8.10}$$

This simply reiterates our earlier discussion concerning dynamic balancing. The energy equation can be derived from Eq. 9.8.6. Take the dot product of **ω** with this equation:

$$\boldsymbol{\omega} \cdot \boldsymbol{\tau} = \boldsymbol{\omega} \cdot \{I\} \frac{d\boldsymbol{\omega}}{dt} + \boldsymbol{\omega} \cdot (\boldsymbol{\omega} \times \{I\}\boldsymbol{\omega})$$

$$\boldsymbol{\omega} \cdot \boldsymbol{\tau} = \boldsymbol{\omega} \cdot \{I\} \frac{d\boldsymbol{\omega}}{dt} \tag{9.8.11}$$

The last term is zero because the expression in parentheses is perpendicular to **ω**. For symmetric tensors, the order of multiplication does not matter, so

$$\boldsymbol{\omega} \cdot \boldsymbol{\tau} = \boldsymbol{\omega} \cdot \{I\} \frac{d\boldsymbol{\omega}}{dt}$$

$$= \frac{d\boldsymbol{\omega}}{dt} \cdot \{I\}\boldsymbol{\omega}$$

$$= \frac{1}{2} \frac{d'}{dt} (\boldsymbol{\omega} \cdot \{I\}\boldsymbol{\omega})$$

$$= \frac{1}{2} \frac{d'}{dt} (\boldsymbol{\omega} \cdot \mathbf{L})$$

$$\boldsymbol{\omega} \cdot \boldsymbol{\tau} = \frac{d}{dt} T_{\text{rot}} \tag{9.8.12}$$

where we have used Eq. 9.7.6 for the rotational kinetic energy and the idea that d/dt and d'/dt are identical when applied to a scalar quantity.

Rotation about a fixed axis is so common that it is easy to incorrectly generalize descriptions of such motion for the case of "free rotation." For rotation about a fixed axis, the angular momentum and angular velocity are parallel. Since the moment of inertia is a tensor, that need not be true in general. This point, which is worth repeating, is best demonstrated in the movements of a high diver or acrobat as shown in Figures 9.8.1 and 9.8.2. There a diver begins a somersault rotating about an axis that is a principal axis of his body. That axis is perpendicular to the length of his body and might be described as "passing through his hips." Once the diver leaves the board, his external torque is zero so his angular *momentum* must remain constant. But by turning himself as shown in the figures, the angular *momentum* no longer aligns with one of the principal axes. To keep the angular *momentum* constant, this causes the angular *velocity* to change. The angular *velocity* now has a component along the length of the diver's body—the diver *twists* while somersaulting!

We shall now solve Euler's equations for the special case of $\tau = 0$. This could be a rotating body in orbit in the Space Shuttle, a rotating body in free fall, or a rotating body supported at its center of mass. The origin will be taken to be at the center of mass.

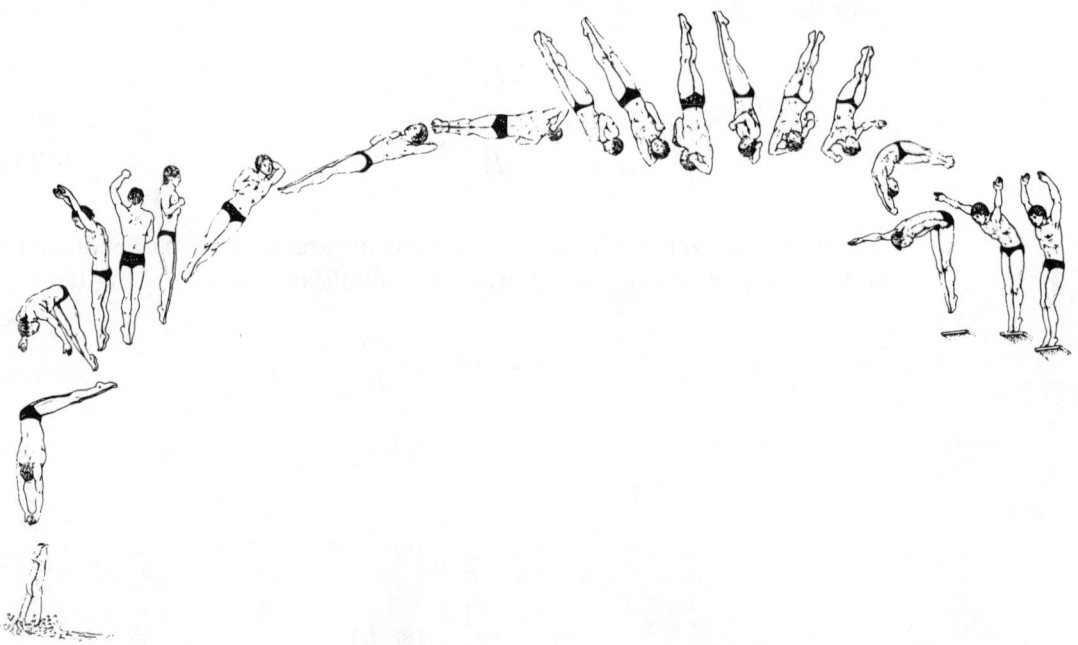

Figure 9.8.1 A forward one-and-one-half somersault with three twists. (From "The Physics of Somersaulting and Twisting" by Cliff Frohlich, © 1980 by Scientific American, Inc. All rights reserved.)

Figure 9.8.2 A forward two-and-one-half somersault with two twists. (From "The Physics of Somersaulting and Twisting" by Cliff Frohlich, © 1980 by Scientific American, Inc. All rights reserved.)

First consider a *symmetric top*. Due to its symmetry, two of the principal moments of inertia are the same. If we take x_3 to be the symmetry axis, then $I_1 = I_2$. This simplifies the third Euler equation to

$$I_3 \dot{\omega}_3 = 0$$

or

$$\omega_3 = \text{const} \tag{9.8.13}$$

Of course, if the body is rotating about x_3, the symmetry axis, then the entire angular velocity remains constant. But this equation says that for *any* rotation, the component of angular velocity along the symmetry axis, ω_3, remains constant.

The other Euler equation, then, becomes

$$\begin{aligned} \dot{\omega}_1 + \gamma \omega_3 \omega_2 &= 0 \\ \dot{\omega}_2 - \gamma \omega_3 \omega_1 &= 0 \end{aligned} \tag{9.8.14}$$

where

$$\gamma \equiv \frac{I_3 - I_1}{I_1} \tag{9.8.15}$$

These coupled first-order differential equations are similar to the ones at the end of Chapter 5 regarding motion in a magnetic field. They can be solved by differentiating

$$\ddot{\omega}_1 + \gamma \omega_3 \dot{\omega}_2 = 0$$
$$\ddot{\omega}_2 - \gamma \omega_3 \dot{\omega}_1 = 0$$
(9.8.16)

and then eliminating the first derivatives by using Eq. 9.8.14:

$$\ddot{\omega}_1 = -(\gamma \omega_3)^2 \omega_1$$
$$\ddot{\omega}_2 = -(\gamma \omega_3)^2 \omega_2$$
(9.8.17)

And these are now the equations of a simple harmonic oscillator whose frequency is $\gamma \omega_3$. We can recall the solution from Chapter 3 as

$$\omega_1 = C_1 \cos((\gamma \omega_3)t + \phi_1)$$
$$\omega_2 = C_2 \cos((\gamma \omega_3)t + \phi_2)$$
(9.8.18)

Since each of Eqs. 9.8.17 is a *second*-order differential equation, there are *two* arbitrary constants in each solution—C_1, ϕ_1 and C_2, ϕ_2. But we started with Eqs. 9.8.14, which were two *first*-order differential equations. Substituting our results back into these equations gives

$$C_1 = C_2$$

and
(9.8.19)

$$\phi_1 = \phi_2 + \frac{\pi}{2}$$

Thus, we may rewrite Eqs. 9.8.18 as

$$\omega_1 = C \cos[(\gamma \omega_3)t + \phi]$$
$$\omega_2 = C \sin[(\gamma \omega_3)t + \phi]$$
(9.8.20)

That is, the components ω_1 and ω_2 trace out a circle. Therefore, the angular velocity vector $\boldsymbol{\omega}$ precesses in a cone about the symmetry axis along which ω_3 remains constant.

For $\gamma > 0$ (that is, for $I_3 > I_1$), the precession of $\boldsymbol{\omega}$ about ω_3 is in the same sense as ω_3. For $\gamma < 0$, $(I_3 < I_1)$, in the opposite sense. Since

$$\omega = \sqrt{\omega_1^2 + \omega_2^2 + \omega_3^2}$$
(9.8.21)

we can use Eq. 9.8.20 to write

$$\omega = \sqrt{\omega_3^2 + C^2}$$

where $C^2 = \omega_1^2 + \omega_2^2$. Of course, ω_3, C, ϕ, and, thus, ω are all determined by the initial conditions. This precession of $\boldsymbol{\omega}$ about ω_3 or the symmetry axis is

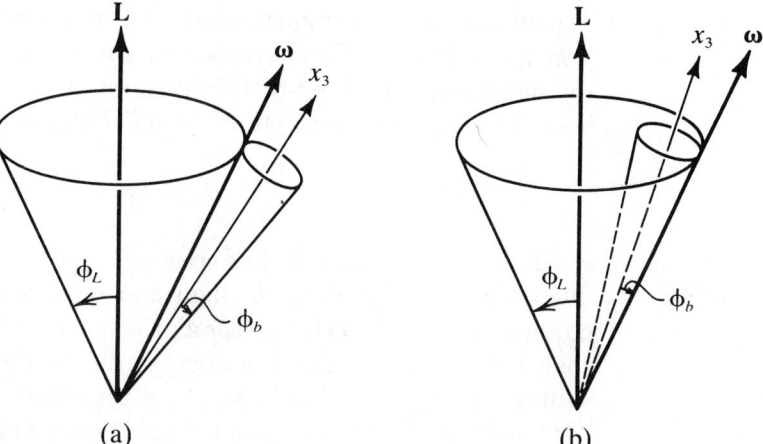

Figure 9.8.3 Free rotations of a top.

as viewed from the body-fixed frame. In this BODY frame, the half angle of this so-called *body cone* of precession is given by

$$\phi_b = \tan^{-1} \frac{C}{\omega_3} \qquad (9.8.22)$$

and is shown in Figure 9.8.3.

What is observed from the fixed LAB frame? There, the angular momentum **L** is fixed in direction with the BODY frame rotating about it. The angle between **ω** and **L** can be obtained from the definition of the dot product. Calling this ϕ_{LAB}, we have

$$\cos \phi_{\text{LAB}} = \frac{\boldsymbol{\omega} \cdot \mathbf{L}}{\omega L} = \frac{\boldsymbol{\omega} \cdot \{I\} \boldsymbol{\omega}}{\omega L} = \frac{2 T_{\text{rot}}}{\omega L} \qquad (9.8.23)$$

and this angle remains constant. This is the half-angle of a *laboratory cone* showing the precession of the angular velocity **ω** about the constant angular momentum **L** as viewed in the LAB frame. This is also shown in Figure 9.8.3.

This can be interpreted by describing the body cone as rolling without slipping around the LAB cone. Their line of contact is, of course, the direction of the angular velocity **ω**. From Eqs. 9.8.20, the angular frequency of precession of **ω** about the symmetry axis as seen from the body frame is simply

$$\Omega_b = \gamma \omega_3 \qquad (9.8.24)$$

The angular frequency of precession of ω about the angular momentum L in the LAB frame is given by

$$\Omega_L = \gamma \omega_3 \frac{\sin \phi_b}{\sin \phi_L} \qquad (9.8.25)$$

The motion of a *nonsymmetric top*—that is, a rigid body with nondegenerate principal moments of inertia—provides some very interesting results. Again, we shall construct a fixed-body reference frame to coincide with the principal axes. We order the principal moments of inertia such that

$$I_3 > I_2 > I_1 \tag{9.8.26}$$

For a rotation precisely about any one of the principal axes nothing interesting occurs; $\boldsymbol{\omega}$ and $\mathbf{L}$ align and remain constant. Therefore, let us consider the case for $\boldsymbol{\omega}$ lying *near* a principal axis, but not along it. That is, we shall *perturb* the rotation from being exactly aligned along a principal axis. Will the resulting motion be stable or unstable?

First picture the motion that would occur *if the angular velocity remained constant* in the body frame. To do this, picture an arrow shot through a box. The box rotates about the arrow like a chicken on a barbeque skewer. *That is the motion of constant angular velocity.* Let us begin by looking at $\boldsymbol{\omega}$ very near x_3. That is, $\boldsymbol{\omega} = \omega_1 \hat{\mathbf{e}}_1 + \omega_2 \hat{\mathbf{e}}_2 + \omega_3 \hat{\mathbf{e}}_3$ and

$$\omega_3 \gg \omega_1$$
$$\omega_3 \gg \omega_2 \tag{9.8.27}$$

We are still restricting our attention to a free rotator with $\boldsymbol{\tau} = 0$. Therefore, the last of Euler's equations 9.8.9 shows

$$\omega_3 = \text{constant} \tag{9.8.28}$$

to first order in ω_1 and ω_2. Keeping ω_3 constant, we can then solve for ω_1 and ω_2 just as before to get

$$\omega_1 = C\sqrt{I_2(I_3 - I_2)} \cos(\gamma\omega_3 t + \phi)$$
$$\omega_2 = C\sqrt{I_1(I_3 - I_1)} \sin(\gamma\omega_3 t + \phi) \tag{9.8.29}$$

[*sin* written above *cos* in second equation as correction]

where

$$\gamma \equiv \sqrt{\frac{(I_3 - I_1)(I_3 - I_2)}{I_1 I_2}} \tag{9.8.30}$$

and the arbitrary constants A and ϕ are determined by the initial conditions. Thus, ω_1 and ω_2 describe an ellipse so the vector $\boldsymbol{\omega}$ precesses in a cone of elliptical cross section about the x_3-axis corresponding to the *largest* principal moment of inertia. What does this motion look like? Think back to the arrow piercing a box. The arrow remains close to the ω_3 axis but moves as viewed by an observer on the box. That is just an additional wobble when viewed by the laboratory frame. Of course, the Chandler wobble of our Earth is an example of this motion.

A similar result can be obtained for an angular velocity $\boldsymbol{\omega}$ initially close to the x_1-axis corresponding to the *smallest* principal moment of inertia. But the rotation near the x_2-axis offers unusual results. This axis has the *intermediate* value of the moments of inertia. For that case, $\boldsymbol{\omega} = \omega_1 \hat{\mathbf{e}}_1 + \omega_2 \hat{\mathbf{e}}_2 + \omega_3 \hat{\mathbf{e}}_3$ with

$$\omega_2 \gg \omega_1$$
$$\omega_2 \gg \omega_3$$
(9.8.31)

The same analysis as before provides

$$\ddot{\omega}_1 = \left[\frac{(I_3 - I_2)(I_2 - I_1)}{I_1 I_2} \omega_2^2\right]\omega_1$$

or (9.8.32)

$$\ddot{\omega}_3 = \left[\frac{(I_1 - I_2)(I_2 - I_3)}{I_2 I_3} \omega_2^2\right]\omega_3$$

Note that these are *not* the equations of simple harmonic oscillators. The terms in brackets are positive! The solutions to these equations are exponential functions. Rotation about the axis corresponding to the intermediate principal moment of inertia is unstable! Demonstrate this yourself by tossing a tennis racquet or book in the air and attempting to spin it about each principal axis. The result is easily observed and is quite surprising!

Chandler Wobble: An example of precession of the rotation axis as viewed from the BODY frame is provided by our own Earth. Earth's rotation axis precesses about our north pole in a circle with radius of about 10 m and with a period of some 430 days. Since latitude is dependent upon this axis of rotation, this causes measurable changes in latitude. These changes in latitude were first discovered and understood by the astronomer S. C. Chandler in 1891.

The more familar precession of Earth's axis, about a cone with half-angle of 23°, is the result of external torques due to the Sun and moon. The large angle involved here means that the north pole traces out a large circle. Polaris, the pole star, will not be the pole star far in the future; nor was it far in the past. The period for this motion is about 26,000 years. As the rotation axis changes, the equatorial plane of Earth also changes. The intersections of this equatorial plane with Earth's orbital plane are known as the *equinoxes*, so this motion is known as the precession of equinoxes.

PROBLEMS

9.1 Find the center of mass of the following:
  (a) A thin, uniform wire bent into a semicircle of radius R.
  (b) A quadrant of a circle of radius R cut out of a plane.
  (c) An octant of a solid sphere of radius R.

9.2 Find the center of mass of the following:
  (a) A thin, uniform wire bent into an "L" with two legs of equal length and at right angles.
  (b) A thin, uniform sheet of metal cut into a triangle whose sides are 3, 4, and 5 cm.
  (c) A uniform hemispherical shell of radius R.

9.3 Find the center of mass of the following:
  (a) The area of the xy plane bounded by the parabola $y = ax^2$ and the line $y = b$.
  (b) The volume bounded by the parabola of revolution $y = a(x^2 + z^2)$ and the plane $y = b$.

9.4 Find the center of mass of a solid, (otherwise) uniform sphere of radius A that has within it a spherical cavity of radius B, centered a distance C from the center of the larger sphere. $A > B + C$.

9.5 A thin, uniform wire bent into a semicircle hangs on a rough peg. The coefficient of friction between wire and peg is μ. Find the maximum angle θ between horizontal and the ends of the semicircle when it is on the verge of slipping.

9.6 A uniform, solid hemisphere rests on the floor and against a vertical wall with the round part of it touching both. The coefficient of friction μ is identical for the hemisphere-wall and hemisphere-floor surfaces. Find the angle θ between the flat side and vertical when it is on the verge of slipping.

9.7 A uniform, solid hemisphere rests on a plane inclined an angle θ from the horizontal. The rounded side is in contact with the plane. The coefficient of friction between the two surfaces is μ. Find the angle θ between the flat side and vertical when it is on the verge of slipping.

9.8 Find the center of mass of a sphere of radius a centered on the origin and made of a layered material whose density is given by

$$\rho = \rho(z) = \rho_0\left(1 + \frac{z}{a}\right)$$

9.9 Explicitly find the moment of inertia of the following:
  (a) A long, thin rod of mass M and length L for rotation about its center of mass.

(b) The same long, thin rod of mass M and length L for rotation about one end.
(c) A hoop or ring of mass M and radius R for rotation about its center of mass with axis perpendicular to the ring.
(d) The same hoop or ring of mass M and radius R for rotation about an axis passing through its edge and perpendicular to the plane of the ring.
(e) The same hoop or ring for rotation about an axis tangent to the edge.

9.10 Explicitly find the moment of inertia for the following:
(a) A solid disk of mass M and radius R for rotation about its center of mass on an axis perpendicular to the plane of the disk.
(b) The same disk for rotation about an axis passing through its edge and perpendicular to the plane of the disk.
(c) The same disk for rotation about an axis tangent to its edge.

9.11 Explicitly find the moment of inertia for the following:
(a) A sphere of mass M and radius R for rotation about an axis passing through its center of mass.
(b) A cylinder of height H, radius R, and mass M for rotation about the symmetry axis.
(c) A cube of mass M and edge length L for rotation about an axis perpendicular to a face and passing through the center of mass.

9.12 Find the moment of inertia for rotation about a symmetry axis for a sphere of radius a whose density is given by

$$\rho = \rho(r) = (\rho_0)\left(1 - \frac{r}{a}\right)$$

What is its moment of inertia if, instead

$$\rho = \rho(r) = (\rho_0)\left(\frac{r}{a}\right)$$

$\rho = \frac{M}{V}$

9.13 Find the moment of inertia for rotation about its symmetry axis for the frustum of a cone of radii a and b and height h. The mass of the cone is m.

9.14 A square whose side is l lies in the xy plane with its center at the origin and a diagonal lying on the x-axis. The triangular portion above this diagonal has a mass per unit area density of σ_1; the portion below, σ_2. Find the center of mass and the moments of inertia for rotation about the x-, y-, and z-axes, and about an axis passing through the center of mass and perpendicular to the square (the "geographic center" of the square is located at the origin).

9.15 Find the moment of inertia for a square lamina of mass M and edge length S for rotation about:
(a) An axis through the center of mass and perpendicular to the plane of the lamina.
(b) An axis through one corner and perpendicular to the plane of the lamina.

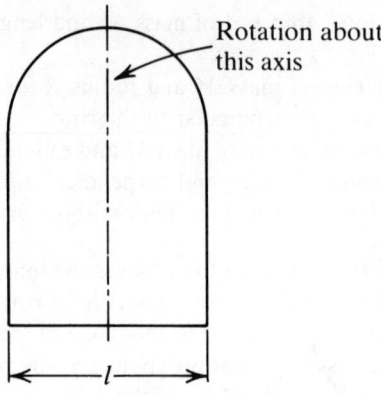

9.16 A piece of sheet metal is cut in the shape of a semicircle attached to a square. Find the moment of inertia for rotation about its "long axis" as shown in the drawing. The total mass is M and the length of a side of the square is l.

9.17 A rigid body consists of six particles, each of mass m, fixed to the ends of three rigid, massless rods of length $2a$, $2b$, and $2c$. The rods are held perpendicular to one another at their midpoints. Find the inertia tensor for this body in terms of axes along the rods,. Show that these are principal axes.

9.18 Show that the largest principal moment of inertia is less than or equal to the sum of the other two.

9.19 Find the inertia tensor for a square lamina of side L and mass M. Use a coordinate system whose origin is located at the center of the lamina with x- and y-axes perpendicular to the sides of the square.

9.20 Find the angular momentum and kinetic energy for rotation of a uniform square lamina about a diagonal.

9.21 Find the inertia tensor for a rectangular lamina with sides of L and W and mass M. Use a coordinate system whose origin is located at the center of the lamina with x- and y-axes perpendicular to the sides of length L and W, respectively.

9.22 Consider a tennis racquet made of a hoop of radius a and a handle of length $2a$. Adjust the mass densities so that the center of mass is at the joint of hoop and handle. Find the racquet's principal axes and its principal moments of inertia.

9.23 Find the inertia tensor for a uniform rectangular solid with mass M and dimensions L, W, and H. Use a coordinate system with its origin at the center of the solid and axes perpendicular to the faces of the solid.

9.24 Find the inertia tensor for a uniform rectangular solid with mass M and dimensions L, W, and H. Use a coordinate system with its origin at one corner and its axes along the edges.

9.25 Set up the cubic equation for finding the principal moments of inertia for the rectangular solid and coordinate system of Problem 9.24.

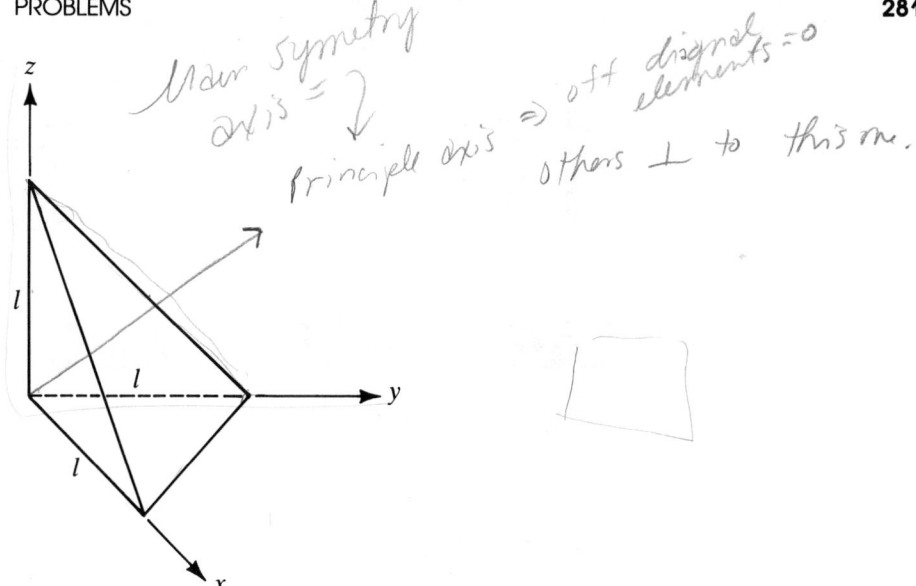

9.26 Consider a right triangular solid of length l along the x-, y-, and z-axes as in the drawing. Its total mass is M. Calculate its inertia tensor for these axes.

9.27 Find the principal moments of inertia and principal axes for the right triangular solid in Problem 9.26.

9.28 Find the inertia tensor for a uniform cube of side length L and mass M in terms of a coordinate system whose origin is located at the center of the cube and whose axes are perpendicular to the faces of the cube.

9.29 Find the moment of inertia for rotation about the z-axis for a right elliptical cylinder of mass M. The cylinder is bounded by plan ends $az = -c$ and $z = +c$. The cylinder's wall is an elliptical surface defined by $(x/a) + (y/b) = 1$.

9.30 Find the principal moments of inertia and the principal axes for a solid, uniform ellipsoid of mass M with axes of $2a > 2b > 2c$.

10

LAGRANGIAN MECHANICS

10.1 Introduction

Thus far, we have solved for the motion of objects or systems of objects by directly applying and solving Newton's Second Law, which can be written as

$$\mathbf{F} = m\mathbf{a} \tag{10.1.1}$$

or

$$\mathbf{F} = \frac{d\mathbf{p}}{dt} \tag{10.1.2}$$

This is a *simple* form for the basis of all mechanics. We have applied it in seemingly ever more complicated situations.

Shortly, we shall write this in a more *complicated* form. It often seems that this is done as a mathematical exercise or an attempt to provide some esoteric elegance to the utilitarian methods of ordinary Newtonian mechanics. But there is method is this madness! Equations 10.1.1 or 10.1.2 are simple and straightforward—for many cases. But for other situations, they rapidly become messy and unmanageable! Even writing the expression for the force of, say, a bead sliding along a bent wire becomes difficult. Solution of the resulting equation using $\mathbf{F} = m\mathbf{a}$ for this case may well be impossible.

Therefore, it is quite worthwhile to spend some considerable effort in reformulating the *ideas* held in Newtonian mechanics so we can solve otherwise intractable problems. Remember, this is only a reformulation so, as we have done before, we shall check the results by applying them to already familiar examples. There is no new information or new areas of validity. We will simply restate Newtonian mechanics in another *form*.

10.2 Generalized Coordinates

In rectangular coordinates, Eq. 10.1.1 becomes

$$F_x = m\ddot{x}$$
$$F_y = m\ddot{y} \tag{10.2.1}$$
$$F_z = m\ddot{z}$$

Ideas seem simpler here. Yet we have encountered situations in which $\mathbf{F} = m\mathbf{a}$ became far simpler in a different coordinate system. For example, central forces would be quite difficult to handle with rectangular coordinates. Therefore, in Chapter 7, the discussion of central forces was carried out in *polar coordinates*. The transformation from rectangular to polar coordinates is

$$r = \sqrt{x^2 + y^2}$$
$$\theta = \tan^{-1}\left(\frac{y}{x}\right) \tag{10.2.2}$$

while the inverse transformation is

$$x = r \cos \theta$$
$$y = r \sin \theta \tag{10.2.3}$$

In a similar manner, we may describe the state or configuration of a system by some *generalized coordinates* $q_1, q_2, \ldots$. These may be positions, angular displacements, whatever. Their time derivatives $\dot{q}_1, \dot{q}_2, \dot{q}_3, \ldots$ are known as *generalized velocities*. For a system of N particles there is, in general, a need for $3N$ coordinates to fully specify the state or configuration of the system. If there are constraints—such as the system is a rigid body, or the N particles are constrained to a surface, or some of them are connected—then less than $3N$ coordinates are needed. Thus, in general, there are a minimum number n of generalized coordinates:

$$q_1, q_2, q_3, \ldots, q_n$$

which will entirely specify the state or configuration of the system. The system is described as *holonomic* if each of the generalized coordinates can vary independently of the others. For that case, there are said to be *n degrees of freedom* for the system. We shall only consider such systems.[1]

The generalized coordinates can be written as some function of the rectangular coordinates and perhaps time. That is,

$$q_1 = q_1(x_1, x_2, x_3, \ldots, x_{3N}, t)$$
$$\vdots \qquad\qquad\qquad\qquad\qquad (10.2.4)$$
$$q_n = q_n(x_1, x_2, x_3, \ldots, x_{3N}, t)$$

where the tensor notation of $x = x_1$, $y = x_2$, $z = x_3$ has been extended so that x_i represents all rectangular coordinates. Likewise, the rectangular coordinates may be written as

$$x_1 = x_1(q_1, q_2, \ldots, q_n, t)$$
$$\vdots \qquad\qquad\qquad\qquad\qquad (10.2.5)$$
$$x_{3N} = x_{3N}(q_1, q_2, \ldots, q_n, t)$$

For a holonomic system, $n = 3N$. The number of degrees of freedom must remain the same. The time-dependence appears only if one of the coordinate systems is moving relative to the other.

Now let the system change from one state or configuration to another nearby. The generalized coordinates change from q_1 to $q_1 + \delta q_1$, from q_2 to $q_2 + \delta q_2$, and so on, while the rectangular coordinates change from x_1 to $x_1 + \delta x_1$, from x_2 to $x_2 + \delta x_2$, and so on. These changes are readily related by looking at the differential form of Eqs. 10.2.4 and 10.2.5:

$$\delta q_1 = \frac{\partial q_1}{\partial x_1} \delta x_1 + \frac{\partial q}{\partial x_2} \delta x_2 + \cdots + \frac{\partial q_1}{\partial x_{3N}} \delta x_{3N}$$
$$\vdots$$

or, more compactly,

$$\delta q_k = \sum_{i=1}^{3N} \frac{\partial q_k}{\partial x_i} \delta x_i \qquad (10.2.6)$$

Likewise,

$$\delta x_1 = \frac{\partial x_i}{\partial q_i} \delta q_1 + \frac{\partial x_i}{\partial q_2} \delta q_2 + \cdots + \frac{\partial x_i}{\partial q_n} \delta q_n$$
$$\vdots$$

or, more compactly,

$$\delta x_i = \sum_{k=1}^{n} \frac{\partial x_i}{\partial q_k} \delta q_k \qquad (10.2.7)$$

---

[1] An example of a nonholonomic system is a sphere or wheel that rolls without slipping. Not all coordinates necessary to specify its orientation may vary independently.

Before going on, let us apply this idea to a specific example. For central forces we choose polar coordinates. Let

$$q_1 = r$$
$$q_2 = \theta \qquad (10.2.8)$$

The transformation equations 10.2.4 and 10.2.5 reduce to the special cases of Eqs. 10.2.2 and 10.2.3 previously discussed. Thus,

$$\delta q_1 = \delta r = \frac{\partial r}{\partial x} \delta x + \frac{\partial r}{\partial y} \delta y$$

$$= \frac{x \delta x}{\sqrt{x^2 + y^2}} + \frac{y \delta y}{\sqrt{x^2 + y^2}}$$

$$\delta q_1 = \delta r = \frac{x \delta x + y \delta y}{\sqrt{x^2 + y^2}} \qquad (10.2.9)$$

$$\delta q_2 = \delta \theta = \frac{\partial \theta}{\partial x} \delta x + \frac{\partial \theta}{\partial y} \delta y$$

$$= \frac{1}{\left(1 + \left(\frac{y}{x}\right)^2\right)} \left(-\frac{y}{x^2}\right) \delta x + \frac{1}{\left(1 + \left(\frac{y}{x}\right)^2\right)} \left(\frac{1}{x}\right) \delta y$$

$$\delta q_2 = \delta \theta = \frac{-y \delta x + x \delta y}{x^2 + y^2} \qquad (10.2.10)$$

And likewise,

$$\delta x = \frac{\partial x}{\partial r} \delta r + \frac{\partial x}{\partial \theta} \delta \theta = \cos \theta \delta r - r \sin \theta \delta \theta \qquad (10.2.11)$$

$$\delta y = \frac{\partial y}{\partial r} \delta r + \frac{\partial y}{\partial \theta} \delta \theta = \sin \theta \delta r + r \cos \theta \delta \theta \qquad (10.2.12)$$

10.3 Generalized Forces

As a particle moves from position $\mathbf{r}$ to $\mathbf{r} + \delta \mathbf{r}$, the small amount of work δW that is done is given by

$$\delta W = \mathbf{F} \cdot \delta \mathbf{r} = F_x \delta x + F_y \delta y + F_z \delta z \qquad (10.3.1)$$

In our extended tensor notation we can write this as

$$\delta W = \sum_i F_i \delta x_i \qquad (10.3.2)$$

For a single particle, i goes from 1 to 3. For N particles, i goes from 1 to $3N$.

In terms of the generalized coordinates, we may use Eq. 10.2.7 to write this as

$$\delta W = \sum_i \left(F_i \sum_k \frac{\partial x_i}{\partial q_k} \delta q_k \right)$$

$$\delta W = \sum_k \left(\sum_i F_i \frac{\partial x_i}{\partial q_k} \right) \delta q_k \tag{10.3.3}$$

This can be written as

$$\delta W = \sum_k Q_k \delta q_k \tag{10.3.4}$$

where

$$Q_k = \sum_i F_i \frac{\partial x_i}{\partial q_k} \tag{10.3.5}$$

is known as the *generalized force*. If q_k is a linear distance, then the associated generalized force Q_k is, indeed, a force. If q_k is an angular displacement, then Q_k turns out to be a torque. In all cases $Q_k \delta q_k$ is the work done as generalized coordinate q_k increases to $q_k + \delta q_k$.

For a conservative force, in Chapter 5, the rectangular components of the force were shown to be the negative partial derivatives of a potential energy function V. In our present notation, $\mathbf{F} = -\nabla V$ becomes

$$F_i = -\frac{\partial V}{\partial x_i} \tag{10.3.6}$$

Thus, the generalized force may be written as

$$Q_k = \sum_i \left(-\frac{\partial V}{\partial x_i} \right) \frac{\partial x_i}{\partial q_k} = -\sum_i \frac{\partial V}{\partial x_i} \frac{\partial x_i}{\partial q_k} \tag{10.3.7}$$

But this is just the usual form of the partial derivative of any function V with respect to variable q_k. Hence,

$$Q_k = -\frac{\partial V}{\partial q_k} \tag{10.3.8}$$

And this is precisely the same form as Eq. 10.3.6 for the rectangular case.

If part of the generalized force cannot be derived from a potential energy function; that is, if it is not conservative, it must be handled separately as

$$Q_k = Q'_k - \frac{\partial V}{\partial q_k} \tag{10.3.9}$$

where Q'_k is the nonconservative part of the generalized force.

10.4 Lagrange's Equations

Now the goal is to find the equation of motion in terms of these generalized coordinates q_k. Direct substitution into Newton's Second Law is messy. A better method is to look at the kinetic energy. In rectangular coordinates this is expressible as

$$T = \sum_i \frac{1}{2} m_i \dot{x}_i^2 \qquad (10.4.1)$$

Recalling Eqs. 10.2.5,

$$x_i = x_i(q_1, q_2, \ldots, q_n, t) \qquad (10.4.2)$$

we can write

$$\dot{x}_i = \frac{dx_i}{dt} = \sum_k \frac{\partial x_i}{\partial q_k} \frac{\partial q_k}{\partial t} + \frac{\partial x_i}{\partial t}$$

or, more simply,

$$\dot{x}_i = \sum_k \frac{\partial x_i}{\partial q_k} \dot{q}_k + \frac{\partial x_i}{\partial t} \qquad (10.4.3)$$

Further, we can see

$$\frac{\partial x_i}{\partial q_k} = \frac{\partial \dot{x}_i}{\partial \dot{q}_k} \qquad (10.4.4)$$

Multiplying by $\dot{x}_i$ and differentiating with respect to t yields

$$\frac{d}{dt}\left(\dot{x}_i \frac{\partial \dot{x}_i}{\partial \dot{q}_k}\right) = \frac{d}{dt}\left(\dot{x}_i \frac{\partial x_i}{\partial q_k}\right)$$

$$\frac{d}{dt}\left(\frac{\partial}{\partial \dot{q}_k}\left(\frac{1}{2}\dot{x}_i^2\right)\right) = \ddot{x}_i \frac{\partial x_i}{\partial q_k} + \dot{x}_i \frac{\partial}{\partial q_k}\left(\frac{dx_i}{dt}\right) \qquad (10.4.5)$$

$$\frac{d}{dt}\left(\frac{\partial}{\partial \dot{q}_k}\left(\frac{1}{2}\dot{x}_i^2\right)\right) = \ddot{x}_i \frac{\partial x_i}{\partial q_k} + \dot{x}_i \frac{\partial \dot{x}_i}{\partial q_k}$$

Further multiplication by m_i gives

$$\frac{d}{dt}\left(\frac{\partial}{\partial \dot{q}_k}\left(\frac{1}{2} m_i \dot{x}_i^2\right)\right) = (m_i \ddot{x}_i) \frac{\partial x_i}{\partial q_k} + m_i \dot{x}_i \frac{\partial \dot{x}_i}{\partial q_k}$$

or

$$\frac{d}{dt} \frac{\partial}{\partial \dot{q}_k}\left(\frac{1}{2} m_i \dot{x}_i^2\right) = F_i \frac{\partial x_i}{\partial q_k} + \frac{\partial}{\partial q_k}\left(\frac{1}{2} m_i \dot{x}_i^2\right)$$

The terms in parenthesis will become the kinetic energy if summed over i. Summing every term over i, then, yields

$$\frac{d}{dt}\frac{\partial T}{\partial \dot{q}_k} = \sum_i \left(F_i \frac{\partial x_i}{\partial q_k}\right) + \frac{\partial T}{\partial q_k} \qquad (10.4.6)$$

From Eq. 10.3.5, the summation is just the generalized force Q_k. Therefore,

$$\frac{d}{dt}\frac{\partial T}{\partial \dot{q}_k} = Q_k + \frac{\partial T}{\partial q_k} \qquad (10.4.7)$$

This set of differential equations in terms of the generalized coordinates are known as *Lagrange's equations*. Solutions to these determine the motion of a body or system, just as solutions to $\mathbf{F} = m\mathbf{a}$.

For conservative systems, we may invoke Eq. 10.3.8 to rewrite Lagrange's equations as

$$\frac{d}{dt}\frac{\partial T}{\partial \dot{q}_k} = \frac{\partial T}{\partial q_k} - \frac{\partial V}{\partial q_k} \qquad (10.4.8)$$

We may further simplify the *form* of this by defining a function

$$L \equiv T - V \qquad (10.4.9)$$

known simply as the *Lagrangian*. Any potential energy function depends only upon coordinates (generalized as well as rectangular) and is independent of velocities (generalized as well as rectangular). That is,

$$\frac{\partial V}{\partial \dot{q}_k} = 0 \qquad (10.4.10)$$

Therefore,

$$\frac{\partial L}{\partial \dot{q}_k} = \frac{\partial T}{\partial \dot{q}_k} \qquad (10.4.11)$$

and

$$\frac{\partial L}{\partial q_k} = \frac{\partial T}{\partial q_k} - \frac{\partial V}{\partial q_k} \qquad (10.4.12)$$

so Lagrange's equations may be written elegantly as

$$\frac{d}{dt}\frac{\partial L}{\partial \dot{q}_k} = \frac{\partial L}{\partial q_k} \qquad (10.4.13)$$

For nonconservative systems, the less elegant form of Eq. 10.4.7 is still valid. Or, we may explicitly write

$$\frac{d}{dt}\frac{\partial L}{\partial \dot{q}_k} = Q'_k + \frac{\partial L}{\partial q_k} \qquad (10.4.14)$$

10.5 Elementary Examples

When developing or formulating something new like this, it is always worthwhile to apply the results first to simple—perhaps even trivial—situations that we already know how to solve. In this way, we can increase our confidence in the reliability of the new formulation, idea, or development. Therefore, we begin by applying Lagrange's equations of motion, Eq. 10.4.12, to rather simple situations.

EXAMPLE 10.5.1 Consider a system of N particles with masses $m_1, m_2, m_3, \ldots, m_N$, located by "ordinary" Cartesian coordinates, $x_1, y_1, z_1, x_2, y_2, z_2, \ldots, x_N, y_N, z_N$, acted upon by "ordinary" forces derivable from a potential energy function V. Show that Lagrange's equations of motion reduce directly to Newton's Second Law ($\mathbf{F} = m\mathbf{a}$).

The kinetic energy is given by

$$T = \sum_{i=1}^{N} \tfrac{1}{2} m_i (\dot{x}_i^2 + \dot{y}_i^2 + \dot{z}_i^2) \tag{10.5.1}$$

The Lagrangian function is

$$L \equiv T - V \tag{10.4.9}$$

The kinetic energy is now independent of the positions, q_k, and the potential energy is independent of the velocities, $\dot{q}_k$. Therefore,

$$\frac{\partial L}{\partial q_k} = -\frac{\partial V}{\partial q_k}$$

for a generalized coordinate q_k. For a specific case, say, $q_k = x_i$, then

$$\frac{\partial L}{\partial q_k} = -\frac{\partial V}{\partial x_i} = F_{x_i} \tag{10.5.2}$$

with similar equations for y_i and z_i, of course. Likewise,

$$\frac{\partial L}{\partial \dot{q}_k} \rightarrow \frac{\partial L}{\partial \dot{x}_i} = m_i \dot{x}_i \tag{10.5.3}$$

Thus, Lagrange's equation yields

$$\frac{d}{dt}(m_i \dot{x}_i) = F_{x_i}$$

or

$$m_i \ddot{x}_i = F_{x_i}$$

with similar results

$$m_i \ddot{y}_i = F_{y_i}$$
$$m_i \ddot{z}_i = F_{z_i} \qquad (10.5.4)$$

And this is clearly the familiar form of Newton's Second Law.

EXAMPLE 10.5.2 Use Lagrange's equations to write the equation of motion for a *simple harmonic oscillator*.

The kinetic energy is given by

$$T = \tfrac{1}{2} m \dot{x}^2 \qquad (10.5.5)$$

and the potential energy is

$$V = \tfrac{1}{2} k x^2 \qquad (10.5.6)$$

So the Lagrangian can be written as

$$L = T - V = \tfrac{1}{2} m \dot{x}^2 - \tfrac{1}{2} k x^2 \qquad (10.5.7)$$

Thus,

$$\frac{\partial L}{\partial x} = -kx \quad \text{and} \quad \frac{\partial L}{\partial \dot{x}} = m\dot{x} \qquad (10.5.8)$$

So Lagrange's equation becomes

$$\frac{d}{dt}\left(\frac{\partial L}{\partial \dot{x}}\right) = \frac{\partial L}{\partial x}$$

$$m\ddot{x} = -kx \qquad (10.5.9)$$

which is the familiar form of Newton's Second Law for a simple harmonic oscillator.

EXAMPLE 10.5.3 Consider a body of mass m moving in a plane and located by polar coordinates. Let $r = q_1$ and $\theta = q_2$. The kinetic energy is given by

$$T = \tfrac{1}{2} m (\dot{r}^2 + r^2 \dot{\theta}^2) \qquad (10.5.10)$$

The generalized forces are given by

$$Q_1 = -\frac{\partial V}{\partial q_1} = -\frac{\partial V}{\partial r}$$

and

$$Q_2 = -\frac{\partial V}{\partial q_2} = -\frac{\partial V}{\partial \theta} \qquad (10.5.11)$$

Since $q_1 = r$ measures a linear distance, the corresponding generalized force is actually a *force*; that is, $Q_1 = F_r$. And since $q_2 = \theta$ measures an angular distance, its corresponding generalized force is a *torque*; that is, $Q_2 = \tau_\theta$. We can write this torque as the distance from an origin multiplied by a component of force in "the θ-direction." That is, $Q_2 = \tau_\theta = rF_\theta$.

Lagrange's equations, then, provide the following:

$$L = T - V = \tfrac{1}{2}m(\dot{r}^2 - r^2\dot\theta^2) - V$$

$$\frac{\partial L}{\partial q_1} = \frac{\partial L}{\partial \dot r} = mr\dot\theta^2 + F_r \quad \text{and} \quad \frac{\partial L}{\partial q_2} = \frac{\partial L}{\partial \theta} = +rF_\theta$$

$$\frac{\partial L}{\partial \dot q_1} = \frac{\partial L}{\partial \dot r} = m\dot r \quad \text{and} \quad \frac{\partial L}{\partial \dot q_2} = \frac{\partial L}{\partial \dot\theta} = mr^2\dot\theta \qquad (10.5.12)$$

$$\frac{d}{dt}\frac{\partial L}{\partial \dot r} = \frac{\partial L}{\partial r} \qquad\qquad \frac{d}{dt}\frac{\partial L}{\partial \dot\theta} = \frac{\partial L}{\partial \theta}$$

$$m\ddot r = mr\dot\theta^2 + F_r \quad \text{and} \quad \frac{d}{dt}(mr^2\dot\theta) = rF_\theta$$

And these are just the equations we have already derived in Chapter 7. For central fields, $F_\theta = 0$, so the quantity $(mr^2\dot\theta)$ must remain constant. This quantity, as you should recall, is the angular momentum. The expression $mr\dot\theta^2$ that appears in the radial equation is sometimes called a "centrifugal force." Again Lagrange's equations provide the same information as Newton's Second Law.

EXAMPLE 10.5.4 Apply Lagrange's equations to an Atwood's machine. Two weights with masses m_1 and m_2 are connected by a light, inextensible string of length l that passes over a pulley of radius r and moment of inertia I supported on a frictionless axis, as illustrated in Figure 10.5.1. A single coordinate x specifies the position of both masses while its time

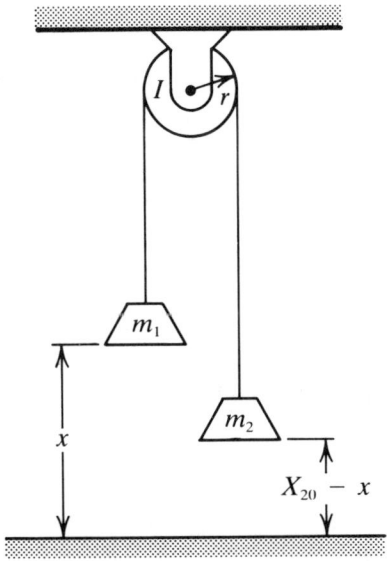

Figure 10.5.1 An Atwood's machine.

derivative $\dot{x}$ describes the speed of both masses and the rim of the pulley. We assume the string does not slip around the pulley. The angular speed of the pulley, then, is $\omega = \dot{x}/r$.

Thus, the kinetic energy is

$$T = \tfrac{1}{2}m_1\dot{x}^2 + \tfrac{1}{2}m_2\dot{x}^2 + \tfrac{1}{2}I\left(\frac{\dot{x}}{r}\right)^2 \tag{10.5.13}$$

and the potential energy is

$$V = +m_1 gx + m_2 g(X_{20} - x) \tag{10.5.14}$$

where X_{20} depends upon the arbitrary choice of level 0 from which to measure potential energy. But, then, any potential energy is defined only to within an additive constant. Thus, the Lagrangian is

$$L = \frac{1}{2}\left(m_1 + m_2 + \frac{I}{r^2}\right)\dot{x}^2 + g(m_2 - m_1)x + m_2 g X_{20} \tag{10.5.15}$$

This last term, the constant, will not appear in any derivatives. Thus,

$$\frac{\partial L}{\partial x} = g(m_2 - m_1) \tag{10.5.16}$$

and

$$\frac{\partial L}{\partial \dot{x}} = \left(m_1 + m_2 + \frac{I}{r^2}\right)\dot{x} \tag{10.5.17}$$

So Lagrange's equation yields

$$\frac{d}{dt}\frac{\partial L}{\partial \dot{x}} = \frac{\partial L}{\partial x}$$

$$\left(m_1 + m_2 + \frac{I}{r^2}\right)\ddot{x} = g(m_2 - m_1)$$

or

$$\ddot{x} = \frac{g(m_2 - m_1)}{m_1 + m_2 + (I/r^2)} \tag{10.5.18}$$

This is the acceleration of the system. For $m_2 > m_1$, the acceleration is positive and x increases (m_1 moves up) when released from rest. For $m_1 > m_2$, the acceleration is negative.

10.6 Systems with Constraints

The example of an Atwood's machine in Section 10.5 involved a system with *constraints*. We might expect that we need at least two coordinates to specify the locations of both the masses, and a third to specify the orientation of the pulley; that is, three coordinates in all. But the inextensible string of length l that connects them provides a constraint. The lack of slipping between string and pulley provides another constraint.

We can formally write these constraints in terms of velocities as

$$\dot{x} = r\dot{\theta}$$

and (10.6.1)

$$\dot{x}_2 = -\dot{x}_1 = -\dot{x}$$

where r, θ, x_2, and x_1 are defined in Figure 10.6.1. Note that x now measures the distance *down*. This is convenient, but it means that the potential energy of mass m_1 must now carry a minus sign,

$$V_1 = -mgx \qquad (10.6.2)$$

Equations 10.6.1 can readily be integrated to

$$x = r\theta + c$$

and (10.6.3)

$$x_2 = C - x$$

The term C, of course, introduces an arbitrary constant into the potential energy V_2 corresponding to m_2. Constraints are given in terms of velocities (or derivatives). If they can then be integrated as here to provide equations relating the generalized coordinates, then they reduce the number of coordinates necessary to specify a system. Such constraints are called *holonomic*; each reduces the number of degrees of freedom by one. Three coordinates

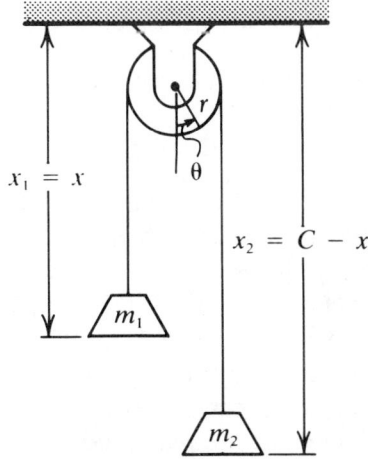

Figure 10.6.1 An Atwood's machine.

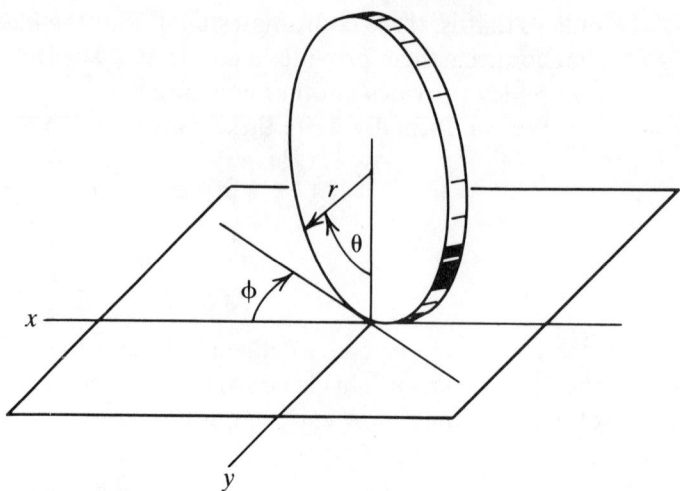

Figure 10.6.2 A rolling hoop.

would be necessary to specify the vertical locations of two masses and the orientation of a pulley. But the holonomic constraints of Eqs. 10.6.3 reduce this to only one. Thus, our Atwood's machine is completely specified by the single coordinate x.

Most constraints of interest in physics are holonomic, so most of our effort shall be directed there. But you may be wondering now how there could be restraints that do *not* reduce the number of degrees of freedom. An example of such a nonholonomic constraint is illustrated in Figure 10.6.2, which illustrates a hoop (or ring, disk, or wheel) rolling along a plane, remaining perpendicular to the plane. Coordinates x and y locate the point of contact; ϕ gives the orientation of the wheel and θ the rotation about its axis. As a constraint, let the wheel *roll* without slipping (even though ϕ is free to vary as the orientation changes about a vertical axis through the point of contact). Thus, the component of the velocity of the point of contact *perpendicular* to the plane of the wheel must vanish. Namely,

$$\dot{x} \sin \phi + \dot{y} \cos \phi = 0 \qquad (10.6.4)$$

Likewise, the component parallel to the plane of the wheel must be

$$\dot{x} \cos \phi - \dot{y} \sin \phi = r\dot{\theta} \qquad (10.6.5)$$

These provide constraints. But they cannot be integrated to provide algebraic equations relating the coordinates themselves. This means that all four coordinates are necessary to specify the system.

Yet they do not vary independently! How can we explain this apparent contradiction? Let the wheel be located at any point given by x and y. If you want a different value for the angle θ, roll the wheel about a circle of appropriate radius to bring it back to its original x, y position after rotating about its own axis to give the desired value for θ. ϕ is then changed at will.

10.7 Applications

We have applied Lagrange's equations to simple systems to both check their correctness and give us confidence in them. Now, let us proceed to apply them to more complex systems. The solution of **F** = m**a** for most of these systems is far more difficult.

EXAMPLE 10.7.1 Consider the *compound Atwood's machine* shown schematically in Figure 10.7.1. A body of mass m_1 is attached to a light, inextensible string. The string passes over a pulley of radius r_1 and moment of inertia I_1. The other end of this string is attached to another pulley of radius r_2 and moment of inertia I_2 and mass M. Bodies of mass m_2 and m_3 are attached to the ends of a light, inextensible string that passes over this pulley. There is no slipping between either string and pulley. This system is fully described by coordinates x_1 and x_2. Constants c_1 and c_2 are determined by the lengths of the two strings, the size of the mounts holding the pulley, and our choice for origins for x_1 and x_2. They shall appear only in the potential energy—and there as arbitrary constants.

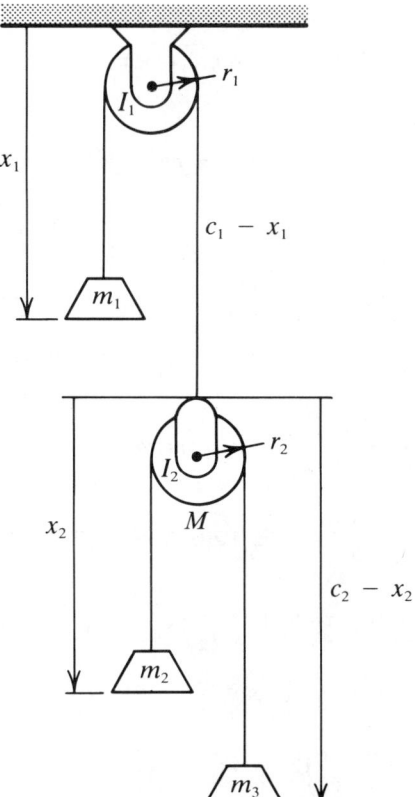

Figure 10.7.1 A *compound* Atwood's machine.

The kinetic energy is given by

$$T = \frac{1}{2}m_1\dot{x}_1^2 + \frac{1}{2}I_1\left(\frac{\dot{x}_1}{r_1}\right)^2 + \frac{1}{2}m_2(-\dot{x}_1 + \dot{x}_2)^2$$
$$+ \frac{1}{2}M\dot{x}_1^2 + \frac{1}{2}I_2\left(\frac{\dot{x}_2}{r_2}\right)^2 + \frac{1}{2}m_3(-\dot{x}_1 - \dot{x}_2)^2$$

Or, simplifying, this becomes

$$T = \frac{1}{2}\left(m_1 + m_2 + m_3 + M + \frac{I_1}{r_1^2}\right)\dot{x}_1^2 + \frac{1}{2}\left(m_2 + m_3 + \frac{I_2}{r_2^2}\right)\dot{x}_2^2$$
$$+ (m_3 - m_2)\dot{x}_1\dot{x}_2 \tag{10.7.1}$$

where the angular speed of the fixed pulley is $\omega_1 = \dot{x}_1/r_1$ and the angular speed of the moveable one is $\omega_2 = \dot{x}_2/r_2$. The moveable pulley is translating as well as rotating so it has translational kinetic energy of $\frac{1}{2}M\dot{x}_1^2$. Thus, $(-\dot{x}_1 + \dot{x}_2)$ and $(-\dot{x}_1 - \dot{x}_2)$ are the velocity of m_2 and m_3 relative to the fixed pulley.

The potential energy is given by

$$V = -m_1 g x_1 - m_2 g(c_1 - x_1 + x_2) - m_3 g(c_1 - x_1 + c_2 - x_2)$$
$$- Mg(c_1 - x_1) \tag{10.7.2}$$

since $V = -mgx$ for distances x measured *downward* as we have here. Simplifying, this gives

$$V = -[(m_1 - m_2 - m_3 - M)x_1 + (m_2 - m_3)x_2]g + C \tag{10.7.3}$$

where

$$C = -[(m_2 + m_3 + M)c_1 + m_3 c_2]g \tag{10.7.4}$$

is just an additive constant.
Thus

$$L = T - V$$

$$\frac{\partial L}{\partial x_1} = -\frac{\partial V}{\partial x_1} = (m_1 - m_2 - m_3 - M)g$$

$$\frac{\partial L}{\partial x_2} = -\frac{\partial V}{\partial x_2} = (m_2 - m_3)g$$

$$\frac{\partial L}{\partial \dot{x}_1} = \left(m_1 + m_2 + m_3 + M + \frac{I_1}{r_1^2}\right)\dot{x}_1 + (m_3 - m_2)\dot{x}_2$$

$$\frac{\partial L}{\partial \dot{x}_2} = \left(m_2 + m_3 + \frac{I_2}{r_2^2}\right)\dot{x}_2 + (m_3 - m_2)\dot{x}_1$$

Thus, Lagrange's equations give

$$\left(m_1 + m_2 + m_3 + M + \frac{I_1}{r_1^2}\right)\ddot{x}_1 + (m_3 - m_2)\ddot{x}_2 = (m_1 - m_2 - m_3 - M)g \tag{10.7.5}$$

and

$$\left(m_2 + m_3 + \frac{I_2}{r_2^2}\right)\ddot{x}_2 + (m_3 - m_2)\ddot{x}_1 = (m_2 - m_3)g \tag{10.7.6}$$

which form a simple set of simultaneous equations for $\ddot{x}_1$ and $\ddot{x}_2$. Further algebraic manipulation provides solutions for $\ddot{x}_1$ and $\ddot{x}_2$.

While these results could certainly have been obtained from a direct application of $\mathbf{F} = m\mathbf{a}$, there is far less pain involved when Lagrange's equations are used.

10.8 Ignorable Coordinates

If a coordinate q_k does not explicitly appear in the expression for kinetic and potential energy, then

$$\frac{\partial L}{\partial q_k} = 0 \tag{10.8.1}$$

Such a coordinate is understandably called an *ignorable coordinate*. This means, of course, that the other side of Lagrange's equation must also vanish. That is,

$$\frac{d}{dt}\frac{\partial L}{\partial \dot{q}_k} = 0 \tag{10.8.2}$$

Hence,

$$\frac{\partial L}{\partial \dot{q}_k} = \text{constant} \tag{10.8.3}$$

This quantity, then, is an *integral of the motion*. Any such constant quantity is of special significance, as we have earlier seen with energy, momentum, and angular momentum.

To see what this quantity is, let us now apply this to the Cartesian coordinates x, y, and z. The kinetic energy is given by

$$T = \tfrac{1}{2}(\dot{x}^2 + \dot{y}^2 + \dot{z}^2) \tag{10.8.4}$$

Suppose that the potential energy and, hence, the Lagrangian, are independent of, say, x. That is,

$$\frac{\partial L}{\partial x} = 0 \tag{10.8.5}$$

Therefore,

$$\frac{d}{dt}\frac{\partial L}{\partial \dot{x}} = \frac{d}{dt}(m\dot{x}) = 0 \tag{10.8.6}$$

or

$$\frac{\partial L}{\partial \dot{x}} = m\dot{x} = \text{constant} \tag{10.8.7}$$

And this is just the momentum. For the Lagrangian (and potential energy) to be independent of the coordinate x, there must be no force in the x direction. Hence, the momentum in the x direction will remain constant.

Thus, for a coordinate q_k, we shall define the corresponding generalized momentum,

$$P_k \equiv \frac{\partial L}{\partial \dot{q}_k} \tag{10.8.8}$$

And this generalized momentum P_k is a constant if q_k is an ignorable coordinate.

10.9 Lagrangian Mechanics and a Rotating Top

A rotating top or gyroscope is an excellent example of Lagrangian mechanics in action. The choice of coordinates used to describe the orientation of a body in space is neither trivial nor unique, and one set of coordinates that has proved quite useful was proposed by Euler; hence, they are known as the *Euler angles*. These angles are best described by starting with a gimbal-mounted gyroscope as shown in Figure 10.9.1. Such a gimbal mount is also known as *Cardan's suspension*.

Select a fixed laboratory frame XYZ as shown in Figure 10.9.1. The orientation of the gyroscope with respect to this frame can then be specified by three angles ϕ, θ, and ψ. Any orientation, then, is specified by the following:

1. A rotation of the outer gimbal through an angle ϕ about the axis AA'.
2. A rotation of the inner gimbal through an angle θ about the axis BB'.
3. A rotation of the gyroscope's rotor through an angle ψ about the axis CC'.

Figure 10.9.2 shows the Euler angles θ, ϕ, and ψ in a more conventional manner. The intersection of the $x_1 x_2$ plane with the XY plane is known as the *line of nodes*. The line of nodes is rotated on angle ϕ from the X-axis. You should realize that this notation and whether θ, ϕ, and ψ are measured clockwise or counterclockwise is *not* universal. The present notation and directions are common, but be careful when you read about the Euler angles in other books.

Angular *displacements* do not commute under addition. That is, a rotation of $\theta = \pi/2$ followed by one of $\phi = \pi/2$ is not the same as $\phi = \pi/2$

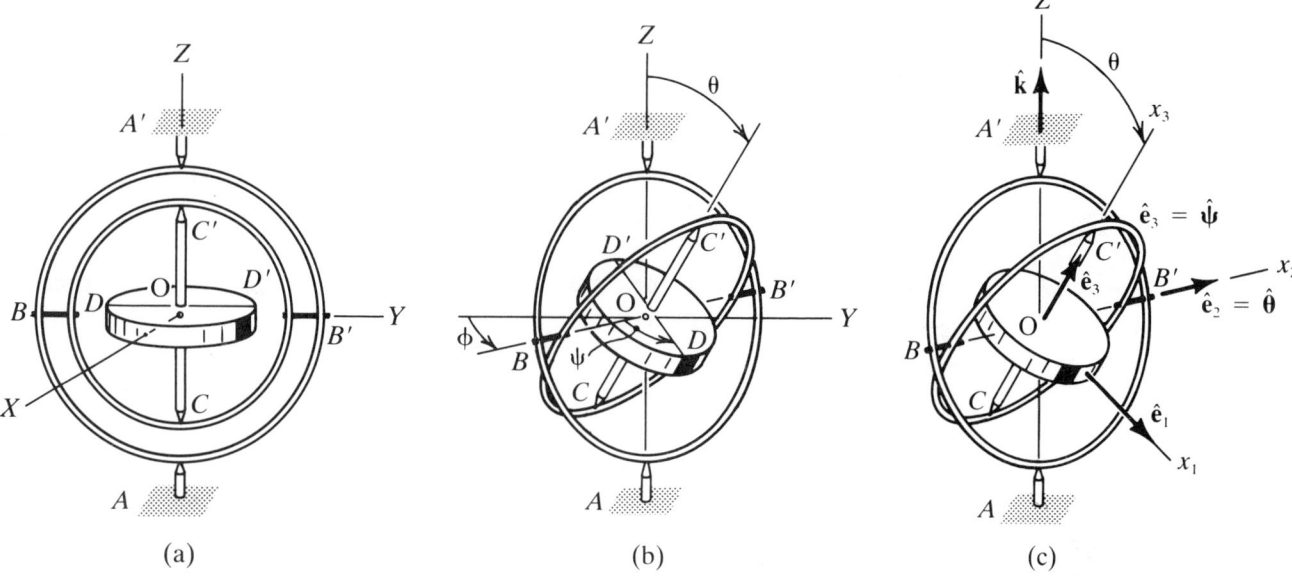

Figure 10.9.1 Gimbal-mounted gyroscope.

followed by $\theta = \pi/2$ (you should convince yourself of this by rotating a book or box through these rotations before continuing). Therefore, angular displacements cannot be treated as ordinary vectors. Nonetheless, angular *velocities* are ordinary vectors and, therefore, may be added as ordinary vectors. Proof of this is straightforward and provides a worthwhile reminder of earlier ideas about rotating coordinate systems developed in Chapter 5.

Consider a system $x_1 y_1 z_1$ rotating with angular velocity $\boldsymbol{\omega}_1$ relative to system $x_0 y_0 z_0$, and a system $x_2 y_2 z_2$ rotating with angular velocity $\boldsymbol{\omega}_2$ relative to system $x_1 y_1 z_1$. We shall show that $x_2 y_2 z_2$ rotates relative to $x_0 y_0 z_0$ with an angular velocity given by $\boldsymbol{\omega}_1 + \boldsymbol{\omega}_2$.

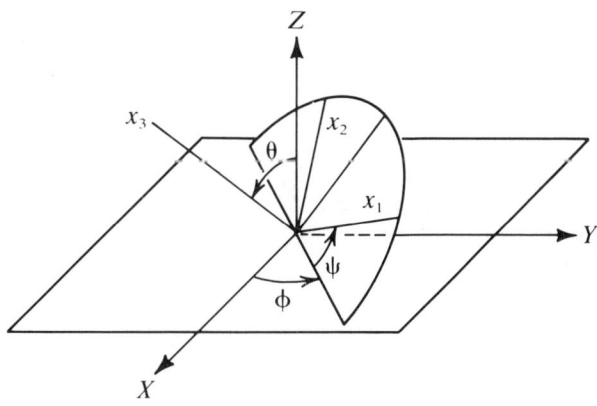

Figure 10.9.2 Euler's angles.

Let **A** be any vector at rest in x_2, y_2, z_2. That is,

$$\left.\frac{d\mathbf{A}}{dt}\right|_2 = 0 \tag{10.9.1}$$

where

$$\left.\frac{d\mathbf{A}}{dt}\right|_2$$

is the time derivative measured in $x_2 y_2 z_2$. Then, by Eq. 6.6.2, the time derivative measured by $x_1 y_1 z_1$,

$$\left.\frac{d\mathbf{A}}{dt}\right|_1$$

is given by

$$\left.\frac{d\mathbf{A}}{dt}\right|_1 = \boldsymbol{\omega}_2 \times \mathbf{A} \tag{10.9.2}$$

Applying Eq. 6.6.2 once again, we can find the time derivative measured by $x_0 y_0 z_0$,

$$\left.\frac{d\mathbf{A}}{dt}\right|_0$$

to be

$$\left.\frac{d\mathbf{A}}{dt}\right|_0 = \left.\frac{d\mathbf{A}}{dt}\right|_1 + \boldsymbol{\omega}_1 \times \mathbf{A}$$

$$= \boldsymbol{\omega}_2 \times \mathbf{A} + \boldsymbol{\omega}_1 \times \mathbf{A}$$

$$\left.\frac{d\mathbf{A}}{dt}\right|_0 = (\boldsymbol{\omega}_2 + \boldsymbol{\omega}_1) \times \mathbf{A} \tag{10.9.3}$$

Thus, $\boldsymbol{\omega}_1 + \boldsymbol{\omega}_2$ must be the angular velocity of $x_2 y_2 z_2$ with respect to $x_0 y_0 z_0$. Angular velocities add as ordinary vectors (because that's what they are).

Now consider the axes and unit vectors of Figures 10.9.1(c) and 10.9.2. XYZ define the fixed laboratory frame; $\hat{\mathbf{e}}_\theta, \hat{\mathbf{e}}_\phi, \hat{\mathbf{e}}_\psi$ define directions for rotations of the Euler angles; x_1, x_2, x_3 or $\hat{\mathbf{e}}_1 \hat{\mathbf{e}}_2, \hat{\mathbf{e}}_3$ define the *principal axes* of the rotor. The angular velocity of the rotor is given by

$$\boldsymbol{\omega} = \dot{\theta}\hat{\mathbf{e}}_\theta + \dot{\phi}\hat{\mathbf{e}}_\phi + \dot{\psi}\hat{\mathbf{e}}_\psi \tag{10.9.4}$$

These are convenient axes, but they are not orthogonal. We can change to the principal axes by writing

$$\hat{\mathbf{e}}_\theta = \hat{\mathbf{e}}_2$$
$$\hat{\mathbf{e}}_\phi = -\sin\hat{\mathbf{e}}_1 + \cos\theta\hat{\mathbf{e}}_3 \tag{10.9.5}$$
$$\hat{\mathbf{e}}_\psi = \hat{\mathbf{e}}_3$$

Then
$$\boldsymbol{\omega} = -\dot{\phi}\sin\theta\hat{\mathbf{e}}_1 + \dot{\theta}\hat{\mathbf{e}}_2 + (\dot{\psi} + \dot{\phi}\cos\theta)\hat{\mathbf{e}}_3 \qquad (10.9.6)$$

In terms of components along the principal axes, the angular velocity is
$$\boldsymbol{\omega} = \omega_1\hat{\mathbf{e}}_1 + \omega_2\hat{\mathbf{e}}_2 + \omega_3\hat{\mathbf{e}}_3 \qquad (10.9.7)$$

Thus,
$$\begin{aligned}\omega_1 &= -\dot{\phi}\sin\theta \\ \omega_2 &= \dot{\theta} \\ \omega_3 &= \dot{\psi} + \dot{\phi}\cos\theta\end{aligned} \qquad (10.9.8)$$

The kinetic energy of a rotating body, given by Eq. 9.8.7, can also be written as
$$T = \tfrac{1}{2}I_1\omega_1^2 + \tfrac{1}{2}I_2\omega_2^2 + \tfrac{1}{2}I_3\omega_3^2 \qquad (10.9.9)$$

where I_1, I_2, and I_3 are the principal moments of inertia and ω_1, ω_2, and ω_3 are the components of the angular velocity in the direction of the principal axes. For a *symmetric* rotor (as we would expect for any typical top or gyroscope), two of the principal moments of inertia are equal. We shall set

$$I_1 = I_2 = I \qquad (10.9.10)$$

and I_3 will be the moment of inertia about the symmetry axis. For this case, the kinetic energy can easily be written with aid of Eq. 10.9.8 as

$$T = \tfrac{1}{2}I\dot{\theta}^2 + I\dot{\phi}^2\sin^2\theta + \tfrac{1}{2}I_3(\dot{\psi} + \dot{\phi}\cos\theta)^2 \qquad (10.9.11)$$

For such a symmetric top or gyroscope that moves about some fixed point of support as shown in Figure 10.9.3, the potential energy can be written as

$$V = mgl\cos\theta \qquad (10.9.12)$$

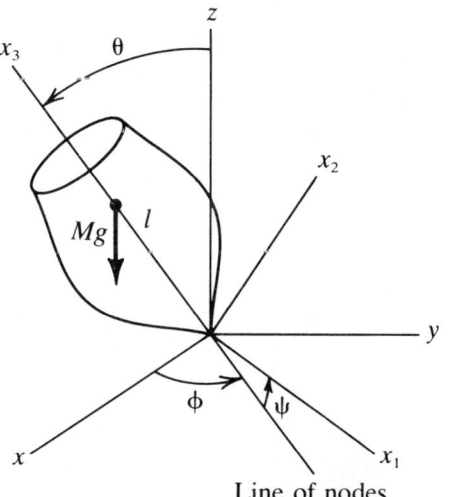

Figure 10.9.3 A spinning top.

where l is the distance between the center of mass and the point of support. Then the Lagrangian may be written as

$$L = \tfrac{1}{2}I\dot{\theta}^2 + \tfrac{1}{2}I\dot{\phi}^2 \sin^2\theta + \tfrac{1}{2}I_3(\dot{\psi} + \dot{\phi}\cos\theta)^2 - mgl\cos\theta \quad (10.9.13)$$

Since the coordinates ψ and ϕ do not appear in the Lagrangian (that is, ψ and ϕ are ignorable coordinates), their corresponding generalized momenta P_ψ and P_ϕ are constants of the motion. Furthermore, the total energy is conserved. That is,

$$\frac{dP_\psi}{dt} = \frac{\partial L}{\partial \psi} = 0$$

$$\frac{dP_\phi}{dt} = \frac{\partial L}{\partial \phi} = 0 \quad (10.9.14)$$

$$\frac{dE}{dt} = -\frac{\partial L}{\partial t} = 0$$

Therefore,

$$P_\psi = \frac{\partial L}{\partial \dot{\psi}} = I_3(\dot{\psi} + \dot{\phi}\cos\theta) \quad (10.9.15)$$

$$P_\phi = \frac{\partial L}{\partial \dot{\phi}} = I\dot{\phi}\sin^2\theta + I_3\cos\theta(\dot{\psi} + \dot{\phi}\cos\theta) \quad (10.9.16)$$

$$E = T + V = \tfrac{1}{2}I\dot{\theta}^2 + \tfrac{1}{2}\dot{\phi}^2\sin^2\theta$$
$$+ \tfrac{1}{2}I_3(\dot{\psi} + \dot{\phi}\cos\theta)^2 + mgl\cos\theta \quad (10.9.17)$$

all remain constant. These equations provide three coupled differential equations for the three angular velocities $\dot{\theta}$, $\dot{\phi}$, and $\dot{\psi}$, so a solution exists—in principle. Detailed solutions for general cases are best done by numerical integration techniques. Nonetheless, we can understand the characteristics of the motion without a complete analytical solution.

Equations 10.9.15 and 10.9.16 for P_ψ and P_ϕ do not contain $\dot{\theta}$ so they may be solved first for

$$\dot{\phi} = \frac{P_\phi - P_\psi \cos\theta}{I\sin^2\theta} \quad (10.9.18)$$

and

$$\dot{\psi} = \frac{P_\psi}{I_3} - \frac{P_\phi - P_\psi \cos\theta}{I\sin^2\theta}\cos\theta \quad (10.9.19)$$

With Eq. 10.9.17 for the energy E, these then yield

$$E = \tfrac{1}{2}I\dot{\theta}^2 + \frac{(P_\phi - P_\psi\cos\theta)^2}{2I\sin^2\theta} + \frac{P_\psi^2}{2I_3} + mgl\cos\theta \quad (10.9.20)$$

This expression may be reduced by one term if the reference point for the potential energy is reduced by defining another "energy" E' given by

$$E' = E - \frac{P_\psi^2}{2I_3} \tag{10.9.21}$$

Solving this equation gives the time dependence of θ, the angle between the symmetry axis and the z-axis in the laboratory frame. In general this solution will require numerical methods or even more elusive elliptical integrals. But this equation *looks* like an effective kinetic energy term of

$$T' = \tfrac{1}{2}I\dot\theta^2 \tag{10.9.22}$$

and an effective potential energy term of

$$V' = \frac{(P_\phi - P_\psi \cos\theta)^2}{2I \sin^2\theta} + mgl \cos\theta \tag{10.9.23}$$

So the early methods developed in Chapter 2 for solving one-dimensional motion are immediately applicable here. Thus, Eq. 2.6.16 immediately becomes

$$\dot\theta = \sqrt{\frac{2}{I}(E' - V')} \tag{10.9.24}$$

which can, in principle, be solved for

$$t = t(\theta) = \sqrt{\frac{I}{2}} \int_{\theta_0}^{\theta} \frac{d\theta}{\sqrt{E' - V'}} \tag{10.9.25}$$

which can then be solved to yield $\theta = \theta(t)$. Just as in the initial discussion of energy in the case of ordinary one-dimensional motion, it will usually be very difficult to explicitly solve for $\theta = \theta(t)$, but much information about the motion can be observed and understood nonetheless.

Figure 10.9.4 shows a graph of effective potential energy $V'(\theta)$ as a function of θ. Because of the sine function in the denominator, V' becomes

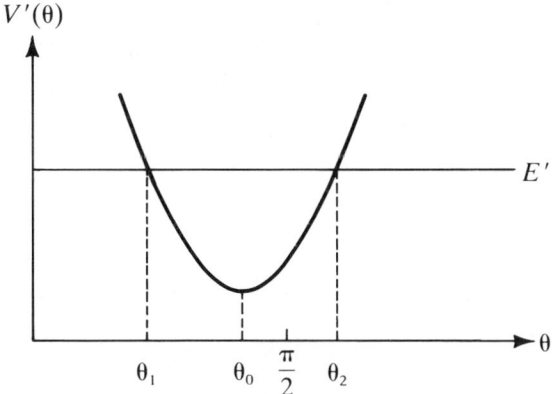

Figure 10.9.4 Effective "potential" for nutating top.

infinite for $\theta = 0$ or π, so there must be a minimum somewhere between these two. The location of this minimum, θ_0, can be found by solving

$$\left.\frac{dV'}{d\theta}\right|_{\theta_0} = -mgl \sin\theta_0 - \frac{(P_\phi - P_\psi \cos\theta_0)(P_\psi - P_\phi \cos\theta_0)}{I \sin^3 \theta_0}$$

$$= 0 \qquad (10.9.26)$$

In general, θ will move back and forth between two extremes θ_1 and θ_2, as shown in Figure 10.9.3. This variation in θ is referred to as *nutation*. At θ_1 and θ_2, the effective kinetic energy is zero or

$$E' = V'(\theta_1) = V'(\theta_2) \qquad (10.9.27)$$

If, however, the effective energy is such that $E' = V'(\theta_0)$, θ remains constant at θ_0 and the top processes with constant angular velocity:

$$\dot\phi_0 = \frac{P_\phi - P_\psi \cos\theta_0}{I \sin^2 \theta_0} \qquad (10.9.28)$$

from Eq. 10.9.18.

10.10 Hamilton's Equations

Let's go back to Eq. 10.4.13, Lagrange's equation,

$$\frac{d}{dt}\frac{\partial L}{\partial \dot q_k} = \frac{\partial L}{\partial q_k} \qquad (10.4.13)$$

which describes the motion of a particle or system as fully and completely as Newton's Second Law, $\mathbf{F} = m\mathbf{a}$. Equation 10.8.8 defined a generalized momentum,

$$P_k = \frac{\partial L}{\partial \dot q_k} \qquad (10.8.8)$$

We shall now write equations of motion—containing the *same information* as Newton's Second Law or Lagrange's equation—in terms of the momenta P_k and coordinates q_k. Direct application of the chain rule results in

$$dL = \sum_k \left(\frac{\partial L}{\partial \dot q_k} d\dot q_k + \frac{\partial L}{\partial q_k} dq_k\right) + \frac{\partial L}{\partial t} dt \qquad (10.10.1)$$

Using Eq. 10.8.8, this easily becomes

$$dL = \sum_k (P_k \, d\dot q_k + \dot P_k \, dq_k) + \frac{\partial L}{\partial t} dt \qquad (10.10.2)$$

We now define a function H by

$$H \equiv \sum_k P_k \dot{q}_k - L \tag{10.10.3}$$

Then

$$dH = \sum_k (P_k \, d\dot{q}_k + \dot{q}_k \, dP_k) - dL$$

$$= \sum_k (P_k \, d\dot{q}_k + \dot{q}_k \, dP_k - P_k \, d\dot{q}_k - \dot{P}_k \, dq_k) - \frac{\partial L}{\partial t} dt$$

$$dH = \sum_k (\dot{q}_k \, dP_k - \dot{P}_k \, dq_k) - \frac{\partial L}{\partial t} dt \tag{10.10.4}$$

By inspection, then, we can see that

$$\dot{q}_k = \frac{\partial H}{\partial P_k} \tag{10.10.5}$$

and

$$\dot{P}_k = -\frac{\partial H}{\partial q_k} \tag{10.10.6}$$

and

$$\frac{\partial H}{\partial t} = -\frac{\partial L}{\partial t} \tag{10.10.7}$$

Equations 10.10.5 and 10.10.6 are known as *Hamilton's equations*. They are the equations of motion we sought in terms of coordinates q_k and momenta P_k. We now have twice as many equations as before, but each one is a first-order instead of second-order differential equation. The function H is called the *Hamiltonian* of the system. For time-dependent potential functions V, H is just the total energy. From Eq. 10.10.7, the total energy (or Hamiltonian) is a constant of the motion for that case.

As with the formulation of Lagrange's equations, this new formulation of mechanics in terms of Hamilton's equations contains no new information. It is just another form of the same information contained in the more familiar Laws of Motion formulated by Newton. They must reduce to the same equations. To see this, let us apply Hamilton's equations to the simple case of a simple harmonic oscillator. For that,

$$H = E = \frac{P^2}{2m} + \frac{kx^2}{2} \tag{10.10.8}$$

Hamilton's equations 10.10.5 and 10.10.6 yield

$$\dot{x} = \frac{P}{m} \tag{10.10.9}$$

and
$$\dot{P} = -kx \qquad (10.10.10)$$

Of course, the former is just a restatement of the definition of momentum. And the latter is but Newton's Second Law.

Although it seems that little is gained by using Hamilton's equations, they prove quite useful in statistical mechanics and quantum mechanics. Further, like Lagrange's equations, these are written in terms of *generalized* coordinates (and momenta), which often prove useful.

10.11 Hamilton's Principle

Newton's Law—like any basic law of physics—may be viewed as a postulate or principle. Its acceptance rests upon experimental verification. That is always the nature of physical laws. Similarly, Hamilton proposed a variational principle, which may be paraphrased as:

The actual motion followed by a mechanical system as it moves from a starting point to a final point within a given time interval will be the motion that provides an extremum for the time integral of the Lagrangian.

Mathematically, that awkward-sounding principle can be written as

$$\delta \left[\int_{t_1}^{t_2} L \, dt \right] = 0 \qquad (10.11.1)$$

where δ indicates a small variation in the motion. An "extremum" can be a minimum, maximum, or point of inflection. The "motion" of a mechanical system can be described by the path of a "system particle" as it moves through configuration space. The configuration of a system at any time is defined by giving the values of all n generalized coordinates $q_1, \ldots, q_n$. This corresponds to a particular point in an n-dimensional hyperspace in which the q_k's are the components measured along the n coordinate axes. This is known as configuration space. Time is a parameter of the path through configuration space. It should be made clear that the path of motion in configuration space need not, in general, have any resemblance to the "real" path of any particle within the system; each point on the path in configuration space represents the complete system at that particular time.

Hamilton's Variational Principle can be visualized graphically in Figures 10.11.1 and 10.11.2. The motion begins at time t_1 with a certain value of the Lagrangian defined by the particular values of the generalized coordinates

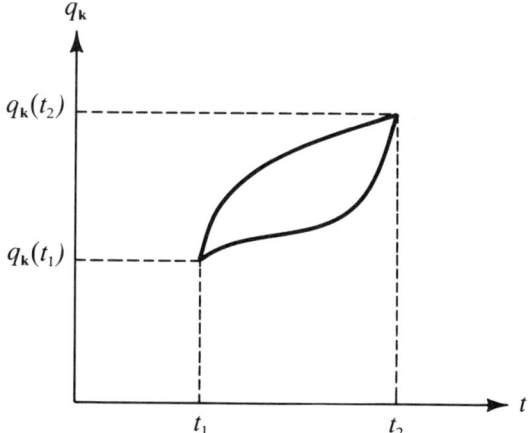

Figure 10.11.1
Variation in motion.

$q_1(t_1), \ldots, q_n(t_1)$ and $\dot{q}_1(t_1), \ldots, \dot{q}_n(t_1)$. At some later time t_2, the generalized coordinates and velocities have particular values of $q_1(t_2), \ldots, q_n(t_2)$ and $\dot{q}_1(t_1), \ldots, \dot{q}_n(t_2)$. Hamilton's Principle states that the motion between these two configurations will be that which causes the integral

$$\int_{t_1}^{t_2} L \, dt$$

to be an extremum. Figure 10.11.1 shows two possible paths for variations in a single coordinate q_k. This is essentially a "two-dimensional projection" from the n-dimensional path in configuration space. Figure 10.11.2 shows three possible paths from t_1 to t_2 that could be taken by a single point particle constrained to move in the xy plane. For such a particle, the xy plane is its configuration space.

We shall now show that Hamilton's Variational Principle leads to Lagrange's equation and, therefore, is equivalent to the other formulations of

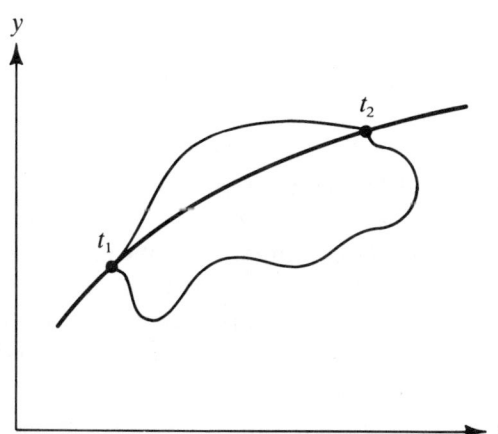

Figure 10.11.2 Path of a "system point" in a two-dimensional configuration space.

the Laws of Motion. To do this, we can calculate the variation directly by writing the Lagrangian explicitly as a function of the q_k's and $\dot{q}_k$'s. Thus,

$$\delta\left[\int_{t_1}^{t_2} L \, dt\right] = \delta\left[\int_{t_1}^{t_2} L(q_k, \dot{q}_k) \, dt\right]$$

$$= \int_{t_1}^{t_2} \delta L(q_k, \dot{q}_k) \, dt$$

$$= \int_{t_1}^{t_2} \sum_k \left(\frac{\partial L}{\partial q_k} \delta q_k + \frac{\partial L}{\partial \dot{q}_k} \delta \dot{q}_k\right) dt = 0 \qquad (10.11.2)$$

But

$$\delta \dot{q}_k = \frac{d}{dt} \delta q_k \qquad (10.11.3)$$

so the last term can be integrated by parts to give

$$\int_{t_1}^{t_2} \frac{\partial L}{\partial \dot{q}_k} \delta \dot{q} \, dt = \int_{t_1}^{t_2} \frac{\partial L}{\partial \dot{q}_k} d(\delta q_k)$$

$$= \left[\frac{\partial L}{\partial \dot{q}_k} \delta q_k\right]_{t_1}^{t_2} - \int_{t_1}^{t_2} \delta q_k \left(\frac{d}{dt} \frac{\partial L}{\partial \dot{q}_k}\right) dt \qquad (10.11.4)$$

The first term on the right must vanish since $\delta q_k = 0$ at both t_1 and t_2 because of the limits to the motion at the start and finish. Thus,

$$\delta\left[\int_{t_1}^{t_2} L \, dt\right] = \int_{t_1}^{t_2} \sum_k \left[\frac{\partial L}{\partial q_k} - \frac{d}{dt} \frac{\partial L}{\partial \dot{q}_k}\right] \delta q_k \, dt = 0 \qquad (10.11.5)$$

The q_k's are independent variables so their variations δq_k are certainly independent. Therefore, each term in the summation must vanish. That is,

$$\frac{\partial L}{\partial q_k} - \frac{d}{dt} \frac{\partial L}{\partial \dot{q}_k} = 0 \qquad (10.11.6)$$

And those are Lagrange's equations!

It is interesting to recount briefly some of the history or background of minimum or variational principles. In the second century B.C., Hero of Alexandria found that the reflection of light could be explained in terms of a minimum principle; namely, a ray of light traveling from one point to another via reflection from a plane mirror always takes the *shortest* path, as shown in Figure 10.11.3. Hero's minimum principle can be extended to a variational or extremum principle. For reflection by an elliptical mirror from one focus to another, the length of the path doesn't change; the length is an inflection point. This is illustrated in Figure 10.11.4. For reflection from a spherical mirror, the length of the path that the ray actually takes is a maximum! This is illustrated in Figure 10.11.5.

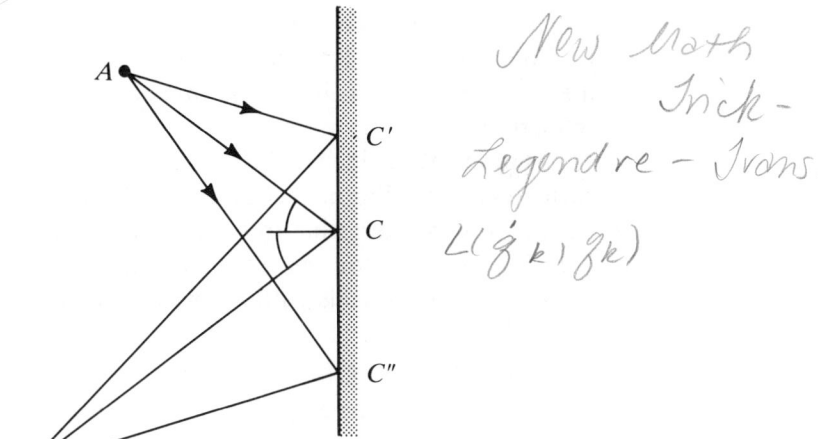

*New Math Trick —
Legendre — Transformation
$L(\dot{q}_k, q_k)$*

Figure 10.11.3 Minimum path length for reflection from a plane mirror. Path ACB is shorter than either $AC'B$ or $AC''B$.

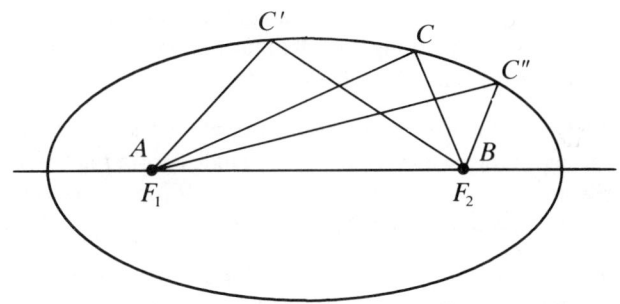

Figure 10.11.4 For an elliptical mirror, the path lengths for all rays leaving one focal point and arriving at the other are identical. $ACB = AC'B = AC''B$.

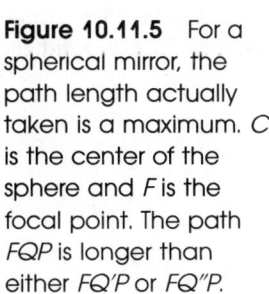

Figure 10.11.5 For a spherical mirror, the path length actually taken is a maximum. C is the center of the sphere and F is the focal point. The path FQP is longer than either $FQ'P$ or $FQ''P$.

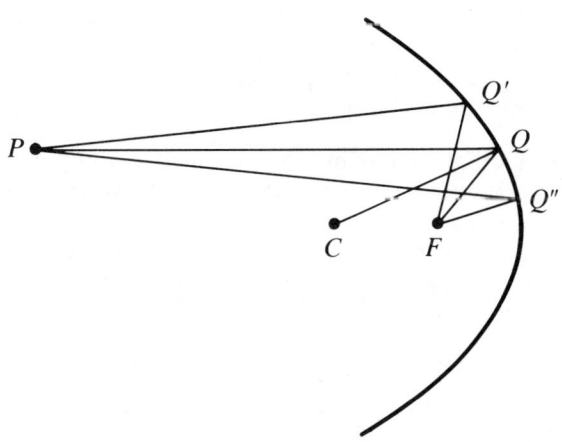

In the seventeenth century, Fermat reformulated this idea of path *length* to a requirement of least *time*. When applied to refraction through various media, Fermat's Principle leads immediately to Snell's Law for the angle of refraction.

Hamilton's Principle, which describes mechanics, is an extension of earlier variational principles due to Hamilton's own work in optics and, thus, waves. So it is interesting to see the extension carried full circle. Hamilton's Principle and Hamilton's equations are of vital importance in quantum mechanics, which deals with the wave nature of matter.

PROBLEMS

10.1 Use Lagrange's equations to find the motion of an object dropped from rest. Extend this to finding the motion of a projectile. Neglect air resistance.

10.2 Use Lagrange's equations to find the motion of a uniform disk rolling down an inclined plane without slipping.

10.3 Write Lagrange's equations for a weighted bicycle wheel ($I = mr^2$) that rolls without slipping along a horizontal surface while acted upon by a horizontal force F acting at the axis in the plane of the wheel.

10.4 A block of mass m slides down a frictionless wedge of mass M. The upper surface of the wedge is inclined an angle θ from the horizontal, and the bottom surface sits on a horizontal, frictionless surface. Find the acceleration of the block and wedge using Lagrange's equations of motion.

10.5 Find expressions for the kinetic energy and the potential energy of a simple pendulum in terms of each of the following used as a generalized coordinate:
  (a) Angular displacement.
  (b) Horizontal displacement.
  (c) Vertical displacement.

10.6 Use Lagrange's equations to write the equation of motion for a simple pendulum using each of the following as a generalized coordinate:
  (a) Angular displacement.
  (b) Horizontal displacement.
  (c) Vertical displacement.

10.7 A uniform bar of mass m and length l is hinged at one end and held horizontal by a spring attached to the other end. The spring has spring constant k. Find the

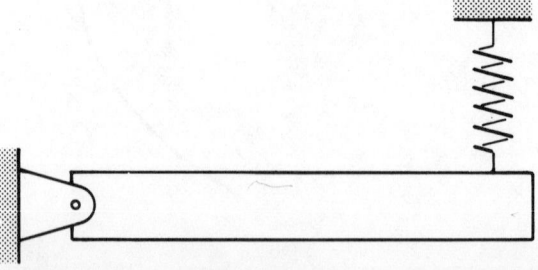

stretch of the spring when the bar is horizontal, in equilibrium. Write the Lagrangian of the system and find the frequency of small oscillations about this equilibrium.

10.8 A uniform rod of length l and mass m is supported by a pivot located a distance h from one end. It is free to swing in a vertical plane. Ignore friction and solve for the motion.

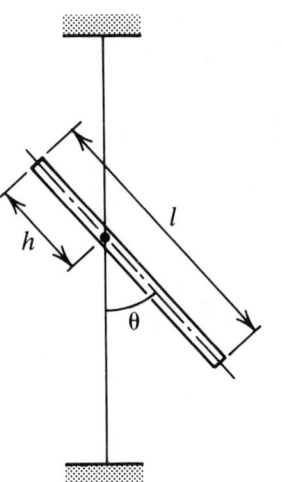

10.9 Use Lagrange's equations to write the equation of motion for a ladder that is set at rest against a frictionless floor and wall.

10.10 Use Lagrange's equations to solve for the motion of a door that has its axis of rotation tilted slightly from vertical. Find the period for small oscillations in terms of the mass and width of the door and its angle from vertical.

10.11 A block of mass m sits on a horizontal, frictionless table. It is attached by a massless string to another block of mass M. The string passes over a pulley of moment of inertia I. Use Lagrange's equations to solve for the motion of this system.

10.12 Replace the massless string in the previous problem with a string of mass μ and length l. Mass M is initially a distance l_0 below the pulley. Find the speed of mass m when it reaches the pulley.

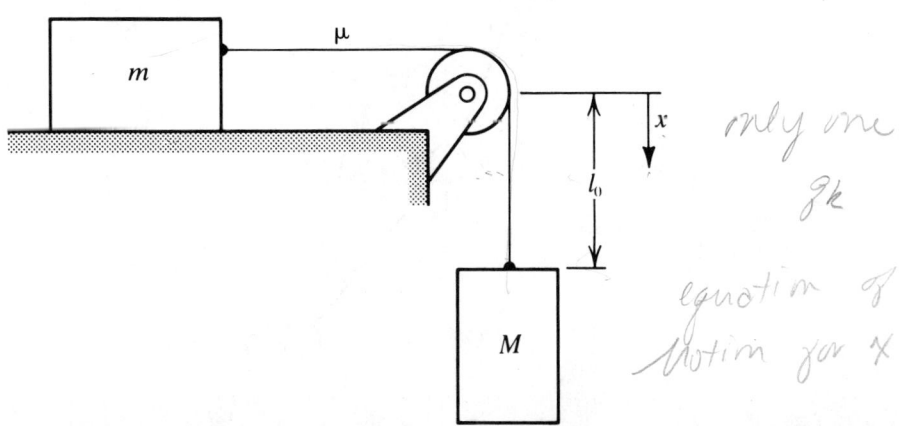

10.13 Use Lagrange's equations to solve for the motion of a simple Atwood's machine consisting of masses m_1 and m_2 attached to a lightweight string hung over a lightweight, frictionless pulley.

10.14 Consider a double-compound Atwood's machine. Objects of mass m_1 and m_2 are attached to the ends of a string passing over pulley P_1. Objects of mass m_3 and m_4 are attached to the ends of a string passing over pulley P_2. Pulleys P_1 and P_2 are attached to the ends of a string passing over a fixed pulley, P_3. All strings are inextensible and all pulleys are of negligible mass. Find the accelerations of the masses.

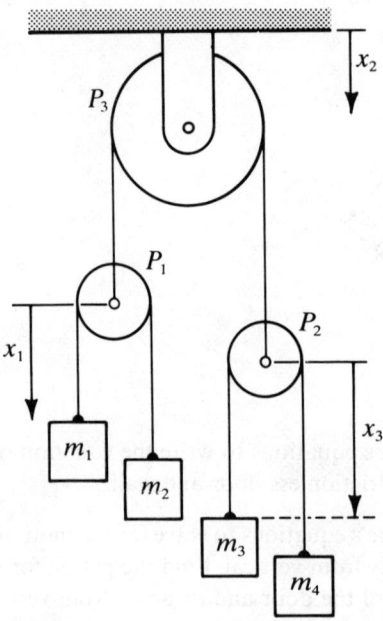

10.15 A double pendulum is constructed of two simple pendulums swinging in the same plane. A cord of length l is attached to a small object of mass m, making a simple pendulum. Another cord of length l hangs from this mass and is attached to another

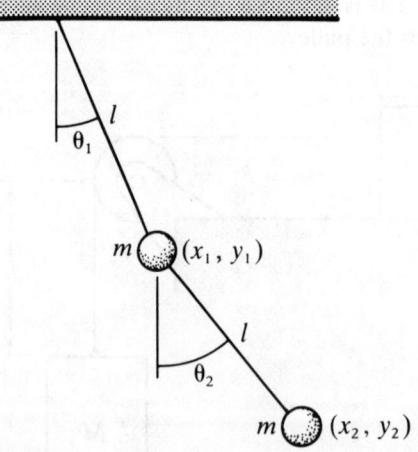

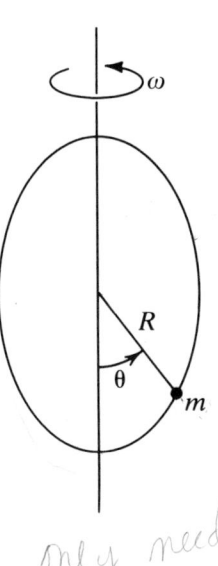

small object of mass m. Write Lagrange's equations of motion for this system. Solve for the motion using the approximation of small oscillations. (This is an interesting problem *to look at* for unequal masses on strings of unequal length.)

10.16 A bead of mass m is free to slide on a circular hoop of radius R. The hoop rotates about a vertical diameter with constant angular velocity ω. Write the Lagrange equation for this system.

10.17 Two identical masses are connected by a spring of spring constant k. An identical spring attaches one mass to a fixed wall. The system is free to move on a frictionless, horizontal surface. Use Lagrange's equations to write the equation of motion for each mass.

10.18 Write the Lagrange equations of motion using polar coordinates r and θ. Express the force in its polar components F_r and F_θ.

10.19 Write the Hamiltonian function for a particle of mass m moving in one dimension in a region with potential energy function V. Show that Hamilton's equations of motion reduce to Newton's Second Law.

10.20 Consider a particle constrained to move on a cylindrical surface

$$x^2 + y^2 + z^2 = R^2$$

that experiences an attractive, linear, central force,

$$F = -kr = -k(x\hat{i} + y\hat{j} + z\hat{k})$$

Write the Hamiltonian for this particle and find the equations of motion.

10.21 For any two functions of (generalized) coordinates and momenta, *Poisson brackets* can be defined by

$$\{g, h\} \equiv \sum_k \left(\frac{\partial g}{\partial q_k}\frac{\partial h}{\partial p_k} - \frac{\partial g}{\partial p_k}\frac{\partial h}{\partial q_k}\right)$$

for $g = g(q_k, p_k)$ and $h = h(q_k, p_k)$. Verify the following:

(a) $\dfrac{dg}{dt} = \{g, H\} + \dfrac{\partial g}{\partial t}$

(b) $\dot{q}_1 = \{q_1, H\}; \quad \dot{p}_1 = \{p_1, H\}$

(c) $\{p_k, p_1\} = \{q_k, q_1\} = 0$

(d) $\{q_k, p_1\} = \delta_{kl} = \begin{cases} 0 & \text{for } k \neq 1 \\ 1 & \text{for } k = 1 \end{cases}$

H is the Hamiltonian function. Poisson brackets are of great use and assistance in understanding the transition from classical mechanics to quantum mechanics. [For more information on this transition using Poisson brackets, see *Nonrelativistic Mechanics* by Robert J. Finklestein (Menlo Park, Calif.: The Benjamin/Cummings Publishing Company, 1973).]

11
STATICS

11.1 Introduction

Statics is the study of forces on (and within) a body when it is in equilibrium. Specifically, we shall look in detail at bodies that are at rest. Of course another inertial observer moving with constant velocity will see this body not at rest but moving with constant velocity. That's all right. Both observers will agree that the body is in equilibrium—the net force and the net torque are both zero.

Statics is but the careful application of Newton's First Law—or Newton's Second Law ($\mathbf{F} = m\mathbf{a}$) when the acceleration is zero. Physicists usually pay more attention to accelerated motion simply because it is more dynamic. A body just sitting still is, after all, rather static (and relatively uninteresting). But statics is an important application of many of the physical principles we've already used. And mechanical and civil engineers spend a sizable portion of their careers using the ideas of statics.

This short chapter will introduce you to a few of the ideas and techniques you can use in statics. It is by no means exhaustive.

11.2 Plane Trusses

Until now we have been concerned with the forces acting *on* a body. Often we were able to consider the forces acting on a *point* particle. But even for an extended body, we were concerned with the external forces acting on it. Now our attention will be focused on the forces *within* a body—on the *internal* forces. We shall consider the equilibrium of structures made of several connected pieces and inquire about the forces exerted by one piece on others.

Because structures can be very complex—like the framing of a building, for example—we shall restrict our attention to *plane structures*, in particular, *plane trusses*. Briefly put, a truss is a structure composed of straight members joined together by *pins* at their *ends*. Further, any loads will be applied only to the pins at the joints.

Figure 11.2.1 shows a simple structure that might be used to support a mailbox. It isn't a truss. Pins may well hold it together at points A, B, and C, and the mailbox exerts a force downward at point D. But the horizontal member CBD does not have forces exerted only at its ends. If no forces are exerted at E, we might ignore the portion CE. But forces (and torques!) are exerted at point G, where the post is driven into the ground.

All right, if that isn't a truss, then what is a truss? Figure 11.2.2(a) shows a simple truss. Note that members AB and BC are separate, distinct pieces held

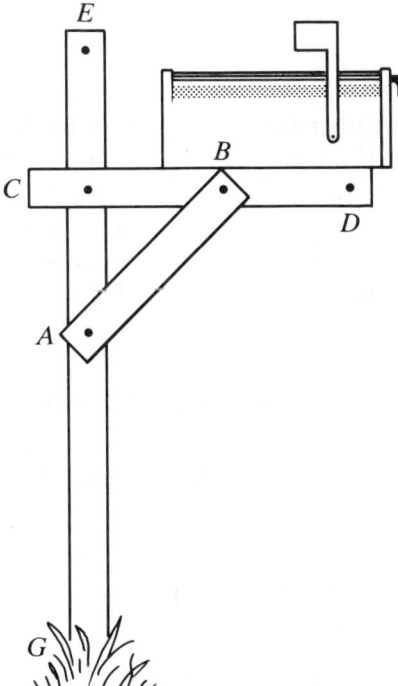

Figure 11.2.1 A simple structure to hold a mailbox.

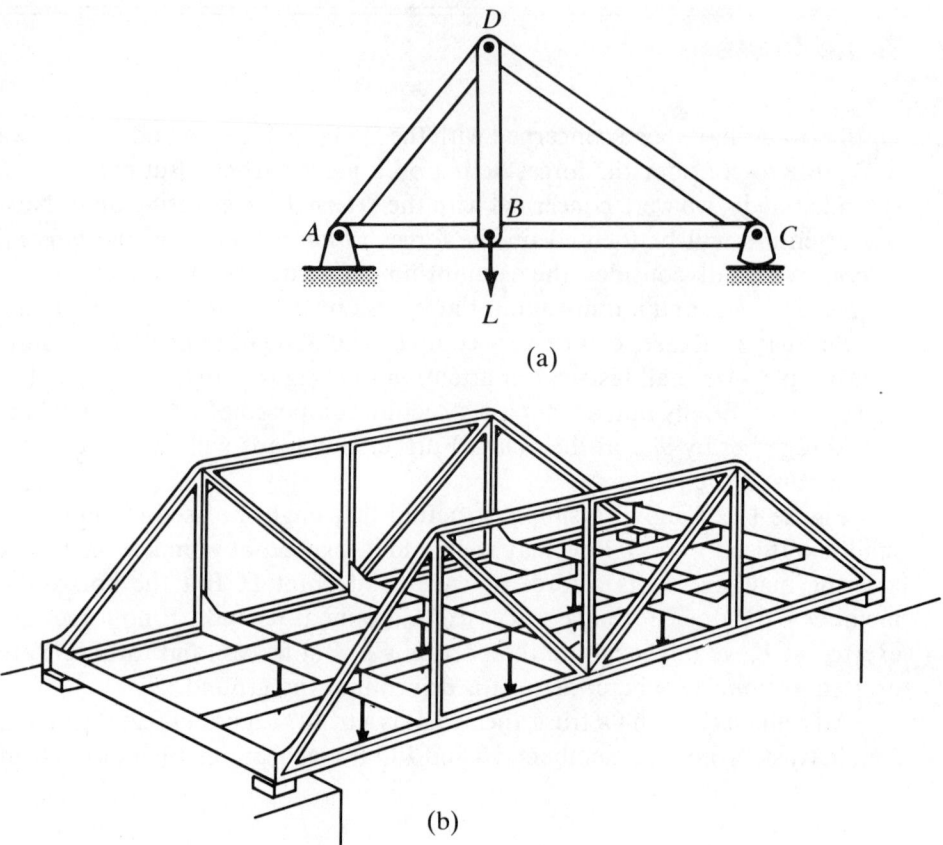

Figure 11.2.2 (a) A simple truss; (b) Simple trusses are common in bridges.

together (and to member DB) by a pin at B. Before we look at the truss, look at its supports at A and C. The support at A is fixed—it can exert both a horizontal and a vertical force on the truss. But the support at C is a rocker—it can exert only a vertical force.

There are two conditions of equilibrium. First, the vector sum of all the forces acting on a body must be zero. That is,

$$\sum \mathbf{F} = 0 \qquad (11.2.1)$$

For a plane truss, all our forces will also be in this same plane so we may write this as

$$\sum F_x = 0$$
$$\sum F_y = 0 \qquad (11.2.2)$$

The second condition is that the sum of all the torques must vanish. This, too, is really a vector equation:

$$\sum \tau = 0 \qquad (11.2.3)$$

SECTION 11.2 / PLANE TRUSSES

But since $\tau = \mathbf{r} \times \mathbf{F}$, for a plane truss and forces confined to this same plane, we can simplify this to the algebraic equation

$$\sum \tau = 0 \tag{11.2.4}$$

where we consider counterclockwise torques as positive and clockwise torques as negative. Or, we may write this as

$$\sum \tau_{\circlearrowleft} = \sum \tau_{\circlearrowright} \tag{11.2.4a}$$

Of course, the torques due to all the forces must be calculated with respect to a single origin, but we are free to choose any point whatsoever to be that origin.

Consider the truss in Figure 11.2.2(a) as a whole. There is a downward load **L** as shown. Forces—as yet undetermined—are exerted by supports at A and C. Our two conditions of equilibrium

$$\sum F_x = 0$$
$$\sum F_y = 0 \tag{11.2.2}$$

and

$$\sum \tau = 0 \tag{11.2.4}$$

provide three equations from which we can determine only three unknowns. Thus we can determine only three components to describe the forces at A and C. As drawn, the force at A can have both vertical and horizontal components while the force at C will have only a vertical component.

What if you built a bridge like that in Figure 11.2.2 (b) *but* with a firm support like that at A supporting both ends? Then the forces at A and C—and throughout the truss—would depend upon the actual construction. Are the supports at A and C pushing the ends together or pulling them apart? The construction is *indeterminate*, and such a truss is said to be *improperly constrained*. Because there are more unknowns than equations, we shall avoid indeterminate situations.

Now that we understand the *support* of a truss, we shall take a closer look at trusses themselves. The simplest stable truss is three members arranged in a triangle, as shown in Figure 11.2.3. Also shown there is an open rectangular

Figure 11.2.3 A three-membered, triangular truss is stable but a four-membered truss collapses.

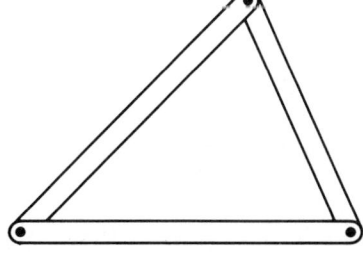
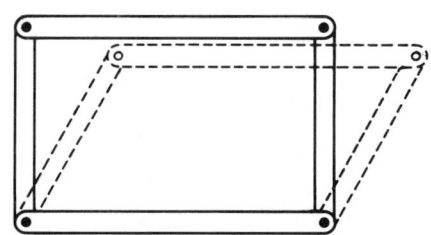

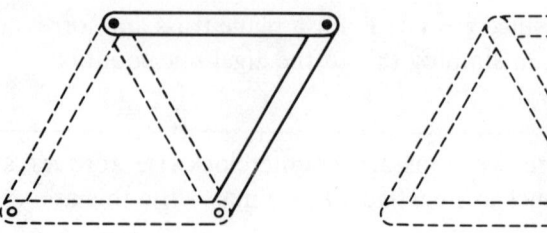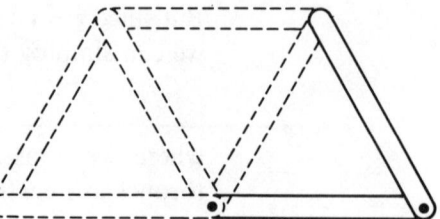

Figure 11.2.4 Two members and one joint added onto the simple triangle.

frame that collapses readily. Trusses that are constructed of a basic, triangular truss, with two members and one joint added, are known as *simple trusses*. Two more members and one joint can then be added again and again, as sketched in Figure 11.2.4. This relationship can easily be expressed as

$$m + 3 = 2j \tag{11.2.5}$$

where m is the number of members and j is the number of joints. If $m + 3 > 2j$, then some of the members are *redundant* and the forces in all the members cannot be determined. If $m + 3 < 2j$, the structure will collapse, as with the rectangular frame of Figure 11.2.3. Even for more complicated, *compound trusses*, ones that cannot be built up from a triangular truss by adding two members and a pin successively, Eq. 11.2.5 still obtains.

In the trusses we shall analyze, the members are held together with pins in their ends. This means that the forces exerted on a member must lie along the axis of the member as shown in Figure 11.2.5. If the pins exert forces that try to pull the two ends apart, we say the member is under *tension*, or that it experiences a *tensile force*. Likewise, if the pins exert forces that try to push the two ends together, we say the member is under *compression*, or that it experiences a *compressive force*. These are the only two cases we shall encounter. They will be labeled T and C, respectively.

To find the *reaction* forces—the forces exerted on the truss by its supports—we treat the truss as a unit. Consider the truss of Figure

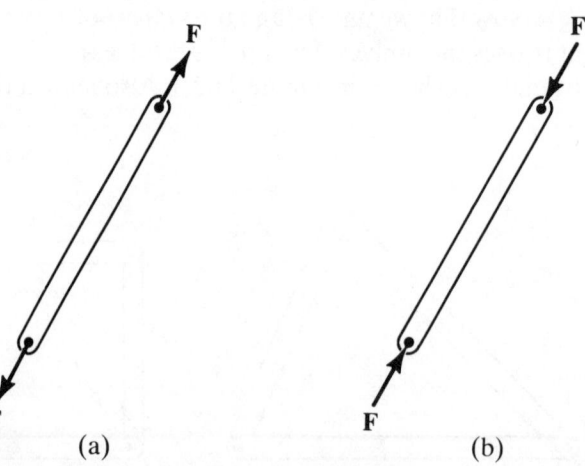

Figure 11.2.5 Members of a truss in (a) tension and (b) compression.

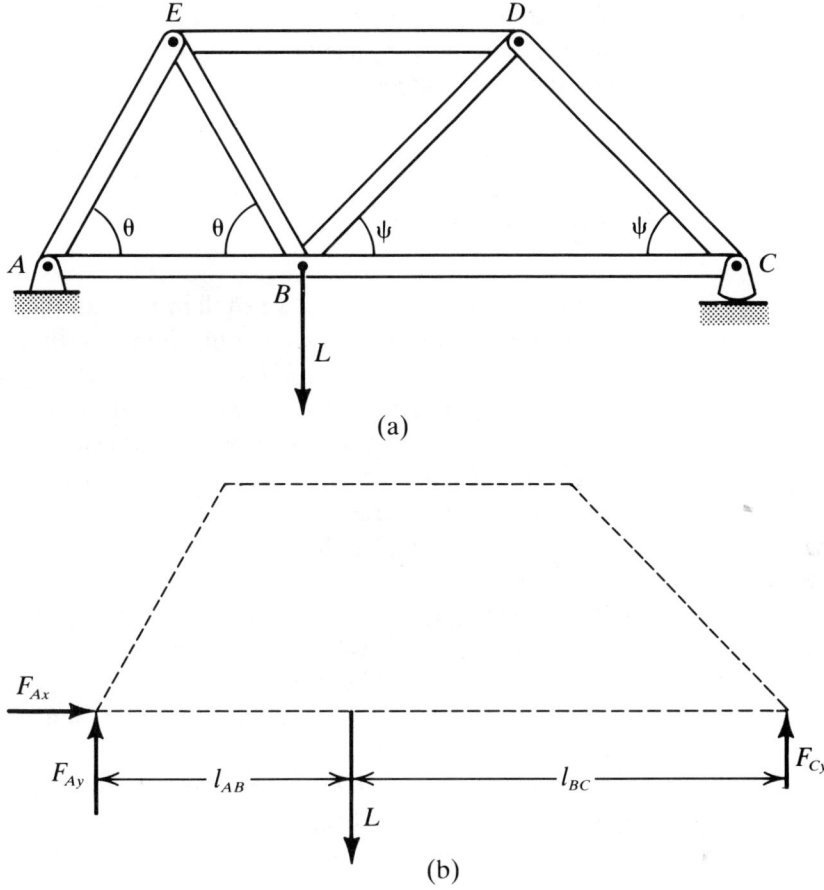

Figure 11.2.6 Example of a truss with load **L**.

11.2.6—part (a) shows the whole truss, while part (b) shows the forces acting on "something." The first condition of equilibrium ($\Sigma F = 0$) gives us

$$F_{AX} = 0$$

$$F_{Ay} + F_{Cy} = L$$

The second condition of equilibrium—zero net torque—yields

$$F_{Ay} l_{AB} = F_{Cy} l_{BC}$$

where we have measured torques about point B. l_{AB} and l_{BC} are the *lengths* of members AB and BC. F_{Ax}, F_{Ay}, and F_{Cy} are the appropriate components of the reaction forces at A and C. Combining these, we have

$$F_{Ay} = L \frac{l_{BC}}{l_{AB} + l_{BC}}$$

and

$$F_{Cy} = L \frac{l_{AB}}{l_{AB} + l_{BC}}$$

All the *external* forces on the truss are now known. How do we determine the force experienced by each individual member? Two very powerful techniques are the *method of joints* and the *method of sections*, which we will now explain.

11.3 Method of Joints

Using the *method of joints*, we shall look at all the forces applied to the pin at each joint. The pin is in equilibrium, so these forces must satisfy the conditions of equilibrium. Of course, since these forces will always be concurrent, there is no information in the torque equation. That means that only two unknown forces can be determined at each joint.

This method is best explained by example, so let us apply this method to the truss we saw earlier in Figure 11.2.6. Let us concentrate our attention on joint A. Figure 11.3.1 shows all the forces acting on joint A. There are two unknown forces $\mathbf{F}_{AB}$ and $\mathbf{F}_{AE}$, the forces due to members AB and AE, respectively. We can solve for these two, but note that *if* there were *three* unknown forces, we would not be able to determine any of them. Instead, we would have to begin with another joint. The horizontal reaction force $\mathbf{F}_{Ax}$ is dotted in Figure 11.3.1 because earlier we found it to be zero; but, in other circumstances, this force might well be present. Note that the unknown forces $\mathbf{F}_{AB}$ and $\mathbf{F}_{AE}$ have *arbitrarily* been drawn pulling away from the pin. That is, they have *arbitrarily* been drawn as *tensile* forces. Whether they are actually in tension or compression remains to be seen. Applying the equilibrium force condition, Eq. 11.2.2, immediately yields

$$F_{AB} + F_{AE} \cos \theta = 0$$

and

$$F_{Ay} + F_{AE} \sin \theta = 0$$

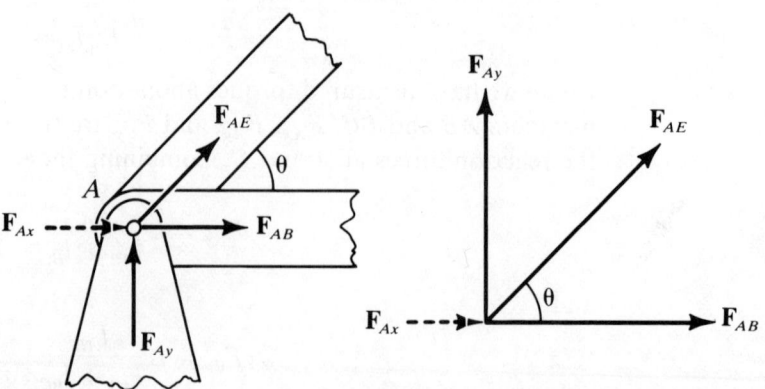

Figure 11.3.1 Forces on pin at joint A.

These two equations are readily solved to obtain

$$F_{AE} = -\frac{F_{Ay}}{\sin \theta}$$

and

$$F_{AB} = F_{Ay} \cot \theta$$

Note that F_{AE} came out negative. That means that our initial "guess" at its direction ("sense" is a better word) was incorrect. The forces are really like those in Figure 11.3.2. F_{AB} is in tension and F_{AE} is in compression. We may write this as

$$F_{AB} = [F_{Ay} \cot \theta], T$$

and

$$F_{AE} = \left[\frac{F_{Ay}}{\sin \theta}\right], C$$

where the expressions in brackets are just the magnitudes of the forces. Using T and C to label tensile and compressive forces turns out to be a very useful technique. Notice that these real forces now can be added graphically to produce a *closed* vector triangle (more than three forces would produce a closed vector polygon). The initial choice for the sense of each force is up to you. Some foresight (or experience) will enable you to choose correctly. An incorrect choice simply shows up as a negative sign in the answer.

Now, returning to the truss of Figure 11.2.6, we know the load L, the reaction forces at A and C, *and*, now, the forces in the members AB and AE. So we proceed on to another joint.

If we now look at joint B, there will be *three* unknown forces—$\mathbf{F}_{BE}$, $\mathbf{F}_{BD}$, and $\mathbf{F}_{BC}$. But we have only two equations. We can't tackle joint B yet. We could solve for the forces at joint C. Or joint E.

Let us look at joint E. Figure 11.3.3 shows the forces applied to the pin at joint E. $\mathbf{F}_{EA}$ is known. By Newton's Third Law, this is equal in magnitude but opposite in direction to force $\mathbf{F}_{AE}$. It is the compressive force that member EA exerts on pin E. The magnitudes—or lengths—of forces F_{ED} and F_{EB} are *not*

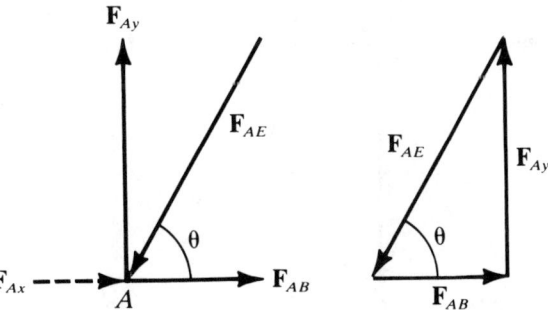

Figure 11.3.2 The real forces on joint A.

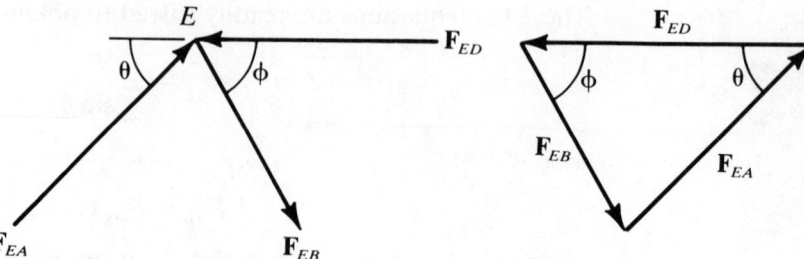

Figure 11.3.3 Forces on pin at joint E.

known. Only their directions are known. Note that F_{EB} has been drawn as a tensile force, and F_{ED} as compressive. Now proceed with Eq. 11.2.2 for equilibrium:

$$F_{EA} \cos \theta + F_{EB} \cos \phi - F_{ED} = 0$$

$$F_{EA} \sin \theta - F_{EB} \sin \phi = 0$$

and the solution is

$$F_{EB} = F_{EA} \frac{\sin \theta}{\sin \phi} = \frac{F_{Ay}}{\sin \phi}, T$$

$$F_{ED} = F_{EA}(\cos \theta + \sin \theta \cot \phi)$$

$$= F_{Ay}(\cot \theta + \cot \phi), C$$

Since F_{EB} and F_{ED} turned out to be positive, the forces are tensile and compressive, respectively, because of the initial choice in the diagram. Note, in Figure 11.3.3, that these forces form a closed triangle for vector addition.

Now we can proceed to joint B. The forces on joint B are shown in Figure 11.3.4. There are *five* forces on this joint, but only two are unknown, F_{BD} and

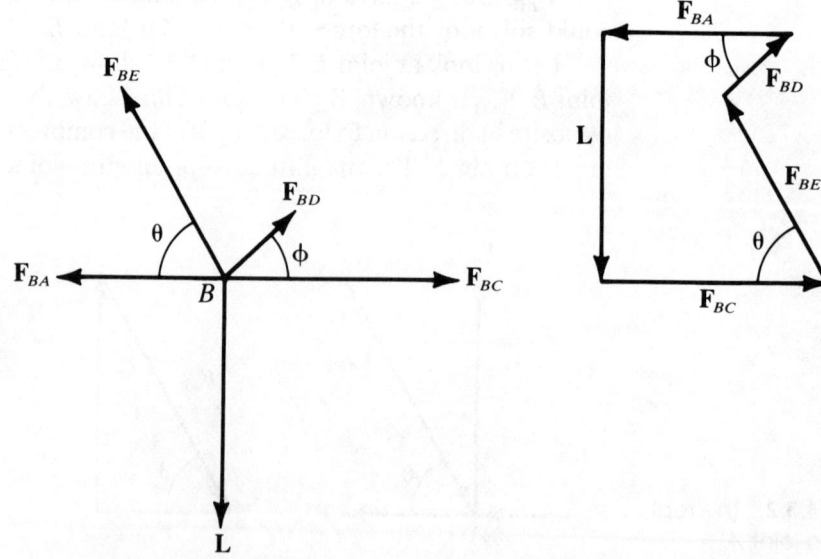

Figure 11.3.4 Forces on pin at joint B.

SECTION 11.3 / METHOD OF JOINTS

F_{BC}. Both have been drawn as tensile forces. Direct application of Eq. 11.2.2, the condition of equilibrium, yields

$$-F_{BA} - F_{BE}\cos\theta + F_{BD}\cos\phi + F_{BC} = 0$$

and

$$F_{BE}\sin\theta + F_{BD}\sin\phi - L = 0$$

and the solution for the two unknown forces is

$$F_{BD} = \frac{L - F_{BE}\sin\theta}{\sin\phi} = \frac{L - F_{Ay}(\sin\theta/\sin\phi)}{\sin\phi}$$

$$F_{BC} = F_{Ay}\left[\cot\theta + \frac{\cos\theta}{\sin\phi} + \frac{\sin\theta\cos\phi}{\sin^2\phi}\right] - L\frac{\cos\phi}{\sin\phi}$$

We shall treat both of these forces as tension although that will depend upon the actual values of θ and ϕ. Note, again, that all these forces produce a closed polygon for vector addition.

Only the force in member CD remains unknown. Figure 11.3.5 shows the forces on joint C. The condition of equilibrium demands that

$$F_{CD}\cos\phi - F_{CB} = 0$$

and

$$F_{Cy} - F_{CD}\sin\phi = 0$$

Notice that we now have two equations and only one unknown. We can readily solve for the unknown force F_{CB} by

$$F_{CD} = \frac{F_{CB}}{\cos\phi} = \frac{1}{\cos\phi}\left[F_{Ay}\left(\cot\theta + \frac{\cos\theta}{\sin\phi} + \frac{\sin\theta\cos\phi}{\sin^2\phi}\right) - L\frac{\cos\phi}{\sin\phi}\right]$$

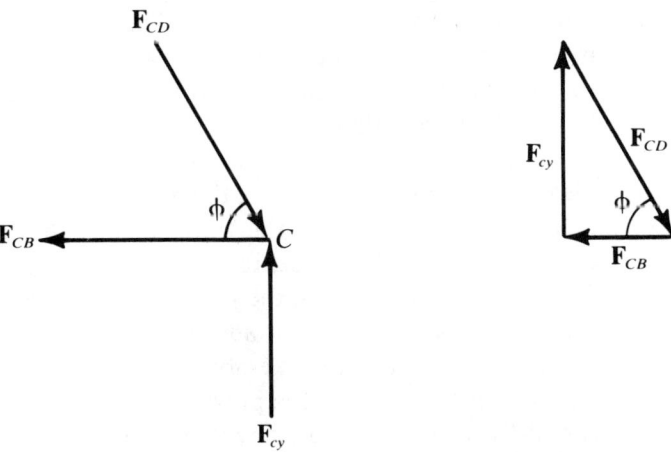

Figure 11.3.5 Forces on pin at joint C.

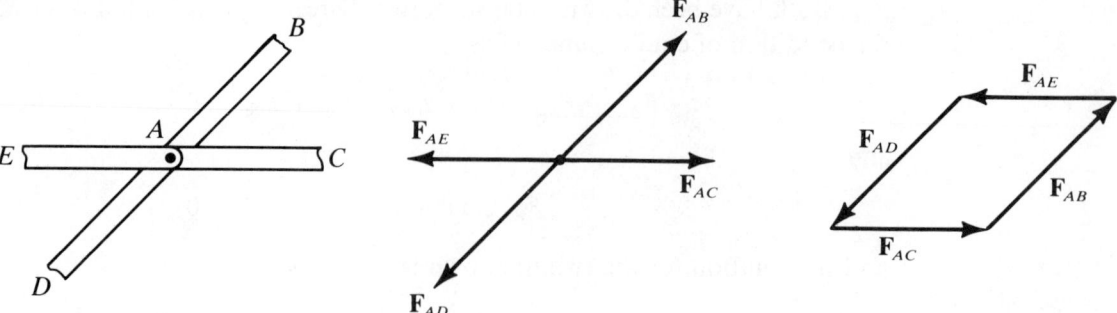

Figure 11.3.6
Members at a joint lying along intersecting straight lines.

A positive value means member *CD* is in compression; negative, in tension. We now turn to the other possible solution:

$$F_{CD} = \frac{F_{Cy}}{\sin \phi}$$

This provides a check on our work and *must* give the same numerical result. Further application of the method of joints to joint *D* where all the forces are known provides another check. These checks should always be carried out.

Thus, the method of joints involves the systematic analysis of the forces in every member of a truss. We can use information obtained at one joint at an adjoining joint through Newton's Third Law. Since the forces at each joint are concurrent, there is no need (or possibility) to use the torque equilibrium equation. Only the straightforward application of $\Sigma \mathbf{F} = 0$ is used. The only drawback to this method is that it can become tedious if you only want to know about the force in a *particular* member of a truss of many members.

We now turn to some loading conditions that warrant special consideration. Once these special loading conditions are understood, a joint satisfying these conditions can be analyzed by *inspection*! Look first at the joint and forces sketched in Figure 11.3.6. There, four members lying along two intersecting straight lines are held by a pin at joint *A*. The forces must add up to zero. The vector addition polygon must be a parallelogram. Thus, under this loading condition, *the forces in opposite members must be equal.*

Figure 11.3.7 shows four additional special loading conditions, although they are all really corollaries of the previous case. In case (a), two members lie along a straight line and the third member lies along the same line as an external load **L**. Of course, this simply replaces the fourth member and its force by the load **L**. Thus, *the forces in the two in-line members must be equal and the force in the other member must equal the load L.*

Case (b) shows three members at a joint, with two lying along the same line. This is the same as the previous case with $L = 0$. Thus, *the forces in the two in-line members must be equal and the force in the other member is zero.*

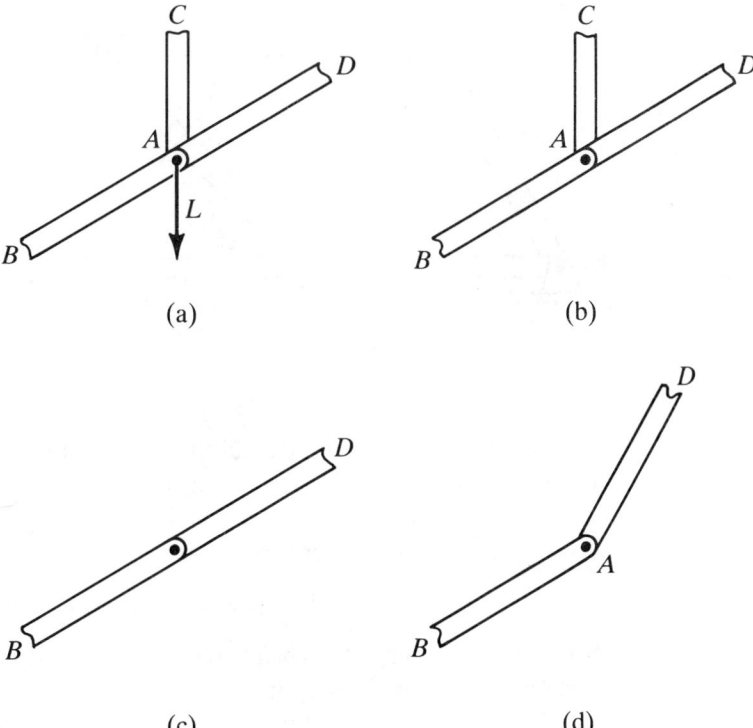

Figure 11.3.7 More special loading conditions.

Case (c) shows two members lying in a straight line. *The forces in these two in-line members must be equal.*

Case (d) shows a joint with two members that are *not* in-line. The two forces exerted on the pin by these two members must *vectorially* add to zero. This is possible only if *the forces in both members are zero.*

11.4 Method of Sections

The *method of sections* allows us to solve for the force in a single, particular member far more directly than the method of joints. In this method, a *section* is passed through the truss. We then apply both equilibrium conditions to either portion of the truss. The forces within the members cut apart by the section then become external forces. As before, this is best explained by example.

Consider once more the truss of Figure 11.2.6. Let's assume that we are interested only in member *ED* and the force in it. Then, as shown by the dotted line in Figure 11.4.1, we pass a section through the truss, ensuring that the section passes through the member in which we are interested. Now look at the two pieces of the truss formed by this section; they are sketched in Figure 11.4.2.

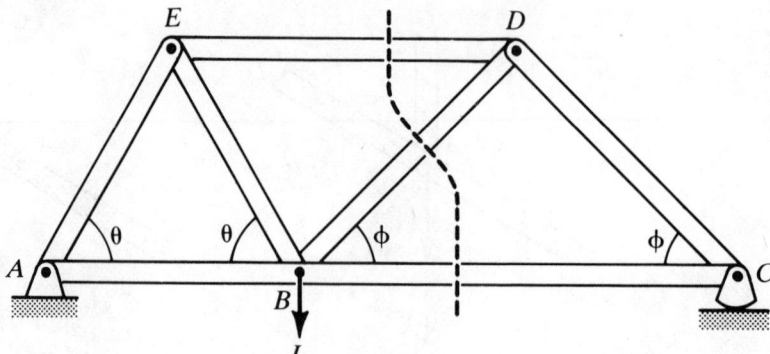

Figure 11.4.1 Passing a section through a truss.

For these pieces, the internal forces within the members are now external forces. This point is quite important but is often not apparent. To be explicit, look at the piece on the right of Figure 11.4.2. Apply the conditions of equilibrium. That the net force is zero means

$$\mathbf{F}_{Cy} + \mathbf{F}_{DE} + \mathbf{F}_{DB} + \mathbf{F}_{CB} = 0$$

Note that the three unknown forces have been drawn *arbitrarily* as forces of tension. That may or may not be true.

But of even more interest is the second condition of equilibrium—that the net torque must vanish. And this is true for torques calculated about *any* point. This idea is quite important. In particular, calculate the torque about point B as shown in Figure 11.4.3. Why point B? We are trying to solve for the single force F_{DE}. We have no interest in the other unknown forces, F_{DB} and F_{CB}. These two forces pass through point B. So their torque with respect to point B is zero. Thus, the second condition of equilibrium, Eq. 11.2.4, reduces to $l_{BC} F_{Cy} + l_{\perp} F_{DE} = 0$, where $l_{\perp}$, the moment arm for force $\mathbf{F}_{DE}$, is

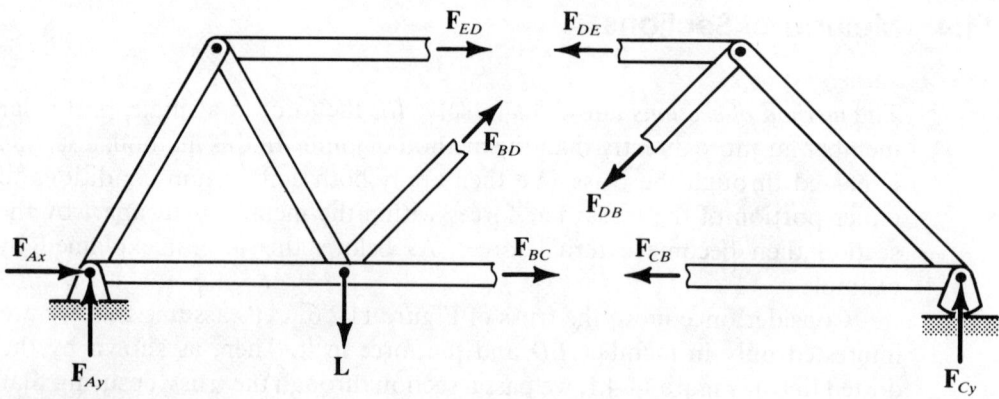

Figure 11.4.2 Forces shown explicitly on the two pieces of a truss.

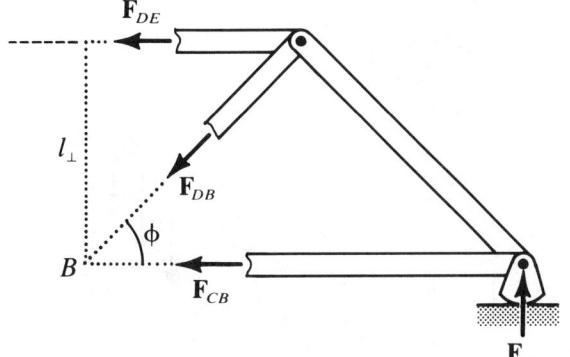

Figure 11.4.3 Torque may be calculated about any point.

the height of the truss. Thus, we can *immediately* solve for the desired force, F_{DE}:

$$F_{DE} = -F_{cy}\left(\frac{l_{BC}}{l_\perp}\right)$$

Note that F_{DE} is negative. That merely means that member DE is actually in *compression*.

Choosing any point other than B for calculating torques would have given us an equation containing all three forces. With the two force equations we could have solved for all three. But if we are only interested in one force, then a judicious choice of origin for calculating torques is a great help. This is even useful if more than one force is to be determined. Unless a complete, member-by-member analysis is required, the additional power of the method of sections usually makes it preferable to the more direct method of joints.

11.5 Cables under Distributed Loads

Consider a cable as shown in Figure 11.5.1. It is supported at its ends A and B. Some sort of load is distributed along the length of the cable as indicated by the arrows. We want to find the *tension* in the cable at any point and the shape that the cable makes.

In discussing cables, we shall make a couple of reasonable assumptions. First, we will deal only with *flexible* cables. That means that at any point the tension lies along a tangent to the cable. Light strings and chains offer ready examples of such flexibility, but most cables under most loads also exhibit this flexibility. We shall also assume that the weight of the cable is negligible compared to the additional load—or that the weight of the cable itself *is* the only load.

We shall use the notation of Figure 11.5.1 throughout this discussion of cables. A and B are the endpoint supports, C is the *lowest* point of the cable, and D is just any point in general. To find out more about the tension in the cable at this general point D, look at Figure 11.5.2. The portion of the cable

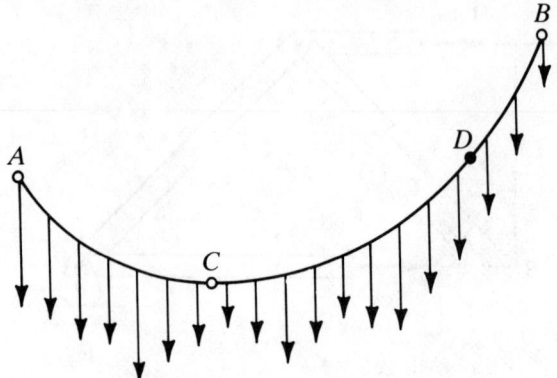

Figure 11.5.1 A cable with a distributed load.

from the lowest point C up to our point of interest D is shown there with all the forces acting on it shown.

We shall call the tension at the lowest point of the cable $\mathbf{T_0}$. It is horizontal. The vector $\mathbf{T}$ is the tension at point D and it points along the line tangent to the cable. The vector $\mathbf{W}$ is the *sum* of all the load supported by the cable between these two points. Since the cable is in equilibrium, these three forces must add up to zero. That is,

$$\mathbf{T} + \mathbf{T_0} + \mathbf{W} = 0 \tag{11.5.1}$$

From this vector addition triangle, we can obtain several useful relations. Namely,

$$T \cos \theta = T_0 \tag{11.5.2}$$

$$T \sin \theta = W \tag{11.5.3}$$

$$\tan \theta = \frac{W}{T_0} \tag{11.5.4}$$

$$T = \sqrt{T_0^2 + W^2} \tag{11.5.5}$$

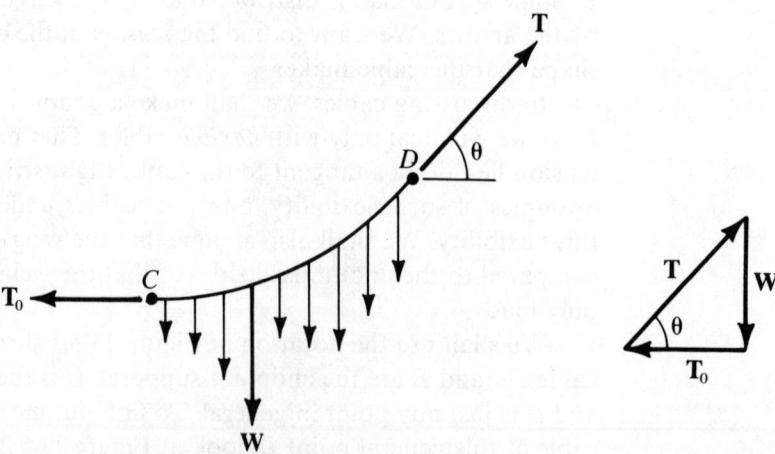

Figure 11.5.2 Forces on a portion of the cable.

The first of these equations says that the horizontal component of the tension within a cable is constant. The second states that the vertical component is equal to the load between the lowest point of the cable and the point of interest.

11.6 Parabolic Cables

Let us now consider cables supporting a load that is *uniformly distributed horizontally*—a suspension bridge is a prime example. Such a cable and load are sketched in Figure 11.6.1. We will denote the load per unit length by w (w has units of newtons per meter). Remember, this must be a *horizontal* length.

For convenience, choose a coordinate frame with origin at the lowest point of the cable C, as shown in Figure 11.6.2. The total load carried is $W = wx$. We can use Eqs. 11.5.4 and 11.5.5 to write

$$T = \sqrt{T_0^2 + w^2 x^2} \qquad (11.6.1)$$

and

$$\tan \theta = \frac{w}{T_0} x \qquad (11.6.2)$$

Due to the nature of the loading, the equivalent total load W is located at a horizontal position of $x/2$. Calculate torques about D and use the torque equation, Eq. 11.2.4, to write

$$(wx)\left(\frac{x}{2}\right) = T_0 y \qquad (11.6.3)$$

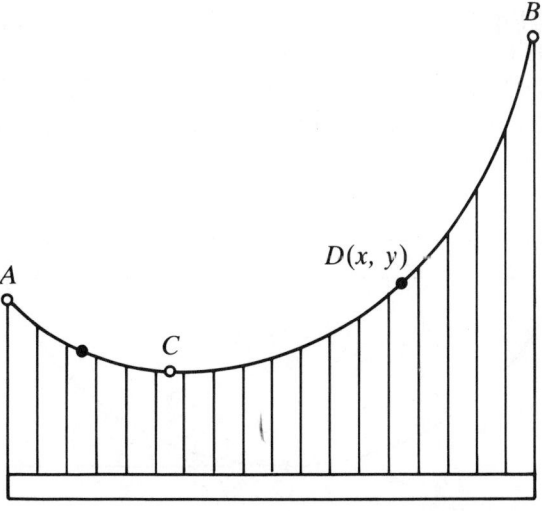

Figure 11.6.1 Cable with uniform horizontal loading.

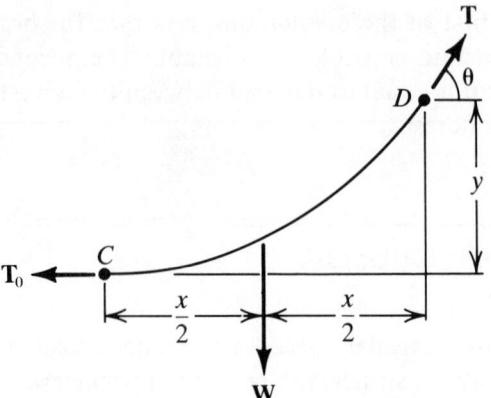

Figure 11.6.2 Portion of cable with uniform horizontal loading.

This is readily solved to provide an equation for $y = y(x)$, the *shape* of the cable:

$$y = \frac{w}{2T_0} x^2 \qquad (11.6.4)$$

Thus a cable under uniform horizontal loading hangs in the shape of a parabola.

If the endpoint supports A and B are at the same vertical level, then their horizontal separation is called the *span* of the cable. The vertical distance between either and the lowest point C is the *sag*. This is shown in Figure 11.6.3; L is the *span*; h, the sag. If these are known, they may be substituted into Eq. 11.6.4 to determine T_0, the minimum tension in the cable:

$$T_0 = \frac{wL^2}{8h} \qquad (11.6.5)$$

Now Eq. 11.6.1 can be used to find the tension at any point in the cable.

If the supports are not at the same height, a complete solution is more subtle. Remember that our coordinate has its origin at C, the lowest point. If, for instance, the elevations, y_A and y_B, of points A and B above C are given

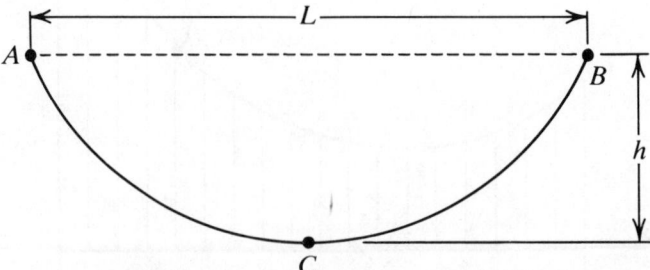

Figure 11.6.3 L is the *span* of the cable; h, the *sag*.

SECTION 11.6 / PARABOLIC CABLES

and their horizontal distance, $x_B - x_A = L$, are given, then we can use Eq. 11.6.4 to write

$$y_A = \frac{w}{2T_0} x_A^2, \qquad y_B = \frac{w}{2T_0} x_B^2$$

or

$$x_A = -\sqrt{\frac{2T_0 y_A}{w}}, \qquad x_B = +\sqrt{\frac{2T_0 y_B}{w}}$$

So that

$$L = x_B - x_A = \sqrt{\frac{2T_0}{w}} (\sqrt{y_B} + \sqrt{y_A})$$

If we set

$$d = \left[\frac{\sqrt{y_B} + \sqrt{y_A}}{2}\right]^2$$

then

$$T_0 = \frac{wL^2}{8d} \tag{11.6.6}$$

Thus, d becomes some sort of a "square mean root" average of the sag and resembles the true sag h for the symmetric case.

Determining the total length of the cable S seems like a worthwhile venture. The length ds of a small portion of the cable is related to its horizontal and vertical dimensions dx and dy by

$$ds^2 = dx^2 + dy^2 \tag{11.6.7}$$

or

$$ds = \sqrt{1 + \left(\frac{dy}{dx}\right)^2}\, dx \tag{11.6.8}$$

From Eq. 11.6.4 we know that

$$\frac{dy}{dx} = \frac{x}{T_0} \tag{11.6.9}$$

Thus, the length of cable from, say, point $C(x = 0)$ to point $B(x = x_B)$ may be written as

$$S_B = \int_0^{x_B} \sqrt{1 + \left(\frac{dy}{dx}\right)^2}\, dx \tag{11.6.10}$$

$$S_B = \int_0^{x_B} \sqrt{1 + \left(\frac{w}{T_0}\right)^2 x^2}\, dx \tag{11.6.11}$$

This can be evaluated explicitly from integral tables to get

$$S_B = \frac{T_0}{2w}\left[\frac{wx_B}{T_0}\sqrt{1 + \frac{wx_B}{T_0}} + \ln\left(\frac{wx_B}{T_0} + \sqrt{1 + \frac{wx_B}{T_0}}\right)\right] \quad (11.6.12)$$

A similar relation holds, or course, for S_A on the other side. Given x_B, we can solve for S_B. But we cannot as readily do the reverse (this is a transcendental equation). Numerical methods are then necessary. So it will be useful in some cases to look for an approximation.

Using the binomial expansion,

$$(1 + u)^{1/2} = 1 + \frac{1}{2}u - \frac{1}{8}u^2 + \cdots$$

we can rewrite Eq. 11.6.11 as

$$S_B = \int_0^{x_B}\left[1 + \frac{1}{2}\left(\frac{w}{T_0}\right)^2 x^2 - \frac{1}{8}\left(\frac{w}{T_0}\right)^4 x^4 + \cdots\right]dx$$

$$S_B = x_B\left[1 + \frac{1}{6}\left(\frac{w}{T_0}\right)^2 x_B^2 - \frac{1}{40}\left(\frac{w}{T_0}\right)^4 x_B^4 + \cdots\right] \quad (11.6.13)$$

Or, recalling from Eq. 11.6.4 that $w/T_0 = 2y_B/x_B^2$, this can be written as

$$S_B = x_B\left[1 + \frac{2}{3}\left(\frac{y_B}{x_B}\right)^2 - \frac{2}{5}\left(\frac{y_B}{x_B}\right)^4 + \cdots\right] \quad (11.6.14)$$

For most cases $y_B \ll x_B$, so only the first two terms need to be considered.

11.7 Catenary Cables

Now consider a cable supporting a load distributed *uniformly along its length*, for example, a cable supporting its own weight. Such a cable is sketched in Figure 11.7.1. Again we shall use w to denote the load per unit length. But

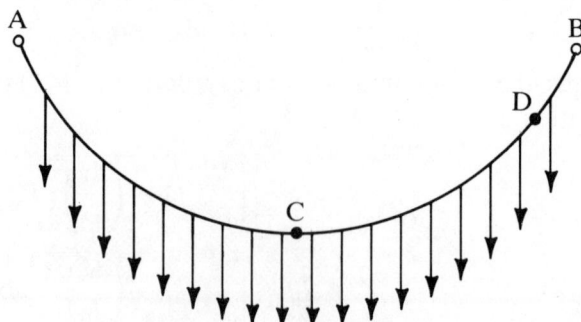

Figure 11.7.1 Cable with load uniformly along its own length.

SECTION 11.7 / CATENARY CABLES

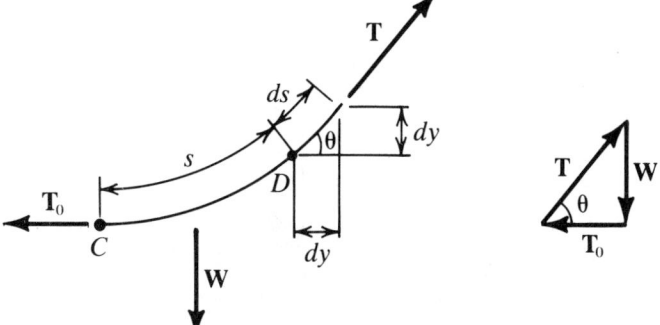

Figure 11.7.2 Portion of cable loaded uniformly along its own length.

now it is the load per unit length *measured along the cable*. We call this length s.

Figure 11.7.2 shows a portion of the cable from its lowest point, C, to some other point, D. The total load carried is $W = ws$. Equation 11.5.5 for the tension now has the form

$$T = \sqrt{T_0^2 + w^2 s^2} \tag{11.7.1}$$

To simplify later equations, we shall introduce a constant $c = T_0/w$. Then we can write T_0 in a *form* similar to W. That is,

$$T_0 = wc, \qquad W = ws \tag{11.7.2}$$

Therefore,

$$T = w\sqrt{c^2 + s^2} \tag{11.7.3}$$

Figure 11.7.2 shows a piece of the cable with all the forces acting on it. We cannot immediately use this to determine $y = y(x)$ as we did in the previous section since we do not know the horizontal location of **W**. To find a relation between horizontal distance x and distance along the cable s, we begin with

$$dx = ds \cos \theta \tag{11.7.4}$$

which can be seen from the figure. Equations 11.5.2, 11.7.2, and 11.7.3 then give

$$dx = \frac{T_0}{T} ds$$

$$= w \frac{wc \, ds}{\sqrt{c^2 + s^2}}$$

$$dx = \frac{ds}{\sqrt{1 + \frac{s^2}{c^2}}} \tag{11.7.5}$$

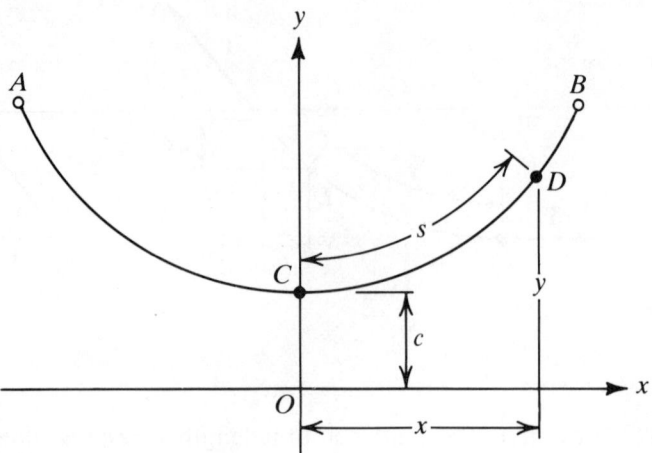

Figure 11.7.3
Coordinate system.

Figure 11.7.3 shows the details of the coordinate system we shall use. As before, the y-axis passes through the lowest point C. But now the origin is chosen to be a distance $c = T_0/w$ below point C. This arbitrary choice of origin is made with some foresight to simplify future expressions. Using an integral table, we can now integrate Eq. 11.7.5 to obtain

$$x = \int_0^s \frac{ds}{\sqrt{1 + \frac{s^2}{c^2}}} = c\left[\sinh^{-1}\frac{s}{c}\right]_0^s = c\sinh^{-1}\frac{s}{c}$$

Or we can write this more conveniently as

$$s = c \sinh \frac{x}{c} \tag{11.7.6}$$

The function sinh ( ) is called the hyperbolic sine function and is defined by

$$\sinh u = \frac{e^u - e^{-u}}{2} \tag{11.7.7}$$

As you might suspect, there is a companion function, cosh ( ) called the hyperbolic cosine function, which is defined by

$$\cosh u = \frac{e^u + e^{-u}}{2} \tag{11.7.8}$$

A few characteristics of these hyperbolic functions are worth noting here. Specifically

$$\frac{d}{du}\sinh u = \cosh u \tag{11.7.9}$$

$$\frac{d}{du}\cosh u = \sinh u \tag{11.7.10}$$

$$\sinh 0 = 0$$

$$\cosh 0 = 1$$

$$\cosh^2 u - \sinh^2 u = 1 \qquad (11.7.11)$$

We have $s = s(x)$ or $x = x(s)$, but would like $y = y(x)$ as an equation for the curve made by the cable. From Figure 11.7.2, we can write

$$dy = dx \tan \theta \qquad (11.7.12)$$

From the vector addition triangle there we can write

$$\tan \theta = \frac{W}{T_0} \qquad (11.7.13)$$

Using Equations 11.7.2 and 11.7.6 we then have

$$dy = \frac{W}{T_0} dx$$

$$= \frac{s}{c} dx$$

$$dy = \sinh \frac{x}{c} dx \qquad (11.7.14)$$

With the coordinate system of Figure 11.7.3 and our earlier discussion, this can be integrated to give

$$y = c \cosh \frac{x}{c} \qquad (11.7.15)$$

This curve, $y = y(x) = c \cosh x/c$ is called a *catenary* and c is known as the *parameter* of the catenary.

Squaring Eqs. 11.7.6 and 11.7.15 and using Eq. 11.7.12, we can write

$$y^2 - s^2 = c^2 \qquad (11.7.16)$$

which allows us to restate Eq. 11.7.3 as

$$T = wy \qquad (11.7.17)$$

That is, the tension at any point in the cable is proportional to the vertical distance y above the x-axis we have constructed.

11.8 Cables with Concentrated Loads

Consider that a flexible cable of negligible weight attached to A and B on which some external force **F** acts at point C. We shall assume that this force pulls the cable taut (this is equivalent to the cable having negligible weight),

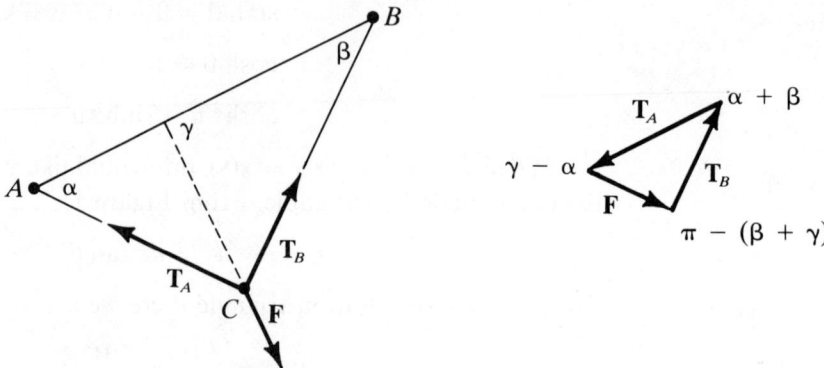

Figure 11.8.1
Concentrated load on a cable.

as sketched in Figure 11.8.1. By the law of cosines, the angles α and β of that figure may be determined to be

$$\cos \alpha = \frac{l_{AB}^2 + l_{AC}^2 - l_{BC}^2}{2 l_{AB} l_{AC}} \quad \text{and} \quad \cos \beta = \frac{l_{AB}^2 + l_{BC}^2 - l_{AC}^2}{2 l_{AB} l_{BC}} \quad (11.8.1)$$

where l_{AB}, l_{BC}, and l_{AC} are the distances between points A and B, B and C, and A and C, respectively.

From the vector addition triangle in the figure, the tensions in the cable can be determined. If $\mathbf{T}_A$ is the tension in the portion of the cable from A to C and $\mathbf{T}_B$, from C to B, then the law of sines allows us to write

$$T_A = F \frac{\sin(\beta + \gamma)}{\sin(\alpha + \beta)} \quad (11.8.2)$$

and

$$T_B = F \frac{\sin(\gamma - \alpha)}{\sin(\alpha + \beta)} \quad (11.8.3)$$

That is, the distances between points A, B, and C on the cable determine the angles—the geometry—of the situation. And these angles, in turn, determine the tensions in the cable. But for an extensible, or stretchable, cable we cannot stop here. For the tension in the cable determines the length of the cable! How can we handle such a situation?

The elongation, extension, or stretch of many bodies—and of even more if we keep the elongation small—is found to be proportional to the tensile force acting upon it. This is Hooke's Law and may be written as

$$\delta l = K l^0 T \quad (11.8.4)$$

where l^0 is the unstretched length and T is the tensile force causing the elongation. The term k is a proportionality constant; in fact, it is the same k that we wrote in Eq. 1.1.1 when we discussed a stretched spring.

Then the new, stretched length is really

$$l = l^0(1 + kT) \quad (11.8.5)$$

So the values for l_{Ac}, l_{BC} in Eq. 11.8.1 should really be

$$l_{AC} = l_{AC}^0(1 + kT_A) \quad \text{and} \quad l_{BC} = l_{BC}^0(1 + kT_B) \tag{11.8.6}$$

You can see the complication—now the lengths depend upon the tensions. But the tensions depend upon the lengths.

These can be solved by iterative approximations. Assume first that the cable does not stretch. Using l_{AC} and l_{BC}, solve for angles α, β, and γ. Then use these angles to solve for tensions T_A and T_B. next, use these tensions to get new values for cable lengths l_{AC} and l_{BC}, and, finally, use these lengths for new angles and tensions. The values should converge on the correct values after a few iterations.

PROBLEMS

11.1 Classify each of the accompanying structures as completely, partially, or improperly constrained; if completely constrained, further classify as internally determinate or indeterminate.

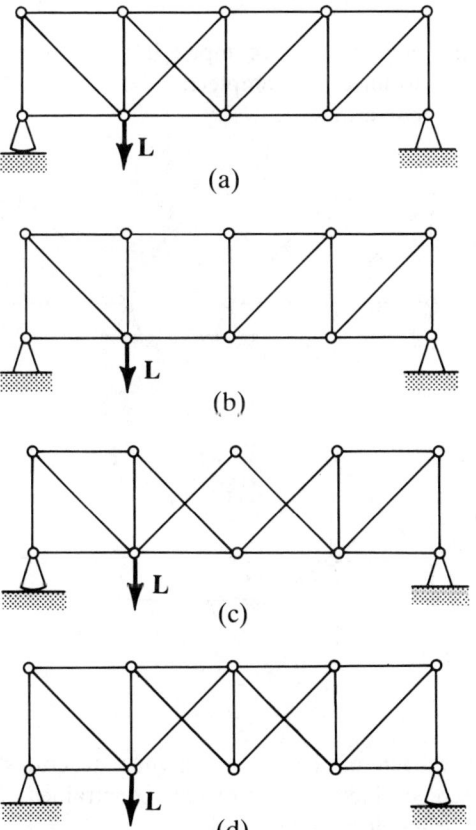

11.2 Classify each of the accompanying structures as completely, partially, or improperly constrained; if completely constrained, further classify as internally determinate or indeterminate.

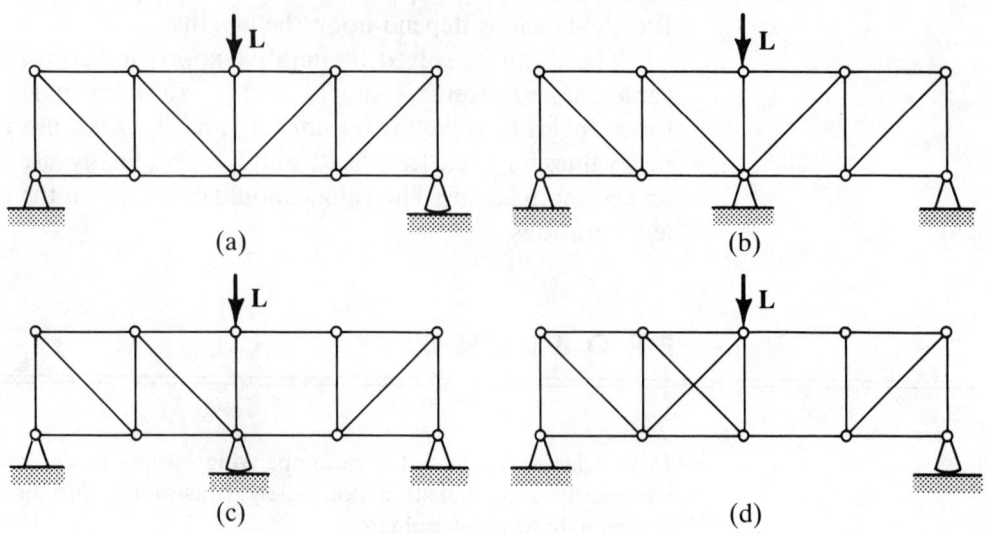

11.3 Classify each of the accompanying structures as completely, partially, or improperly constrained; if completely constrained, further classify as internally determinate or indeterminate.

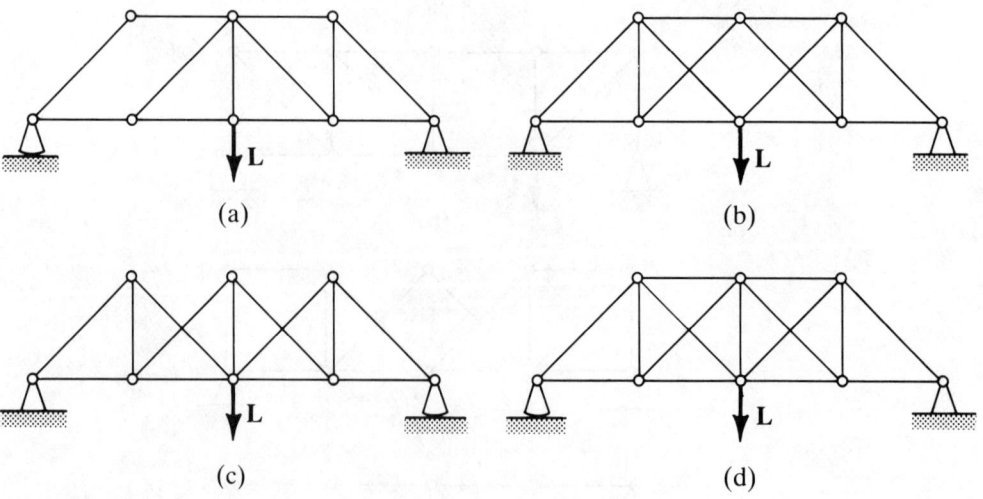

11.4 Classify each of the accompanying structures as completely, partially, or improperly constrained; if completely constrained, further classify as internally determinate or indeterminate.

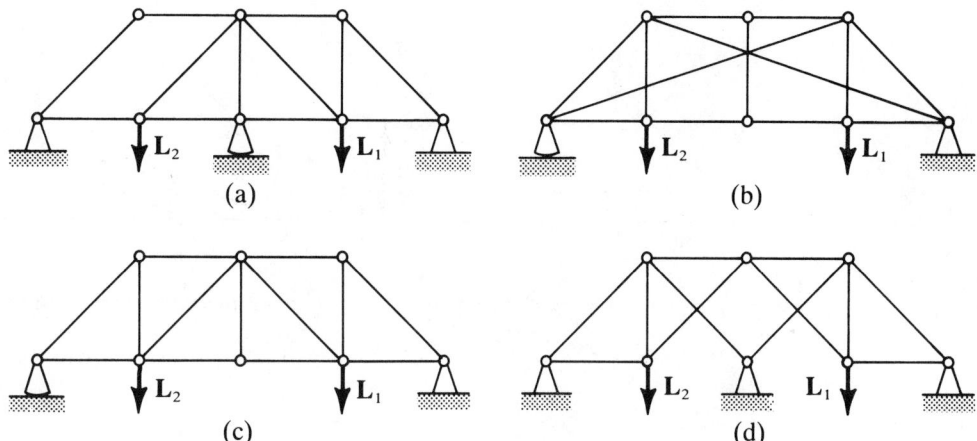

(a) (b) (c) (d)

11.5 Using the method of joints, determine the force in kilonewtons (kN) in each member (and whether it is tensile or compressive) for the truss shown.

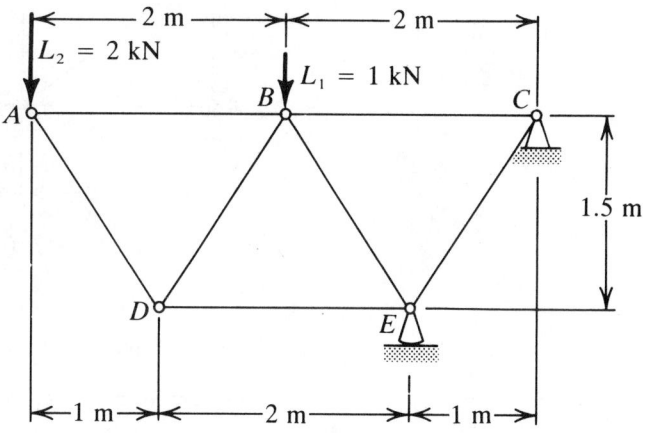

11.6 Using the method of joints, determine the force in each member of the trusses shown.

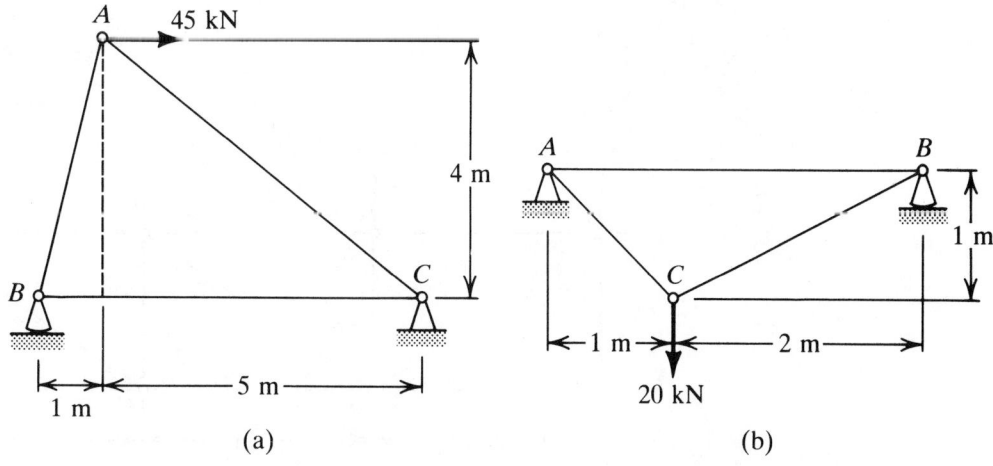

(a) (b)

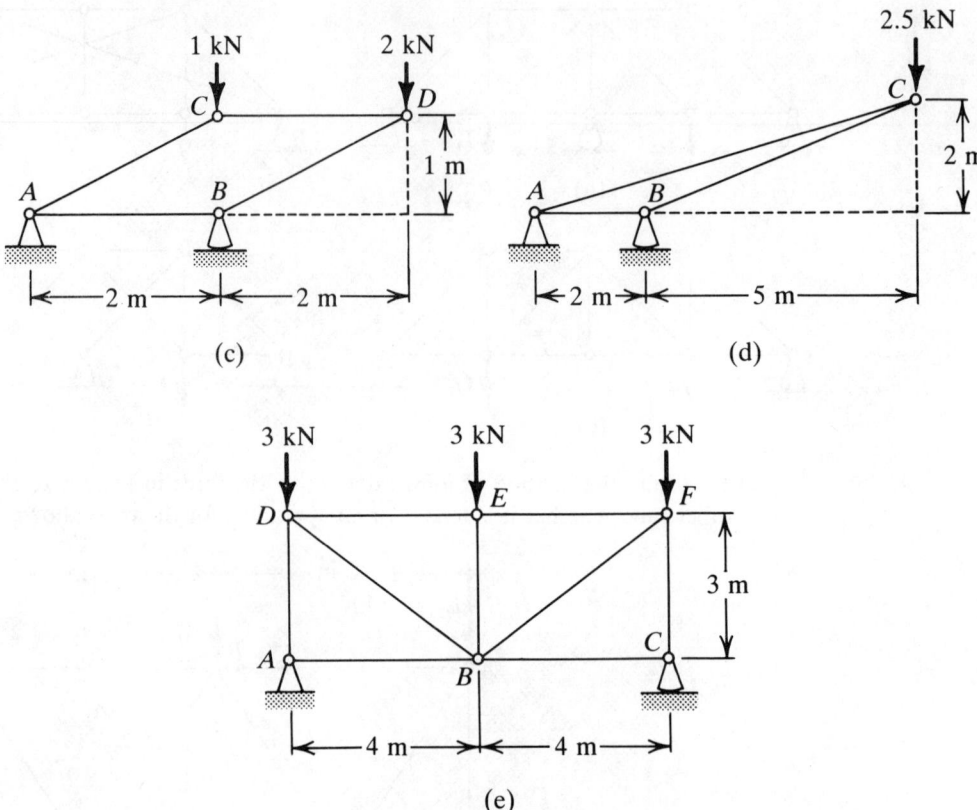

11.7 Find the force in members BA and GA of the truss shown.

11.8 Find the force in members EF and CF of the truss shown.

11.9 Find the force in members BF and BC of the truss shown.

11.10 Find the force in members EC and EF of the truss shown.

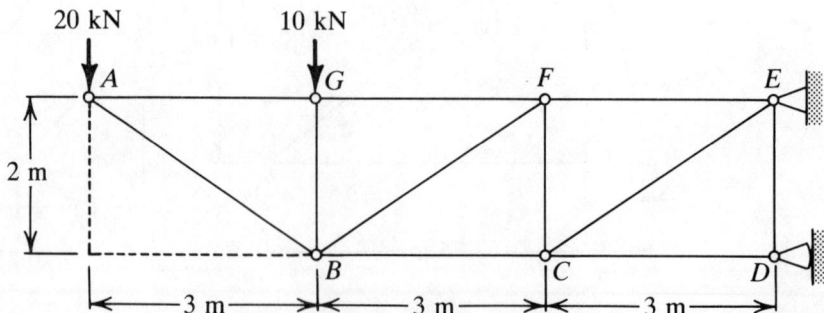

PROBLEMS

11.11 Find the force in members *GH* and *HI* in the truss shown.

11.12 Find the force in members *EF* and *FD* in the truss shown.

11.13 Find the force in members *BD* and *CD* in the truss shown.

11.14 Find the force in members *EA* and *EC* in the truss shown.

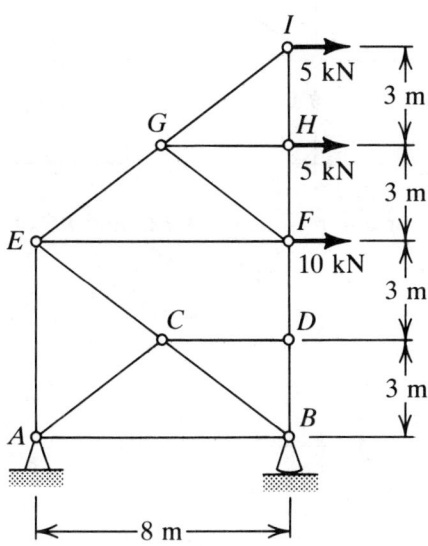

11.15 Find the force in members *HG*, *BG*, and *BC* of the stadium truss shown.

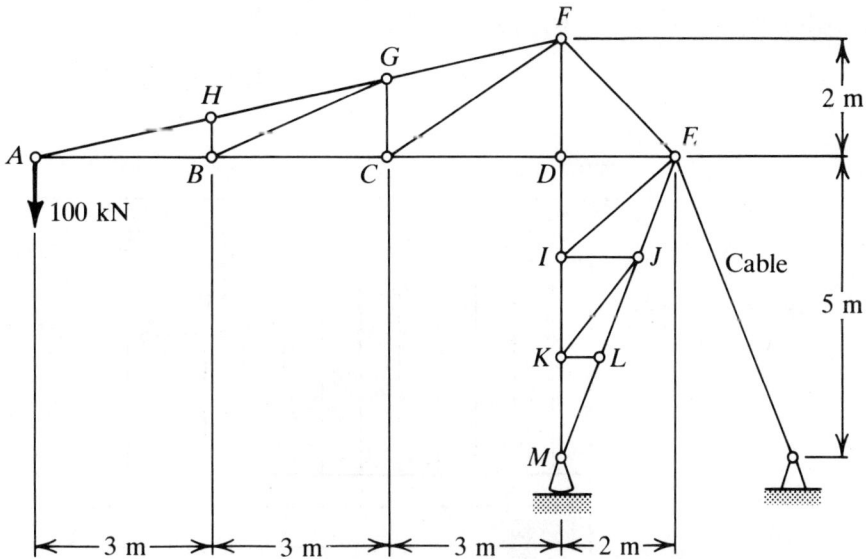

11.16 Find the force in members EF, BF, and BC in the truss shown for $L_1 = 100$ kN and $L_2 = 150$ kN.

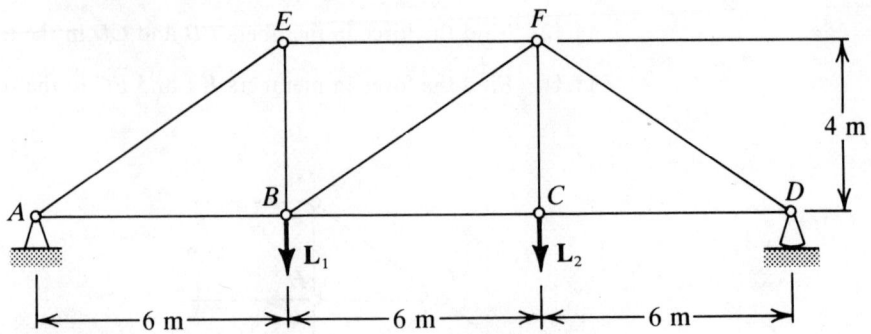

11.17 Find the force in members ED, EC, and BC of the truss shown.

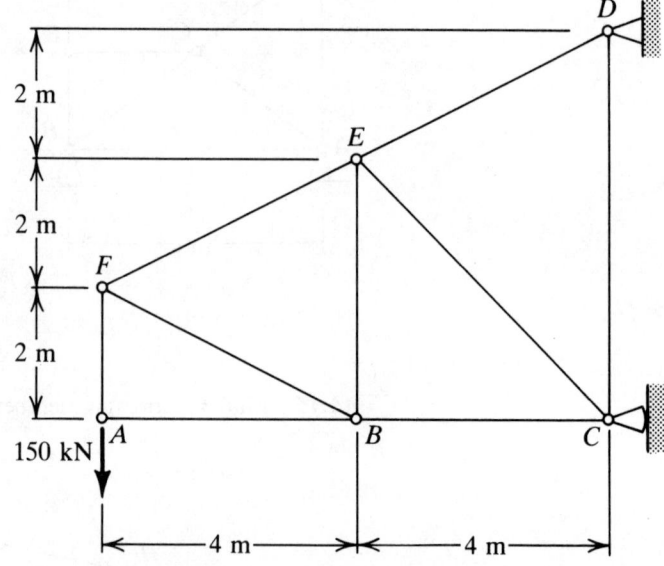

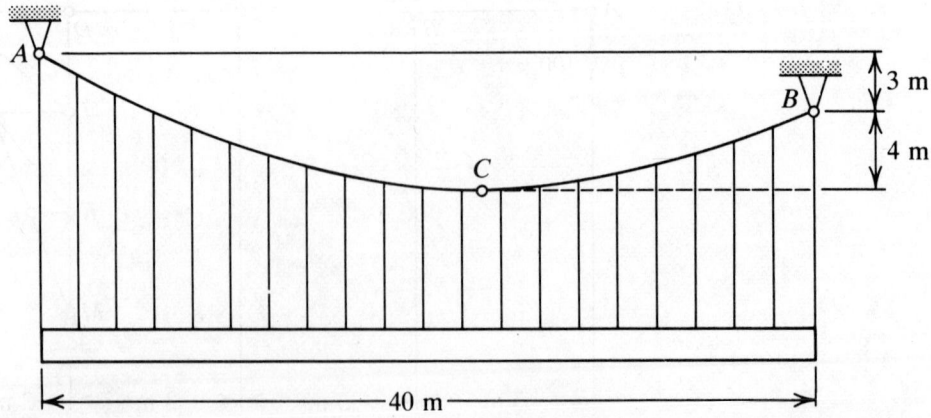

11.18 A cable, attached at A and B as shown in the sketch, supports a load of 200 N/m of uniform horizontal distribution. Find the tension in the cable at points A and B.

11.19 A steam pipe runs between two buildings and is supported by a cable as shown. The steam pipe has a linear mass density of 115 kg/m. Neglect the mass of the cable. Determine the horizontal location of the lowest point of the cable and the tension in the cable where it is attached to each building.

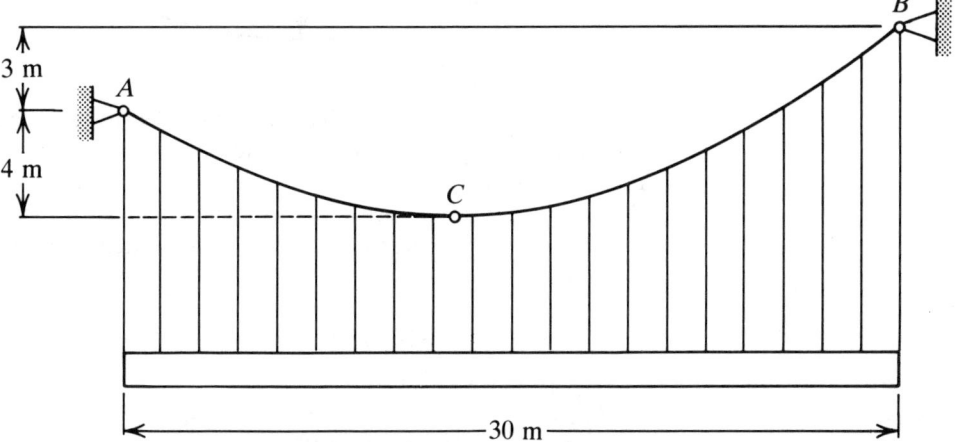

11.20 A flexible cable is attached to a fixed support at A, passes over a frictionless pulley at B, and then supports a mass M as shown in the sketch. The span of the cable is 20 m and the sag is 0.5 m. The mass per unit length of the cable is 0.8 kg/m. Neglect the weight of the cable from the pulley to the suspended weight. Determine the amount of mass suspended, M. Find the slope of the cable at support A.

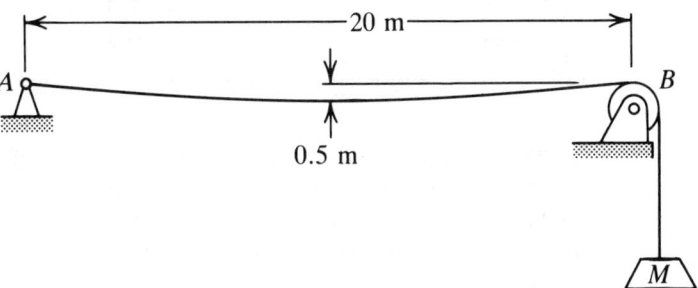

11.21 A 75-m surveying tape has a mass of 2 kg. The tape is stretched between two points of the same elevation until the tension at either end is 80 N. Find the *horizontal* distance between the ends of the tape. Find the amount of sag.

11.22 A 200-m electric power transmission cable with mass of 37 kg is strung between two towers each 20 m high. The lowest point of the cable is then found to be 10 m above the ground. Find the weight of the cable and the horizontal distance between the towers if the minimum tension in the cable is 900 N.

11.23 A uniform cable with linear mass density of 0.23 kg/m is suspended as shown. Find the minimum and maximum values of tension in the cable. Find the length of the cable. (This solution involves a transcendental equation that is best handled numerically.)

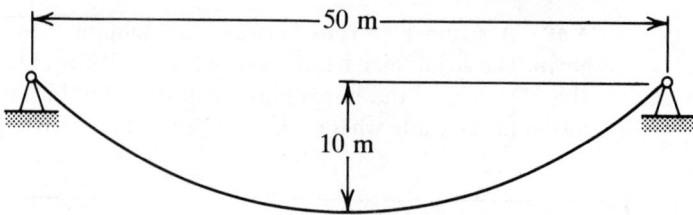

*11.24 An electric power line with mass density of 0.25 kg/m is strung between transmission towers as shown in the accompanying sketch. Since the middle tower, at B, is slender, it is desired that the resultant horizontal force on that tower vanish. Determine the sag h necessary between A and B to meet this requirement. (This solution involves a transcendental equation that is best solved using numerical methods.)

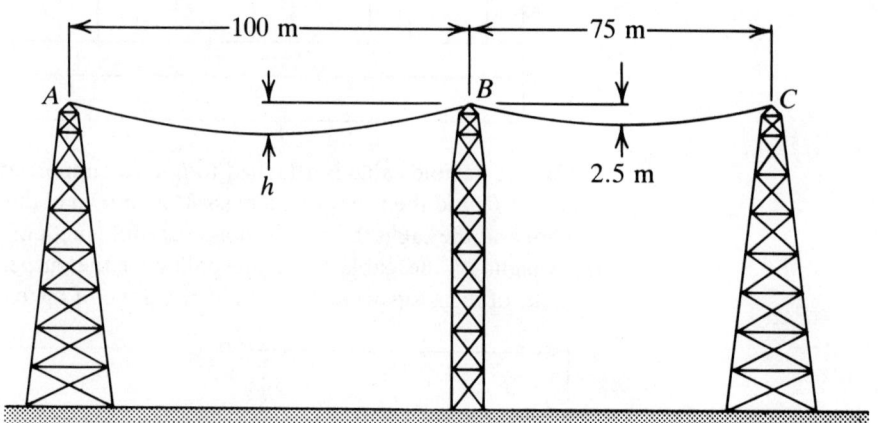

*11.25 A rubber band with an unstretched length of 20 cm is attached to points A and B. 5 cm from end B, a 200-g weight is attached. Assume that the original rubber band obeys Hooke's Law with a "spring constant" of 200 N/m. Find the tension T_{AC} and T_{CB} in each portion of the rubber band.

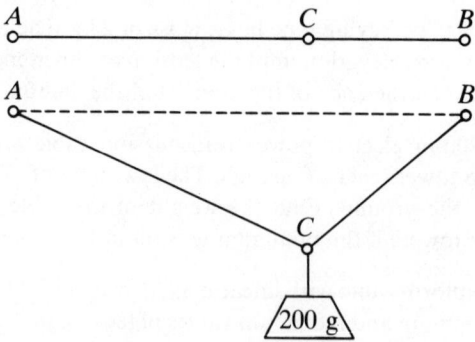

*11.26 A rubber band has an unstretched length of 20 cm and is fastened at A and B, which are also 20 cm apart. Then a pulley (to assure symmetry) supporting a 200-g weight is placed on the rubber band. Assume that the unstretched rubber band has a "spring constant" of 200 N/m (remember that each half will then have twice that value as its spring constant). (Actually, a rubber band does not obey Hooke's Law very well; so we are being somewhat cavalier to assign a spring *constant* to a rubber band. But just use it as an assumption.) Find h, the distance the load will pull the rubber band down from its unloaded position.

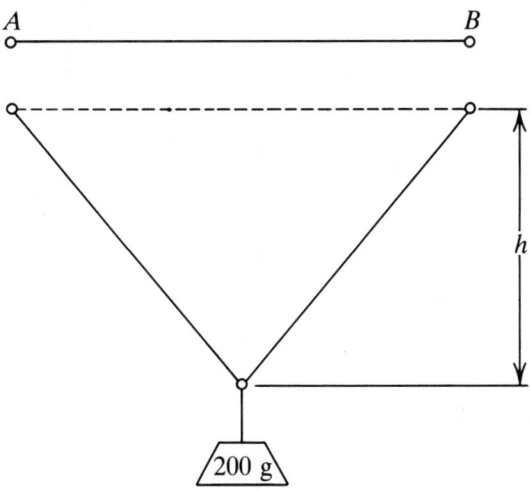

*11.27 Aluminum cables AC and BC with unstretched lengths of 10 m and 15 m are attached to supports at A and B 15 m apart. A load of 5 kN is placed at point C where they are joined. Find the tension T_{AC} and T_{BC} in each cable if the two cables obey Hooke's law with spring constants of $K_{AC} = 1.67 \times 10^4$ N/m and $K_{BC} = 2.5 \times 10^4$ N/m.

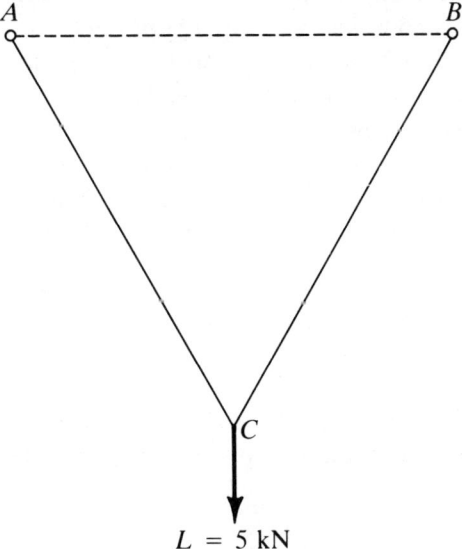

12

FLUID MECHANICS

12.1 Introduction

As in all of mechanics, we can focus our attention on fluids *at rest* (which we call *hydrostatics*) or we can study fluids *in motion* (which we call *hydrodynamics*). A *fluid* is defined as a substance that does not have a fixed shape and that can flow. Therefore, both liquids and gases are fluids. But there are also important differences between liquids and gases. Liquids are not readily compressed; we can usually consider their volumes or densities as fixed. Gases can be compressed far more easily. Liquids have a distinct surface whereas gases completely fill up a container.

We will find Newton's Second Law, **F** = m**a**, and all the earlier ideas of energy conservation to be quite useful when discussing fluids. Of course this should be so, as a fluid is merely a collection of particles (but such a large collection that it is not of much use to ask for the details of energy, force, acceleration, or mass of each individual particle).

12.2 Hydrostatics: Fluids at Rest

The idea of *pressure* is useful, even essential, in discussing fluids. Pressure is the force exerted on an area divided by that area:

$$p = \frac{F}{A} \qquad (12.2.1)$$

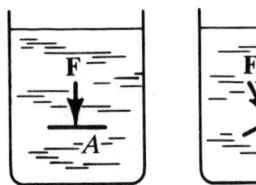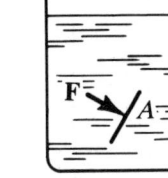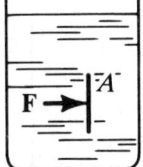

Figure 12.2.1 Pressure.

Pressure, however, is not a vector quantity. Experimentally, it is observed that the magnitude of the force exerted on some small area is the same for any direction or orientation of that area. This is shown schematically in Figure 12.2.1. You have experienced this if you have dived into a swimming pool or the ocean—when you dive you feel the increased pressure on all parts of your body. We may say that *a fluid exerts pressure equally in all directions*.

The force exerted by a fluid at rest is always *perpendicular* to the area on which the force is exerted. Suppose this were not true. Then a fluid at rest could exert both a perpendicular component of force $F_\perp$ and a parallel component $F_\parallel$ on an area A of its container, as sketched in Figure 12.2.2. By Newton's Third Law, then, the container would exert equal and opposite forces on the surface of the fluid. Since fluids—by their definition—can flow, this parallel component $F_\parallel$ would cause the fluid to flow; it would not be at rest. Thus, a fluid at rest can exert only perpendicular forces. While this is not true for fluids in motion—viscous or frictional drag forces are parallel components—it is often useful to consider only perpendicular components for even the case of moving fluids.

Swimmers and divers also know that *pressures at equal depths are the same*. The pressure 2 m below the surface of a swimming pool is the same whether a diver is on the bottom of a 2-m pool or 2 m deep in a 5-m diving well. It is the same if he is 2 m deep in one of the Great Lakes. The pressure increases with depth, of course; but depends only upon the depth from the atmosphere-water surface. It is independent of the distance to the bottom. It is independent of the *size* of the pool or lake!

Having stated these important characteristics of hydrostatics, let us return and look at some of them more closely.

Consider a *very small* triangular prism of fluid as shown in Figure 12.2.3. The prism has a cross section of an isosceles triangle and its ends are parallel

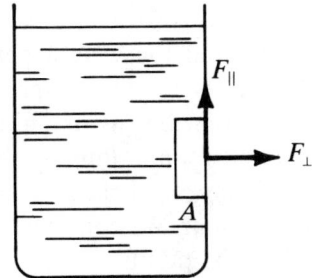

Figure 12.2.2 Fluids at rest exert only perpendicular forces.

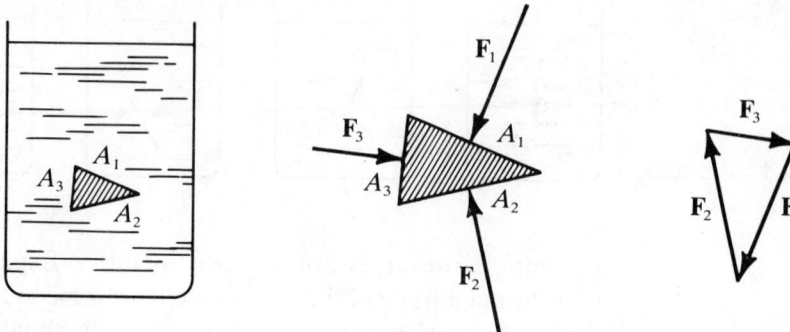

Figure 12.2.3 Forces on a small prism-shaped volume of fluid.

to each other and perpendicular to the three sides of the prism. These sides have areas A_1, A_2, and A_3. Thus the prism has acting on it forces $\mathbf{F}_1$, $\mathbf{F}_2$, and $\mathbf{F}_3$, as shown—*plus* two additional forces acting on the ends. By symmetry, these two forces must identically cancel. Since the fluid—including this small volume of it—is in equilibrium, the net force on it must be zero. That is,

$$\mathbf{F}_1 + \mathbf{F}_2 + \mathbf{F}_3 = 0 \quad (12.2.2)$$

as shown in the figure.

By construction of the prism and the relationship of force and pressure given in Eq. 12.2.1, this vector addition triangle must also be isosceles. Therefore,

$$F_1 = F_2 \quad (12.2.3)$$

Also, the two corresponding pressures must be identical. But this construction was entirely independent of the orientation of the prism or the angle between the sides. Therefore, we can conclude that the pressure is the same in *any* direction:

$$p = \frac{F}{A} = \frac{F_1}{A_1} = \frac{F_2}{A_2} \quad (12.2.4)$$

A fluid also has weight, which we have not yet considered. We write the weight per unit volume as $\mathbf{w}$; then

$$\mathbf{w} = \rho \mathbf{g} \quad (12.2.5)$$

where ρ is the density, the mass per unit volume; that is, the *force* per unit volume due to gravity. If the fluid is (or can be considered) incompressible, then the density ρ is constant and the weight per unit volume $\mathbf{w}$ remains constant throughout the fluid. If the fluid is compressible (for example, a gas), the direction of $\mathbf{w}$ remains constant but its magnitude changes.

Consider two points P_1 and P_2 in a fluid as shown in Figure 12.2.4. To account for possible variations in $\mathbf{w}$, let them be nearby (so we can treat $\mathbf{w}$ as a constant) and separated by vector $d\mathbf{r}$. We construct a right circular cylinder of cross section dA about vector $d\mathbf{r}$; points P_1 and P_2 lie on the faces of this

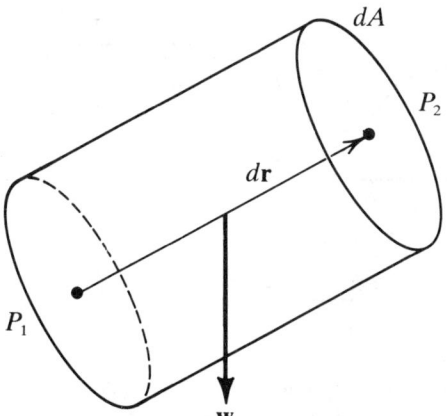

Figure 12.2.4 Pressure variation in a fluid.

cylinder. Forces—due to pressure and to gravity—are exerted all over this cylinder and in various directions. Look at only the components of forces *along the direction of d*$\mathbf{r}$:

$$\mathbf{w} \cdot d\mathbf{r}\, dA + p_1\, dA - p_2\, dA = 0 \qquad (12.2.6)$$

where $dV = dr\, dA$ is the volume within the cylinder and p_1 and p_2 are the pressures at P_1 and P_2, respectively. Thus, the difference in pressure for these two nearby points separated by $d\mathbf{r}$ is just

$$dp = p_2 - p_1 = \mathbf{w} \cdot d\mathbf{r} \qquad (12.2.7)$$

We can then readily integrate to find the difference in pressure between *any* two points. For two points located by $\mathbf{r}_1$ and $\mathbf{r}_2$, the difference in pressure is

$$p_2 - p_1 = \int_{\mathbf{r}_1}^{\mathbf{r}_2} \mathbf{w} \cdot d\mathbf{r} \qquad (12.2.8)$$

This integral is a *line* integral, of course.

This important relationship means that the pressure *difference* between two points in a fluid is determined by the gravitational forces on the fluid and the spatial relations of the two points themselves. Any change in pressure at one point in a fluid at rest will be transmitted to every other point in the fluid. This fact—known as Pascal's Law—is the basis of hydraulic jacks, hydraulic servo-control systems, and hydraulic suspension.

We shall find the pressure at a depth h below the surface of a liquid. Assume that the density remains constant. Consider points A, B, C, and D in Figure 12.2.5. All are in a plane at depth h below the surface; point O is on the surface. This figure may as well represent a swimming pool. Then the pressure at point O on the surface is merely atmospheric pressure: $p_O = 1$ atm.

From Eq. 12.2.8 we know

$$p_A - p_O = \int_{\mathbf{r}_O}^{\mathbf{r}_A} \mathbf{w} \cdot d\mathbf{r} = \int_{\mathbf{r}_O}^{\mathbf{r}_A} \rho \mathbf{g} \cdot d\mathbf{r} = \rho g h$$

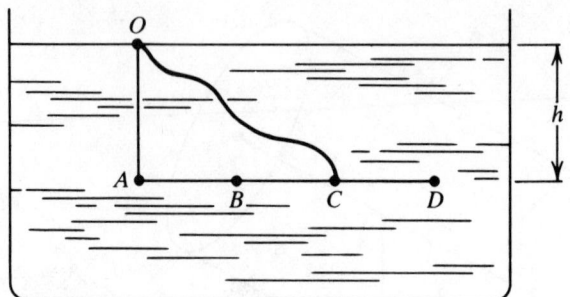

Figure 12.2.5 Pressure at depth h in a liquid.

For point A, directly below O, it's easy to see that $p_A - p_O = \rho g h$ since $h = |\mathbf{r}_A - \mathbf{r}_O|$. But, for *each* other point B, C, or D, the line integral has exactly the same value. This is easiest to see by looking at a path from O to A and then from A to B or C or D. But the result is exactly the same for any path you care to use—like the one sketched from O to C.

Thus, we can write

$$p(h) = p_O + \rho g h \qquad (12.2.9)$$

where p_O is the pressure at some point O and $p(h)$ is then the pressure anyplace in the fluid a depth h below point O. To use this form, of course, we must be able to consider the density as a constant.

To see this another way, look at the rectangular prism of fluid drawn in Figure 12.2.6. Since the entire fluid is at rest, the net force on the prism must be zero. By symmetry, the forces on the sides of the volume must cancel. The upward force, the force on the bottom of the prism due to pressure, is $p(h)A$. The downward forces, the weight of the fluid and the force on the top due to pressure, are $\rho g h A$ and $p_O A$. Thus,

$$p(h)A = p_O A + \rho g h A$$

And, dividing by the area A gives Eq. 12.2.9 again.

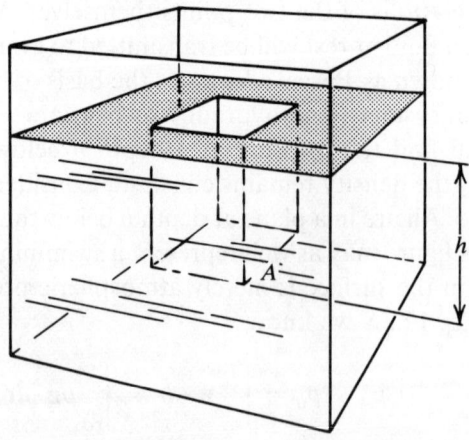

Figure 12.2.6 Pressure at depth h.

Thus, you can see that the *increase* in pressure at depth h is the additional weight of a column of fluid of height h divided by the area of that column.

All of this means that the pressure and, hence, the force on the bottom of a diver—or any body submerged in a fluid—is greater than that on top. This means that a body submerged in a fluid experiences a net upward, or buoyant, force due to the fluid surrounding it.

To be more quantitative, look at the weight of the fluid contained within some volume V. That is,

$$\mathbf{W} = \iiint_V \mathbf{w} \, dV \tag{12.2.10}$$

For a fluid at rest, this is precisely balanced by the forces due to pressure by the surrounding fluid on the surface surrounding this volume:

$$\mathbf{F}_b = \iint_S \mathbf{n} p \, dS \tag{12.2.11}$$

Since the fluid is at rest, these two forces must balance each other:

$$\mathbf{F}_b = -\mathbf{W} = -\iiint_v \mathbf{w} \, dV \tag{12.2.12}$$

Thus, the buoyant force on a volume in a fluid is equal to the weight of the fluid inside that volume. Known as Archimedes' Principle, this can also be stated as: *the buoyant force on a body immersed in a fluid is equal to the weight of the fluid displaced.* This buoyant force is due to the pressure of the fluid surrounding a volume and is the same whether that volume contains more fluid or matter of some other type.

We live at the bottom of an ocean of air. Since we are accustomed to the ordinary pressure of 1 atm in which we live, we seldom think of it at all and it is often convenient to calibrate pressure gauges to read *zero* at ordinary atmospheric pressure. Such gauges essentially read the $\rho g h$ term, called the *gauge pressure*. On the other hand, $p_o + \rho g h$—the real, actual, authentic pressure present—is called the *absolute pressure*.

Pressure is a force divided by an area, so it must have units of N/m². This is given the special name of *pascal*, abbreviated Pa. Because a pascal is a rather small pressure, kilopascals are very useful:

$$10^3 \text{ N/m}^2 = 1 \text{ kPa} \tag{12.2.13}$$

Now, what is ordinary, everyday atmospheric pressure? Figure 12.2.7 shows a mercury barometer. A long tube is completely filled with mercury. The open end is then capped and submerged in a container of mercury. When the cap—or thumb—is removed from the tube, mercury will flow out of the tube

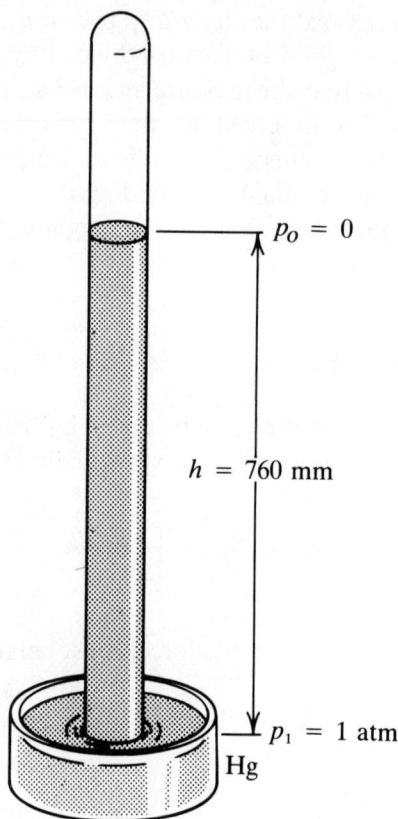

Figure 12.2.7 Mercury barometer.

until the top of the mercury column is about 760 mm above the surface of the mercury below. The top of the column has zero pressure—there's a vacuum there. And the surface of the mercury in the container below is at one atmosphere—it's open to the air. Thus, Eq. 12.2.9 immediately yields

$$p_1 - p_0 = 1 \text{ atm} - 0 = \rho_{Hg} g h$$

or

$$1 \text{ atm} = \left(13.6 \times 10^3 \, \frac{\text{kg}}{\text{m}^3}\right)\left(9.8 \, \frac{\text{m}}{\text{s}^2}\right)(0.76 \text{ m})$$

$$1 \text{ atm} = 1.01 \times 10^5 \text{ Pa} = 101 \text{ kPa}$$

EXAMPLE 12.2.1 A hydraulic jack uses Pascal's law of undiminished transmission of pressure throughout a fluid. Consider a car, sketched in Figure 12.2.8, with a mass of 1.5 metric tons (1500 kg) sitting on a rack supported by a piston of 15-cm radius. Hydraulic fluid supports the piston. Hydraulic lines connect it to a smaller piston that is connected to a motor and a valve and tank arrangement. The motor can only exert a force of 100 N (about 22 lb) on this second, pumping piston. What must be the radius of this second

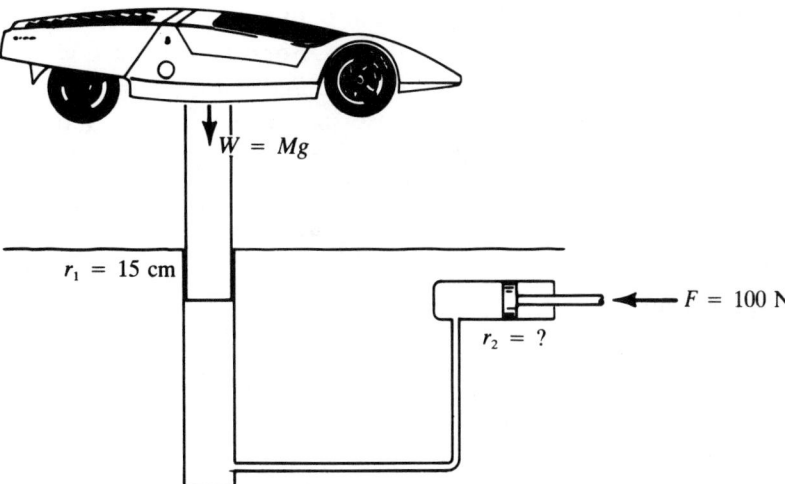

Figure 12.2.8
Hydraulic jack.

piston so that the pressure at its face is the same as that inside the cylinder where the piston is lifting the automobile?

$$P_1 = \frac{(1500 \text{ kg})(9.8 \text{ m/s}^2)}{\pi(.15 \text{ m})^2} = \frac{100 \text{ N}}{\pi r_2^2} = P_2$$

$$r_2 = 0.0124 \text{ m} = 12.4 \text{ mm}$$

This secondary, pumping piston must be small with a radius of 12.4 mm. What does all this mean? With this proper choice of pistons, a very small force of only 100 N is supporting the 15,000-N weight of the automobile! Is this a case of sleight of hand or are we really getting more out than we put in? We're certainly getting out more force. But we won't get out more work. To lift the car a distance l_1, the piston will have to travel a distance l_2. If the fluid is not compressible, then it is easy to see that

$$l_1(\pi r_1^2) = l_2(\pi r_2^2)$$

And we know the pressures are the same:

$$p_1 = \frac{W}{\pi r_1^2} = \frac{F}{\pi r_2^2} = p_2$$

Combining these we have

$$l_1 W = l_2 F$$

The left side is the work *out* in lifting the car; the right side, the work *in* due to the pump. And they're equal. Of course we've neglected any friction and any complications due to the weight of the hydraulic fluid or position of the pump.

But the density is not always constant—especially when dealing with a gas. A change in pressure will usually cause a change in volume and, hence, in density. We can write an equation relating these changes:

$$\frac{dV}{V} = -\frac{dp}{B} \tag{12.2.14}$$

where B is called the bulk modulus. It is merely a proportionality constant that varies with the gas. For an ideal gas, it is constant. For a real gas, it will vary a little as pressure and temperature change. The minus sign is present because an increase in pressure causes a decrease in volume. Using the definition of density, $\rho = M/V$, we can write

$$d\rho = -\frac{M}{V^2} dV = -\left(\frac{M}{V}\right)\frac{dV}{V}$$

or

$$\frac{d\rho}{\rho} = -\frac{dV}{V} \tag{12.2.15}$$

Therefore,

$$\frac{d\rho}{\rho} = -\frac{dp}{B} \tag{12.2.16}$$

An ideal gas obeys an equation of state:

$$pV = nRT \tag{12.2.17}$$

where n is the number of moles of gas being considered, T is the temperature, and the universal gas constant R is equal to

$$R = 8.314 \times 10 \frac{J}{C° \text{ mole}}$$

So we can rewrite the density as

$$\rho = \frac{Mp}{nRT} = \frac{mp}{RT} \tag{12.2.18}$$

where $m = M/n$ is the mass of one mole.

Let us use this information to find the pressure in our atmosphere as a function of altitude z. Assume temperature T remains constant. Exchanging h for $-z$ in Eq. 12.2.9 or going back to more the basic equations 12.2.7 and 12.2.5, we have

$$\frac{dp}{dz} = -\rho g \tag{12.2.19}$$

With Eq. 12.2.18 this is

$$\frac{dp}{dz} = -\left(\frac{mg}{RT}\right)p \tag{12.2.20}$$

Integration then yields an expression for pressure as a function of altitude in an atmosphere at uniform temperature:

$$p = p_0 e^{(-mg/RT)z} \tag{12.2.21}$$

and this is quite different from our earlier expressions involving only a constant density.

12.3 Moving with the Flow

A direct generalization of the ideas of particle motion in Chapter 9 would lead us to follow the motion of individual fluid particles. Given initial conditions $x_0, y_0, z_0, v_{0x}, v_{0y}$, and v_{0z} of individual fluid particles, we would expect to predict the later motion of these individual particles. This approach can be taken. But for most purposes it is more useful to describe the motion of a fluid in terms of its density $\rho(x, y, z, t)$ and velocity $\mathbf{v}(x, y, z, t)$ at each point in space and each instant of time. This is the approach we shall use here. We shall be concerned with what happens at a particular position and time more than with what happens to a particular particle.

But the laws of mechanics apply to these particles and not merely to the space and time through which they flow. Thus, we shall still have occasion to look at the particles themselves. This will become especially evident when we need derivatives of quantities describing the fluid—like the density or pressure. The time rate of change of a quantity measured at a point fixed in space is the partial time derivative (e.g., $\partial p/\partial t$) while the time rate of change of a quantity measured by a particle moving with the fluid is the total time derivative. Using pressure as an example, that means

$$\frac{dp}{dt} = \frac{\partial p}{\partial t} + \frac{\partial p}{\partial x}\frac{dx}{dt} + \frac{\partial p}{\partial y}\frac{dy}{dt} + \frac{\partial p}{\partial z}\frac{dz}{dt} \tag{12.3.1}$$

where dx/dt, dy/dt, and dz/dt are the components of the fluid velocity $\mathbf{v}$.

Using the gradient defined by equation (4.6.8), we can write this as

$$\frac{dp}{dt} = \frac{\partial p}{\partial t} + \mathbf{v} \cdot \nabla p \tag{12.3.2}$$

Of course, a similar equation obtains for any variable of interest. So we may write this in terms of just the operators:

$$\frac{d}{dt} = \frac{\partial}{\partial t} + \mathbf{v} \cdot \nabla \tag{12.3.3}$$

Consider a small volume element δV *that moves along with the fluid*. As the fluid moves, this small volume may expand or contract, but it always contains the same fluid particles. In rectangular coordinates, we can express this volume as

$$\delta V = \delta x \, \delta y \, \delta z \tag{12.3.4}$$

Figure 12.3.1 shows a sketch of such a volume that expands as it moves. The fluid velocity may well be different on the different sides of this volume. If so, this may mean that this volume varies with time. This is easier to see if we look at the sides δx, δy, and δz of this small volume. If the velocity v_x varies

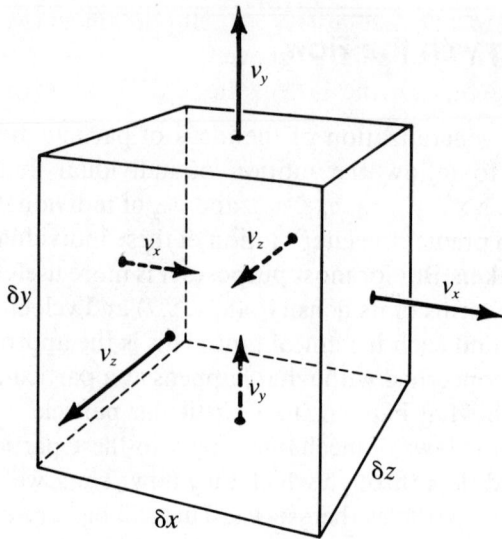

Figure 12.3.1 A volume element $V = xyz$ that expands as it moves.

from one end of δx to the other, then the length δx will change with time. Specifically,

$$\frac{d\,\delta x}{dt} = \frac{\partial v_x}{\partial x}\delta x \tag{12.3.5}$$

And, of course, similar equations hold for δy and δz:

$$\frac{d\,\delta y}{dt} = \frac{\partial v_y}{\partial y}\delta y \tag{12.3.6}$$

and

$$\frac{d\,\delta z}{dt} = \frac{\partial v_z}{\partial z}\delta z \tag{12.3.7}$$

All these are really true only in the limit of each of these sides approaching zero.

These can be put together to find the time rate of change of the entire volume element δV:

$$\frac{d}{dt}\delta V = \frac{d}{dt}(\delta x\,\delta y\,\delta z)$$

$$= \left(\frac{d\,\delta x}{dt}\right)\delta y\,\delta z + \delta x\left(\frac{d}{dt}\delta y\right)\delta z + \delta x\,\delta y\,\frac{d}{dt}\delta z$$

$$= \frac{\partial v_x}{\partial x}\delta x\,\delta y\,\delta z + \delta x\,\frac{\partial v_y}{\partial y}\delta y\,\delta z + \delta x\,\delta y\,\frac{\partial v_z}{\partial z}\delta z$$

$$= \left(\frac{\partial v_x}{\partial x} + \frac{\partial v_y}{\partial y} + \frac{\partial v_z}{\partial z}\right)\delta x\,\delta y\,\delta z$$

$$\frac{d}{dt}\delta V = \nabla\cdot\mathbf{v}\,\delta V \tag{12.3.8}$$

SECTION 12.3 / HYDRODYNAMICS

This just means that $\partial \delta V/\partial t = 0$. But that must clearly be the case for smooth flow past a fixed point in space.

Even though the little volume δV may expand or contract, the mass inside it must remain constant. The mass contained inside δV is

$$\delta m = \rho \, \delta V \tag{12.3.9}$$

For δm to remain constant, its time derivative must vanish:

$$\frac{d}{dt} \delta m = \frac{d}{dt}(\rho \, \delta V) = 0 \tag{12.3.10}$$

We shall carry out this differentiation with the help of Eq. 12.3.8:

$$\begin{aligned}\frac{d}{dt} \delta m &= \frac{d}{dt}(\rho \, \delta V) \\ &= \frac{d\rho}{dt} \delta V + \rho \frac{d}{dt} \delta V \\ &= \frac{d\rho}{dt} \delta V + \rho \nabla \cdot \mathbf{v} \, \partial V = 0\end{aligned} \tag{12.3.11}$$

Or, after dividing by δV,

$$\frac{d\rho}{dt} + \rho \nabla \cdot \mathbf{v} = 0 \tag{12.3.12}$$

Remember, this *total* time derivative must refer to the derivative seen by an observer *moving along with the fluid*. We shall use Eq. 12.3.3 to write this in terms of partial derivatives, referring to a *fixed* observer. Thus,

$$\frac{\partial \rho}{\partial t} + \mathbf{v} \cdot \nabla \rho + \rho \nabla \cdot \mathbf{v} = 0 \tag{12.3.13}$$

Or, combining the last two terms,

$$\frac{\partial \rho}{\partial t} + \nabla \cdot (\rho \mathbf{v}) = 0 \tag{12.3.14}$$

This is the *equation of continuity*. It is essentially a statement of the conservation of matter. You will encounter similar equations of continuity in electromagnetism.

We have made frequent use of the divergence in describing fluid flow. The curl, too, is useful in analyzing fluid flow. Equation 4.6.31 is useful here.

$$\iint_S (\nabla \times \mathbf{V}) \cdot \hat{\mathbf{n}} \, dS = \oint \mathbf{V} \cdot d\mathbf{r} \tag{4.6.31}$$

Figure 12.3.2 shows two examples of flow for which the line integral $\oint \mathbf{V} \cdot d\mathbf{r}$ is clearly nonzero. Thus, the curl of the velocity, $\nabla \times \mathbf{v}$, must also be

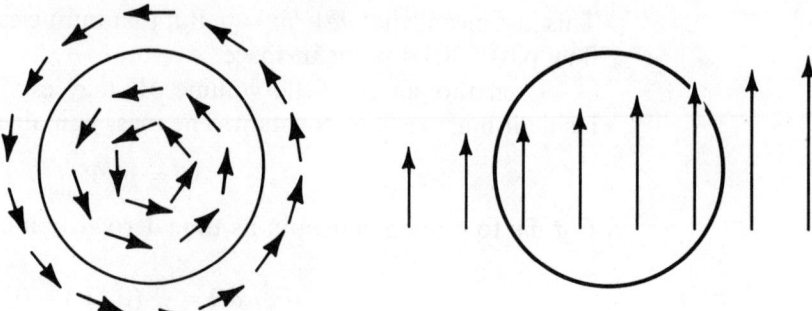

Figure 12.3.2 Two examples of flow with nonzero curl *v*.

nonzero. The example on the left is that of a vortex; on the right, a transverse velocity gradient. If curl **v** vanishes, we call the flow *irrotational flow*.

Consider throwing a small paddle wheel into fluid as it flows. Or watch a leaf on top of a river. For *irrotational flow*, the leaf or paddle wheel moves along without any rotation. But if $\nabla \times \mathbf{v} \neq 0$, then there will be rotation. Let us find the relation between this rotation, the angular velocity $\boldsymbol{\omega}$, and the curl **v**, $\nabla \times \mathbf{v}$, of the fluid flow.

Consider a primed coordinate system rotating with angular velocity $\boldsymbol{\omega}$. There, by Eq. 6.6.12, the velocity **v'** measured in this frame and the velocity **v** measured in the fixed frame are related by

$$\mathbf{v} = \mathbf{v}' + \boldsymbol{\omega} \times \mathbf{r}' \qquad (12.3.15)$$

where **r'** is the position vector in the primed, rotating frame. Take the curl of the velocity

$$\nabla \times \mathbf{v} = \nabla \times \mathbf{v}' + \nabla \times (\boldsymbol{\omega} \times \mathbf{r}')$$
$$= \nabla \times \mathbf{v}' + \boldsymbol{\omega}(\nabla \cdot \mathbf{r}') - (\boldsymbol{\omega} \cdot \nabla)\mathbf{r}'$$
$$= \nabla \times \mathbf{v}' + 3\boldsymbol{\omega} - \boldsymbol{\omega}$$
$$\nabla \times \mathbf{v} = \nabla \times \mathbf{v}' + 2\boldsymbol{\omega} \qquad (12.3.16)$$

(We have evaluated $\nabla \times (\boldsymbol{\omega} \times \mathbf{r}')$ using the vector identity $\mathbf{A} \times (\mathbf{B} \times \mathbf{C}) = (\mathbf{A} \cdot \mathbf{C})\mathbf{B} - (\mathbf{A} \cdot \mathbf{B})\mathbf{C}$. The divergence of a position vector is simply 3 and explicit evaluation of $(\boldsymbol{\omega} \cdot \nabla)\mathbf{r}'$ does, in fact, yield $\boldsymbol{\omega}$.)

We may set $\nabla \times \mathbf{v}' = 0$. This irrotational flow is what an observer would see if he rode along with the paddlewheel or leaf, moving along with the fluid. Then, from Eq. 12.3.16,

$$\boldsymbol{\omega} = \tfrac{1}{2}(\nabla \times \mathbf{v}) \qquad (12.3.17)$$

This $\boldsymbol{\omega}$, then, is the angular velocity of the fluid—or the angular velocity of a small paddlewheel moving freely with the fluid.

12.4 Hydrodynamics

So far we have *described* the motion of a fluid; formally this is called fluid *kinematics*. Now we move on to fluid *dynamics* and study the *causes* of this motion.

You already know by Eq. 12.2.5 that due to gravity there is a force on a volume element δV of

$$\boldsymbol{\omega}\delta V = \rho \mathbf{g}\, \delta V \tag{12.4.1}$$

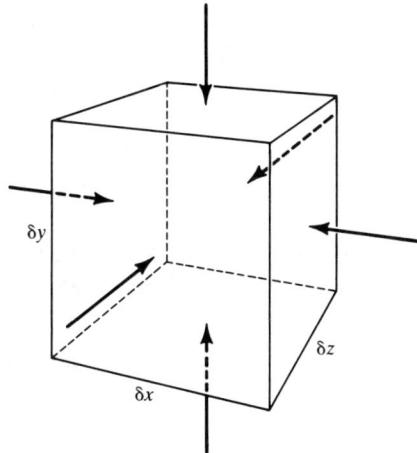

Figure 12.4.1 Forces due to pressure on a volume element V.

But how can you handle the forces due to pressure? Figure 12.4.1 shows our now familiar volume element δV with forces on all sides due to pressure. Of course, pressure may change as we move a distance δx, δy, or δz. And this will cause the forces to change so the net force need not be zero. Figure 12.4.2 shows in more detail the x-components of the forces acting on opposite faces of δV. If we call δF_x the net x-component of force on δV, then

$$\begin{aligned} \delta F_x &= F_x(x) - F_x(x + \delta x) \\ &= [p(x) - p(x + \delta x)]\delta y\, \delta z \\ &= \frac{p(x) - p(x + \delta x)}{\delta x}\, \delta x\, \delta y\, \delta z \end{aligned}$$

$$\delta F_x = -\frac{\partial p}{\partial x}\, \delta V \tag{12.4.2}$$

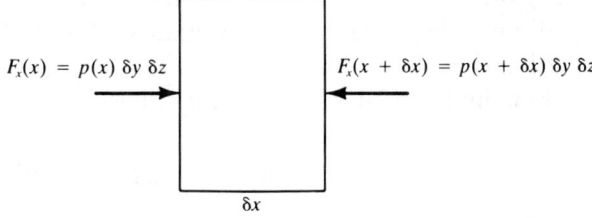

Figure 12.4.2 Detailed look at x-components of forces on faces yz of volume element V.

Combining δF_x, δF_y, and δF_z to make the vector $\delta \mathbf{F}$, we have

$$\delta \mathbf{F} = -\left(\frac{\partial p}{\partial x}\hat{\mathbf{i}} + \frac{\partial p}{\partial y}\hat{\mathbf{j}} + \frac{\partial p}{\partial z}\hat{\mathbf{k}}\right)\delta V$$

or

$$\delta \mathbf{F} = -\nabla p\, \delta V \tag{12.4.3}$$

The net force on volume element δV then is just the sum of Eqs. 12.3.18 and 12.4.3. This and Eq. 12.3.10 allow us to write Newton's Second Law as

$$\rho\, \delta V \frac{d\mathbf{v}}{dt} = \rho \mathbf{g}\, \delta V - \nabla p\, \delta V$$

or, more simply, as

$$\rho \frac{d\mathbf{v}}{dt} + \nabla p = \rho \mathbf{g} \tag{12.4.4}$$

Just as we did with Eqs. 12.3.12 and 12.3.13, we shall write this in terms of the partial derivates that refer to a *fixed* observer. Then

$$\rho\left(\frac{\partial \mathbf{v}}{\partial t} + \mathbf{v}\cdot\nabla\mathbf{v}\right) + \nabla p = \rho \mathbf{g}$$

or

$$\frac{\partial \mathbf{v}}{\partial t} + \mathbf{v}\cdot\nabla\mathbf{v} + \frac{1}{\rho}\nabla p = \mathbf{g} \tag{12.4.5}$$

which is known as Euler's equation of motion for a fluid. Operator equation 12.3.3 applied to a scalar quantity is straightforward. Applied to the pressure, we have Eq. 12.3.2, and the last term, $\mathbf{v}\cdot\nabla p$, is just the dot product of the velocity with the gradient of the pressure. Now, having applied Eq. 12.3.4 to a vector quantity, the velocity, we have an unusual-looking term $\mathbf{v}\cdot\nabla\mathbf{v}$. This term is to be read as an operator $(\mathbf{v}\cdot\nabla)$ operating upon the velocity $\mathbf{v}$. It can be shown that

$$(\mathbf{v}\cdot\nabla)\mathbf{v} = \nabla\left(\frac{1}{2}v^2\right) - \mathbf{v}\times(\nabla\times\mathbf{v})$$

If the density is directly related to the pressure (for example, in an ideal gas) then there are four quantities at every point in space that determine or specify the flow: v_x, v_y, v_z, and p. Euler's equation (Eq. 12.4.5) gives us three equations. The equation of continuity (Eq. 12.3.14) provides the fourth equation.

Rewrite Eq. 12.4.4 and multiply by δV:

$$\rho\, \delta V \frac{d\mathbf{v}}{dt} = (\rho \mathbf{g} - \nabla p)\, \delta V$$

SECTION 12.4 / HYDRODYNAMICS

But since $\delta m = \rho \, \delta V$ is a constant, we may pull $\rho \delta V$ through the differentiation to write

$$\frac{d}{dt}(\rho \mathbf{v} \, \delta V) = (\rho \mathbf{g} - \nabla p) \, \delta V \tag{12.4.6}$$

$\rho \mathbf{v} \, \delta V$ is just the momentum contained in volume element δV and the right-hand side is the net force on that element. This equation is equivalent for fluids to

$$\frac{d\mathbf{p}}{dt} = \sum \mathbf{F}$$

Equation 12.4.6 is true for each point in space. We might want to integrate this equation over some volume to give

$$\frac{d}{dt} \iiint_V \rho \mathbf{v} \, dV = \iiint_V \rho \mathbf{g} \, dV - \iiint_V \nabla p \, dV \tag{12.4.7}$$

It can be shown that

$$\iiint_V \nabla p \, dV = \iint_S \hat{\mathbf{n}} p \, dS$$

This resembles Gauss' Theorem, Eq. 4.6.21, and is a corollary to it. Using this, our equation now becomes

$$\frac{d}{dt} \iiint_V \rho \mathbf{v} \, dV = \iiint_V \rho \mathbf{g} \, dV + \iint_S (-\hat{\mathbf{n}}) \, p \, dS \tag{12.4.8}$$

The integral on the left is the total linear momentum contained within some volume V. The time rate of change of this momentum is equal to the two terms on the right. The first is just the weight of all the fluid in the volume. The second is the sum of all the forces acting on the *surface* of this volume. Of course, this is exactly what we would expect from our earlier discussion of a system of particles in Chapter 8.

We considered conservation of mass and arrived at the equation of continuity. Now we have looked at linear momentum. Energy proved to be of enormous importance with single bodies. Energy will prove just as useful here with fluids.

Take the dot product of $\mathbf{v}$ with both sides of Eq. 12.4.6. Remember that

$$\delta m = \rho \, \delta V = \text{constant}$$

and

$$\frac{d}{dt}(v^2) = \frac{d}{dt}(\mathbf{v} \cdot \mathbf{v}) = 2\mathbf{v} \cdot \frac{d\mathbf{v}}{dt}$$

CHAPTER 12 / FLUID MECHANICS

and you will readily have

$$\frac{d}{dt}\left(\frac{1}{2}\rho v^2 \, \delta V\right) = \mathbf{v} \cdot (\rho \mathbf{g} - \nabla p) \, \delta V \qquad (12.4.9)$$

The term in parenthesis on the left is but the kinetic energy contained in volume δV. The left-hand side, the velocity dotted into the net force, is but the power supplied. This equation says that the increase in kinetic energy of a volume of fluid is equal to the power supplied to it.

It is useful to modify this equation by replacing some of the right-hand terms with potential energy terms on the left. To do this, we can use Eqs. 12.3.4 and 12.3.9:

$$\frac{d}{dt}(p \, \delta V) = \frac{dp}{dt}\delta V + p \frac{d \, \delta V}{dt}$$

$$\frac{d}{dt}(p \, \delta V) = \frac{\partial p}{\partial t}\delta V + \mathbf{v} \cdot \nabla p \, \delta V + p \nabla \cdot \mathbf{v} \, \delta V$$

Thus,

$$-\mathbf{v} \cdot \nabla p \, \delta V = -\frac{d}{dt}(p \, \delta V) + \frac{\partial p}{\partial t}\delta V + p \nabla \cdot \mathbf{v} \, \delta V \qquad (12.4.10)$$

Using the gravitational potential Φ defined in Eq. 7.8.12, we can write the $\mathbf{v} \cdot \rho \mathbf{g} \, \delta V$ term as

$$\mathbf{v} \cdot \rho g \, \delta V = \rho \, \delta V \mathbf{v} \cdot \mathbf{g}$$
$$= -\rho \, \delta V \mathbf{v} \cdot \nabla \Phi$$
$$= -\rho \, \delta V \left(\frac{d\Phi}{dt} - \frac{\partial \Phi}{\partial t}\right)$$

where we have used Eq. 12.3.4 once again. Recalling once more that $\rho \, \delta V$ is constant, we can rewrite this as

$$\mathbf{v} \cdot \rho \mathbf{g} \, \delta V = -\frac{d}{dt}(\rho \Phi \, \delta V) + \rho \, \delta V \frac{\partial \Phi}{\partial t} \qquad (12.4.11)$$

Of course $\partial \Phi / \partial t$ will vanish for all situations except, perhaps, those in astrophysical hydrodynamics involving flow of stellar dust clouds or the like. Nevertheless, we shall carry it along for a while.

Now we can collect all the total time derivative terms together and write

$$\frac{d}{dt}\left[\left(\frac{1}{2}\rho v^2 + p + \rho \Phi\right)\delta V\right] = \left(\frac{\partial p}{\partial t} + \rho \frac{\partial \Phi}{\partial t}\right)\delta V + p \nabla \cdot \mathbf{v} \, \delta V \qquad (12.4.12)$$

where $\frac{1}{2}\rho v^2$ is the kinetic energy density and $\rho \Phi$ is the gravitational potential energy density. We may then identify the pressure p as another potential energy density. As mentioned earlier, $\partial \Phi / \partial t$ will usually vanish. For smooth, streamline flow the pressure at a given point in space will remain constant; so $\partial p/\partial t$ will also vanish. For an incompressible fluid, the equation of continuity requires that

$$\mathbf{V} \cdot \mathbf{v} = 0$$

so the entire right-hand side vanishes, leaving only

$$\frac{1}{2}\rho v^2 + p + \rho \Phi = \text{constant} \qquad (12.4.13)$$

This is known as Bernoulli's equation. But, before we discuss its applications, let us see if it is possible to include compressible fluids as well.

As volume element δV expands under pressure p, it does work on the surrounding fluid. We can write this as

$$dW = p \, d(\delta V) \qquad (12.4.14)$$

The time rate at which this work is done is

$$\frac{dW}{dt} = p\frac{d\,\delta V}{dt} = p\mathbf{V} \cdot \mathbf{v}\,\delta V \qquad (12.4.15)$$

If we consider only so-called homogeneous fluids, that is, ones with density determined only by pressure, then we can define a potential energy associated with this work. We define u as the potential energy density so the potential energy contained in volume element δV is the negative of the work done on the surrounding fluid as its pressure goes from some standard reference pressure p_0 to a final pressure p. That is,

$$u\,\delta V = -\int dW = -\int_{p_0}^{p} p\,d(\delta V) \qquad (12.4.16)$$

Clearly,

$$\frac{d(u\,\delta V)}{dt} = -\frac{dW}{dt}$$

so Eq. 12.4.12 can now be written as

$$\frac{d}{dt}\left[\left(\frac{1}{2}\rho v^2 + p + \rho \Phi + u\right)\delta V\right] = \left(\frac{\partial p}{\partial t} + \rho\frac{\partial \Phi}{\partial t}\right) \qquad (12.4.17)$$

As explained earlier, for most problems of interest the right-hand side vanishes ($\partial p/\partial t$ is nonzero for turbulent flow). So we can write for those cases

$$\frac{d}{dt}\left(\frac{1}{2}\rho v^2 + p + \Phi\rho + u\right)\delta V = 0 \qquad (12.4.18)$$

or

$$\tfrac{1}{2}\rho v^2 + p + \rho\Phi + u = \text{constant} \qquad (12.4.19)$$

And this is Bernoulli's equation. Now, what does it mean?

If we look at various positions in a fluid, we will find that the expression in Eq. 12.4.19 always has the same value. An increase in one term, or an increase in one form of energy, must be accompanied by a decrease someplace else. This is just a statement of energy conservation. To begin with, let us consider a fluid at rest. Thus, $v = 0$. If we further confine our discussion to an incompressible fluid, u disappears. This leaves only

$$p + \rho\Phi = \text{constant}$$

Φ is very general. Near Earth's surface, where g is constant, Φ has the far simpler appearance of

$$\Phi = + gy$$

where y is the height above some reference level. Thus, Bernoulli's equation gives

$$p + \rho g y = \text{constant} \tag{12.4.20}$$

We can understand this better by writing

$$p_1 + \rho g y_1 = p_0 + \rho g y_0$$

or

$$p_1 = p_0 + \rho g (y_0 - y_1)$$

and this is but Eq. 12.2.9 with $y_0 - y_1 = h$.

Or consider flow of an incompressible fluid through a horizontal pipe as shown in Figure 12.4.3. We may then neglect the $\rho\Phi$ and u terms so that

$$\frac{1}{2}\rho v^2 + p = \text{constant} \tag{12.4.21}$$

or

$$\frac{1}{2}\rho v_1^2 + p_1 = \frac{1}{2}\rho v_2^2 + p_2$$

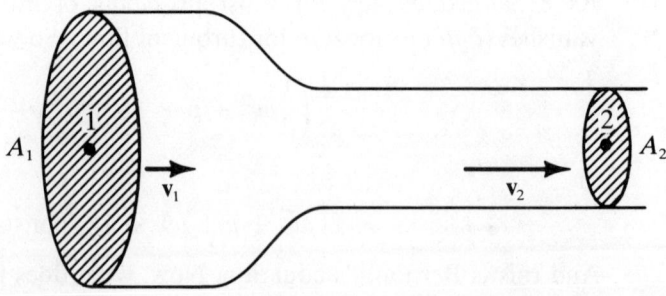

Figure 12.4.3 Pressure is greater at point 1 than at point 2.

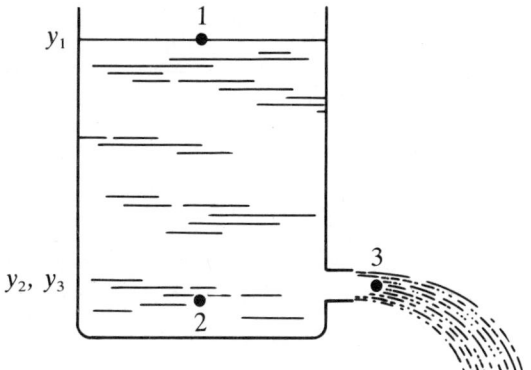

Figure 12.4.4
Torricélli's Theorem.

By the equation of continuity, the cross-section areas and velocities at points 1 and 2 are related by

$$A_1 v_1 = A_2 v_2 \tag{12.4.22}$$

$$p_1 = p_2 + \frac{1}{2}\rho(v_2^2 - v_1^2)$$

$$p_1 = p_2 + \frac{1}{2}\rho v_2^2 \left[1 - \left(\frac{A_1}{A_2}\right)^2\right] \tag{12.4.23}$$

So p_1 is clearly greater than p_2. That is, a decrease in the flow velocity is accompanied by an increase in pressure—and vice versa.

This idea can be used in such diverse applications as perfume atomizers, Venturi flow meters, air speed indicators (Pitot tubes), and airfoil (wing) design.

Apply Bernoulli's equation to the situation sketched in Figure 12.4.4. There a large tank has a small hole punched in it near the bottom. Because the hole is small, the distance from the water level to opening, $y_1 - y_3$, does not change appreciably as we discuss this problem. We assume an incompressible fluid (like water—well, it's nearly incompressible). Then we have

$$\frac{1}{2}\rho v^2 + p + \rho g y = \text{constant} \tag{12.4.24}$$

or

$$\frac{1}{2}\rho v_3^2 + p_3 + \rho g y_3 = \frac{1}{2}\rho v_1^2 + p_1 + \rho g y_1$$

From the statement of the problem, we can set $v_1 = 0$. But what is p_3, the pressure in the flow *outside* the opening? It's the same as p_1, atmospheric pressure. So that leaves only

$$v_3 = \sqrt{2g(y_1 - y_3)} \tag{12.4.25}$$

This is the same velocity a drop of water (or an apple) has if it drops from rest through a distance of $y_1 - y_3$. This is Torricelli's Theorem.

PROBLEMS

12.1 What is the difference in blood pressure between the bottom of the feet and the top of the head for a 1.8-m person who is standing straight up? (Assume that the density of blood is the same as the density of water, $\rho = 1000$ kg/m^3.)

12.2 What is the approximate difference in air pressure between the top and bottom of the CN Tower in Toronto? The tower has a height of 533 m. (Assume a density for air of about 1.3 kg/m^3.)

12.3 A solid aluminum ($\rho = 2700$ kg/m^3) cylinder with a volume of 0.5 m^3 is on the bottom of a swimming pool, 2 m below the surface. What is the buoyant force acting on it? How much work is done by a scuba diver in bringing it to the surface?

12.4 Ice has a density of about 920 kg/m^3. An ice cube with a volume of 8.7 cm^3 floats in a glass of water that is filled to the brim. What volume of water will flow over the edge of the glass as the ice melts?

12.5 A U-tube has mercury ($\rho = 13{,}600$ kg/m^3) alone on the left and water ($\rho = 1000$ kg/m^3) above mercury on the right. The height of the mercury column on the left is 40 cm above the table as shown. The height of the mercury-water interface on the right is 15 cm above the table. Find the pressure at the interface and the height of the water column above this interface. Both ends of the U-tube are open to the air.

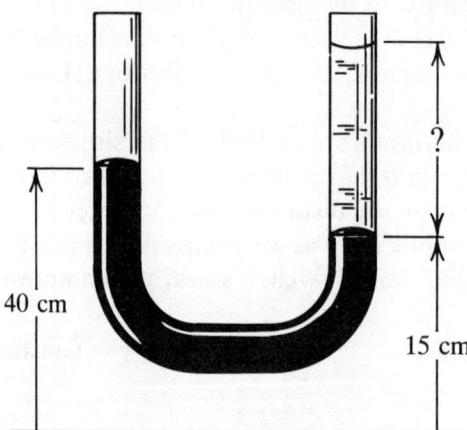

12.6 On a day when the meteorologist gives the barometric pressure as 747 mm of mercury, what is atmospheric pressure in kilopascals?

12.7 On a day when the meteorologist gives the barometric pressure as 763 mm of mercury, what is the maximum height a "perfect" lift pump might raise a column of water?

12.8 The density of the air varies with altitude. Suppose, though, that it were constant. If the *entire* atmosphere had a constant density of 1.3 kg/m^3, what would be the thickness of the atmosphere? This is the density of air at sea level for a temperature of 0°C, the so-called "standard temperature and pressure" (STP).

PROBLEMS

12.9 To demonstrate his principle, Pascal burst a wooden barrel by pouring a small amount of water into a long tube suspended above the barrel. Suspend a long tube of 0.25-cm radius above a 25-cm wooden barrel. Attach the tube to a hole drilled in the barrel top. Fill the tube with water to a height of 12 m (this is probably enough to burst the barrel). Find the mass and weight of the water in the tube. Find the net force on the lid of the barrel.

12.10 A rectangular dam 50-m high and 120-m wide forms a lake with an average depth of 32 m and length of 1.2 km. Find the total force on the dam when water just spills over the top of the dam.

12.11 What pressure is necessary in the water mains to allow a firehose to send water a height of 20 m?

12.12 As wind blows at 30 m/s across the roof of a house, what is the net force on the roof if it has an area of 160 m²?

12.13 A horizontal pipe has a narrow region. Cross sections for both regions are $A_1 = 0.010$ m³ and $A_2 = 0.005$ m³. Under most conditions, water cannot withstand negative pressures (that is, tensile stresses). Hence water begins to boil locally where the absolute pressure drops to zero. This known as *cavitation*. What is the maximum flow rate that can be achieved in the pipe if the pressure at P_1 is 1.5 atmospheres and cavitation does not occur at P_2?

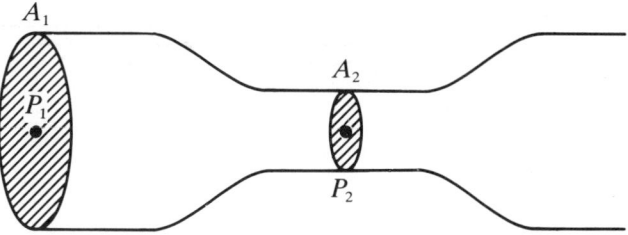

12.14 Water flows through a pipe as shown. Cross-sectional areas are $A_1 = 0.06$ m² and $A_2 = 0.02$ m³ and the difference in levels is $y_2 - y_1 = 2$ m. The mass flow rate

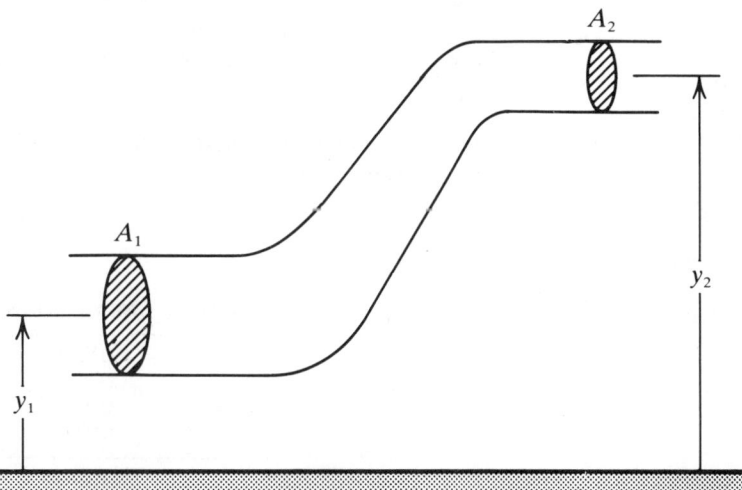

through the pipe is 40 kg/s. Find the speeds v_1 and v_2 of the water across surfaces A_1 and A_2. What is the difference in pressure between the centers of these cross-sectional areas?

12.15 Water spouted upward from Yellowstone's Old Faithful geyser reaches a height of 12 m. Approximately what must be the pressure inside the geyser to allow the water to do this? What is the speed of the water at ground level?

12.16 The air speed of an airplane can be measured using a Pitot tube as shown. Show that the airspeed is given by

$$v = \sqrt{\frac{2(p_1 - p_2)}{\rho}}$$

where p_1 and p_2 are the pressures measured at positions 1 and 2 and ρ is the density of the air.

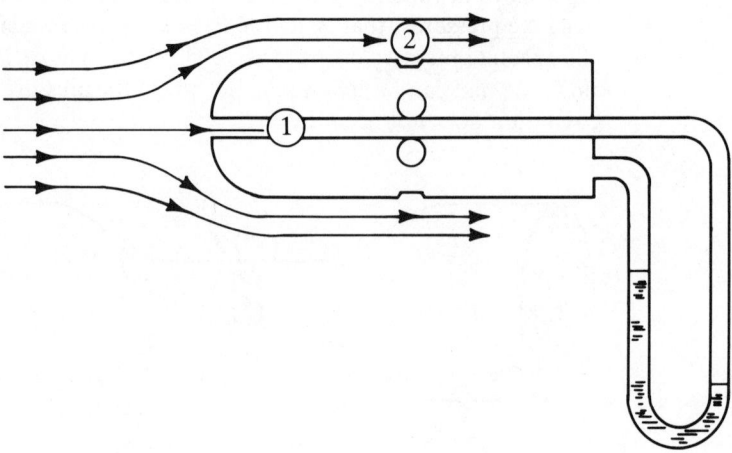

*12.17 A small hole with cross section of 1×10^{-5} m^3 is drilled near the bottom of a cylindrical container of water 20 cm in radius. The container is initially filled to a height of 1 m above the hole. What is the velocity of the water leaving the hole as a function of time?

12.18 The reservoir and piping system shown is initially filled with water to a height of 1 m. Take atmospheric pressure, Patm, to be 101 kPa. Find the pressure at points 1 through 5 just as the water starts to flow.

Position	Cross Sections
2	.010 m²
3	.005 m²
4	.010 m²
5	.010 m²

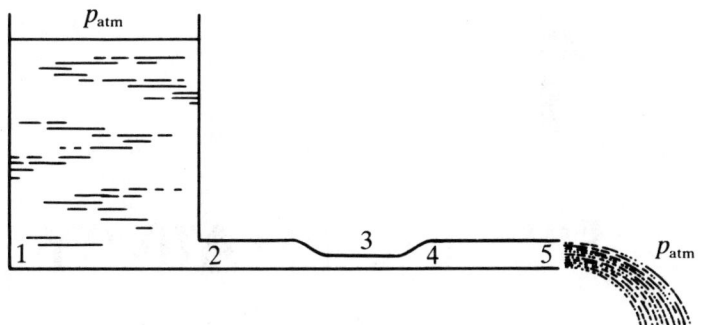

*12.19 For the reservoir and piping as shown, find an expression for the height of water in the reservoir as a function of time, $y(t)$. Find an expression for the mass flow rate as a function of time. What is the pressure at position 3 as a function of time? These may all be done analytically in closed form or a numerical solution (table or graph) may be obtained. This tank has a cross section of 0.25 m².

13

SPECIAL RELATIVITY

13.1 Introduction

Taylor and Wheeler begin their excellent book, *Spacetime Physics*, with a parable of considerable significance:

> Once upon a time there was a Daytime surveyor who measured off the king's lands. He took his directions of north and east from a magnetic compass needle. Eastward directions from the center of the town square he measured in meters.... Northward directions were sacred and were measured in a different unit, in miles.... His records were complete and accurate and were often consulted by the Daytimers.
>   Nighttimers used the services of another surveyor. His north and east directions were based on the North Star. He too measured distances eastward from the center of the town square in meters... and sacred distances north in miles.... His records were complete and accurate. Every corner of a plot appeared in his book with its two coordinates,....*

Our task, in relativity, is to reconcile the observations of two observers —not unlike the surveyors in this fable. But in our case the two observers are

---

* From Edwin F. Taylor and John Archibald Wheeler, *Spacetime Physics* (New York: W. H. Freeman and Company, 1966), p. 1.

moving *relative* to each other. We've already spent some time looking at this situation in Chapter 6, which could have been entitled "Newtonian Relativity" or "Galilean Relativity." There, we looked at very complicated situations—observers moving relative to each other with constant velocity, accelerating observers, rotating observers, even rotating and accelerating observers!

In this chapter we shall restrict our attention to the special case of two observers (and their reference frames) moving relative to each other along a straight line with *constant velocity*. This is the *theory of special relativity*. It was first understood, developed, and explained by Einstein in 1905.

So what is the point, then, of this chapter? If we are able to measure with enough accuracy or if the speeds involved are great enough, we find that the clear, straightforward, obvious relationships we derived in Sections 6.5 and 6.6 simply do not give the right answers. But they do work quite well for ordinary, everyday speeds (like a car at 100 km/hr or even a spacecraft at 30,000 km/hr).

13.2 Galilean Relativity

Consider two reference frames, labeled A and B, which move relative to each other with constant velocity. Let us orient them so they are parallel and their x-axes align with their relative velocity. Further, to simplify matters, let us start our clocks or stopwatches when the origins of the two reference frames coincide. All this is shown schematically in Figure 13.2.1.

Now consider a body located at point P. Observers in both frames record its position and velocity. Of course they get different numbers. But these numbers can be related to each other by using the familiar transformation equations:

$$x_A = x_B + vt \qquad (13.2.1)$$

$$y_A = y_B \qquad (13.2.2)$$

$$z_A = z_B \qquad (13.2.3)$$

$$v_{Ax} = v_{Bx} + v \qquad (13.2.4)$$

$$v_{Ay} = v_{By} \qquad (13.2.5)$$

$$v_{Az} = v_{Bz} \qquad (13.2.6)$$

where x_A, y_A and z_A are the coordinates of point P measured by observer A while x_B, y_B, and z_B are the coordinates measured by observer B; v_{Ax}, v_{Ay}, and v_{Az} are the components of P's velocity measured by A, and v_{Bx}, v_{By} and v_{Bz} are the velocity components measured by B. Everybody measures t for the time

Figure 13.2.1
Reference frames A and B move relative to each other with constant velocity.

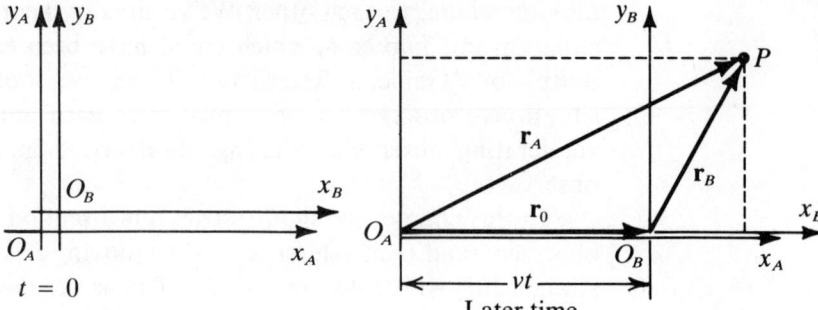

and v is the relative speed of the two observers. These equations are but a restatement of Eqs. 6.5.1 and 6.5.2.

These clear, straightforward, obvious transformation equations have been known and understood since the time of Galileo. All the *laws* of mechanics remain the same for two observers moving relative to each other with constant velocity: both are *inertial frames*. The two observers will record different *numbers* when they watch an experiment. But they will still find that the laws of mechanics ($\mathbf{F} = m\mathbf{a}$, and so on) are satisfied by their data.

Consider the situation sketched in Figure 13.2.2. An apple falls from a tree and is observed by a "stationary observer" A standing under the tree and by a "moving observer" B located on a railroad hand car moving by with

Figure 13.2.2
Observers from two reference frames watch an apple fall.

constant velocity **v**. What does each observer record for the motion of the apple? *A* sees that it falls straight down. *B* sees that it moves along a parabolic trajectory. Each observes different initial conditions and different motion but both agree that the motion observed is due to $\mathbf{F} = m\mathbf{a}$. The laws of mechanics are the same for both observers.

But what of other laws of *physics*? What of the laws of optics? Or of electromagnetism? We shall see.

13.3 Historical Background

In the late nineteenth century, James Clerk Maxwell did for electricity and magnetism much of what Newton had earlier done for mechanics. He consolidated and summarized all of it in four equations, now known as Maxwell's equations. Among other things, these equations predict that a *wave* can be caused or carried by the electric and magnetic fields. And the speed of this wave is given in terms of electric and magnetic properties. For empty space, this speed "just happened" to be astonishingly close to the speed of light. This led to the discovery that light was, indeed, an electromagnetic wave.

Maxwell's equations, then, predicted a speed for this electromagnetic wave called light (or any other electromagnetic wave like radio waves) of $c = 3.00 \times 10^8$ m/s. But if a light wave's velocity were measured by two observers moving relative to each other, who would actually find this for the measured value? The speed of sound is about 330 m/s. But that's in *still* air. A listener moving through still air at 30 m/s will measure the velocity of the sound waves to be 360 m/s as he approaches a ringing bell, and 300 m/s as he leaves it. The speed of sound is 330 m/s only in one particular, preferred reference frame—the reference frame of the still air.

And what of light? For forty years or so surrounding the turn of the century, the best experimental scientists of the world sought to find this preferred reference frame in which Maxwell's equations accurately predicted the speed of light. Their quest for this *luminiferous ether*, as they called it, makes an interesting saga of its own. But the conclusion reached is that there is no such preferred frame—or, rather, that *every* inertial reference frame is just such a preferred frame. The speed of light was experimentally found to be the same for all observers!

This clearly contradicts the velocity transformation equation (Eq. 13.2.4) and the ideas from which it was derived. But the evidence was (and remains) overwhelming. How can this be? Implicit in the development of the Galilean transformations were the ideas of "absolute space" and "absolute time." Both observers *A* and *B* will agree on the distance between any two points and both will agree on what time it is. That was obvious. But it led to the wrong conclusion!

13.4 Einsteinean Relativity/Spacetime Coordinates

Einstein's postulate of relativity can be stated as follows:

All the laws of physics remain the same for all observers moving with constant velocity relative to each other.

Period. That's it. We know it to be true for mechanics, and experiments demonstrate it to be true for optics and electromagnetism. So we shall *begin* with this. We shall *require* this. And this postulate will then force a reformulation of our basic ideas of space and time.

We can no longer *assume* anything. We must be quite specific as to what we mean when we ask for a particular measurement. We must be very specific in what we mean when we talk of distances or of times or of a reference frame.

For a reference frame, consider a huge framework of metersticks and clocks as sketched in Figure 13.4.1. The metersticks are arranged along the x-,

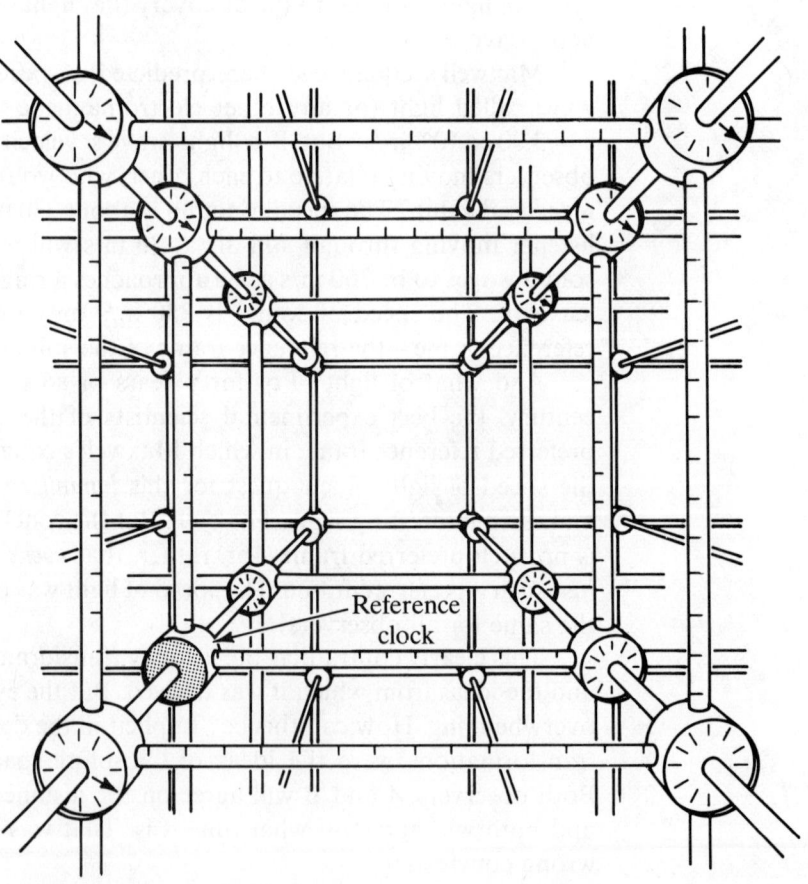

Figure 13.4.1 A reference frame.

y-, and z-axes in order to give the spatial coordinates of any point we desire. Scattered throughout this framework—at each junction of metersticks—are a multitude of time clocks. We may stick a card into any one of these and it will imprint it with the time and location (t and x, y, z). Each stamp of such a clock is an *event*. We shall carry out all our measurements in terms of these events. If a firecracker explodes, or someone sneezes, we can record the event by recording the time of a nearby clock and looking at the nearby metersticks.

Of course all the clocks must run at exactly the same rate. And they must be synchronized. We can't have one clock reading 12:38 while one a meter away reads 10:32. How can they be synchronized? With the Galilean or Newtonian concept of "absolute time" this was a ridiculous question—just set them to read the same time! But now we must give a careful, operational definition of just what we mean by synchronizing these clocks. The easiest way has to do with flashing light signals between the clocks. But we got into this whole mess because of light's peculiar behavior. So using light for the synchronization of the clocks looks like it might bias our results (it doesn't; it's a valid method, too).

Consider for a moment a clock, O, located at, say, the origin of our coordinate system or meterstick framework of Figure 13.4.2 and *one other* clock, C, that we wish to synchronize to this origin clock. Let a messenger on a bicycle travel back and forth between these clocks. We need only require that she travels at constant speed (she carries her own watch and calculates her speed using it) and uses the same path each time. She sticks a card into clock C and then brings this record of C's time back to the origin and clock O for comparison. When she arrives at the origin, clock O will read something *different* from the time earlier recorded at C. But the two clocks are synchronized if O is later by just the amount of time the messenger requires to

Figure 13.4.2
Synchronizing clocks.

make the trip. This method may seem clumsy and tedious, but it is effective and avoids using light itself.

If we can synchronize the clocks with messengers on bicycles, then we can synchronize the clocks with light signals. At, say, high noon—local time—each clock sends out a flash of light. These individual light flashes arrive at the origin at times Δt *after* the origin clock indicates high noon. The set of clocks are synchronized to this origin clock if these times Δt are related to their distances from the origin Δs by

$$\Delta t = \frac{\Delta s}{c} \qquad (13.4.1)$$

where c is the speed of light. Note that this means the various clocks would indicate different times if we could—somehow—see them all at the same time. But we can't do that. In fact, the idea of "the same time" is worth some considerable discussion in its own right. Since light *always* travels at a constant speed, it is especially convenient to use in this synchronization scheme. It isn't necessary to synchronize using light signals, but it is convenient.

This framework of perpendicular metersticks and synchronized clocks then defines our reference frame. Or we might call it a *space-time* coordinate system. We will often simply say "an observer" but, by that, we mean the entire frame or system.

13.5 Simultaneity

We have said that time can no longer be considered absolute. This implies that *simultaneity is relative*. That is, two events that are observed to happen at the *same* time by one observer may be observed to happen at *different* times by another observer moving with respect to the first. That simultaneity is relative can be directly linked to many (or all) of the future results of relativity that we shall study.

Suppose that a long, very fast alien spaceship passes over North America, passing over Philadelphia and Toronto just as street lights are turned on in both cities. This is sketched in Figure 13.5.1. Street lights are turned on in both cities at *exactly* 8:00 P.M. Eastern Daylight Time. The emissions of the first photon from a lamp in each city mark two events. These events are simultaneous to an Earthbound observer. We can determine these events to be simultaneous because they occur at exactly 8:00 P.M. as registered by local clocks that have been synchronized previously.

A more basic—and more useful—way to determine their simultaneity is to station an observer *exactly midway* between Philadelphia and Toronto. This observer will receive the first emission of light from the two cities at *exactly the same time*; they arrive together.

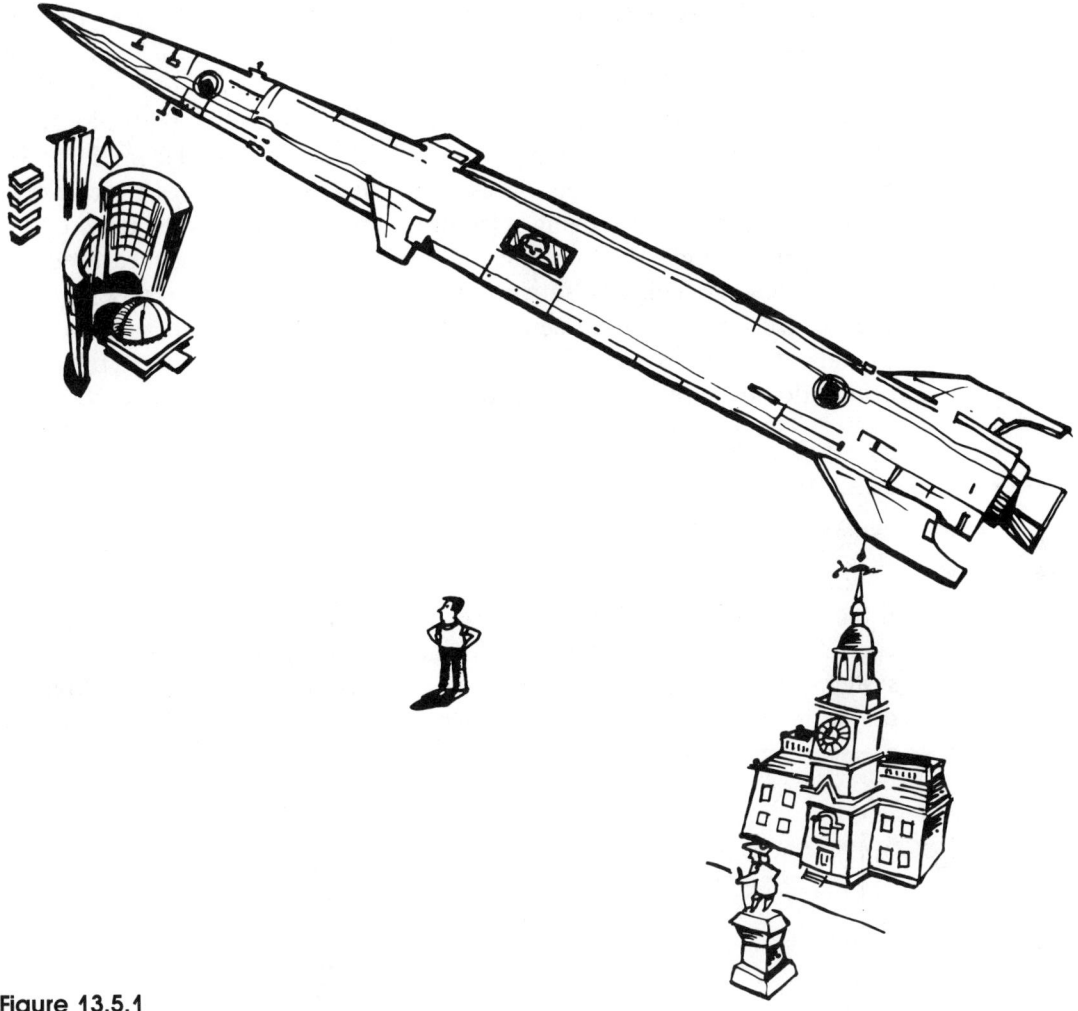

Figure 13.5.1
Simultaneity is relative.

However, observers aboard the northbound spaceship do *not* see lights in both cities come on at exactly the same time. The spaceship observers will see the lights in Toronto come on first. The lights in Philadelphia will come on later.

To see how this can be, let us use the same test for simultaneity that worked so well in the Earth frame. Station an observer on the spaceship midway between Philadelphia and Toronto at 8:00 P.M. (Earth time). This observer is just above the Earthbound observer we used two paragraphs ago. But he doesn't remain there. Rather, he moves toward Toronto, encountering the light from Toronto first while the light from Philadelphia reaches him later.

This explanation shows that simultaneity is relative. But we shall need some mathematical relationships to provide details like *how much* of a difference is observed. In our example, with the two cities 725 km apart, for a spaceship traveling at 90% the speed of light, a difference of 5 ms is to be expected.

This effect is symmetric. For observers moving with the spaceship, all the on-board clocks are carefully synchronized. But when viewed from Earth, the clocks on the front part of the spaceship are late. For both cases, the clocks on the *leading edge* of the moving reference frame are observed to be *late*.

By now you may be thinking, "all right, so that's what is measured; but what's *really* happening?" But you can't ask that. Physics is an observational science. We are stuck with operational definitions. What we measure is all we can talk about in Physics. This is not to imply that things like love, truth, and beauty do not exist or are not important. They're just not physics.

13.6 Lorentz Transformations

If the Galilean transformations of Section 13.2 are not correct, what are the correct transformation equations? Figure 13.6.1(a) shows two coordinate systems aligned with parallel axes and a relative motion along the direction of their x-axes. As their origins momentarily coincide, clocks in *both* systems are set to zero. We can say that frame B moves to the *right* with speed v

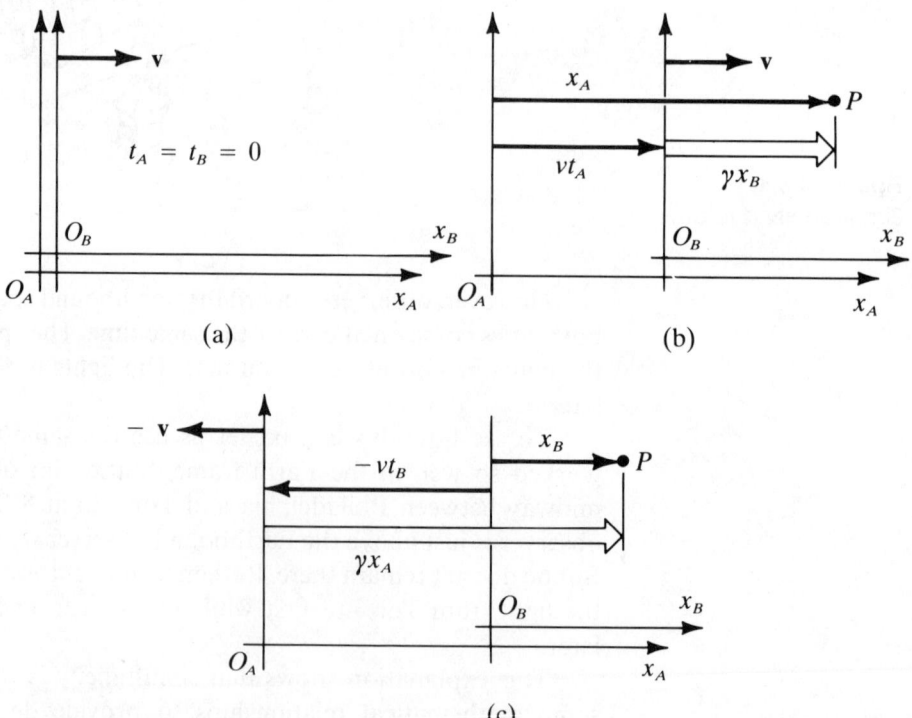

Figure 13.6.1
Distances measured in two moving reference frames.

relative to *A*. Or that *A* moves to the *left* with speed v relative to *B*. Or *B* moves with velocity **v** with respect to *A*. Or *A* moves with velocity $-$**v** with respect to *B*. All four statements say exactly the same.

Sometimes we may for convenience think of one frame as "fixed" and the other as "moving." But that is merely convenience. There is a relative velocity between the two frames. This velocity is shared. Observers in both frames *do* agree on the value for v. If they didn't, one would get a greater value than the other. And symmetry prohibits that.

Let observers in both frames measure the coordinates of some point *P*. How can these values be related? For the present, we shall be concerned only with the *x*-coordinate, the distance along x_A or x_B measured from O_A or O_B. *A* measures a distance x_A from his origin O_A to the point *P*. Likewise, *B* measures a distance x_B from her origin to the point *P*. *A* knows that *B*'s origin O_B has traveled a distance vt since they coincided. Therefore, we might expect

$$x_A = vt + x_B$$

But this is just the Galilean transformation.

Let's be careful about what *A* really measures. We know times and distances appear different for the two observers. So let us replace t with t_A, the *time that A measures*. And there may be some distortion of x_B by the time *A* measures it. Therefore, we can safely write

$$x_A = vt_A + \gamma x_B \qquad (13.6.1)$$

where γ is a "distortion factor" that we must evaluate later. Figure 13.6.1(b) illustrates this addition as seen by *A*. Notice the double vector used to indicate possible distortion of the distance x_B as measured by an observer from the *other* frame.

Figure 13.6.1(c) shows this situation from *B*'s point of view. *B* can clearly measure distances x_B and vt_B. But *B*'s measurement of x_A may be distorted. Therefore, we have

$$\gamma x_A = vt_B + x_B \qquad (13.6.2)$$

These two equations must hold for *any* point *P*. So let *P* be on the crest of a light wave that left the common origins when they were coincident. That is, at $t_A = t_B = 0$, a flashbulb was set off at $x_A = x_B = 0$. Both observers find that light travels with the same speed c. Thus,

$$x_A = ct_A$$

and

$$x_B = ct_B$$

For these values, Eqs. 13.6.1 and 13.6.2 become

$$ct_A = vt_A + \gamma ct_B \quad \text{and} \quad \gamma ct_A = vt_B + ct_B$$

or

$$(c - v)t_A = \gamma ct_B \quad \text{and} \quad \gamma ct_A = (c + v)t_B$$

Dividing one by the other, the times cancel, and we are left with

$$\frac{c-v}{\gamma c} = \frac{\gamma c}{c+v}$$

or

$$\gamma^2 = 1 - \frac{v^2}{c^2}$$

Thus,

$$\gamma = \sqrt{1 - \frac{v^2}{c^2}} \qquad (13.6.3)$$

This allows us to write Eq. 13.6.2 as

$$x_A = \frac{x_B + vt_B}{\sqrt{1 - v^2/c^2}} \qquad (13.6.4)$$

We can readily solve Eq. 13.6.1 for t_A to write

$$t_A = \frac{t_B + (v/c^2)x_B}{\sqrt{1 - v^2/c^2}} \qquad (13.6.5)$$

Or, we can solve for x_B and t_B to get the inverse transformations:

$$x_B = \frac{x_A - vt_A}{\sqrt{1 - v^2/c^2}} \qquad (13.6.6)$$

$$t_B = \frac{t_A - (v/c^2)x_A}{\sqrt{1 - v^2/c^2}} \qquad (13.6.7)$$

These are known as the *Lorentz transformations*. Notice that for $v \ll c$, they reduce to the earlier Galilean transformations—as, indeed, they must. Notice, further, that while the equations for x_B and t_B are easy to arrive at algebraically, we can get them by symmetry. Interchanging the roles of A and B is exactly the same as replacing v with $-v$!

Lorentz transformations have been recast in a most useful form by Professor Robert Brehme.* Define a parameter α by

$$\sin \alpha = \frac{v}{c} \qquad (13.6.8)$$

Figure 13.6.2 shows a right triangle with α as one of its angles. From that you can see that

$$\frac{1}{\sqrt{1 - (v^2/c^2)}} = \frac{c}{c\sqrt{1 - (v_2/c^2)}} = \frac{1}{\cos \alpha} = \sec \alpha \qquad (13.6.9)$$

---

*F. W. Sears and R. W. Brehme, *Introduction to the Theory of Relativity* (Reading, MA: Addison-Wesley Publishing Co., 1968), p. 19.

Figure 13.6.2 Some useful relationships from Pythagoras.

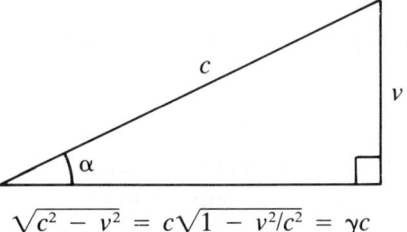

$$\sqrt{c^2 - v^2} = c\sqrt{1 - v^2/c^2} = \gamma c$$

and

$$\frac{v}{c\sqrt{1 - (v^2/c^2)}} = \tan \alpha \qquad (13.6.10)$$

We can use this to write the Lorentz transformations as

$$x_A = x_B \sec \alpha + (ct_B) \tan \alpha \qquad (13.6.11)$$

and

$$(ct_A) = (ct_B) \sec \alpha + x_B \tan \alpha \qquad (13.6.12)$$

This elegant form points out the symmetry between the spatial coordinate X and the temporal coordinate (ct). Of course, we can also write

$$x_B = x_A \sec \alpha - (ct_A) \tan \alpha \qquad (13.6.13)$$

and

$$(ct_B) = (ct_A) \sec \alpha - x_A \tan \alpha \qquad (13.6.14)$$

And this makes numerical calculations quick and easy. Trig functions are far easier to handle than $\sqrt{1 - (v^2/c^2)}$ on a hand-held calculator or computer. Remember, though, α is merely a parameter; nothing is rotated through a real angle α.

So far we have not mentioned the coordinates y and z, perpendicular to the relative velocity vector. They are not altered at all. That is,

$$y_A = y_B \qquad (13.6.15)$$

$$z_A = z_B \qquad (13.6.16)$$

Professor Hans Ohanian gives an elegant and quick proof of these coordinate transformations.* Consider two identically manufactured pipes moving relative to each other—moving in the common direction of their lengths. Consider pipe A "at rest" and pipe B "moving." If this motion causes a contraction in the y_B and z_B directions, then pipe B could pass through pipe A. But the motion of the pipes is symmetric. We may ride along with B and consider it at rest and A moving. Then such a contraction would allow pipe A to pass through pipe B. This is contradictory. Thus, there is no contraction (or elongation) in the y and z directions.

---

* Hans C. Ohanian, *Physics* (New York: W. W. Norton & Company, 1985), vol. 1, p. 422.

13.7 Application of the Lorentz Transformations

You may already know that "moving metersticks shrink in the direction of their motion." This is known as the Lorentz-Fitzgerald contraction. But what does it mean and how does it come about?

First, you must decide how to measure a moving meterstick or a moving rocket ship. It sounds simple enough. But if you will think on this for a moment, you'll realize that it is not something you ordinarily do. Consider a long freight train moving along a track. How would you measure its length? In actual practice, you'd probably *stop* it and then measure its length. But we can't do that; it keeps on moving. We could send two motorcyclists out—one at the front of the engine and the other at the end of the caboose— to make marks at the front and rear of the train *at the same time*. Then, at our leisure, we can measure the distance between *these* marks.

Figure 13.7.1 shows a meterstick at rest in the B frame, moving at velocity **v** with respect to the A frame. The two ends of the meter stick are marked by events E_1 and E_2. This may be exploding two flashbulbs or marking two chalk marks on a blackboard. They must occur *at the same time*—as seen by an observer at rest in the A frame.

Now it is very useful to use the Lorentz transformations for *differences* in coordinates rather than just coordinates. But since the equations are linear, the *differences* in coordinates transform just like the coordinates themselves. That is, we can rewrite Eqs. 13.6.13 and 13.6.14 as

$$\Delta x_B = \Delta x_A \sec \alpha - (c\,\Delta t_A) \tan \alpha \qquad (13.7.1)$$

$$(c\,\Delta t_B) = (c\,\Delta t_A) \sec \alpha - \Delta x_A \tan \alpha \qquad (13.7.2)$$

For our moving meterstick, we have $\Delta t_A = 0$ (the events marking the ends must occur simultaneously) and $\Delta x_B = L_0$ (the length of the stick *at rest*). Then

$$\Delta x_B = \Delta x_A \sec \alpha$$

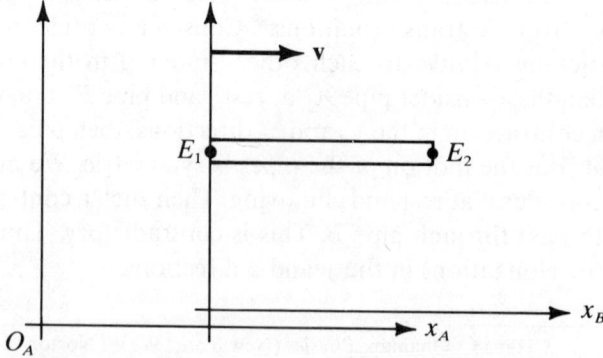

Figure 13.7.1
Measuring the length of a moving meter stick.

or

$$L = L_0 \sqrt{1 - \frac{v^2}{c^2}} \qquad (13.7.3)$$

where we have now used $\Delta x_A = L_0$. This is the Lorentz-Fitzgerald contraction. But notice that observer B may complain that the measurement was done incorrectly. Equation 13.7.2 gives

$$c\,\Delta t_B = 0 - L \tan \alpha$$

or

$$\Delta t_B = -\frac{v^2}{c^2} L_0 \qquad (13.7.4)$$

That is, *observer B will see event E_2 occur before event E_1*—he will insist that A has measured a shorter length because he first marked the *front* of the meterstick, waited until the rear of the stick moved closer to that mark, and then marked the rear. Nonsimultaneity is not an *explanation* of this length contraction. They're both real. They're both the consequence of the lack of an absolute time or space.

How large is this effect? For ordinary speeds, $\sqrt{1 - (v^2/c^2)}$ is as close to unity as you care to calculate. But look at the case of $v = 0.9c$. Then

$$L = 0.44 L_0 \qquad (v = 0.9c)$$

For such extremely relativistic speeds, this effect is quite noticeable! Or, to find a contraction of 10% ($L = 0.9 L_0$), the speed must be

$$v = 0.44c \qquad (L = 0.9 L_0)$$

To observe a contraction of 1% ($L = 0.99 L_0$), we must have a speed of

$$v = 0.14c \qquad (L = 0.99 L_0)$$

You may also already know that "moving clocks run slower." This is known as the Einstein time dilation. But what does it mean and how does it come about?

How would you compare your own watch with the watch of a friend and see if one ran slower than the other? You'd set them side by side and watch them. But you can't do that if one is moving. The best you can do is to compare a *single*, moving clock with *two* synchronized, stationary clocks as sketched in Figure 13.7.2. Place clock $A1$ near A's origin and clock B near B's origin. Both read $t_A = t_B = 0$ as they pass. What are the readings on clocks $A2$ and B as they pass each other?

Clock B is at the origin, so $x_B = 0$. We can call its time reading t_B or Δt_B. Let's call it $\Delta \tau$. Equations 13.6.9 and 13.6.12 then give the reading on clock $A2$.

$$\Delta t_A = \Delta \tau \sec \alpha = \frac{\Delta \tau}{\sqrt{1 - (v^2/c^2)}} \qquad (13.7.5)$$

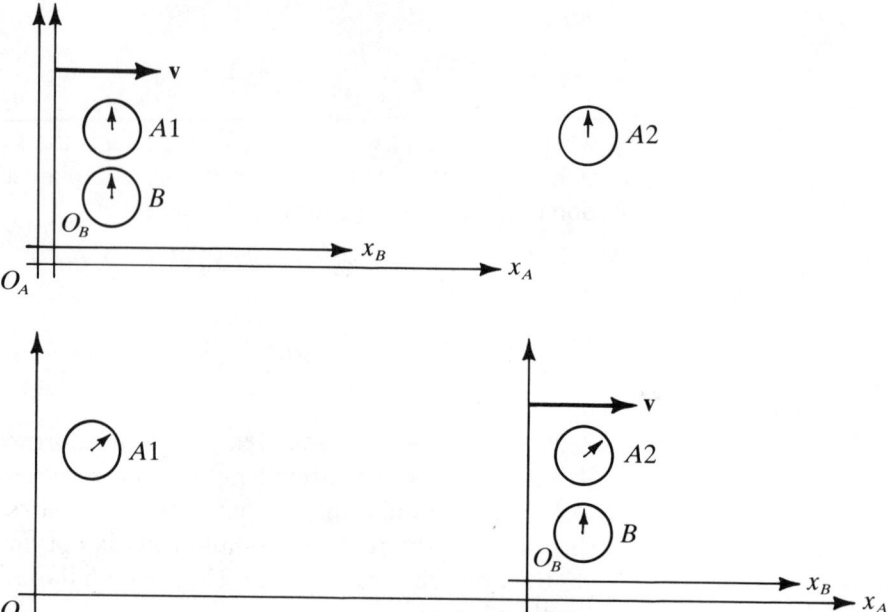

Figure 13.7.2
Measuring the rate of a moving clock.

$\Delta \tau$ is the amount of time between two events *as measured by a single clock present at both events*. We shall call this the *proper time*. Clock B is setting still in the "moving" B frame. Δt_A is the amount of time between two events as measured *by two different clocks*, synchronized and at rest with respect to each other.

Δt_A is larger than $\Delta \tau$. If it takes 10 s for clock B to pass between clocks $A1$ and $A2$, as measured by these two clocks, it may only require 7 s as measured by B. Thus, A will conclude that B's clock is running slowly.

But this effect is symmetric. Let B watch a *single* clock at rest in A but moving with respect to B as sketched in Figure 13.7.3. The "stationary" observer still finds the "moving" clock running slower!

This is not mere mathematical magic. It is an observable phenomenon. Radioactive particles in cosmic rays traveling at extremely high speeds take longer to decay than identical particles at rest in a laboratory.

While the time and distance between any two events will be quite different for two moving observers, is there nothing on which the two will agree? Is there nothing that is characteristic of the two *events themselves* and, thus, independent of the observers? It turns out that we can define an *invariant interval*, Δs, between two events given by

$$\Delta s^2 = (\Delta x^2 + \Delta y^2 + \Delta z^2) - c^2 \, \Delta t^2 \qquad (13.7.6)$$

that will be the same for *all* observers moving at constant speed. If $(\Delta x^2 + \Delta y^2 + \Delta z^2) < c^2 \, \Delta t^2$, then it is more convenient and useful to define $\Delta \tau$ by

$$c^2 \, \Delta \tau^2 = c^2 \, \Delta t^2 - (\Delta x^2 + \Delta y^2 + \Delta z^2) \qquad (13.7.7)$$

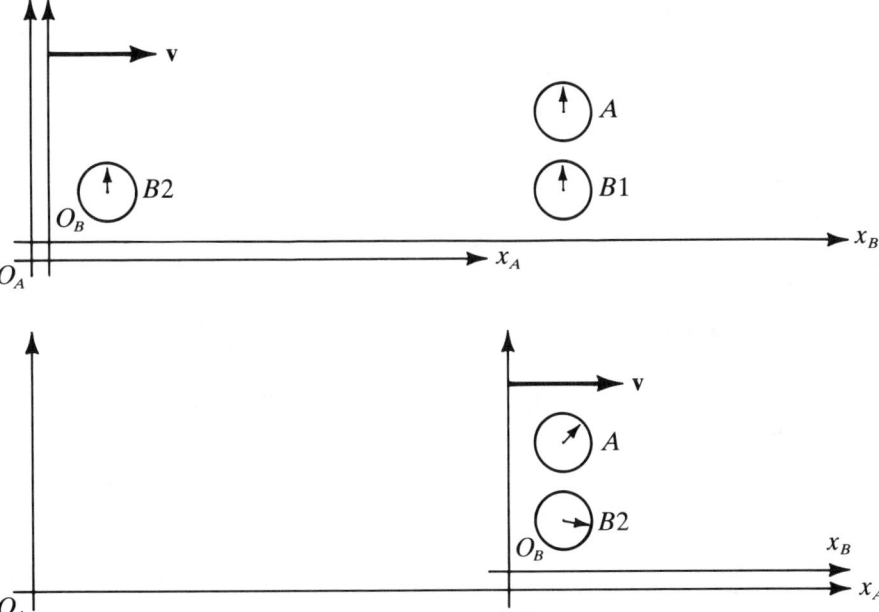

Figure 13.7.3 Time dilation is symmetric.

which will remain invariant for all inertial observers. $\sqrt{\Delta t^2}$ is the *temporal* separation of two events and $\sqrt{(\Delta x^2 + \Delta y^2 + \Delta z^2)}$ is the *spatial* separation of two events. $\Delta s = \sqrt{\Delta s^2}$ turns out to be the spatial separation of the two events in a frame that is moving so that the two events are simultaneous. $\Delta \tau = \sqrt{\Delta \tau^2}$ turns out to be the temporal separation for the frame in which the two events are coincident. Clearly Δs and $\Delta \tau$ cannot both be used in discussing a single pair of events.

13.8 Minkowski Diagrams

The Lorentz transformations we developed earlier relate the space and time coordinates in one coordinate system to those in another in just such a way as to keep the speed of light constant in both. They express the basic characteristics of spacetime. They express the *geometry* of spacetime, if you like. But they are still unfamiliar. Their predictions do violence to our intuition. Now we shall make drawings to express these transformations graphically. These are called *Minkowski diagrams*.

We shall only be concerned with one-dimensional motion. As before, the two frames or observers have a relative motion parallel to the common direction of their *x*-axes. We shall only be concerned with motion in the *x* direction.

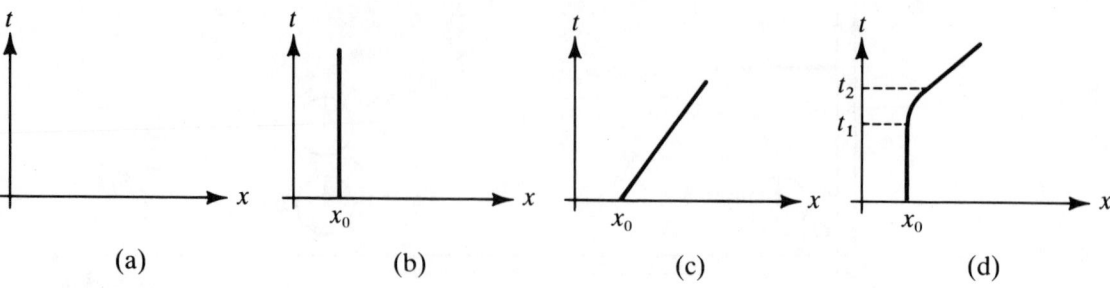

Figure 13.8.1
Minkowski diagrams and worldlines.

By convention, the time axis is chosen to be vertical; the x-axis, horizontal as shown in Figure 13.8.1(a). Figure 13.8.1(b) shows a body that remains at rest at position x_0. Figure 13.8.1(c) shows a body moving to the right at constant speed. Figure 13.8.1(d) indicates a body initially at rest at x_0 that begins to accelerate at time t_1 until it reaches some velocity at t_2 and then continues with constant velocity. Each curve showing the motion of an object is called the *worldline* of that object. Notice that this is just what you did in your first introduction to motion—but now the axes are oriented as shown in Figure 13.8.1 and we've coined the word "worldline" for efficiency and brevity.

Before we continue, let us choose scales for both axes. In the equations of the Lorentz transformations, you will recall that it was the quantity (ct) that exhibited such great symmetry with x. We shall exploit this and plot our worldlines on axes of (ct) and x. Instead of plotting time directly in seconds, we shall plot it in terms of the *distance* light would travel in that time. We can use distances of light-years or light-seconds.

$$1 \text{ light-second} = 1 \text{ } l\text{-sec} = 3 \times 10^8 \text{ m} \tag{13.8.1}$$

With this choice of scales, the worldline of a flash of light will make an angle of 45° with the x-axis as shown in Figure 13.8.2. Speeds less than c are

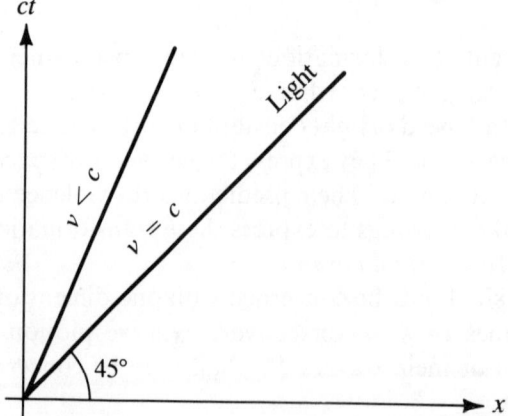

Figure 13.8.2 Plotting ct versus x ensures 45° for the worldline of light.

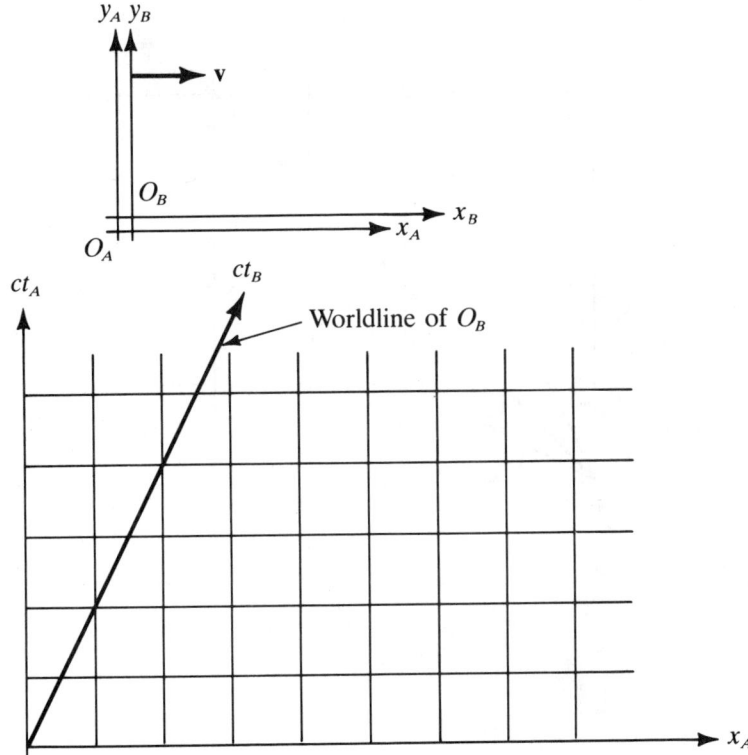

Figure 13.8.3
Worldline of O_B as seen by A.

indicated by worldlines between that of light and the (ct) axis. Most ordinary, everyday speeds correspond to worldlines almost on top of the (ct) axis.

Consider our familiar reference frames A and B moving relative to each other as shown at the top of Figure 13.8.3. The worldline of the origin of the B frame, O_B, as seen in frame A is a straight line as shown there. But that is just a graph of $x_B = 0$. For a graph of (ct_B) versus x_B, this is just the (ct_B)-axis. Its tilt from the (ct_A)-axis is determined by the velocity $\mathbf{v}$. For this particular diagram, $v = 0.5c$.

Now, where is the x_B-axis on this diagram? Is it also rotated counterclockwise, down below the x_A-axis? No, go back to the Lorentz transformation equation (Eq. 13.6.14). The x_B-axis is just a graph of $(ct_B) = 0$, and that requires

$$ct_A = \frac{v}{c} x_A \qquad (13.8.2)$$

Figure 13.8.4 shows the ct_B- and x_B-axes on the ct_A, x_A Minkowski diagram.

Consider the worldlines of the ends of a meterstick aligned in B along the x_B-axis, as shown in Figure 13.8.5. In B's frame, they are $L_0 = 1$ m apart. But in A's frame they are a distance of only

$$L = L_0 \sqrt{1 - \frac{v^2}{c^2}} \qquad (13.7.3)$$

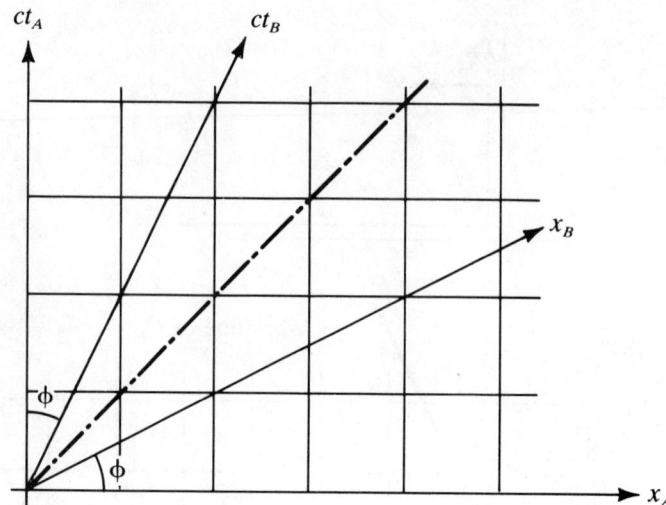

Figure 13.8.4 Axes of moving reference frames.

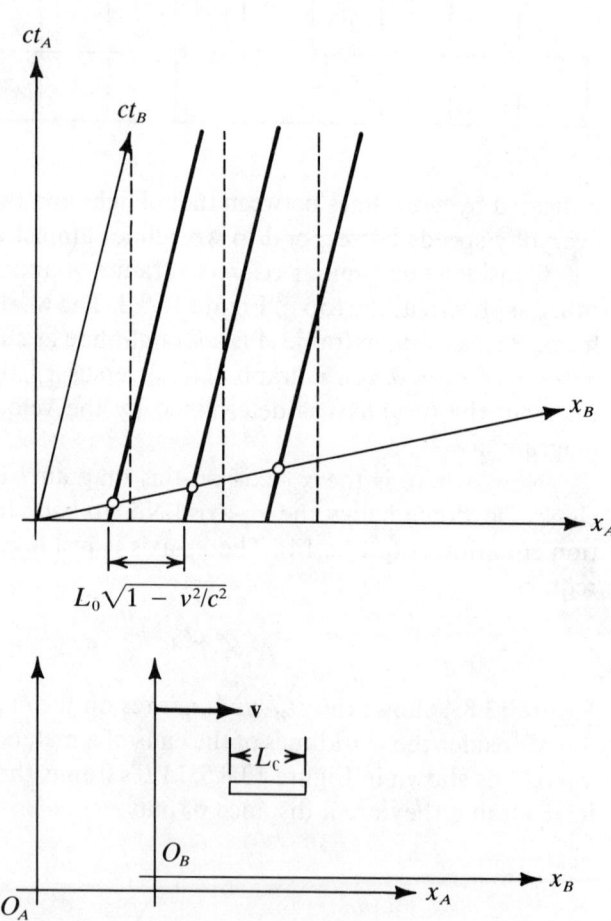

Figure 13.8.5 Spatial coordinates.

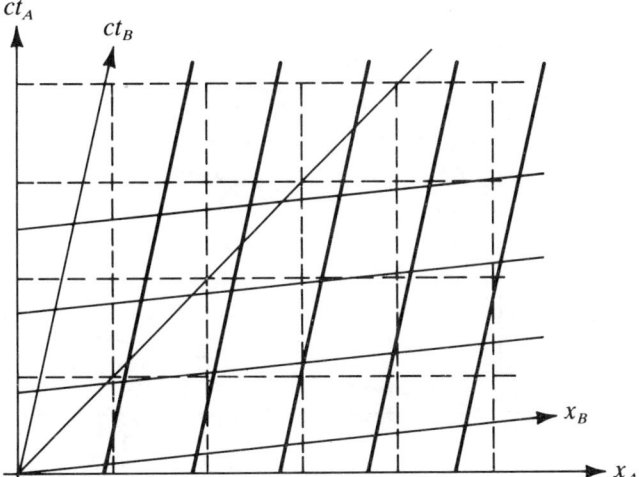

Figure 13.8.6 Superposition of both space time coordinate systems.

apart. These worldlines—and others parallel to them—are *lines of constant spatial coordinate* in the B coordinate frame. Similar lines of constant *time* in B can be drawn to complete our representation of the B coordinate system. This is shown in Figure 13.8.6. Note that the worldline for light evenly bisects both the ct_B-, x_B-axes and the ct_A-, x_A-axes.

Many quantities of interest can be read *directly* from a Minowski diagram. Horizontal lines represent a distance measured in the A-frame, Δx_A. Since it is horizontal, it describes two events that are simultaneous to observer A. A vertical line measures the *time* between two events coincident in A; that is, $c\Delta t_A$. Slanted lines parallel to the x_B-axis measure Δx_B. And slanted lines parallel to the ct_B-axis, $c\Delta t_B$. These four cases handle most of the situations we shall encounter; they are shown in Figure 13.8.7.

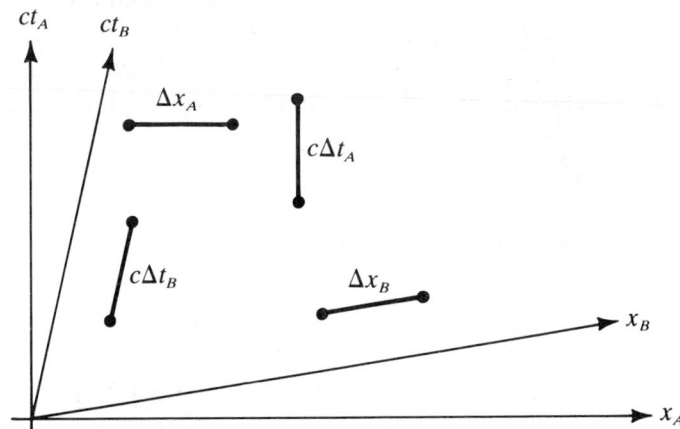

Figure 13.8.7 Several quantities can be read directly from a Minkowski diagram.

Figure 13.8.8
Simultaneity, again.

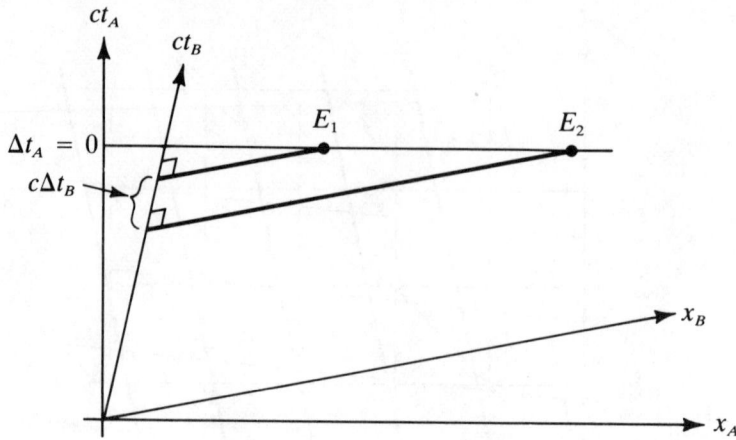

Minkowski diagrams contain the same information as the Lorentz transformations—they are a *graphical* representation of the same relationships. We can become more familiar with their use by looking at some previous examples. We start with the idea that simultaneity is relative.

Figure 13.8.8 shows two events, E_1 and E_2, that are *simultaneous* in frame A. They happen at the same time—as seen by observer A. That means they both appear on a single horizontal line in the diagram. Such horizontal lines—parallel to the x_A-axis and perpendicular to the ct_A-axis—are lines of constant time in A, t_A = constant. Lines of constant time in B, t_B = constant are parallel to the x_B-axis. As such, they are "perpendicular" to the ct_B-axis ("orthogonal" is a better word here). Draw two such lines in the diagram that pass through events E_1 and E_2. Their intersections with the ct_B-axis give the times at which the two events occur—as measured by B! Thus, from the diagram, it is easy to see that B will *not* observe the two events to be simultaneous—E_2 occurs earlier than E_1.

Figure 13.8.9 shows more details for the same situation. Events E_1 and E_2 occur at the same time in frame A and are a distance Δx_A apart there. In frame B, these same events are separated by Δx_B in distance and Δt_B in time. Let these two events measure the ends of, say, a meterstick whose length A wants to measure as it comes speeding by. From Eq. 13.7.3 we know that

$$\Delta x_A = \Delta x_B \sqrt{1 - \frac{v^2}{c^2}} \qquad (13.8.3)$$

Further, from Eq. 13.7.4 we know

$$|c\,\Delta t_B| = \frac{v}{c} \Delta x_B \qquad (13.8.4)$$

In B's frame, Δx_A, Δx_B, and $c\,\Delta t_B$ form a "right" triangle with Δx_A being the hypotenuse, Δx_B the long leg, and $c\,\Delta t_B$ the short leg. For all such "right"

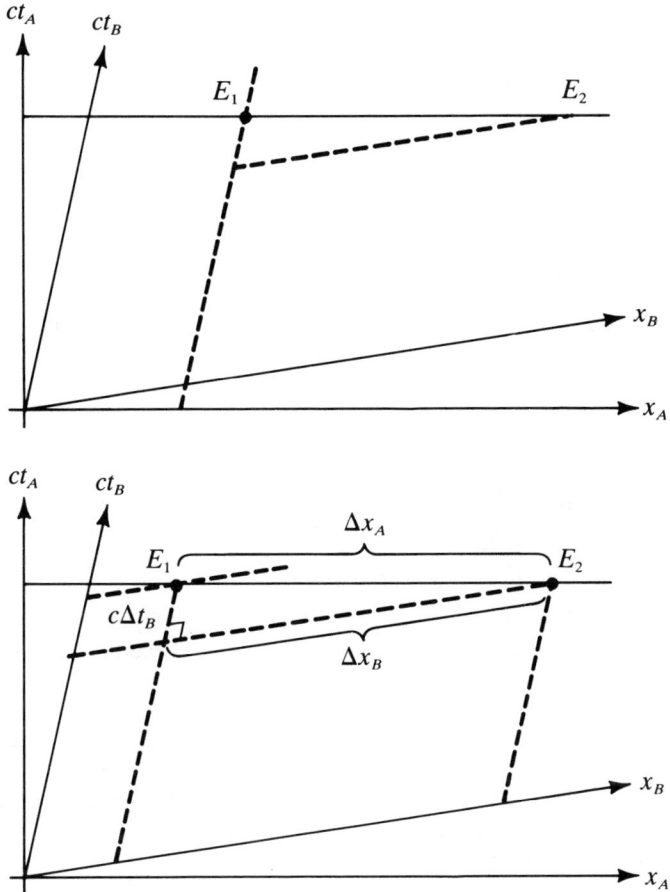

Figure 13.8.9
Relationships between the two frames.

triangles in Minkowski diagrams, we find the disconcerting results that

$$(\text{hypotenuse}) = \sqrt{1 - \frac{v^2}{c^2}} (\text{long leg}) \qquad (13.8.5)$$

$$(\text{short leg}) = \frac{v}{c} (\text{long leg}) \qquad (13.8.6)$$

Note that the hypotenuse is shorter than the long leg!

This strange result is not limited to the skewed right triangle of B's frame. Figure 13.8.10 shows two events E_1 and E_2 that are simultaneous in A's frame. They occur a distance $x_A = vt_A$ apart at time t_A. B sees E_2 occur at time t_B. Equation 13.6.10 and 13.6.14 tell us that $t_B = t_A \sqrt{1 - (v^2/c^2)}$, which is a *smaller* number than t_A. So, even for observer A, the hypotenuse is shorter than the long leg! Equations 13.8.5 and 13.8.6 are true for *both* observers.

Besides illustrating these unusual relations for what appear to be right triangles, Figure 13.8.10 illustrates time dilation. Events E_0 and E_2 can be two

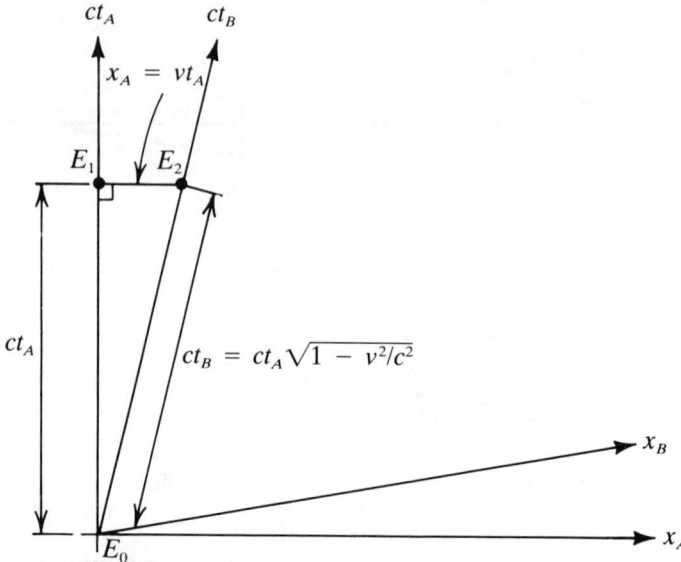

Figure 13.8.10 More relationships between the two frames.

"ticks" of a single clock at rest in B's frame. Observer A measures a *larger* amount of time to elapse between these two events. Again

$$\Delta t_A = \Delta \tau \sec \alpha = \frac{\Delta \tau}{\sqrt{1 - (v^2/c^2)}} \qquad (13.7.5)$$

holds where Δt_A is the time between events measured by two synchronized clocks in A and $\Delta \tau$ is the "proper time"—the time between events measured by a single clock, at rest in B, which is present at both events.

13.9 Velocity Transformation

As a young child, I was fascinated by the following question that my father posed to me: If a machine gun, which fires bullets at 100 miles per hour, is mounted on an airplane that flies 200 miles per hour, will the airplane run into its own bullets? As an experienced physicist, I am still fascinated by similar questions like: If an alien spacecraft fires phaser torpedoes forward with a speed of $0.75c$ (with respect to the spacecraft, of course), while the spacecraft itself has a speed of $0.3c$ with respect to Earth, how fast are the torpedoes observed to travel with respect to Earth?

The obvious, common-sense, intuitive answer is $1.05c$. But velocities greater than c do all sorts of terrible things to the Lorentz transformations. So let's investigate this situation more closely.

Figure 13.9.1 shows our usual case of frames A and B. In frame B, body C is observed to have a speed of v_{CB} in the x direction. What is the speed of C with respect to A, v_{CA}?

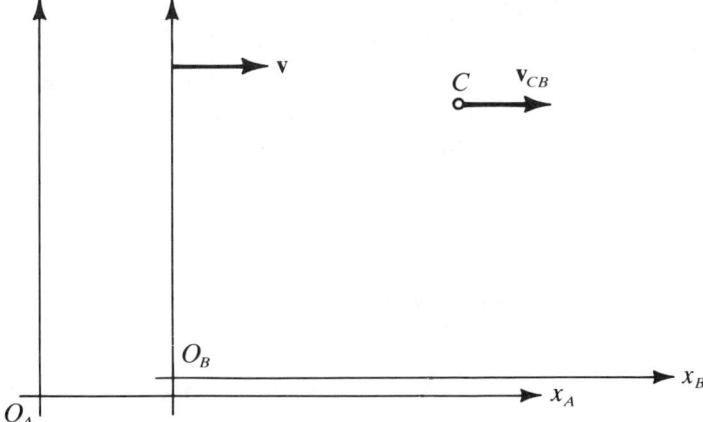

Figure 13.9.1 Velocity transformation.

The velocity measured by B is just dx_B/dt_B and the velocity measured by A is just dx_A/dt_A; so, from the Lorentz transformations equations 13.6.4 and 13.6.5 or 13.6.11 and 13.6.12, we can immediately write

$$v_{cA} = \frac{dx_A}{dt_A} = \frac{dx_B \sec\alpha + (c\,dt_B)\tan\alpha}{\dfrac{1}{c}[(c\,dt_B)\sec\alpha + dx_B \tan\alpha]} \qquad (13.9.1)$$

This looks much more formidable than it is. Multiply numerator and denominator by $\sin\alpha/dt_B$ and recall that $\sin\alpha = v/c$, and this reduces to

$$v_{CA} = \frac{v_{CB} + v}{1 + (vv_{cB}/c^2)}$$

This was for the special case of C moving only in the x direction. We can write this explicitly as

$$v_{xCA} = \frac{v_{xCB} + v}{1 + (vv_{xCB}/c^2)} \qquad (13.9.2)$$

If C has y and z-components of velocities, the corresponding denominators will be identical to that of Eq. 13.9.1, but since $y_A = y_B$ and $z_A = z_B$, the numerators will be simpler. Hence, these components of velocity transform according to

$$v_{yCA} = v_{yCB}\frac{\sqrt{1 - (v^2/c^2)}}{1 + (vv_{xCB}/c^2)} \qquad (13.9.3)$$

and

$$v_{zCA} = v_{zCB}\frac{\sqrt{1 - (v^2/c^2)}}{1 + (vv_{xCB}/c^2)} \qquad (13.9.4)$$

Now we have the information to answer the alien spacecraft-phaser torpedo question posed a little earlier. The phaser torpedo will be observed by A to have a speed of $0.857c$. That's faster than the $0.75c$ in B's frame, of course, but far different than the $1.05c$ predicted by a Galilean transformation. These velocity transformations reflect the fact that no inertial observer will see a real, material object traveling faster than the speed of light. The speed of light is a universal, cosmic speed limit.

13.10 Doppler Effect

We have looked at an alien spacecraft firing phaser torpedoes. What happens, instead, if a searchlight is turned on? *All* the v's then become c's in Eq. 13.9.2. Everybody still gets the same value for the speed of light. But, different observers measure different values for the *frequency*. This is just the relativistic Doppler effect.

A Minkowski diagram like Figure 13.10.1 is especially useful in this situation. The spacecraft and floodlight (or laser or anything that emits photons) leave the A origin at event E_1. Event E_2 is the end of one period at the floodlight. One wavelength or one period of light has been released. In B's frame, the amount of time between these two events is Δt_B or $\Delta \tau$. A would *calculate* that event E_2 occurs at his time t_{A2} where

$$t_{A2} = \frac{\Delta \tau}{\sqrt{1 - v^2/c^2}}$$

But the light from the *end* of this wavelength or period has not yet arrived at A's origin. The propagation of the light back to A is shown by the 45° line between E_2 and E_3. Additional time Δt_{A2} is required for the end of this

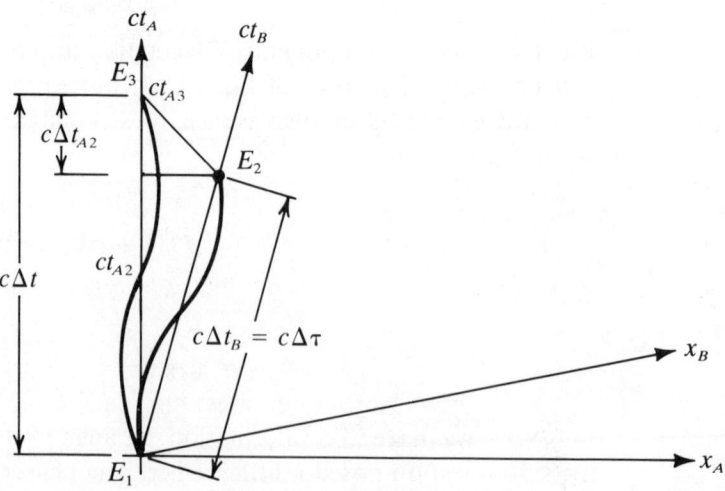

Figure 13.10.1
Relativistic Doppler effect.

wavelength or period to be communicated back to A. The x_A-coordinate of E_2 is vt_{A2}. So $c\,\Delta t_{A2}$ must equal that as well.

$$\Delta t_{A2} = \frac{v}{c} t_{A2}$$

The total time elapsed in A's frame between events E_1 and E_3 is $\Delta t = t_{A2} + \Delta t_{A2}$,

$$\Delta t = \Delta \tau \frac{1 + (v/c)}{\sqrt{1 - (v^2/c^2)}}$$

or

$$\Delta t = \Delta \tau \sqrt{\frac{1 + (v/c)}{1 - (v/c)}} \qquad (13.10.1)$$

This describes the periods. Wavelengths are proportional to periods, so we can write

$$\lambda = \lambda_0 \sqrt{\frac{1 + (v/c)}{1 - (v/c)}} \qquad (13.10.2)$$

Frequencies are inversely proportional to periods ($f = 1/T$), so

$$f = f_0 \sqrt{\frac{1 - (v/c)}{1 + (v/c)}} \qquad (13.10.3)$$

The basic results are the same as the classical Doppler effects—a receding source is "red shifted" toward longer wavelengths and an approaching source is "blue shifted" toward shorter wavelengths. But the details—Eqs. 13.10.1 through 13.10.3—are different than the classical example of, say, the wavelength or frequency of a railroad train whistle as it approaches a listener and then recedes into the distance.

13.11 Momentum and Mass-Energy

We have now investigated relativistic *kinematics*, the description of motion involving ultra-high velocities. But what of relativistic *dynamics*, an explanation of the causes of motion? Newton's Second Law in the form of

$$\mathbf{F} = m\mathbf{a}$$

has been very useful in the past. Can we retain it in this form? No, this predicts a constant acceleration rising from a constant force. And a constant acceleration acting for long enough will give as large a velocity as you might

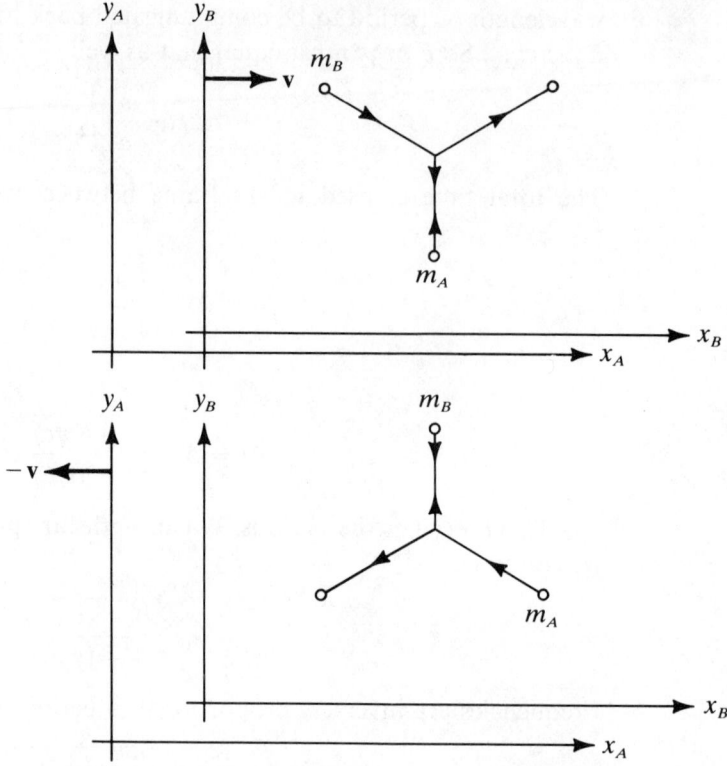

Figure 13.11.1 Two balls colliding.

want. In particular, the velocity can eventually exceed c, the speed of light. And that causes all sorts of difficulties with the Lorentz transformations.

But we have also written Newton's laws in terms of momentum and found the ideas of momentum conservation to be useful—even vital. Perhaps we can retain the ideas of momentum conservation. Previously, we have defined momentum as

$$\mathbf{p} = m\mathbf{v} \tag{13.11.1}$$

Can we retain this? If not, how must we modify it to retain conservation of momentum?

Consider our usual observers A and B, each in his respective reference frame, moving with relative velocity v along their common x-direction. The two observers have identical balls labeled m_A and m_B (i.e., they would be identical if compared side by side, at rest with respect to each other). Thus we shall say

$$m_A = m_B = m$$

Each observer throws his ball along his own y direction so the two balls collide, as sketched in Figure 13.11.1. It is an elastic collision and the balls rebound with their initial speeds.

The situation is completely symmetric. So the speed v_{Ay} that A measures for m_A must be the same as the speed v_{By} that B measures for m_B. Let us call this speed u_0. That is,

$$v_{Ay} = v_{By} = u_0$$

We shall now reserve v for the relative velocity between observers and use u for the velocity of an object. Look closely at what each observer measures. A throws a ball up with mass m and velocity u_0. Thus, A measures its initial y momentum to be

$$p_{A0y} = mu_0$$

The ball returns with velocity $-u_0$ (that is, it is moving down). Now its y-momentum is

$$p_{Afy} = -mu_0$$

So there has been a *change* in the y- component of the momentum of ball A of

$$\Delta p_A = -2mu_0 \qquad (13.11.2)$$

What does A observe to happen to ball B? Ball m_B now has an x-component of velocity equal to $-v$,

$$v_{Bx} = -v$$

as seen by A. According to Eq. 13.9.3, A will observe its y velocity, u_y, to be

$$u_y = -u_0 \sqrt{1 - \frac{v^2}{c^2}} \qquad (13.11.3)$$

So its initial y momentum, as seen by A, is $-mu_y$ or

$$p_{B0y} = -mu_0 \sqrt{1 - \frac{v^2}{c^2}}$$

where the negative sign simply indicates "down." After the collision, the motion in the y direction is reversed so

$$p_{Bfy} = mu_0 \sqrt{1 - \frac{v^2}{c^2}}$$

for a *change* in the y momentum of ball B, as seen by A, of

$$\Delta p_B = 2mu_0 \sqrt{1 - \frac{v^2}{c^2}} \qquad (13.11.4)$$

Clearly this is *not* the opposite of Δp_A as given by Eq. 13.11.3. Therefore, momentum as defined by $p = mv$ is not conserved in relativity!

But momentum conservation is so important. Is there no way it can be salvaged? How can we redefine momentum so that it is conserved in relativity? Our new definition must reduce to $p = mv$ for speeds much less than c since we know this quantity "works" or is conserved for low speeds.

In the situation just considered, we ran into trouble because the y-component of the momentum depended on the x-component of the velocity. We shall seek to avoid that. We know that distances in the y direction are unaffected by velocities in the x direction. But the time Δt necessary to travel a y displacement Δy is affected. To circumvent this problem, let us use the *proper time* $\Delta \tau$ as measured by the moving object itself. Therefore, the quantity $\Delta y / \Delta \tau$ will be the same for all observers. From Eq. 13.7.5 we know that Δt and $\Delta \tau$ differ due to time dilation by

$$\Delta \tau = \Delta t \sqrt{1 - \frac{u^2}{c^2}}$$

That means that

$$\frac{\Delta y}{\Delta \tau} = \frac{\Delta y}{\Delta t} \frac{1}{\sqrt{1 - u^2/c^2}} \qquad (13.11.5)$$

That is, the y-component of

$$\frac{v}{\sqrt{1 - u^2/c^2}}$$

is the same for all observers moving at constant velocity along their common x direction. We shall generalize this and *define* the relativistic momentum by

$$p = \frac{mu}{\sqrt{1 - u^2/c^2}} \qquad (13.11.6)$$

where u is the speed of an object of mass m. Momentum, defined by this equation, is conserved for all observers in all reference frames moving with constant velocity relative to each other.

Notice that we could write the components of momentum as

$$p_x = m \frac{dx}{d\tau}$$

$$p_y = m \frac{dy}{d\tau} \qquad (13.11.7)$$

$$p_z = m \frac{dz}{d\tau}$$

since

$$\frac{dx}{d\tau} = \frac{dx}{dt} \frac{dt}{d\tau}$$

SECTION 13.11 / MOMENTUM AND MASS-ENERGY

and, by Eq. 13.7.5

$$\frac{dt}{d\tau} = \frac{1}{\sqrt{1 - u^2/c^2}} \qquad (13.11.8)$$

which is just the extra factor required in the momentum.

What happens to energy in relativity? We began our initial study of energy by defining the kinetic energy as the work done by a force when it moves an object of mass m from rest to a final speed u. We shall do the same now. We cannot use $\mathbf{F} = m\mathbf{a}$, but the alternate form,

$$F = \frac{dp}{dt} \qquad (13.11.9)$$

is still valid in relativity. Just as before, we begin by defining the work and seeing what changes.

$$\begin{aligned}
KE = W &= \int F\, dx \\
&= \int \frac{d}{dt}\left[\frac{mu}{\sqrt{1 - u^2/c^2}}\right] dx \\
&= \int \frac{d}{dt}\left[\frac{mu}{\sqrt{1 - u^2/c^2}}\right] \frac{dx}{dt}\, dt \\
&= \int u\, \frac{d}{dt}\left[\frac{mu}{\sqrt{1 - u^2/c^2}}\right] dt \\
&= \int_0^u u\, d\left[\frac{mu}{\sqrt{1 - u^2/c^2}}\right] \\
&= \int_0^u u\left\{\left[\frac{m\,du}{\sqrt{1 - u^2/c^2}}\right] + \frac{mu^2}{c^2}\left[\frac{du}{\sqrt{(1 - u^2/c^2)^3}}\right]\right\} \\
&= \int_0^u \left[\frac{mu}{\sqrt{1 - u^2/c^2}}\right] du \\
&= \left.\frac{mc^2}{\sqrt{1 - u^2/c^2}}\right|_0^u
\end{aligned}$$

$$KE = W = \frac{mc^2}{\sqrt{1 - u^2/c^2}} - mc^2 \qquad (13.11.10)$$

How do we interpret this result? First, we must see if this reduces to our familiar $mu^2/2$ for low velocities. Then, what is the significance of the constant term, mc^2, in this equation?

We start by expanding the denominator using the binomial expansion

$$\left(1 - \frac{u^2}{c^2}\right)^{-1/2} = 1 + \frac{1}{2}\frac{u^2}{c^2} + \cdots \quad (13.11.11)$$

Thus, the kinetic energy of Eq. 13.11.10 is

$$KE = mc^2\left(1 + \frac{1}{2}\frac{u^2}{c^2} + \cdots\right) - mc^2$$

And that is just the classical form of

$$KE = \frac{1}{2}mu^2 \quad \text{(for } u \ll c\text{)}$$

for speeds u much less than the speed of light c (so that higher-order terms in u^2/c^2 may be neglected). So our result is *consistent* with what we have already developed for low speeds. That is important for any new theory must be consistent with established, verified theories in regions or situations where both can be used. A new theory may extend the frontiers but it must not contradict experimental results.

Now, how shall we interpret Eq. 13.11.10 for the kinetic energy? The constant term, mc^2, is interpreted as the *rest energy* that an object has simply because it has mass m. The first term is the *total energy*,

$$E = \frac{mc^2}{\sqrt{1 - u^2/c^2}} \quad (13.11.12)$$

Notice that, just as we did for momentum in Eq. 13.11.7, we can also write this as

$$E = mc^2 \frac{dt}{d\tau} \quad (13.11.13)$$

Kinetic energy, the energy of motion, is the *difference* between the total energy of a body and its rest energy. Kinetic energy is the *additional* energy that has been added due to motion. That is reasonable and as we might expect. The unexpected—and very interesting—feature is that energy is to be associated with mass. The rest energy, often written E_0,

$$E_0 = mc^2 \quad (13.11.14)$$

is energy due to the *mass* itself. You may think of mass as concentrated or stored energy. Despite the humorous cartoon depicting Einstein writing $E = ma^2$ and rejecting it and then trying $E = mb^2$ and rejecting it before settling upon $E = mc^2$, this is not something separate. It is not an independent postulate. It is an integral part of relativity. That is, the equivalence of mass and energy is a straightforward result of the Lorentz transformations and the idea that work causes a change in energy.

We have defined momentum by Eq. 13.11.6 and energy by Eq. 13.11.12 in terms of the velocity of a particle. We can eliminate the velocity u by solving

SECTION 13.11 / MOMENTUM AND MASS-ENERGY

for it in one equation and substituting into the other. The result is a very useful relation:
$$E^2 = p^2c^2 + m^2c^4$$
or, the equivalent
$$E^2 - p^2c^2 = m^2c^4 \tag{13.11.15}$$

For very high speeds, such that $pc \gg mc^2$, this reduces to
$$E = pc \tag{13.11.16}$$
which is always true for *massless* particles such as photons.

Now return to Eqs. 13.11.7, 13.11.13, and 13.11.15:

$$p_x = m\frac{dx}{d\tau}$$

$$p_y = m\frac{dy}{d\tau} \tag{13.11.7}$$

$$p_z = m\frac{dz}{d\tau}$$

$$E = mc^2\frac{dt}{d\tau} \tag{13.11.13}$$

$$E^2 - p^2c^2 = m^2c^4 \tag{13.11.15}$$

The mass m, the speed of light c, and the proper time τ are all invariants. So p_x, p_y, p_z, and E/c must transform exactly like x, y, z, and ct. That is, they must satisfy the Lorentz transformations, Eqs. 13.6.4 through 13.6.16. In particular, we can write

$$p_{xA} = \frac{p_{xB} + vE_B/c^2}{\sqrt{1 - v^2/c^2}} \tag{13.11.17}$$

$$p_{yA} = p_{yB}$$

$$p_{zA} = p_{zB}$$

$$E_A = \frac{E_B + vp_{xB}}{\sqrt{1 - v^2/c^2}}$$

Or, in terms of the parameter α that we defined in Eq. 13.6.8, these can be written as

$$p_{xA} = p_{xB}\sec\alpha + \frac{E_B}{c^2}\tan\alpha \tag{13.11.18}$$

$$p_{yA} = p_{yB}$$

$$p_{zA} = p_{zB}$$

$$\frac{E_A}{c^2} = \frac{E_B}{c^2}\sec\alpha + p_{xB}\tan\alpha$$

Note, from Eq. 13.11.15, that mc^2 is the invariant magnitude for the four-vector p_x, p_y, p_z, and E/c just as Δs was the invariant interval for $\Delta x, \Delta y, \Delta z$, and $c\Delta t$ in Eq. 13.7.6.

It is interesting to apply these transformations to photons. Consider a photon moving along the positive x direction viewed by our usual observers in reference frames A and B. Combining Eq. 13.11.16 with the last of Eqs. 13.11.17 or 13.11.18 yields

$$E_A = E_B \frac{1 + v/c}{\sqrt{1 - (v^2/c^2)}}$$

$$E_A = E_B \sqrt{\frac{1 + (v/c)}{1 - (v/c)}} \quad (13.11.19)$$

The photon's energy measured by A is greater than its energy measured by B. Since there is a direct relation between energy and frequency, this turns out to be another expression of the Doppler effect.

Equation 13.10.3 gave the frequence of light observed in a rest frame in terms of the frequency of the same light observed in a moving frame. In terms of our usual frames A and B of Figure 13.11.2, this can be written as

$$f_A = f_B \sqrt{\frac{1 - (v/c)}{1 + (v/c)}} \quad (13.11.20)$$

Combining this with Eq. 13.11.19 yields

$$\frac{E_A}{f_A} = \frac{E_B}{f_B} \quad (13.11.21)$$

This means that the ratio of the *energy* of a photon divided by the *frequency* of the associated radiation is a constant for all observers. We can write this as

$$E = hf \quad (13.11.22)$$

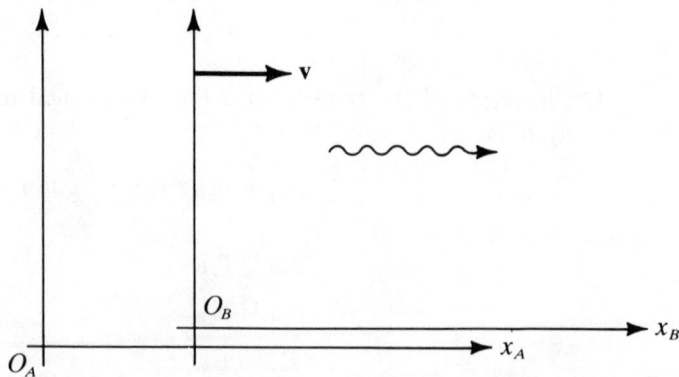

Figure 13.11.2 The energy of a photon is different for different observers.

where *h* is *Planck's constant*. This constant and this idea that Planck first proposed in thermodynamics and that Einstein used in explaining the photoelectric effect (for which he received the Nobel Prize) turns out to be ultimately and intimately tied to all of relativity—the constancy of the speed of light and the geometry of space time.

13.12 Relativistic Mass

But wait! This discussion of relativity can't just end (or can it?). What about the "relativistic mass"? Everybody already "knows" that the mass increases according to

$$m = \frac{m_0}{\sqrt{1 - (v^2/c^2)}}$$

That goes right along with time dilation and length contraction. Doesn't it?

That all depnds. Our definition of the momentum explicitly contains the factor $\sqrt{1 - (v^2/c^2)}$. And the mass is constant and invariant. But, if you decide to define the momentum using something that *looks* more like the classical form

$$p = m^*v \qquad (13.12.1)$$

then the "mass" m^* is not constant or invariant. This mass does increase with velocity according to

$$m^* = \frac{m_0}{\sqrt{1 - (v^2/c^2)}}$$

where m_0 is not the "rest mass," the value measured for the mass when it is at rest. Either formulation is correct.

The important thing about mass in relativity is its equivalence with energy. We can no longer talk about the conservation of mass and the conservation of energy separately. We must now be concerned with the conservation of mass energy. They really describe the same thing.

This equivalence shows up in the binding energies of particles. While it is true for the binding energies of electrons in chemical reactions, it is easier to observe in the binding energies of nucleons in nuclear reactions. As a single simple example, consider two nuclei of heavy hydrogen (two deuterons) that fuse together to form a nucleus of helium (an alpha particle). The sum of the masses of the two deuterons is *greater* than the mass of the product helium nucleus. Mass is not conserved: the final mass is measureably less than the initial mass. But energy is given off in the form of γ rays. And the energy of these γ rays is *precisely* equivalent to the "missing" mass, Δm, according to

$$E_\gamma = \Delta mc^2 \qquad (13.12.2)$$

This equivalence of mass and energy, of course, is the basis of all nuclear power—both fission and fusion. A water molecule has a mass that is somewhat less than the sum of the masses of the oxygen and hydrogen atoms that formed it. So this is also the basis of energy released from chemical reactions—like burning hydrogen, oil, or coal—but in these cases the differences in mass are so small that they are quite difficult to measure.

PROBLEMS

13.1 Frame B moves relative to frame A with a relative speed of $0.8c$ along the common direction of their x-axes. All clocks read zero as the origins coincide. B observes an explosion (an event) on his x-axis at position $x_B = 1.5$ km and time $t_B = 0.001$ s. When and where does A observe this event?

13.2 A flashbulb goes off as the origins of our usual A and B frames coincide. Frames A and B have a relative speed of $0.6c$. The wave front arrives at $x_B = y_B = z_B = 500$ m at time t_B. What is this time t_B? What are the coordinates x_A, y_A, z_A, and t_A corresponding to this event? Use *these* coordinates to calculate the speed of light in A's frame.

13.3 Two events in frame B occur with (x_B, y_B, z_B, t_B) coordinates of E_1(100 m, 0, 0, 3×10^{-7} s) and E_2(200 m, 100 m, 0, 5×10^{-7} s). Frame B moves past frame A so that $\alpha = 45°$. Find the following:
 (a) The coordinates of E_1 and E_2 in frame A.
 (b) The interval, Δs, between E_1 and E_2 using A's coordinates.
 (c) The interval, Δs, between E_1 and E_2 using B's coordinates.

13.4 An alien spaceship traveling at $v = 0.5c$ travels north from Los Angeles toward San Francisco. The Captain and First Officer, standing at the bow and stern, 500 m apart, sneeze at precisely the same time—as observed by their crew. What time difference measured by Earth-based observers is there between their two sneezes?

13.5 A meterstick is at rest in frame B in the xy plane, making an angle of $60°$ with the x-axis. Frame B moves with speed of $0.5c$ relative to frame A along the common direction of their x-axes. What angle does A measure between the traveling meterstick and the x_A-axis? How long is the meterstick measured to be—in A's frame?

13.6 A space traveler whizzes past Earth at a speed of $0.5c$. As *his* clock indicates a passage of 100 seconds of time, how much time elapses on Earth? What length will Earth-based scientists measure for the length of his spaceship if it is 100 m long at rest?

13.7 A spaceship 100 m long (when measured at rest) travels at a speed $0.75c$ relative to Earth. A flashbulb is flashed at the front or bow of the ship. How much time—measured on board—elapses before the light reaches the tail or stern? How much time—measured on Earth—elapses between these two events?

PROBLEMS

13.8 Frame B moves relative to frame A with the speed $0.4c$. When the two origins coincide, clocks at both origins read zero. One hour later—measured by B—a lab assistant at B's origin sneezes.
    (a) According to A, when did the sneeze occur?
    (b) How long—according to A—does it take the signal (or information) from the sneeze to reach A's origin? What does A's clock read then?
    (c) How long—according to B—does it take the signal from the sneeze to reach A's origin?

13.9 How fast must a space traveler travel with respect to Earth so that Earthbound observers will observe ten hours to pass on Earth clocks while only one hour passes for the space traveler (as seen by Earth)?

13.10 How fast must a space traveler travel with respect to Earth so his clock is observed to run at one-half the rate of a clock at rest on Earth?

13.11 The mean decay time of a muon at rest is 2.2×10^{-6} sec. What mean decay time will be measured in a lab for muons traveling at $0.999c$?

13.12 Earth's atmosphere extends (very) roughly 10-km high. How long does this distance appear to a cosmic ray muon traveling at $0.99c$? How much (proper) time elapses as the muon travels this distance?

13.13 A railroad train at rest in a yard is carefully measured to have a proper length of precisely 1.0 km. What will be its length, measured by observers at rest on Earth, if it travels 100 km/hr? If it travels 100,000 km/hr? If it travels 100,000,000 km/hr?

13.14 Certain particles traveling at 2.4×10^8 m/s are observed to have a mean lifetime of 2×10^{-8} s. What is their lifetime at rest?

13.15 A high-speed alien spacecraft whose proper length is 200 m flies into a straight tunnel bored through a mountain. The proper length of the tunnel is 100 m. An observer at rest beside the tunnel observes the spacecraft contracted to a length of 98 m. How fast is the craft going? The tunnel has doors at each end that shut and reopen immediately. The observer at rest beside the tunnel activates both doors at exactly the same time. Thus, he concludes that the spacecraft was momentarily confined inside the tunnel.
    An observer on board the spacecraft, though, observes the tunnel to be contracted. So—for this observer—it is clear the spacecraft could not be confined in the tunnel.
    Describe the two events—the two doors' closings—from this reference frame. Use both Minkowski diagrams and the Lorentz transformations to reconcile the two descriptions.

13.16 Two particles are moving in *opposite* directions along the x-axis. Particle 1 has a speed of $-0.6c$ (to the left); particle 2, $0.8c$ (to the right). What is the speed of particle 2 as seen by 1, v_{21}?

13.17 A speeding motorist is stopped by a city policeman for going through a red light. The motorist offers as an excuse that the red light ($\lambda = 6.50 \times 10^{-7}$ m) appeared green ($\lambda = 5.25 \times 10^{-7}$ m) due to the Doppler effect. If this were true, what was his speed?

13.18 Frame B moves relative to frame A at a speed that gives $\alpha = 45°$. Both sets of clocks read zero when the origins are coincident.
    (a) Find the relative speed of the two frames.
    (b) How far are the two origins separated—as measured in A's frame by x_A—when A's clock shows $t_A = 5$ s?
    (c) What will B's clock located at the origin read then?

13.19 Repeat Problem 13.18 for $\alpha = 30°$.

13.20 Repeat Problem 13.18 for $\alpha = 60°$.

13.21 What is the maximum speed for which the actual, relativistic kinetic energy is within one percent of the classical form, $\frac{1}{2}mv^2$?

13.22 What is the minimum speed for which the relativistic kinetic energy and the total energy vary by less than one percent?

13.23 Show that the speed of a particle is given by

$$v = c\sqrt{1 - \left(\sqrt{\frac{E_0}{E}}\right)^2}$$

where $E_0 = mc^2$ is its rest energy, and E its total energy.

13.24 Derive Eq. 13.11.15.

13.25 How much energy is released if an electron-positron pair (each with mass of 9.1×10^{-31} kg) is annihilated?

13.26 How much energy is released if 1 g of matter is converted entirely into energy?

A

CONVERSATIONAL PASCAL

A.1 Getting Started

In any physics course, it's often easy for the physics of the situation to become obscured by the mathematics. To avoid this (for the present), let's look at a harmonic oscillator and use the simplest mathematical tools possible.

For our harmonic oscillator, we'll look at a mass, m, attached to one end of a spring having spring constant k. The other end of the spring is firmly attached to a rigid support, as shown in Figure A.1.1. When the mass is moved some distance, x, away from equilibrium, the spring exerts a force to restore it—to push or pull it back to equilibrium. This force is found to be proportional to the distance the mass is from equilibrium (the constant of proportionality being the spring constant k). This behavior of a linear restoring force is known as Hooke's Law and, as an equation, can be written as

$$F = -kx \qquad (A.1.1)$$

From Newton's Second Law of Motion we also know

$$F = ma \qquad (A.1.2)$$

where a is the acceleration experienced by mass m when acted upon by force F. Therefore,

$$ma = -kx$$

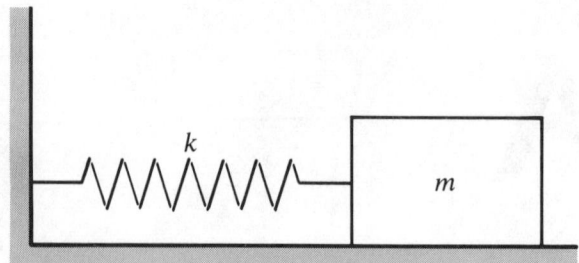

Figure A.1.1 A simple harmonic oscillator.

or

$$a = -\frac{kx}{m} \quad (A.1.3)$$

If the mass is moved to some initial position, x_0, and released with some initial velocity, v_0, Eq. 1.1.3 should enable us to locate the mass at any later time—*in principle*.

For simpler cases, such as $a = a_0$ (a constant), you already know that

$$v = a_0 t + v_0$$
$$x = \tfrac{1}{2} a_0 t^2 + v_0 t + x_0 \quad (A.1.4)$$

And for $a = a(t)$, we can integrate directly to get

$$v = \int a(t)\, dt + v_0$$
$$x = \int v(t)\, dt + x_0 \quad (A.1.5)$$

But our present situation doesn't lend itself to such direct methods. So, we shall omit integration entirely and solve this problem *by hand*! Using a hand-held calculator will make the calculations go much faster.

To begin, label some columns on a sheet of paper as follows:

Time	Position	Velocity	Acceleration
0	x_0	v_0	$-\dfrac{k}{m} x_0$

Usually we know where the mass is and how fast it is going at the beginning. These are the *initial conditions*. We can write x_0 and v_0 for these initial conditions when $t = 0$. Knowing the position allows us to calculate the force and then the acceleration. Acceleration for this mass and spring is $-(k/m)x_0$. All of these values are entered in the previous table.

Although we begin by looking at the motion of a mass attached to a spring—a simple harmonic oscillator—the ideas and techniques we develop

SECTION A.1 / GETTING STARTED

here will be useful in many different systems. This simple harmonic oscillator is just an example, but an important one.

The acceleration changes as x changes. If we observe this system for a very brief time, Δt, so that x doesn't change much, then acceleration doesn't change much either. For that matter, the velocity stays nearly the same during this time, too.

If we let time increase to $t_1 = \Delta t$, where Δt is small enough that the change in acceleration is negligible, we may treat the acceleration as a constant. Then,

$$v(t_1) = v_0 + a\,\Delta t \tag{A.1.6}$$

Likewise, the velocity will be nearly constant over such a small time interval. So we can write

$$x(t_1) = x_0 + v\,\Delta t \tag{A.1.7}$$

Now the data table looks like this:

Time	Position	Velocity	Acceleration
0	x_0	v_0	$-\frac{k}{m}x_0$
t_1	$x(t_1)$	$v(t_1)$	

Now let's try this with some actual numbers. Suppose we use a spring with spring constant $k = 1000$ N/m attached to a body with a mass of $m = 5$ kg. Pull it to one side a distance of 10 cm or $x_0 = 0.10$ m. Let it go from rest so that $v_0 = 0.00$ m/s. As you release it, its acceleration is $-(k/m)x_0 = -(1000/5)0.1 = -20$; that is, -20 m/s². For the time increment Δt, use 0.01. From Eq. 1.1.6, the velocity at $t = 0.01$ s must be $v(0.01) = 0 + (-20)(0.01) = -2$; that is, -0.2 m/s. This velocity and Eq. 1.1.7 allow us to calculate the position at $t = 0.01$ s. We find $x(0.01) = 0.10 + (-0.2)(0.01) = 0.0998$, which is really 0.0998 m.

With actual numbers, then, our data table looks like this:

Time	Position	Velocity	Acceleration
0.00	0.10000	0.00	-20
0.01	0.0998	-0.20	

Since we now know the current (or "new") position, we can calculate the current (or "new") value of the acceleration from

$$a(t_1) = -\frac{k}{m}x(t_1) \tag{A.1.8}$$

and include it in our table.

Let time increase by Δt again, this time to t_2. Just as before, assume that Δt is so small that v and a are almost constant. Then we can write

$$v(t_2) = v(t_1) + a(t_1)\,\Delta t \tag{A.1.9}$$

$$x(t_2) = x(t_1) + v(t_2)\,\Delta t \tag{A.1.10}$$

Or we can write this as

$$v(\text{new}) = v(\text{old}) + a(\text{old})\,\Delta t \tag{A.1.11}$$

$$x(\text{new}) = x(\text{old}) + v(\text{old})\,\Delta t \tag{A.1.12}$$

We can then calculate a new value for the acceleration:

$$a(\text{new}) = -\frac{k}{m}\,x(\text{new}) \tag{A.1.13}$$

Using this method over and over again, we can continue to fill in our table of values. The smaller we make Δt, the more accurate our assumptions (and our results) become. The larger we make Δt, the quicker we're finished, but with less accurate results.

In this way, we reduce this problem to a particular pattern of "plug-and-crank arithmetic." But we fear you will quickly tire of this. So think. Is there anyone on campus you could get to do these simple arithmetical operations and give you the finished table, all neatly filled in?

Of course there is—your friendly campus computer. It speaks several languages, if not English or algebra. So we have to take a crash course in what we'll call "conversational Pascal." This isn't a course in computer programming, so we're not going to learn *Real* Pascal—rather, just enough to solve our current problem.

The spring constant, k, and the mass, m, have numerical values. We can inform the computer of this by telling it

```
K := k
```
(A.1.14)

```
M := m
```
(A.1.15)

K and M are locations—just like a scratch pad, a chalkboard, or an envelope. The variables k and m are the actual numbers we then store in the respective locations—or write on the scratch pad or chalkboard or put in the envelope. The assignment operator in Pascal is " := ", which is not the same as the test for equality or the definition of a constant, " = ". A semicolon separates two Pascal statements.

To make this example more explicit, suppose the spring has a spring constant of 5.23 N/m and is attached to a mass of 0.750 kg. Eqs. A.1.14 and A.1.15 then become

```
K := 5.23
```
(A.1.16)

and

```
M := 0.750
```
(A.1.17)

SECTION A.1 / GETTING STARTED

Please note that the computer has no way of keeping track of units, so that's up to you. The computer works only on the numbers.

When $t = 0$, the initial position is x_0 and the initial velocity is v_0. We can tell the computer this by saying

$$T := 0.0 \tag{A.1.18}$$

$$X := x_0 \tag{A.1.19}$$

$$V := v_0 \tag{A.1.20}$$

where x_0 and v_0 are just numbers—like 1.50 (m) and 2.35 (m/s). T, X, and V are locations inside the computer—or scratch pads, chalkboards, or envelopes, if you please—that now contain the numbers 0.0, x_0, and v_0, respectively. (Incidentally, .0 is not allowed in Pascal; be sure to use 0.0.)

We must now determine a value for Δt. For the time being, let's choose $\Delta t = 0.05$. Tell the computer

$$DT := 0.05; \tag{A.1.21}$$

The location of DT now contains our choice for Δt.

The computer can now calculate the force and acceleration if we tell it

$$F := -K*X; \qquad A := \frac{F}{M} \tag{A.1.22}$$

Note that the asterisk ($*$) is used for multiplication and the slash (/) for division. A semicolon (;) is used to separate Pascal statements. The computer doesn't understand "$A := -KX/M$" (at least, not for what we need done).

To start writing our table, we could now instruct the computer to

$$\text{WRITELN (T, X, V, A)} \tag{A.1.23}$$

Although we need A for the calculations, we're usually more interested in X and V. So, even though the computer will be keeping A's value inside, let's ask for a simplified printout:

$$\text{WRITELN (T, X, V)} \tag{A.1.24}$$

We need just a few more things. In Pascal, even conversational Pascal, we must take care of a few housekeeping chores before we can really begin. First, we must give the program a name and set up the input and output. We do this by

$$\text{PROGRAM OSCIL (INPUT, OUTPUT)} \tag{A.1.25}$$

Then we must tell the computer all the variables we plan to use so room can be set aside for them:

$$\text{VAR M, K, X, V, A, T:REAL} \tag{A.1.26}$$

These are all *real* variables (i.e., they are not integers). Our time increment DT from Eq. A.1.21 is a constant so we shall declare that separately by saying

CONST DT = 0.05 (A.1.27)

Note that there is no colon in front of this equal sign. Now we are ready to begin, and we tell the computer:

BEGIN (A.1.28)

We can now add the system parameters K and M from Eqs. A.1.14 through A.1.17. We can even combine them onto a single line:

K := k;   M := m (A.1.29)

Then we can combine the initial conditions of Eqs. A.1.18 through A.1.19:

T := 0;   X := x_0;   V := v_0 (A.1.30)

We are now ready for the main calculations using the ideas developed in Eqs. A.1.8 through A.1.13, but let's include the force explicitly as in Eq. A.1.22. That looks like

$$F := -K*X;   A := \frac{F}{M};$$ (A.1.31)

V := V + A*DT; (A.1.32)

Whoops! That doesn't seem to be correct unless A*DT is equal to zero. But ":=" in Pascal isn't quite what you would expect. A good English paraphrase of this Pascal statement might be:

"Take the numerical value written at location A and multiply it by the numerical value written at location DT. Now take the numerical value written at location V and add it to this product. Finally, put this *new* value into location V."

The *original* value stored at location V is erased, forgotten, lost, destroyed. The values at A and DT remain.

Likewise, Eq. A.1.12, which we used to calculate the new position, readily translates into

X := X + V*DT (A.1.33)

These are the *new* values of X, V, and A, corresponding to a new time. To find the new time, we can instruct the computer to calculate

T := T + DT (A.1.34)

So that we can see the results, we tell it to

  WRITELN (T, X, V);   (A.1.35)

to get a new set of entries on our data table.

We now need to repeat these calculations over and over again. But instead of making more and more copies of the same statements, we tell the computer to REPEAT these calculations UNTIL, say, $T > 50$. And the program itself must formally end with END. Thus, our entire program now looks like

```
PROGRAM OSCIL (INPUT, OUTPUT);
VAR M, K, V, A, T, F: REAL;
CONST DT = 0.05;
BEGIN
    V := 0; X := 0;
    M := 1.0; K := 2.5;
    T := 0;
    REPEAT
        F := -K*X; A := F/M;
        V := V + A*DT; X := X + V*DT;
        T := T + DT;
        WRITELN (T, X, V)
    UNTIL T > 50
END.
```
(A.1.36)

What do we choose for Δt? For accuracy, we want it small. We might try 0.01; or 0.001 would be even more accurate. But that will generate 1000 or 10,000 entries on our data table! Increasing Δt to 0.1 or 0.5 gives us a reasonable number of entries, but these data are unreliable. How can we have both accuracy (from a small Δt) and a reasonable number of data entries?

What if you were still processing Eqs. A.1.11, A.1.12, and A.1.13 by hand? You could use $\Delta t = 0.001$ for accuracy, but only use the results for $t = 0$, $t = 0.1$, $t = 0.2$, and so on. In this way, all the calculations are made every 0.001 seconds, but the results are *printed* only every 0.1 seconds—one printout every 100 calculations. Now let's tell the computer the same thing.

We first change DT to 0.001 and add a new variable, COUNT, to keep track of how many calculations we've made. (We make COUNT an integer.) Now the beginning of our program is

```
PROGRAM OSCIL (INPUT, OUTPUT);
VAR M, K, X, V, A, T, F:REAL;
    COUNT:INTEGER;
CONST DT = 0.001;
```
(A.1.37)

Instead of writing out the values *every*time, we just increment COUNT by one and when it reaches 100, do something—like write the values and reset

COUNT to zero. We can tell the computer all of this by

```
COUNT := COUNT + 1;
IF COUNT = 100 THEN
    BEGIN
        WRITELN (T, X, V);
        COUNT := 0;
    END;
```
(A.1.38)

Now the complete program looks something like this:

```
PROGRAM OSCIL (INPUT, OUTPUT);
VAR M, K, X, V, A, T, F:REAL;
    COUNT:INTEGER;
CONST DT = 0.001;
BEGIN
    V := 0; X := 1.0;
    M := 1.0: K = 2.5;
    T := 0;
    COUNT := 0;
    REPEAT
        F := -K*X; A := F/M;
        V := V + A*DT; X := X + V*DT;
        T := T + DT;
        COUNT := COUNT + 1;
        IF COUNT = 100 THEN
            BEGIN
                WRITELN (T, X, V);
                COUNT := 0;
            END;
    UNTIL T > 50;
END.
```
(A.1.39)

This programs works quite well. If we interchange the X and V calculations, calculating the new position before the new velocity, then the program would use a numerical integration method developed by Euler over two centuries ago. But the order of calculations in your program greatly increases the accuracy of the calculations, an important result only recently recognized by Professor Cromer.[1]

A.2 Elegant Output

We now have written a thoroughly functional program. It can do calculations and print out an exhaustive data table. Can we ask for anything more? Of course we can!

---

[1] Alan Cromer, "Stable solutions using the Euler approximation," *American Journal of Physics*, Vol. 49 (May 1981), pp. 455-459.

SECTION A.2 / ELEGANT OUTPUT

There are often times when it is helpful to make a remark in the program to remind ourselves of something. We can do this in Pascal because anything enclosed between (* and *) is ignored. For example, we might add a heading like

$$\text{(* CALCULATION OF POSITION FOR HARMONIC OSCILLATOR *)} \tag{A.2.1}$$

to our program. After the REPEAT we might add

$$\text{(* MAJOR CALCULATIONS OCCUR HERE *)} \tag{A.2.2}$$

The computer ignores these lines, but they may be helpful when *we* read the program.

A WRITELN statement in Pascal can be used to print more than just numbers. The statement

$$\text{WRITELN ('HARMONIC CALCULATIONS')} \tag{A.2.3}$$

tells the computer to print everything *within* the single quotations. We can use this feature and print numerical values at the same time, too.

As an example, suppose the spring constant is $k = 125$ N/m and this spring is attached to a mass of $m = 0.75$ kg. We can tell the computer this by

$$K = 125; \tag{A.2.4}$$

$$M = 0.75; \tag{A.2.5}$$

We might then add

$$\text{WRITELN('SPRING CONST = ',K,' MASS = ',M)} \tag{A.2.6}$$

The computer will then write the expression in the first set of quotation marks

SPRING CONST =

write the numerical value stored in location K,

SPRING CONST = 125

and then write the expression in the second set of quotation marks:

SPRING CONST = 125   MASS =

Finally, it will print the numerical value stored in M:

SPRING CONST = 125   MASS = 0.75

Things could be spruced up a bit if we had the units written out as well. To do this we could change the statement to

$$\text{WRITELN('SPRING CONST = ',K,'N/M   MASS = ',M,'KG')} \tag{A.2.7}$$

The material inside the quotation marks will be written out *exactly* as it appears. The material outside the quotation marks will be used as a command to look up and then write a numerical value.

Graphs are often better at displaying information than data tables. For example, it's easier to find the period, or see if it's varying, or determine what's happening to the amplitude in a graph.

So far we have used the WRITELN command, which writes out the information and then provides a *new line*. But Pascal also has a WRITE command that does *not* go on to the next line. We will use it to insert a number of spaces and then write a symbol, forming a rough but useful graph. To DO something a number of times we could use the REPEAT or WHILE commands, but the FOR command was designed for situations just like ours. To print N spaces we can say

$$\text{FOR I} := 1 \text{ TO N DO WRITE (' ');} \tag{A.2.8}$$

After these N spaces are written, we can then write an asterisk ('*') and go on to a new line with

$$\text{WRITELN('*')} \tag{A.2.9}$$

But how do we determine N? We want a graphical representation of the position X. When X is large and positive, we need N to be large and the asterisk is printed on the right of our output. For X near zero, the asterisk should be near the middle, and for X large but negative, toward the left. You are free to choose the parameters to suit yourself, but one possibility is

$$N = 40 + 20*X \tag{A.2.10}$$

which satisfies our needs. When $X = 1.0$, $N = 60$ and the asterisk is on the right. N is 40 when $X = 0$. And for $X = -1.0$, $N = 20$ and the asterisk is on the left. However, Pascal will object to this as written. N must be an *integer* to use it in the FOR statement. And X has already been declared a *real* number. We can correct this mismatch of data types with the TRUNC function that TRUNCates a *real* number and retains only the *integer* part of it by stating

$$N := 40 + \text{TRUNC}(20*X) \tag{A.2.11}$$

where we have added the required colon to the equal sign. Note that $20*\text{TRUNC}(X)$ will give very different, and interesting, results. Remember, we must go back to the beginning of the program and declare variables I and N to be *integers*; this can be done on the same line with COUNT:

$$\text{VAR COUNT, I, N:INTEGER;} \tag{A.2.12}$$

This addition should now result in a reasonable graph with time plotted downward and position horizontally. Rotating the computer's printout by

90° yields a more conventional distance/time graph. It may be useful, though, to still have the time written explicitly. We can do that with the statement

    WRITE(T);   (A.2.13)

just before the FOR statement.

So far, we've only used $F = -K*X$ in our program. This is true for a simple harmonic oscillator and assumes no friction at all. But suppose our apparatus were submerged in water—or corn syrup or molasses. Then there would be a frictional force opposing the motion. This force is proportional to the velocity, so our *net* force would be

$$F := -K*X - C*V \qquad (A.2.14)$$

where C is as yet undefined (it's small for water and very large for molasses). We can go back to the beginning of the program and define C by

    CONST C = c   (A.2.15)

where c is any value we want. What units must c have? We're now describing a *damped harmonic oscillator*.

Ordinarily we will be concerned with the units of anything we use; c must have units of N/(m/s) or kg/s. But we shall ignore such reasonable considerations in this chapter to concentrate on the computer program and the motion it describes.

Adding this velocity-dependent term to the force means we are describing a damped harmonic oscillator.

We can add still another component or term to the force, an external driving force like

$$F = F_0 \sin \omega t \qquad (A.2.16)$$

that varies with time. This is now a *forced harmonic oscillator*.

Once we define F0 and W as

    CONST F0 = f_0;   (A.2.17)

    CONST W = ω;   (A.2.18)

we can rewrite the force equation:

$$F := -K*X - C*V + F0*\text{SIN}(W*T) \qquad (A.2.19)$$

This means that we're trying to move the mass with this external driving force. How will it respond? Certainly increasing $F0$ will increase the amplitude of the motion. But the motion's dependence upon the frequency ω or W is of more interest. What frequency causes the largest amplitude? This short program allows immediate investigation of this driven harmonic oscillator. We'll return to it later and find analytical solutions to this problem.

A.3 Sometimes More Is Really Less

We're finished with the physics of the problem for the time being. After all our modifications, our program now looks like this:

```
(* Harmonic Oscillator Program *)                                    (A.3.1)
PROGRAM OSCIL. (INPUT, OUTPUT)
VAR M, K, X, A, T, F, C, F0, W:REAL;
    COUNT,N,I              :INTEGER;
CONST DT = 0.025;
  BEGIN
      WRITELN ('Harmonic Oscillator Calculations');
      T := 0;
      V := 0; X := 0; (* Initial Conditions *)
      M := 1.0; K := 2.5; (* System Parameters, SHO *)
      WRITELN ('Spring constant = ', K, 'N/M Mass = ',M, 'kg');
      C := 0.20; F0 := 0; W := 1.0; (* Parameters for complex system *)
      COUNT := 0;
      REPEAT
          F := -K*X - C*V + F0*SIN (W*T); A = F/M;
          V := V + A*DT; X := X + V*DT;
          T := T + DT;
          COUNT := COUNT + 1;
          IF COUNT = 10 THEN
          BEGIN
              WRITE (T);
              N := 40 + TRUNC (20*X);
              FOR I := 1 TO N DO WRITE (' ');
              WRITELN ('*');
              COUNT := 0;
          END;
      UNTIL T > 50;
  END.
```

So far, we've had to change the entries in the program every time we needed to change any of these values. There must be something we can do to decrease this workload (of course there is!). Rather than assigning each value in an explicit statement, we can use READ statements.

If we tell the computer

```
READLN (A);
```

SECTION A.3 / SOMETIMES MORE IS REALLY LESS

it will stop everything and wait until we type in a number and press ⟨ENTER⟩ before it goes on. We can combine several variables into a single READ statement like

READLN (A, B, C);

but a READ statement by itself can make it difficult to remember just what is being requested. It is a good idea to put in a WRITE statement to *ask* for the information. Such a statement is called a *prompt*.

Instead of explicitly writing

$$M := 1.0; \; K := 2.5; \; (* \text{ System Parameters, SHO } *) \tag{A.3.2}$$

to assign values for the mass and spring constant, we might use

WRITE ('What is the MASS of the oscillator?'); (A.3.3)
READLN(M);
WRITE('And the SPRING CONSTANT?');
READNN(K);

We can replace the assignment of values to V and X by

WRITELN('Now for the initial conditions ...'); (A.3.4)
WRITE('What is the initial position?');
READLN(X);
WRITE('And the initial velocity?');
READLN(V);

If the damping coefficient C is needed for a damped harmonic oscillator, we can use something like

WRITE ('What damping coefficient should I use?'); (A.3.5)
READLN(C);

For the forced harmonic oscillator, we can ask the computer to READ values for the external driving force $F0$ and its angular frequency W with statements like

WRITE ('What is the magnitude of the external force?'); (A.3.6)
READLN (F0);
WRITE ('And what is its angular frequency?');
READLN (W);

Now we can have the computer READ any information we choose. But at present the computer will READ the information, use it once, and END. We need some way to direct the computer back to the beginning READ statements again and again. We can do this by nesting most of our program inside a WHILE...DO loop like

```
PROGRAM OSCIL(INPUT, OUTPUT);                                    (A.3.7)
VAR M, D, X, V, A, T, F, L, F0, W; REAL;
    COUNT, N, I           ; INTEGER;
CONST DT = 0.001;
BEGIN
WHILE TRUE DO
BEGIN
    WRITE ('What is the MASS of the oscillator?');
    READLN (M);
    WRITE ('And the SPRING CONSTANT?');
    READLN (K);
    WRITELN ('And now for the initial conditions...');
    WRITE ('What is the initial POSITION?');
    READLN (X);
    WRITE ('And the initial VELOCITY?');
    READLN (V);
    ⋮
    T := 0;
    COUNT := 0;
    REPEAT
        F := −K*X; A := F/M;
        (* F := −K*X − C*V + F0*SIN(W*T) if needed *)
        V := V + A*DT; X = X + V*DT;
        T = T + DT;
        COUNT = COUNT + 1
        IF COUNT = 10 THEN
        BEGIN
            WRITE(T);
            N := 40 + TRUNC(20*x);
            FOR I := 1 TO N DO WRITE (' ');
            WRITELN('*');
            COUNT := 0;
        END;
    UNTIL T > 10;
END;
END.                                                             (A.3.8)
```

We can add WRITE prompts and READ statements for any other variable we need if we want to investigate a damped or forced harmonic oscillator. This program works, but it is like the sorcerer's apprentice in that there is no way to stop it. We can always stop it with a BREAK key or the equivalent; if we want a "cleaner" method, we can check the value of the mass M and continue only WHILE $M > 0$. That is, the program will stop if we enter negative value for the mass.

To do this, we can check the value of M by

```
BEGIN                                                              (A.3.9)
    WRITE ('What is the MASS?');
    READLN (M);
    WHILE M > 0 DO
        BEGIN
        WRITE ('And the SPRING CONSTANT?');
        READLN (K);
           ⋮
```

We now have a very useable program involving real physics, and it uses all four major constructs available to Pascal:

```
WHILE ··· DO                                                       (A.3.10)
FOR ··· TO ··· DO
REPEAT ··· UNTIL
IF ··· THEN ...
```

A.4 Into The Wild Blue...

We can now accurately (and elegantly) calculate the rather complex motion of an oscillator. As you shall see later, this is far from a trivial problem. It's the computer and Pascal that make our calculations fairly easy. Because of that, *you* can expend your current energies on understanding the physics of the motion—how it moves and why.

Now let's apply our powerful tool to some other situations. First, consider the very familiar case of free fall with no air resistance. This is a case of constant acceleration and the results are described by

$$v = v_0 + gt \tag{A.4.1}$$

$$x = x_0 + v_0 t - \tfrac{1}{2} g t^2 \tag{A.4.2}$$

with which you should be very familiar. Such motion is a result of a force

$$F = -mg \tag{A.4.3}$$

that is proportional to the mass of the body. (This was one of Galileo's big contributions to physics.) To use our Pascal program, all we need do is change the force statements to

```
F := -M*G                                                          (A.4.4)
```

We must define G earlier ($G = 9.8$ for MKS units) and do not need to define K, C, $F0$, or W since they have been removed from our force equation.

We can now turn to air resistance and find that for small, smooth objects at low speeds it is nearly proportional to the speed of the object. We accommodate for this by changing the force statement to

$$F := -M*G - C*V \qquad (A.4.5)$$

This slight change is trivial as far as the computer is concerned, but it makes our force a much closer approximation of real forces encountered in nature. As you shall see later, this slight change greatly increases the complexity of the analytic solution in closed form.

With such a force, it becomes very interesting to look at the behavior of the speed. We therefore write out the speed again. If dropped from rest, the speed increases initially, but reaches (or approaches) some value and then remains constant. This is called the *terminal speed* of the object. A constant speed means zero acceleration or zero net force, so

$$F = -mg - cv_t = 0 \qquad (A.4.6)$$

$$v_t = -\frac{mg}{c} \qquad (A.4.7)$$

where v_t is the terminal speed. Terminal speed for a man falling without a parachute might be about 120 mph. A parachute increases the drag (increases C) so the terminal speed is (greatly) decreased. Even before deploying a parachute, a sky diver can change his speed by extending his arms and legs; this changes his drag (and, thus, C) and, in turn, the terminal speed.

Air resistance is complex. As we shall see later, considering it as a force proportional to velocity (or speed) is a reasonable approximation and allows us to solve the problem without too much difficulty using calculus. For some objects—and especially for higher speeds—a force proportional to the *square* of the velocity is a better approximation.

Our first attempt might be to change our definition of the force to

$$F := -M*G - C2*V*V \qquad (A.4.8)$$

where $C2$ is our proportionality constant. $C2$ is a coefficient just like C for the linear case, but it has different units. Therefore, we can avoid later difficulties later by remembering that it is different and not writing it simply as C. But wait just a minute. Whether V is positive when the object is going upward or negative when the object is going downward, the term $-C2*V*V$ is always negative (assuming $C2 > 0$). A negative force is *downward*. Air resistance, being a frictional force, is always opposed to the motion. When V is positive, the air resistance force is negative, and vice versa.

We could take care of this with IF statements checking on whether V is directed upward or downward. But a shorter method is to use the absolute value function, ABS( ), available in Pascal. With that we can correctly define our force by

$$F := -M*G - C2*V*ABS(V) \qquad (A.4.9)$$

ABS (V) is the absolute value of V. It is V whenever V is positive and $-V$ whenever V is negative. That gives the correct sign relations we need on V and the air resistance force.

Again, if we look at velocity as a function of time, we will find a *terminal speed* as in the earlier case. As then, a terminal speed occurs when the net force is zero (the downward gravitational component is exactly balanced by the upward air resistance component). Thus,

$$F = -mg + c_2 v_t^2 = 0 \qquad (A.4.10)$$

$$v_t^2 = \frac{mg}{c_2} \qquad (A.4.11)$$

$$v_t = \sqrt{\frac{mg}{c_2}} \qquad (A.4.12)$$

So far, all of our programs have handled motion in one dimension only. But it is easy to alter a program to handle projectile motion—or any motion in two dimensions under the influence of any two-dimensional force. The general case of three dimensions can easily be handled as well. Instead of dealing only with $F = ma$, we must look at $F_x = ma_x$ and $F_y = ma_y$.

To look at simple projectile motion, we replace our single force statement by two,

FX := 0; $\qquad (A.4.13)$

FY := $-$M*G;

The acceleration of gravity G and mass M will have to be defined earlier. Instead of calculating a single acceleration from $A := F/M$, we must now calculate the two components of the acceleration,

AX := FX/M; $\qquad (A.4.14)$

AY := FY/M;

The x- and y-components of position and velocity must be calculated separately by

X := X + VX*DT; $\qquad (A.4.15)$

Y := Y + VY*DT;

VX := VX + AX*DT; $\qquad (A.4.16)$

VY := VY + AY*DT;

Remember that DT can be changed at will to increase or decrease the accuracy of the calculations. And COUNT can be changed if you want more or fewer data printed. (Always be sure that *you* control the computer—*not* vice versa.)

B
CALCULUS REVIEW

B.1 Differential Calculus

We define acceleration as the time rate of change of the velocity and velocity as the time rate of change of the position. Mathematically, we can write these as

$$a = \frac{\Delta v}{\Delta t} \quad \text{and} \quad v = \frac{\Delta x}{\Delta t} \tag{B.1.1}$$

where the "Δ" means "change in."

But how large a value of Δt shall we use? We can describe the *instantaneous* acceleration or the *instantaneous* velocity as the value we obtain as we use an ever smaller value for Δt—as Δt approaches 0. We might write that mathematically as

$$a = \lim_{\Delta t \to 0} \frac{\Delta v}{\Delta t} \quad \text{and} \quad v = \lim_{\Delta t \to 0} \frac{\Delta x}{\Delta t} \tag{B.1.2}$$

Such a limiting process is the basis for *differential calculus*. The *derivative* of any function $f(x)$ is defined by

$$\frac{df(x)}{dx} \equiv \lim_{\Delta x \to 0} \frac{f(x + \Delta x) - f(x)}{\Delta x} \tag{B.1.3}$$

To make this more meaningful, let $f(x) = x^2$ as an example. Then,

$$\frac{d(x^2)}{dx} = \lim_{\Delta x \to 0} \frac{(x + \Delta x)^2 - x^2}{\Delta x}$$

$$= \lim_{\Delta x \to 0} \frac{[x^2 + 2x\,\Delta x + (\Delta x)^2] - x^2}{\Delta x}$$

$$= \lim_{\Delta x \to 0} \frac{2x\,\Delta x + (\Delta x)^2}{\Delta x}$$

$$= \lim_{\Delta x \to 0} (2x + \Delta x)$$

$$\frac{d}{dx}(x^2) = 2x \tag{B.1.4}$$

That is, the derivative of x^2 is $2x$.

But a derivative is far more than just a relationship between two functions. Figure B.1.1 shows a sketch of the function $f(x) = x^2$. When we apply Equation B.1.3 to find the derivative of a function, we're really finding the slope of a line. As an example, find the derivative of x^2 for $x = 1$. Figure B.1.1 shows what we would need to do for $\Delta x = 2, 1,$ and 0.5.

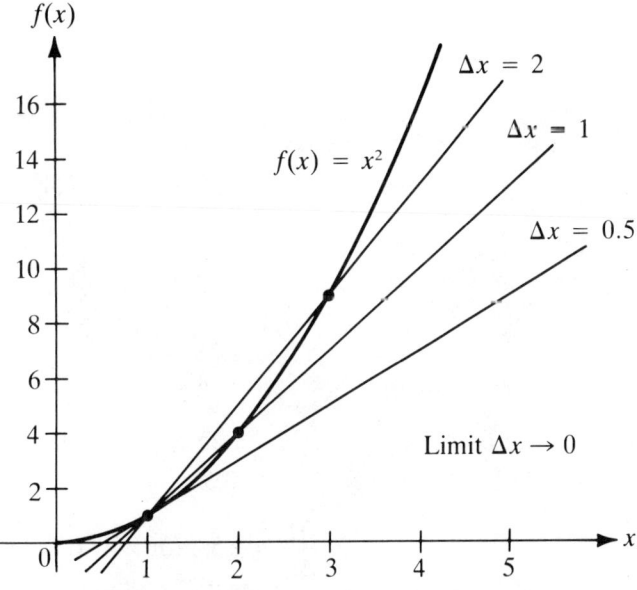

Figure B.1.1

For $f(x) = x^2$, $x = 1$ and $\Delta x = 2$, we have

$$\frac{df}{dx} \to \frac{\Delta f}{\Delta x} = \frac{f(x + \Delta x) - f(x)}{\Delta x}$$

$$= \frac{f(1 + 2) - f(1)}{2}$$

$$= \frac{f(3) - f(1)}{2}$$

$$= \frac{9 - 1}{2}$$

$$= \frac{8}{2}$$

$$\frac{\Delta f}{\Delta x} = 4 \quad \text{for } x = 1, \Delta x = 2 \tag{B.1.5}$$

For $x = 1$, we can evaluate $\Delta f / \Delta x$ by

$$\frac{\Delta f}{\Delta x} = \frac{f(x + \Delta x) - f(x)}{\Delta x}$$

$$= \frac{f(1 + 1) - f(1)}{1}$$

$$= \frac{f(2) - f(1)}{1}$$

$$= \frac{4 - 1}{1}$$

$$\frac{\Delta f}{\Delta x} = 3 \quad \text{for } x = 1, \Delta x = 1 \tag{B.1.6}$$

For $\Delta x = 0.5$, we have

$$\frac{\Delta f}{\Delta x} = \frac{f(x + \Delta x) - f(x)}{\Delta x}$$

$$= \frac{f(1 + 0.5) - f(1)}{0.5}$$

$$= \frac{f(1.5) - f(1)}{0.5}$$

$$= \frac{2.25 - 1}{0.5}$$

$$= \frac{1.25}{0.5}$$

$$\frac{\Delta f}{\Delta x} = 2.5 \quad \text{for } x = 1, \Delta x = 0.5 \tag{B.1.7}$$

SECTION B.1 / DIFFERENTIAL CALCULUS

For $\Delta x = 0.25$, we have the following:

$$\frac{\Delta f}{\Delta x} = \frac{f(x + \Delta x) - f(x)}{\Delta x}$$

$$= \frac{f(1 + 0.25) - f(1)}{0.25}$$

$$= \frac{f(1.25) - f(1)}{0.25}$$

$$= \frac{1.5625 - 1.0}{0.25}$$

$$= \frac{0.5625}{0.25}$$

$$\frac{\Delta f}{\Delta x} = 2.25 \quad \text{for } x = 1, \Delta x = 0.25 \tag{B.1.8}$$

We can now make a table of $\Delta f / \Delta x$ versus Δx:

Δx	$\frac{\Delta f}{\Delta x}$
2.0	4
1.0	3
0.5	2.5
0.25	2.25

If we would continue with even smaller values of Δx, we would find the values of $\Delta f / \Delta x$ getting even closer to 2.0 as given by Eq. B.1.4.

With this in mind and looking at Figure B.1.7, we can see that *the derivative of a function is the slope of a line drawn tangent to that function at the point in question. A derivative is the slope of a line.*

Look at the function $f(x) = x^n$. To find the derivative, we'll need $f(x + \Delta x) = (x + \Delta x)^n$. We can best see this by using binomial expansion:

$$(x + y)^n = x^n + nx^{n-1}y + \frac{n(n-1)}{2!}x^{n-2}y^2 + \frac{n(n-1)(n-2)}{3!}x^{n-3}y^3 + \cdots \tag{B.1.9}$$

$$(x + \Delta x)^n = x^n + nx^{n-1}\Delta x + \frac{n(n-1)}{2!}x^{n-2}(\Delta x)^2 + \cdots \tag{B.1.10}$$

Now to find the derivative of $f(x) = x^n$, we have

$$\frac{d}{dx}(x^n) = \lim_{\Delta x \to 0} \frac{(x + \Delta x)^n - x^n}{\Delta x}$$

$$= \lim_{\Delta x \to 0} \frac{\left[x^n + nx^{n-1}\Delta x + \frac{n(n-1)}{2!}x^{n-2}(\Delta x)^2 + \cdots\right] - x^n}{\Delta x}$$

$$= \lim_{\Delta x \to 0}\left[nx^{n-1} + \frac{n(n-1)}{2!}x^{n-2}(\Delta x) + \cdots\right]$$

$$\frac{d}{dx}(x^n) = nx^{n-1} \tag{B.1.11}$$

This relation, which should already be quite familiar, is one of the most used in differential calculus.

We can use this relation to find the derivatives of the trigonometric functions by first noting that they can be expressed as a power series. For instance,

$$\sin x = x - \frac{x^3}{3!} + \frac{x^5}{5!} - \frac{x^7}{7!} + \cdots \tag{B.1.12}$$

$$\frac{d}{dx}(\sin x) = \frac{d}{dx}\left(x - \frac{x^3}{3!} + \frac{x^5}{5!} - \frac{x^7}{7!} + \cdots\right)$$

$$= \frac{dx}{dx} - \frac{1}{3!}\frac{dx^3}{dx} + \frac{1}{5!}\frac{dx^5}{dx} - \frac{1}{7!}\frac{dx^7}{dx} + \cdots$$

$$= 1 - \frac{1}{3!}(3x^2) + \frac{1}{5!}(5x^4) - \frac{1}{7!}(7x^6) + \cdots$$

$$\frac{d \sin x}{dx} = 1 - \frac{x^2}{2!} + \frac{x^4}{4!} - \frac{x^6}{6!} + \cdots$$

But this power series is just the cosine:

$$\cos x = 1 - \frac{x^2}{2!} + \frac{x^4}{4!} - \frac{x^6}{6!} + \cdots \tag{B.1.13}$$

Hence,

$$\frac{d}{dx}\sin x = \cos x \tag{B.1.14}$$

Likewise,

$$\frac{d}{dx}\cos x = \frac{d}{dx}\left(1 - \frac{x^2}{2!} + \frac{x^4}{4!} - \frac{x^6}{6!} + \cdots\right)$$

$$= \frac{d}{dx}(1) - \frac{1}{2!}\frac{d}{dx}(x^2) + \frac{1}{4!}\frac{d}{dx}(x^4) - \frac{1}{6!}\frac{d}{dx}(x^6) + \cdots$$

$$= 0 - \frac{1}{2!}(2x) + \frac{1}{4!}(4x^3) - \frac{1}{6!}(6x^5) + \cdots$$

$$= -\left(x - \frac{x^3}{3!} + \frac{x^5}{5!} - \cdots\right)$$

$$\frac{d}{dx}\cos x = -\sin x \tag{B.1.15}$$

Equations B.1.14 and B.1.15 will also prove quite useful throughout this course.

Also of use is the derivative of the *product* of two functions, like $h(x) = f(x) \cdot g(x)$. Direct application of Eq. B.1.3, our definition of a derivative, gives us the desired result.

$$\frac{dh}{dx} = \frac{d}{dx}[f(x)g(x)]$$

$$= \lim_{\Delta x \to 0} \frac{f(x + \Delta x)g(x + \Delta x) - f(x)g(x)}{\Delta x}$$

$$= \lim_{\Delta x \to 0} \frac{\left[f(x) + \frac{df}{dx}\Delta x\right]\left[g(x) + \frac{dg}{dx}\Delta x\right] - f(x)g(x)}{\Delta x}$$

$$= \lim_{\Delta x \to 0} \frac{f(x)g(x) + \frac{df}{dx}g\,\Delta x + f\frac{dg}{dx}\Delta x + \frac{df}{dx}\frac{dg}{dx}(\Delta x)^2 - f(x)g(x)}{\Delta x}$$

$$= \lim_{\Delta x \to 0}\left[\frac{df}{dx}g + f\frac{dg}{dx} + \frac{df}{dx}\frac{dg}{dx}\Delta x\right]$$

$$\frac{dh}{dx} = \frac{d}{dx}(fg) = \frac{df}{dx}g + f\frac{dg}{dx} \tag{B.1.16}$$

B.2 Integral Calculus

If $x = x(t)$ is *given*, then it is a simple and straightforward procedure to take the derivative to obtain the velocity function $v = v(t) = dx(t)/dt$. This can then be differentiated once more to obtain the acceleration, and multiplied by mass to determine the force, F. All of this is simple and straightforward. However, if we know $x(t)$ completely, there would hardly be any reason to pursue the matter further since $x(t)$ completely describes the motion. What usually is the case is that we *know* the force, F, and we want to solve for the motion, $x(t)$. This is much more difficult to do. We will need more tools—namely *integral calculus*.

An integral is sometimes called an antiderivative. If

$$f(u) = \frac{dF(u)}{du}$$

then

$$F(u) = \int f(u)\,du$$

by definition. That is, the *integral* of a function is another function whose derivative is the original function. To be specific, the above defines an *indefinite integral*. A *definite integral* is written with limits (values or symbols) at either end of the integral sign and—as its name implies—has a definite value (*unless one of the limits is itself a variable*).

$$\int_a^b f(u)\,du = F(b) - F(a) \tag{B.2.1}$$

would be a definite number—the difference in the values of F evaluated at the upper limit b and at the lower limit a. One of the limits, however, can be a variable:

$$\int_a^u f(u)\,du = F(u) - F(a) \tag{B.2.2}$$

Our result is now a function of u. The u *inside* the integral sign is a "dummy variable" or "variable of integration" and should not also be used in our limit. While the notation of Eq. B.2.2 is common, it is technically incorrect and would be written more correctly as

$$\int_a^u f(u')\,du' = F(u) - F(a) \tag{B.2.3}$$

where a different variable, u', has been used as the dummy variable inside the integral.

The integral is useful to physicists because it is essentially an *infinite sum*. We used the computer in Chapter 1 and Appendix A to determine velocity or position by

$$x = x_0 + \sum_1 v_i \Delta t \tag{B.2.4}$$

$$v = v_0 + \sum a_i \Delta t \tag{B.2.5}$$

with accuracy increasing as Δt became smaller. In the limit as Δt becomes infinitesimally small, this sum becomes an integral:

$$\int v(t)\,dt = \lim_{\Delta t \to 0} \sum v_i \Delta t \tag{B.2.6}$$

This can be seen clearly by taking a simple function, say $f(x) = x$, and drawing a graph of this as in Figure B.2.1.

To find the area under this curve from 0 to 4, represent this by the area of a series of rectangles as in

$$A = \sum_i f(x_i)\,\Delta x \tag{B.2.7}$$

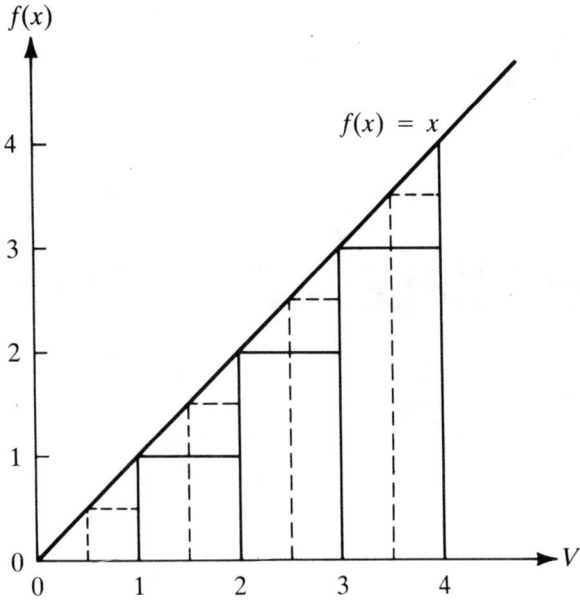

Figure B.2.1

where $x_i = 0, \Delta x, 2\Delta x, 3\Delta x \cdots$. As shown in Figure B.2.1 for $\Delta x = 1$, we have

$$A = (0)(1) + (1)(1) + (2)(1) + (3)(1) = 6 \tag{B.2.8}$$

As shown by the broken lines for $\Delta x = 0.5$, we have

$$A = (0)(.5) + (.5)(.5) + (1.0)(.5) + (1.5)(.5)$$
$$+ (2.0)(.5) + (2.5)(.5) + (3.0)(.5) + (3.5)(.5)$$
$$= 7 \tag{B.2.9}$$

If we continued with smaller and smaller values for Δx, our answer would approach the value of 8, which we know to be the true area from simple plane geometry. We could also find the area under this curve by evaluating a definite integral as in:

$$A = \int_0^4 f(x)\,dx = \frac{1}{2}x^2 \Big|_0^4 = \frac{1}{2}(4)^2 - 0 = 8 \tag{B.2.10}$$

where the upper and lower limits on the integral form the boundaries of our area. Thus, a definite integral is the "area" under the curve representing the function bounded on two sides by the limits.

C
MULTIPLE INTEGRALS

Integrals in two or three dimensions occur throughout physics. They are important—even crucial—to understanding much of advanced physics. The ideas involved are simple enough, yet experience shows that many, many students have difficulty with their explicit evaluations. The choice of limits proves especially difficult.

To explain or review these multiple integrals, let us use as an example Eqs. 9.1.4 and 9.1.5 that give the position of the center of mass of an extended body:

$$X = \frac{1}{M} \iiint_V x\rho \, dV$$

$$Y = \frac{1}{M} \iiint_V y\rho \, dV \qquad (9.1.4)$$

$$Z = \frac{1}{M} \iiint_V z\rho \, dV$$

$$M = \iiint_V \rho \, dV \qquad (9.1.5)$$

APPENDIX C / MULTIPLE INTEGRALS

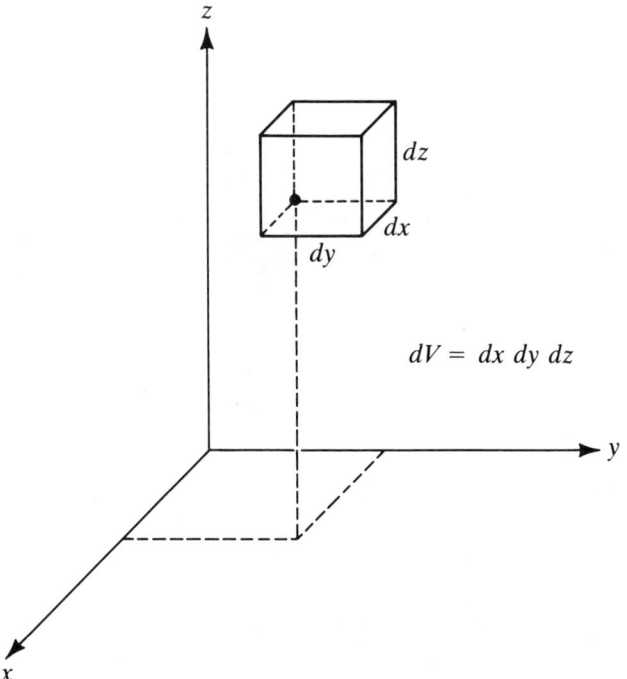

Figure C.1 A representative differential volume element in Cartesian coordinates.

In rectangular or Cartesian coordinates the infinitesimal volume element is

$$dV = dx\, dy\, dz \tag{C.1}$$

It is worthwhile to consider *why* this is so for this simple case in which the result is either quite familiar or intuitively obvious.

A set of rectangular coordinate axes are shown in Figure C.1. The back corner of the cube (indicated by a heavy dot) locates position (x, y, z). Now, one by one, increment each of the coordinates by an infinitesimal amount and see what volume is created. As x increases to $x + dx$, a *line* is created of length dx. Next increase, say, z to $z + dz$ (the order in which we look at the coordinates shouldn't make any difference). Now the line has swept out a *surface* of area $dx\, dz$. As y now increases to $y + dy$, this surface generates a *volume*, $dx\, dz\, dy$, and this is simply a detailed development of Eq. C.1.

EXAMPLE C.1 Find the center of mass of the wedge shown in Figure C.2. The dimensions are shown there and it is assumed to be of uniform density.

Before applying Eq. 9.1.4 to find the center of mass, it is better to find the total mass from Eq. 9.1.5. This provides a chance to look carefully at the limits of integration and the result can be easily checked. Since this wedge is just half of a rectangular solid, the total mass must be

$$M = \frac{1}{2} \rho abc \tag{C.2}$$

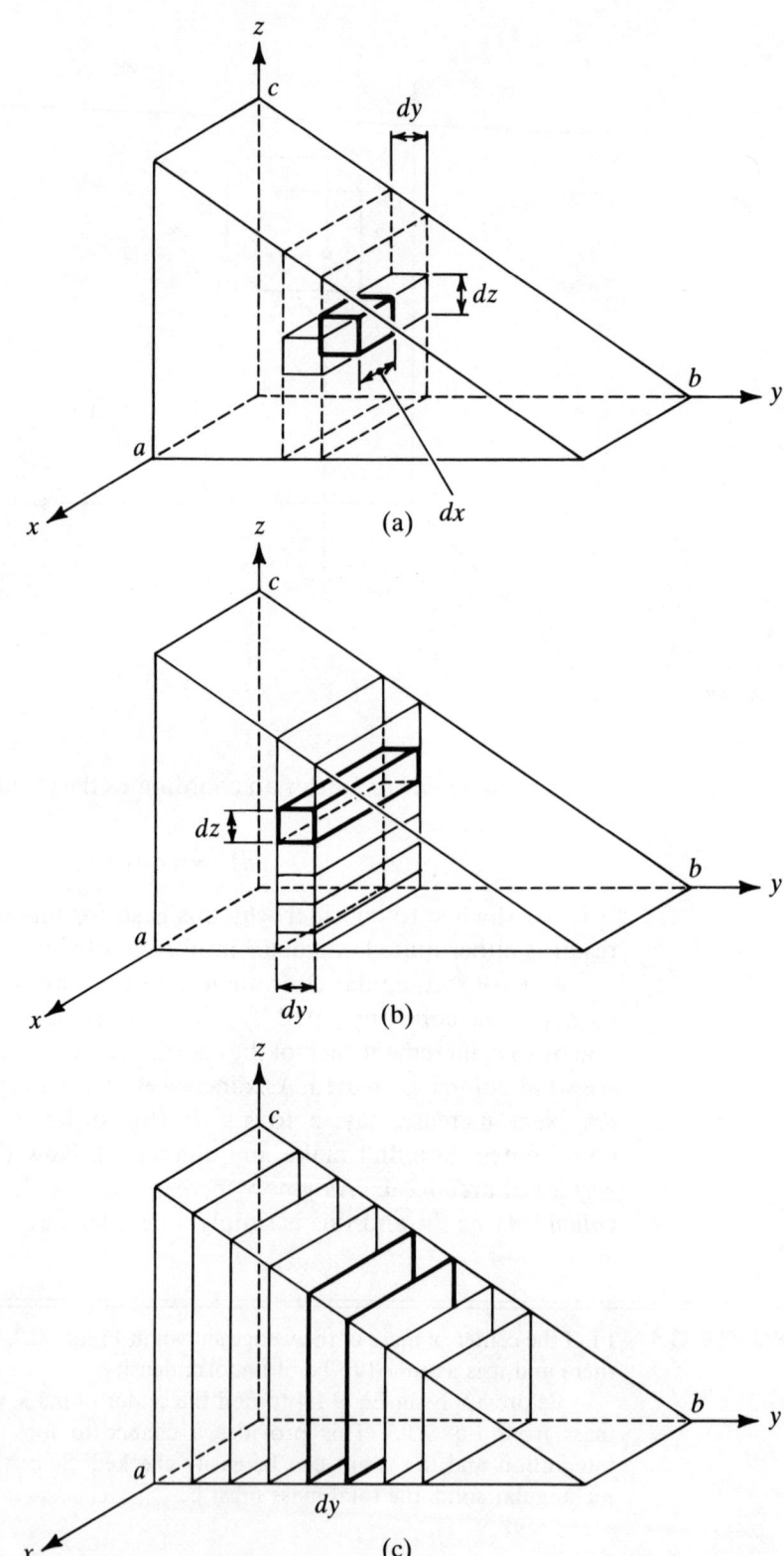

Figure C.2 (a) Integration along *x* generates a small rectangular solid; (b) Integration along *z* generates a thin slab; (c) Integration along *y* generates the entire volume.

APPENDIX C / MULTIPLE INTEGRALS

Applying Eq. 9.1.5 directly, we have

$$M = \iiint_V \rho \, dV = \rho \iiint dx \, dz \, dy$$

Since ρ is a constant, it can be moved through the integral signs, and the result is independent of the order in which the integrations are carried out over the three variables.

The integration over x moves the small cube $dV = dx \, dy \, dz$ along the x direction, generating a small representative volume element, shaped as a *rectangular solid* as shown in Figure C.2(a). This rectangular solid has one limit at $x = 0$ and the other at $x = a$. Thus, these are the limits of integration:

$$M = \rho \int \left[\int \left(\int_0^a dx \right) dz \right] dy$$

$$(\;\;) = x \Big|_0^a = a - 0 = a$$

$$M = \rho a \int \left[\int dz \right] dy$$

Now the integration becomes more interesting. The integration over z moves this slender rectangular solid from one limit to the other in the z direction and generates a *slab* as shown in Figure C.2(b).

But what are these limits? From the bottom at $z = 0$ to the top. At $z = c$? No! The z value for the top of this slab is a function of y. A little thought will show that the equation of the top of the wedge is

$$z = -\frac{c}{b} y + c$$

and that value of z is now the upper limit.

$$M = \rho a \int \left[\int_0^{-(c/b)y + c} dz \right] dy$$

$$= \rho a \int \left[z \Big|_0^{-(c/b)y + c} \right] dy$$

$$= \rho a \int \left(-\frac{c}{b} y + c \right) dy$$

$$= -\frac{\rho a c}{b} \int y \, dy + \rho a c \int dy$$

APPENDIX C / MULTIPLE INTEGRALS

The integration over y moves the thin slab from $y = 0$ to $y = b$ and, thus, sweeps through the entire volume of the wedge as shown in Figure C.2(c).

$$M = -\frac{\rho ac}{b} \int_0^b y\, dy + \rho ac \int_0^b dy$$

$$= -\frac{\rho ac}{b} \frac{1}{2} y^2 \Big|_0^b + \rho ac\, y \Big|_0^b$$

$$= -\frac{1}{2} \rho abc + \rho abc$$

$$M = \frac{1}{2} \rho abc$$

And this is the result we expected from Eq. C.2. This same technique and choice of limits can now be used to determine the coordinates of the center of mass:

$$X = \frac{1}{M} \iiint_v x\rho\, dv \qquad (9.1.4)$$

$$= \frac{\rho}{\frac{1}{2}\rho abc} \int_y \left[\int_z \left(\int_{x=0}^a x\, dx \right) dz \right] dy$$

$$= \frac{2}{abc} \int_y \left[\int_z \frac{1}{2} x^2 \Big|_0^a dz \right] dy$$

$$= \frac{2}{abc} \cdot \frac{a^2}{2} \int_y \left[\int_{z=0}^{-(c/b)y+c} dz \right] dy$$

$$= \frac{a}{bc} \int_y z \Big|_0^{-(c/b)y+c} dy$$

$$= \frac{a}{bc} \int_{y=0}^b \left(-\frac{c}{b} y + c \right) dy$$

$$= \frac{a}{bc} \left[-\frac{c}{b} \int_0^b y\, dy + c \int_0^b dy \right]$$

$$= \frac{a}{b} \left(-\frac{1}{b}\frac{1}{2} y^2 \Big|_0^b + y \Big|_0^b \right)$$

$$= \frac{a}{b} \left(-\frac{1}{2} b + b \right)$$

$$X = \frac{1}{2} a \qquad (C.3)$$

The y-coordinate of the center of mass is found in just the same way:

$$Y = \frac{1}{M} \iiint_v y\rho\, dv \qquad (9.1.4)$$

$$= \frac{\rho}{\frac{1}{2}\rho abc} \int_y y \left[\int_z \left(\int_{x=0}^a dx \right) dz \right] dy$$

APPENDIX C / MULTIPLE INTEGRALS

As far as the integrations over x and z are concerned, y may be treated as a *constant* and moved through the integral signs.

$$Y = \frac{2}{abc} \cdot a \int_y y \left[\int_{z=0}^{-(c/b)y+c} dz \right] dy$$

$$= \frac{2}{bc} \int_{y=0}^{b} y\left(-\frac{c}{b}y + c\right) dy$$

$$= \frac{2}{bc}\left(-\frac{c}{b}\int_0^b y^2\, dy + c\int_0^b y\, dy\right)$$

$$= \frac{2}{b}\left(-\frac{1}{b}\frac{1}{3}y^3\Big|_0^b + \frac{1}{2}y^2\Big|_0^b\right)$$

$$= \frac{2}{b}\left(-\frac{1}{3}b^2 + \frac{1}{2}b^2\right)$$

$$Y = \frac{1}{3}b \tag{C.4}$$

And the z-coordinate of the center of mass is found in the same way:

$$Z = \frac{1}{M}\iiint_v z\rho\, dv$$

$$= \frac{\rho}{\frac{1}{2}\rho abc}\int_y\left[\int_z z\left(\int_0^a dx\right) dz\right] dy$$

For the integration over x, z behaves as a constant and, therefore, is pulled out of the x-integral.

$$Z = \frac{2}{bc}\int_y\left[\int_{z=0}^{-(c/b)y+c} z\, dz\right] dy$$

$$= \frac{2}{bc}\int_y \frac{1}{2}z^2\Big|_0^{-(c/b)y+c} dy$$

$$= \frac{1}{bc}\int_{y=0}^b \left(-\frac{c}{b}y+c\right)^2 dy$$

$$= \frac{1}{bc}\int_0^b \left(\frac{c^2}{b^2}y^2 - \frac{2c^2}{b}y + c^2\right) dy$$

$$= \frac{c}{b}\left(\frac{1}{b^2}\int_0^b y^2\, dy - \frac{2}{b}\int_0^b y\, dy + \int_0^b dy\right)$$

$$= \frac{c}{b}\left(\frac{1}{b^2}\cdot\frac{1}{3}b^3 - \frac{2}{b}\cdot\frac{1}{2}b^2 + b\right)$$

$$= c\left(\frac{1}{3} - 1 + 1\right)$$

$$Z = \frac{1}{3}c \tag{C.5}$$

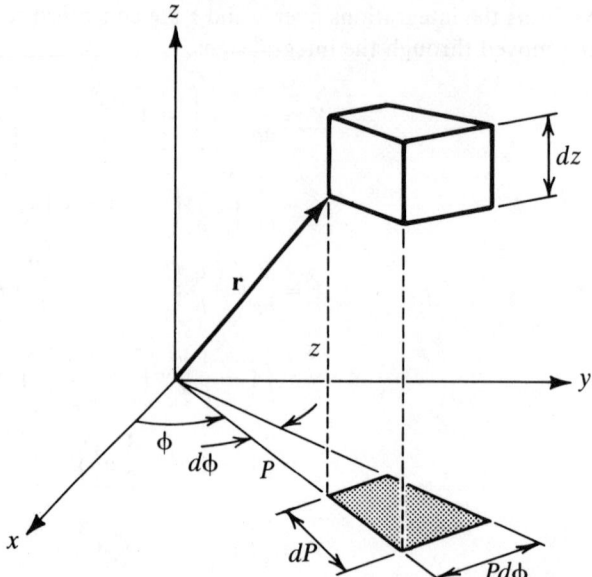

Figure C.3 A representative differential volume element in cylindrical coordinates.

Thus, the center of mass of the wedge shown in Figure C.1 is located at

$$(X, Y, Z) = \left(\frac{a}{2}, \frac{b}{3}, \frac{c}{3}\right)$$

What form does dV take in other coordinate systems? Figure C.3 shows a position vector **r** in *cylindrical* coordinates. The location of a point is given by specifying P, the perpendicular distance from the z-axis; ϕ, the angle of rotation from the x-axis; and z, the height above the xy plane. The Greek letter ρ is ordinarily used for the perpendicular distance from the z-axis, but since we are now using ρ for the density, we shall use P to avoid confusion.

The infinitesimal volume dV is the volume generated when each of the coordinates increases a small amount. This is illustrated in Figure C.3. As ϕ increases to $\phi + d\phi$, a *line* is generated of length $P\,d\phi$. Notice that the length of this line depends upon the value of another coordinate! Nothing like this happened with rectangular coordinates. Increasing P to $P + dP$ causes this line to generate a surface of *area*, $P\,d\phi\,dP$. Increasing z to $z + dz$ moves this area through a *volume*, $P\,d\varphi\,dP\,dz$. Thus, for cylindrical coordinates,

$$dv = P\,dP\,dz\,d\phi \tag{C.6}$$

EXAMPLE C.2 Find the center of mass of the cone shown in Figure C.4 (the height is h and the radius of the base is a). As usual, the cone is of uniform density.

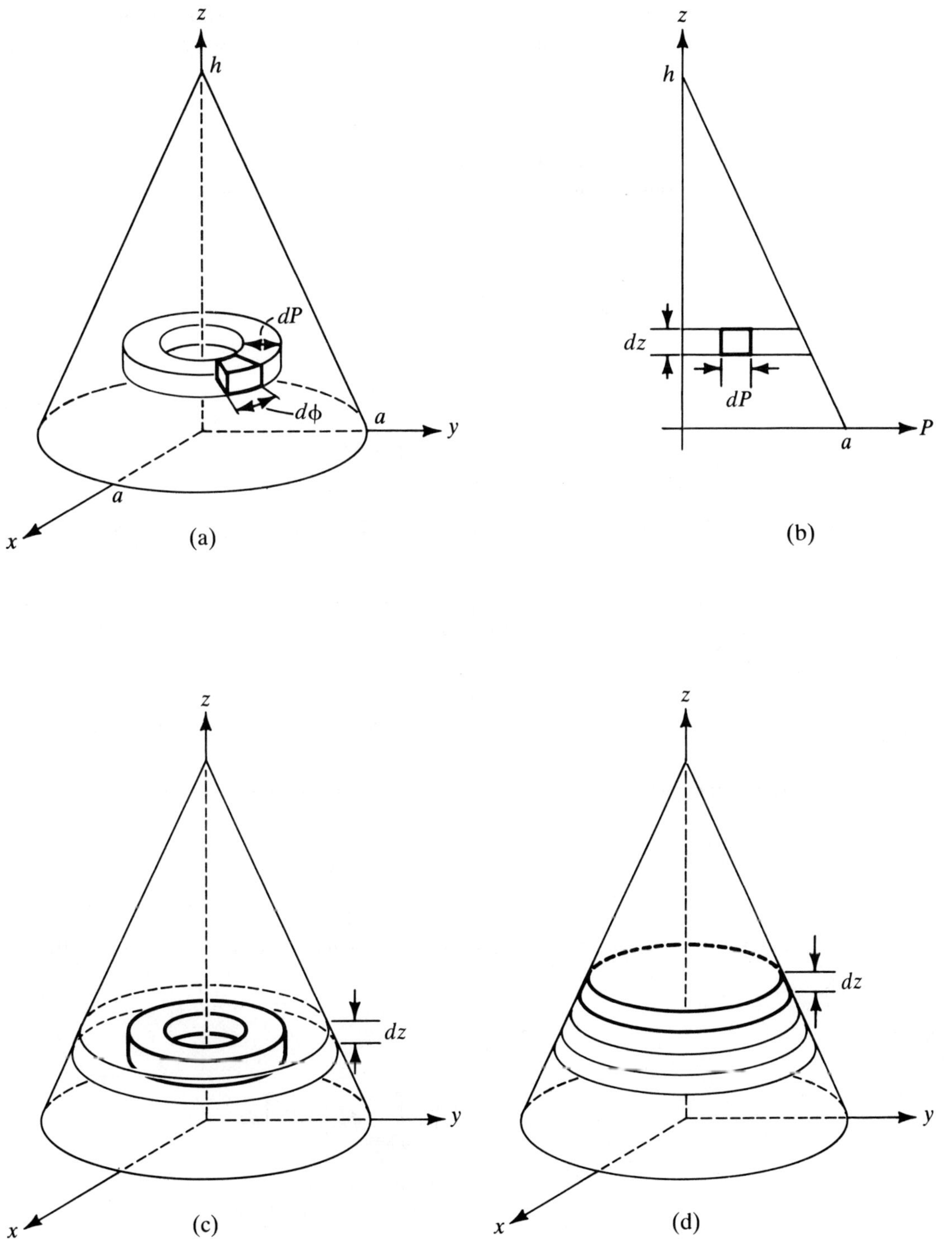

APPENDIX C / MULTIPLE INTEGRALS

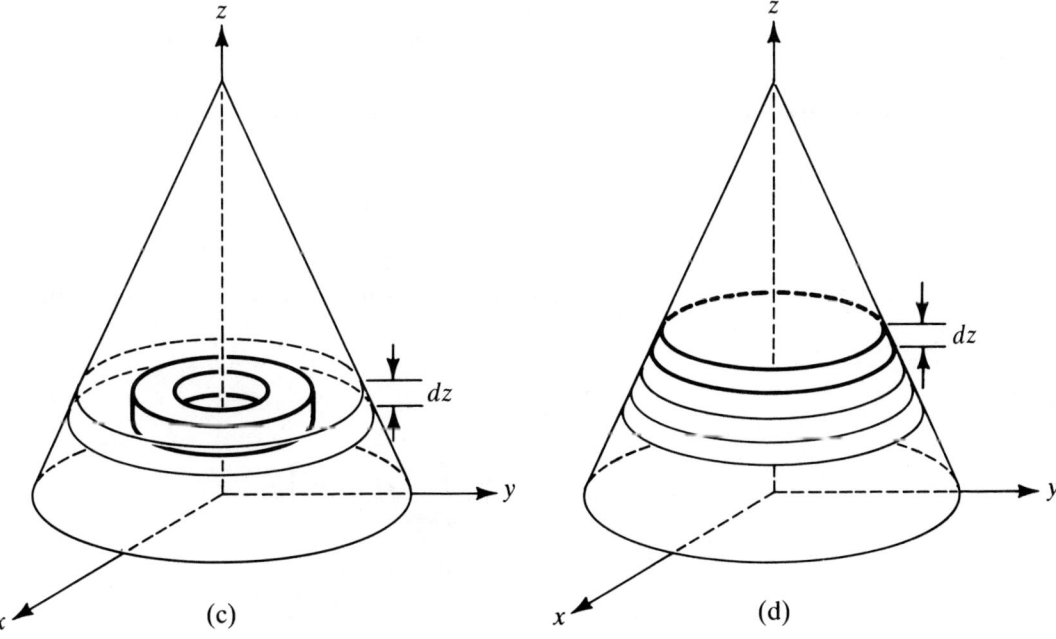

Figure C.4 (a) Integration over ϕ generates a ring; (b) Integration over P changes the radius of the ring; (c) Thus, a thin disc is generated; (d) Integration over z generates the entire volume.

As shown in Figure C.4(a), an integration over ϕ causes our small volume to generate a ring. We shall do this integration first. *From symmetry we can readily see that*

$$X = Y = 0$$

Nonetheless, let us calculate this explicitly.

From geometry (or a handbook) we find that the volume of a cone is

$$V = \frac{\pi}{3} a^2 h$$

or

$$M = \frac{\pi}{3} a^2 h \rho \tag{C.7}$$

$$X = \frac{1}{M} \iiint_V x \rho \, dV$$

But $x = P \cos \phi$ must be used since the variables of integration are P, ϕ, and z:

$$X = \frac{1}{M} \rho \int_z \left[\int_P \left(\int_{\phi=0}^{2\pi} P \cos \phi \, d\phi \right) P \, dP \right] dz$$

And

$$\int_0^{2\pi} \cos \phi \, d\phi = 0$$

so $X = 0$ explicitly. Likewise, an explicit evaluation of Y will involve

$$\int_0^{2\pi} \sin \phi \, d\phi = 0$$

and $Y = 0$ explicitly. But that still leaves the evaluation of Z, which is interesting.

$$Z = \frac{1}{M} \iiint_V z \rho \, dV$$

$$= \frac{1}{M} \rho \int_z z \left[\int_P \left(\int_{\phi=0}^{2\pi} d\phi \right) P \, dP \right] dz$$

$$= \frac{1}{M} 2\pi \rho \int_z z \left[\int_{P=0}^{P = -(a/h)z + a} P \, dP \right] dz$$

The limits on this integral over P are crucial. Note that this is not presently an integral over P between 0 and a! Figure C.4(b) shows a cross section in the zP plane. Just as with the wedge of the earlier example, the upper limit on P is a function of z. When we complete the integration over P, the "ring" will have generated a circular *slice* of thickness dz. But, continuing the integration, we have

$$Z = \frac{2\pi\rho}{M} \int_z z \left[\frac{1}{2}\left(-\frac{a}{h}z + a\right)^2 \right] dz$$

$$= \frac{2\pi\rho}{M} \int_z z \left[\frac{a^2}{h^2} z^2 - \frac{2a^2}{h} z + a^2 \right] dz$$

$$= \frac{\pi\rho a^2}{M} \left[\frac{1}{h^2} \int_{z=0}^{h} z^3 \, dz - \frac{2}{h} \int_0^h z^2 \, dz + \int_0^h z \, dz \right]$$

$$= \frac{\pi\rho a^2}{M} \left[\frac{1}{h^2} \cdot \frac{1}{4} h^4 - \frac{2}{h} \cdot \frac{1}{3} h^3 + \frac{1}{2} h^2 \right]$$

$$= \frac{\pi\rho a^2 h^2}{M} \left(\frac{1}{4} - \frac{2}{3} + \frac{1}{2} \right)$$

$$= \frac{\pi\rho a^2 h^2/12}{\pi\rho a^2 h/3}$$

$$Z = \frac{1}{4} h \tag{C.8}$$

The final integration over z sweeps the representative circular slice throughout the *entire volume*, as shown in Figure C.4(d).

Figure C.5 shows a point located by spherical coordinates r, θ, and ϕ, where r is the distance from the origin, θ is the rotation from the z-axis, and ϕ is the rotation from the x-axis. As before, dV is the volume generated by an infinitesimal increase in each of these coordinates. Increasing θ to $\theta + d\theta$ moves a point through a line of length $r\, d\theta$. Increasing ϕ to $\phi + d\phi$ moves a

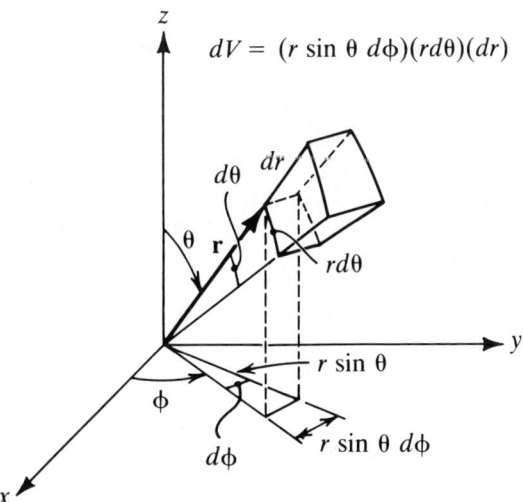

Figure C.5 A representative differential volume element in spherical coordinates.

point through a distance of $r \sin \theta \, d\phi$, as shown in the figure. Both increases, then, generate a surface whose area is $(r \, d\theta)(r \sin \theta \, d\phi) = r^2 \sin \theta \, d\theta \, d\phi$. An increase in r to $r + dr$ then moves this surface a distance dr to generate the volume

$$dV = r^2 \sin \theta \, d\theta \, d\phi \, dr \tag{C.9}$$

for spherical coordinates.

EXAMPLE C.3 Find the center of mass of a hemisphere of radius a as shown in Figure C.6. By symmetry, we may immediately set

$$X = Y = 0 \tag{C.10}$$

This will not be done explicitly as was done in the previous example. The mass of a hemisphere is

$$M = \frac{2}{3} \pi a^3 \rho$$

where ρ is, of course, the density (which we assume to be uniform).

The z-coordinate of the center of mass is found, as always, from Eq. 9.1.4:

$$Z = \frac{1}{M} \iiint_V z\rho \, dV$$

$$= \frac{1}{M} \rho \int_r \left[\int_\theta \left(\int_\phi z \, d\phi \right) \sin \theta \, d\theta \right] r^2 \, dr$$

And, just as earlier, we must write z in terms of the variables of integration before we can proceed:

$$z = r \cos \theta$$

$$Z = \frac{1}{M} \rho \int_r \left[\int_\theta \left(\int_\phi r \cos \theta \, d\phi \right) \sin \theta \, d\theta \right] r^2 \, dr$$

$$= \frac{1}{M} \rho \int_r \left[\int_\theta \left(\int_{\phi=0}^{2\pi} d\phi \right) \cos \theta \sin \theta \, d\theta \right] r^3 \, dr$$

$$= \frac{2\pi\rho}{M} \int_r \left[\int_{\theta=0}^{\pi/2} \cos \theta \sin \theta \, d\theta \right] r^3 \, dr$$

$$= \frac{2\pi\rho}{M} \cdot \frac{1}{2} \int_{r=0}^{a} r^3 \, dr$$

$$= \frac{\pi\rho}{M} \cdot \frac{1}{4} a^4$$

$$= \frac{\pi\rho a^4/4}{2\pi\rho a^3/3}$$

$$Z = \frac{3}{8} a \tag{C.11}$$

APPENDIX C / MULTIPLE INTEGRALS

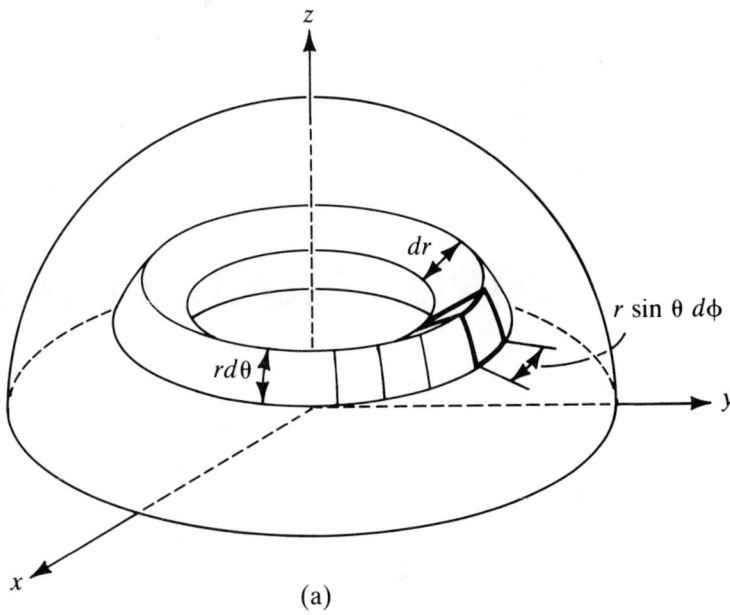

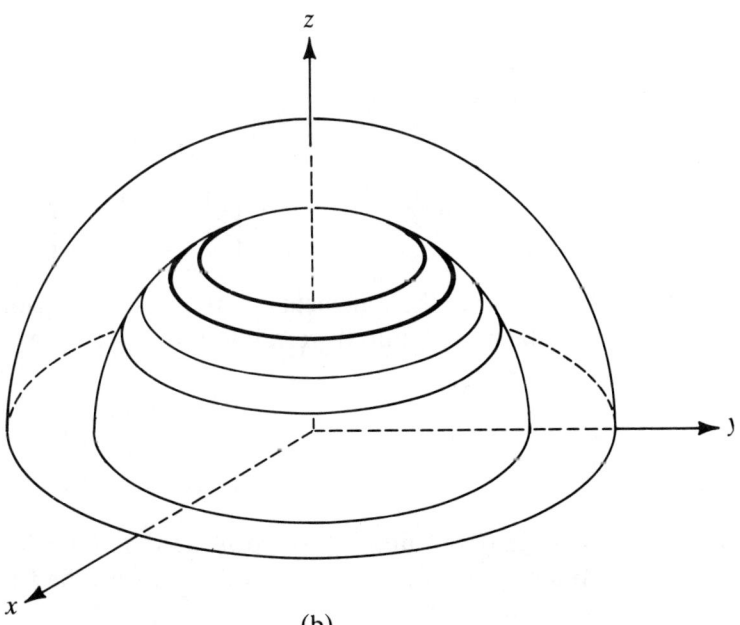

Figure C.6 (a) Integration over ϕ again generates a ring; (b) Integration over θ generates a thin hemispherical shell.

Figure C.6(a) indicates the integration over ϕ. The result is a *ring*. Further integration over θ generates a thin representative spherical shell of radius r and thickness dr, as shown in Figure C.6(b). Final integration over r generates the entire volume.

D
MATRIX MULTIPLICATION

A *matrix* is a rectangular array of elements—usually numbers, like

$$\begin{bmatrix} 1 & 5 & 7 & 2 \\ 12 & 7 & 9 & 10 \\ 10 & 2 & 4 & 8 \end{bmatrix}$$

We can write these elements symbolically as A_{ij} for the element in the ith row and the jth column. If we call the matrix $[A]$, we could write it as

$$[A] = \begin{bmatrix} A_{11} & A_{12} & A_{13} & A_{14} \\ A_{21} & A_{22} & A_{23} & A_{24} \\ A_{31} & A_{32} & A_{33} & A_{34} \end{bmatrix}$$

Note that the number of rows need not be the same as the number of columns. However, most of the matrices we shall encounter in this text will be 3×3 matrices.

Two matrices may be *multiplied* together. The result will be another matrix. We can multiply matrix $[A]$ with matrix $[B]$, as

$$[A][B] = [C]$$

if (and only if) the number of *columns* in $[A]$ is the same as the number of *rows* in $[B]$. It is undefined otherwise. An element in $[C]$ is the sum of the

APPENDIX D / MATRIX MULTIPLICATION

products of the elements in the corresponding row in [A] and the corresponding column in [B]. Pictorially, we might show this as

$$[C] = [A][B] = \begin{bmatrix} A_{11} & A_{12} & A_{13} \\ A_{21} & A_{22} & A_{23} \\ A_{31} & A_{32} & A_{33} \\ A_{41} & A_{42} & A_{43} \end{bmatrix} \begin{bmatrix} B_{11} & B_{12} \\ B_{21} & B_{22} \\ B_{31} & B_{32} \end{bmatrix} = \begin{bmatrix} C_{11} & C_{12} \\ C_{21} & C_{22} \\ C_{31} & C_{32} \\ C_{41} & C_{42} \end{bmatrix}$$

A numerical example may also be useful. Multiply the two matrices

$$[A] = \begin{bmatrix} 1 & 2 & 3 \\ 4 & 5 & 6 \\ 7 & 8 & 9 \end{bmatrix} \quad \text{and} \quad [B] = \begin{bmatrix} 2 & 4 & 6 \\ 8 & 10 & -8 \\ -6 & -4 & -2 \end{bmatrix}$$

$$[A][B] = \begin{bmatrix} 1 & 2 & 3 \\ 4 & 5 & 6 \\ 7 & 8 & 9 \end{bmatrix} \begin{bmatrix} 2 & 4 & 6 \\ 8 & 10 & -8 \\ -6 & -4 & -2 \end{bmatrix}$$

$$= \begin{bmatrix} 1 \times 2 + 2 \times 8 + 3 \times (-6) & 1 \times 4 + 2 \times 10 + 3 \times (-4) & 1 \times 6 + 2 \times (-8) + 3 \times (-2) \\ 4 \times 2 + 5 \times 8 + 6 \times (-6) & 4 \times 4 + 5 \times 10 + 6 \times (-4) & 4 \times 6 + 5 \times (-8) + 6 \times (-2) \\ 7 \times 2 + 8 \times 8 + 9 \times (-6) & 7 \times 4 + 8 \times 10 + 9 \times (-4) & 7 \times 6 + 8 \times (-8) + 9 \times (-2) \end{bmatrix}$$

$$= \begin{bmatrix} 2 + 16 - 18 & 4 + 20 - 12 & 6 - 16 - 6 \\ 8 + 40 - 36 & 16 + 50 - 24 & 24 - 40 - 12 \\ 14 + 64 - 54 & 28 + 80 + 36 & 42 - 64 - 18 \end{bmatrix}$$

$$= \begin{bmatrix} 0 & 12 & -16 \\ 12 & 42 & -28 \\ 24 & 72 & -40 \end{bmatrix}$$

Symbolically, we can write the individual elements of $[C] = [A][B]$ as

$$C_{ij} = \sum_{k=1}^{N} A_{ik} B_{kj} \tag{D.1}$$

In fact, this notation is so common in some fields of physics that the summation sign is *omitted* and this is simply written as

$$C_{ij} = A_{ik} B_{kj} \tag{D.2}$$

with the *implicit understanding* that there is to be a *summation* over k since it is a *repeated subscript*.

Notice that the *order* of the two matrices in the multiplication is very important. In fact, if the number of rows is different from the number of columns, $[B][A]$ may be undefined even though the multiplication $[A][B]$

can be carried out. Matrix multiplication does *not* commute. That is, in general

$$[A][B] \neq [B][A] \tag{D.3}$$

Try this for yourself with the numerical examples of $[A]$ and $[B]$ given above. Carry out the multiplication for $[B][A]$. The result will be very different than the matrix we found for $[A][B]$. Only for special cases will that not be true.

INDEX

A

Acceleration, 1, 22, 407
Acceleration constant, 2, 25, 408
Air resistance, 117
Allowed regions, 35
Amplitude, 56, 66
Angle, apsidal, 203, 205
Angular frequency, 55
Angular momentum, 179, 189, 247
Aphelion, 187, 197
Apogee, 187
Apsidal angle, 203, 205
Archimedes' Principle, 351
Arrival velocity, 197
Atwood's machine, 291, 293, 295
Axes, principal, 262
Axis, 251

B

Balancing
    dynamic, 263
    static, 263
Barometer, 352
BASIC, 1
Bernoulli's equation, 363, 365
Body cone, 275
Body frame, 270
Boosts, gravitational. *See* Gravitational boosts
Brahe, Tycho, 189

C

Cables, 327
Calculus, 424
    vector, 91
Capacitor, 69
Cardan's suspension, 298
Catenary, 332
Center of mass, 223, 246
    frame, 236
Central forces, 176
Centrifugal barrier, 182
Centrifugal force, 159, 165, 182, 191
Centrifugal potential energy, 182
Centripetal acceleration, 164, 165
Centripetal force, 164, 165
Chain rule, 96
Chandler wobble, 277
Collisions, 231
    elastic, 233
    head-on, 233
Commensurable frequencies, 120
Condensor, 69
Conic sections, 185
Conservation
    of angular momentum, 1
    of energy, 33, 122
    of momentum, 225
Conservative forces, 121, 130
Constraints, 292
    holonomic, 294
Coordinates
    cylindrical, 147
    fixed, 154
    generalized, 283
    ignorable, 297
    inertial frame, 154
    LAB frame, 154
    moving, 153
    noninertial frame, 154
    plane polar, 144
    polar, 145, 283
    rotating, 156
    spacetime, 374
    spherical, 148
Coordinate system, 23, 89, 144
    rotation transformation, 91
Coriolis force, 159, 166
Cross product, 87
Curl, 95, 103, 177

D

Damped harmonic oscillator, 58, 417
Damping. *See* Overdamping and Underdamping

Del operator, 97
Density, 247
Derivative, 424
Determinant, 88
Differential equations, 51, 52, 185
Differential operators, vector, 95
Directional derivative, 97
Displacement, 22
Divergence, 95, 99
  theorem, 100
Doppler effect, 394
Dot product, 86
Driven harmonic oscillator. *See* Forced harmonic oscillator
Dynamic balancing, 263
Dynamics, 23, 166

E

Eccentricity, 185
Effective potential, 187
  energy, 182
Effective spring constant, 58, 201
Eigenvalues, 267
Eigenvectors, 267
Einstein time dilation, 383
Electromagnetic fields, 133
Elliptic integrals, 253
Energy
  conservation, 34
  kinetic. *See* Kinetic energy
  and mass, 395
  potential. *See* Potential energy
Equilibrium, 57, 201, 254
Escape velocity, 194
Ether, luminiferous, 373
Euler's angles, 269, 298
Euler's equations, 269
Euler's relations, 51
Exponentials, 51

F

Fictitious forces, 155
Field
  electromagnetic, 133, 210
  gravitation, 205
Flow, irrotational, 358
Fluids, 346. *See also* Hydrostatics and Hydrodynamics
Forbidden regions, 35
Force, 1, 407

Force (*continued*)
  central, 176
  centrifugal, 159, 165
  centripetal, 164, 165, 191
  conservative. *See* Conservative forces
  constant, 25
  Coriolis, 159
  fictitious, 155
  generalized, 285
  inertial, 155
  position-dependent, 33
  separable, 115
  time-dependent, 31
  velocity-dependent, 39
Forced harmonic oscillator, 64
Foucault pendulum, 168
Frame, center of mass, 236
Frequency, 56
  commensurable, 120
  radial oscillations, 202
  resonance, 66
  See also Angular frequency
Friction, 11, 15, 28, 422

G

Galilean relativity, 371
Gauss's Theorem, 100, 361
Generalized coordinates, 283
Generalized forces, 285
Gimbal mounting, 298
Gradient, 95
Gravitational boosts, 191
Gravitational braking, 199
Gravitational field, 205
Gravitational potential, 206
  energy, 362
Gravity, 204
Gyroscope, 254, 298

H

Hamiltonian, 305
Hamilton's equations, 304
Hamilton's Principle, 306
Harmonic oscillator, 1, 50, 254, 407
  three-dimensional, 119
Hero of Alexandria, 308
Hero's minimum principle, 308
Hooke's Law, 1, 50
Hydrodynamics, 355
Hydrostatics, 346

INDEX

I

Ignorable coordinates, 297
Inductor, 69
Inertia, 23
    moment of. *See* Moment of inertia
Inertial forces, 155
Intertia tensor, 260
Initial conditions, 2, 26, 53, 60, 408
Inner product, 86
Integral, 429
    line. *See* Line integrals
    of the motion, 181, 182, 297
    multiple, 432
Inverse-square force, 182
Irrotational flow, 358

K

Kepler, Johannes, 189
Kepler's Laws, 189
Kinematics, 21
Kinetic energy, 33, 268, 362
Kirchhoff's rules, 70

L

LAB cone, 275
LAB frame, 270
Lagrange's equations, 287, 304, 308
Lagrangian, 288
Lagrangian mechanics, 282
Latitude, 166, 171
Laws of motion, 1, 23, 24, 25, 180, 224, 282, 304, 305, 308, 314, 346, 347, 395, 407
L'Hôpital's Rule, 194
Light, constant speed of, 373
Line integrals, 123, 130
Lissajous figures, 121
Loads, 314
Lorentz-Fitzgerald contraction, 383
Lorentz transformations, 378
Luminiferous ether, 373

M

Mass
    center of. *See* Center of mass
    and energy, 395
    reduced, 238

Mass (*continued*)
    variable, 227
Matrices, 444
Maxwell's equations, 373
Members of a truss, 317
Method
    of joints, 320
    of sections, 325
Minkowski diagrams, 385
Moment
    arm, 89
    of inertia, 251, 254, 256, 262
Momentum, 225, 395
Motion
    constant acceleration, 2, 22, 25, 408
    one-dimensional, 21
    projectile. *See* Projectile motion
    three-dimensional, 114
    See also Laws of motion

N

Natural frequency, 65
Newton's Laws of Motion. *See* Laws of motion
North Pole, 169
Nutation, 304

O

Orbital transfers, 191
Orbits. *See* Central forces
Oscillation
    radial, 200
    radius of, 74
Oscillator. *See* Simple harmonic oscillator
Overdamping, 59

P

Parallel-axis theorem, 257
Partial derivative, 96, 130, 132
Particles, system of, 222
Pascal, 407
Pascal's Law, 349, 352
Pendulum
    Foucault, 168
    physical, 73, 253
    simple, 71
Perigee, 187
Perihelion, 187, 197

Permittivity, 184
Perpendicular-axis theorem, 258
Phase angle, 65, 67
Physical pendulum, 253
Pitot tubes, 365
Plane truss. *See* Truss
Polar coordinates, 283
Position, 22
Potential, gravitational, 206
Potential energy, 34, 121, 177
Power series, 56
Precession, 256
Pressure, 347, 363
Principal axes, 262
PRINT statements, 6
Product of inertia, 262
Projectile motion, 116
Proper time, 398

R

Radial oscillations, 200
Radius of oscillation, 74
Reduced mass, 238
Reference frame. *See* Coordinate system
Relativistic mass, 403
Relativity, 370
    Doppler effect, 394
    Einstein time dilation, 383
    energy, 395, 400
    Galilean, 371
    laws of motion, 395
    light, constant speed of, 373
    Lorentz-Fitzgerald contraction, 383
    Lorentz transformations, 378
    luminiferous ether, 373
    mass, 403
    mass and energy, 395
    Minkowski diagrams, 385
    momentum, 395
    proper time, 398
    rest energy, 400
    rest mass, 400, 403
    simultaneity, 376
    spacetime, 374
    special, 370
    velocity transformations, 392
    worldlines, 386
Resistor, 69
Resonance, 66
Resonance frequency, 66
Rest energy, 400

Rest mass, 400, 403
Right-hand rule, 87
Rigid bodies, 246
Rotating top, 298
Rotation, 251
Rotational force, 252
Rotational mass, 252
Rutherford scattering, 210

S

Sag of cable, 330
Scalar product, 86
Scattering, 210
Scattering angle, 211
Scattering cross section, 212
Sections, method of, 325
Separable forces, 115
Shock absorbers, 63
Simple harmonic oscillator, 52. *See also* Harmonic oscillator
Simultaneity, 376
Somersault, 254, 272
Spacetime, 374
Span of cable, 330
Special relativity, 370
Spring constant, 1, 50, 407
    effective, 58, 201
Stable orbits, 202
Static balancing, 263
Statics, 164, 314
Steady state, 64
Stokes's Theorem, 103, 130, 357

T

TAB function, 10
Taylor series, 56
Tension, 327
Tensor, 260. *See also* Inertia tensor
Terminal speed, 15, 42, 422
Top, rotating, 298
Torque, 74, 88, 179, 248, 252
Torricelli's Theorem, 365
Transfer velocity, 196
Transformation matrix, 91
Transformations, velocity, 392
Transient, 64
Truss, 315
Turning points, 35

U

Underdamping, 60
Universal gravitation constant, 184

V

Variation Principle, Hamilton's, 306
Vector addition, 79, 81
    parallelogram method, 81
Vector algebra, 79
Vector calculus, 91
Vector components, 82
Vector differential operators, 95
Vector magnitude, 83
Vector multiplication, 86
    cross product, 87
    dot product, 86
    by a scalar, 80

Vector multiplication (*continued*)
    scalar product, 86
    vector product, 87
Vector polygon, 81
Vector product, 87
Vectors, 79
    unit, 82
Vector subtraction, 84
Vector triangle, 81
Velocity, 22
    arrival, 197
    escape, 194
    transfer, 196
    transformations, 392

W

Work, 34
Worldlines, 386

H.E.
no → Althoff
yes → whiting